Forthcoming from The MIT Press

Plastics for Architects and Builders
by Albert G. H. Dietz
The objective of this "primer" on plastics is to acquaint the architect, designer, builder, and contractor with polymers and their potentialities in building, covering not only structural aspects, but interior, decorative, and lighting applications as well. The user of this book will obtain basic, essential knowledge of the nature and the applicability of these versatile materials. While it does not attempt to cover the entire field, it provides a firm groundwork in an area that could well profit from a knowledge of fundamental facts; equipped with this information the architect or builder can avoid trial-and-error methods in the employment of the plastic medium in building.

The book begins with an extensive series of illustrations, including a four-color section, thus providing, in this way, a visual introduction or sample "case history" of the many uses of plastics in modern architecture. The text classifies polymers into simple groupings of application in building, describes their general properties, and sets forth composite materials and assemblies based on plastics, taking up the principal methods of fabrication as they affect uses. The parts of a building and the ways in which plastics can be employed are described in detail.

The text is liberally supported by sketches and tables, clearly and usefully presented. The generic names and properties of the principal plastics currently in structural use and their applications in construction work and utility appliances are summarized in comprehensive form.

The MIT Press
Massachusetts Institute of Technology
Cambridge, Massachusetts 02142

COMPOSITE ENGINEERING LAMINATES

COMPOSITE ENGINEERING LAMINATES

Edited by Albert G. H. Dietz

THE MIT PRESS
Massachusetts Institute of Technology
Cambridge, Massachusetts, and London, England

Preface

Composite laminated materials for engineering purposes have experienced great and rapid development in the 20 years since *Engineering Laminates,** the predecessor of this book, appeared. It is therefore appropriate to examine this field anew.

This is not a second edition. The major part of the material is new, but, where appropriate, information contained in the previous book has been incorporated.

Although most of the material is new, the general objective is still much the same as was stated in the Preface to the former book:

"Laminates of engineering significance form the subject matter of this book. This includes composites of essentially all one class of material such as wood or metal, composites of several widely different materials such as plastic and fiber, composites essentially the same density throughout and composites of lightweight cores and denser stronger skins commonly referred to as sandwich structures. A given material is included in this book if it is useful from an engineering standpoint and if it is significantly different in its composite properties or use from its constituents. ... Laminates in which the surface material is essentially only a protective coating are excluded. Plated, galvanized, and similar thinly coated materials are, consequently, ruled out. Evidently, the dividing line between coated materials and laminated materials is not a sharp one, and it may be questioned whether some of the material excluded might not properly have been included, and whether some materials included might not be considered to fall in the class of coatings. Unquestionably, in many laminates one of the principal functions of the surface materials is protection.

"Although this book is devoted to laminates, it necessarily includes a good deal of information respecting the properties of the base materials of which the laminates are composed. Furthermore, it discusses subjects which are not peculiar to laminates alone. Thus the chapter on hard surfacing by fusion welding contains a discussion of hardness and abrasion resistance ... and the various sections dealing with metal-clad laminates include information respecting the corrosion resistance and other properties of the cladding metals.

"An attempt has been made throughout to emphasize the engineering properties of the materials. Where the authors have thought it desirable, they have included discussion of manufacturing and fabricating procedures, applications, and economics. In general, however, these discussions are subordinate to the exposition of engineering properties."

In a symposium such as this some overlapping among closely related chapters is unavoidable. In some instances such overlapping has been welcomed, as, for instance, among Chapters 1, 6, and 8, which in many ways complement each other. In general, however, duplication has been avoided as much as possible.

In light of changes that have occurred in the engineering uses of composite laminates in the past twenty years, some chapters of the former book have been dropped, some have been consolidated, and one has been divided. In the arrangement of the book, Chapters 1 and 2 are general in content, Chapters 3 through 7 are devoted to nonmetallic laminates, Chapters 8 and 9 are not restricted as to materials, and the balance concern mainly metallic materials, although the final chapter involves both glass and steel.

Many new authors have joined a number of those who contributed to the former book. Yehuda Stavsky joins N. J. Hoff in large-scale revision of Chapter 1; J. M. Carney and J. L. Welsh, Jr., collaborate with David R. Countryman in rewriting Chapter 4; Joseph R. Ryan revises Chapter 7 by the late George B. Watkins; D. V. Rosato joins R. T. Schwartz in completely rewriting Chapter 8; and Raymond M. Sears joins Unto U. Savolainen in Chapter 10. Charles Hemming, Chapter 9, and Howard S. Avery, Chapter 14, are the two former authors who have themselves reworked their chapters. Completely new are Richard Blomquist (Chapter 2), Russell P. Wibbens (Chapter 3), M. G. Young (Chapter 5), L. J. Broutman (Chapter 6), Robert H. Brown (Chapter 11), Jack B. Morgan (Chapter 12), Albert Hoersch, Jr. (Chapter 13), H. S. Ingham, Jr., succeeding his father (Chapter 15), and William B. Crandall (Chapter 16).

The individual authors have worked devotedly to prepare this material, in time snatched from crowded schedules. The publishers have been most helpful in every way. Such virtues as the book possesses should be credited these individuals.

ALBERT G. H. DIETZ

Cambridge, Massachusetts
February 1969

---

* Albert G. H. Dietz, *Engineering Laminates*, John Wiley & Sons, Inc., New York, 1949.

Contents

Introduction

Laminated composite engineering materials are employed in increasing volume and in more and more diverse fields because (1) they combine the properties of their component parts to obtain composite properties which may be new or unique, or (2) they make it easier or less costly to obtain certain properties than is possible with "solid" materials. The behavior of thermostat metals depends on the combination of dissimilar metals; safety glass owes its value to the combination of quite different materials; structural sandwiches possess high weight-strength ratios; clad metals combine strong or inexpensive cores with highly corrosion-resistant faces; laminated timbers can be made in sizes and shapes unattainable in solid timber; plastics-based laminates combine strength, flexibility, chemical and electrical properties, and abrasion resistance. The list can be greatly extended.

Laminating to improve the properties of materials or to combine several materials into one is not new, but the rapid expansion of the principle into a great variety of different applications is relatively recent and is accelerating at an increasingly rapid pace. To a large degree, therefore, it represents something new in the field of materials.

Early examples of laminates include glued wood, in the form of parallel-laminated members and plywood, often considered to be quite new. In Figure 1 is shown a piece of ancient Egyptian laminated wood, now in the possession of The Metropolitan Museum of Art, New York, which was found at Thebes and belongs to the Eighteenth Dynasty (about 1500 B.C.). At the bottom are five thin pieces glued-laminated to the heavier piece. At the top is a piece of glued veneer. Figure 2 shows a sarcophagous now in the Boston Museum of Fine Arts. It consists of fine wood veneers overlaid on a substrate of heavier wood. Such combinations of several layers of wood are found in a variety of ancient Egyptian items. The Romans used plywood for fine furniture. The greatly increased use of plywood and its application to engineering structures is, however, a recent phenomenon, largely made possible by improved synthetic adhesives.

Figure 1 Fragment of Egyptian Laminated Wood of the Eighteenth Dynasty (about 1500 B.C.) Found at Thebes. Five thin pieces are glued-laminated at the bottom to the heavier piece. A piece of veneer is glued at the top. (Courtesy Metropolitan Museum of Art, New York)

Figure 2 Egyptian Sarcophagous of Fine Wood Veneers Laminated to Heavier Wood Substrate. (Courtesy Boston Museum of Fine Arts)

The makers of arms and armor,* in the Near and Far East as well as in Europe, understood the value of laminated metal. In Figure 3 is an example of fifteenth century German armor (circa 1475) showing a typical laminated structure of alternating layers of steel and iron. These craftsmen understood intuitively and by experience the subtle principles of blunting and arresting fracture, a subject of fundamental research by present-day investigators.

Outstanding examples of fine craftsmanship in laminated metal are to be found in the famed Damascus swords and in the consummate skill and art of the Japanese swordmaker. The

* The information concerning armor was made available by Mr. Stephen V. Grancsay, curator of arms and armor, the Metropolitan Museum of Art, New York. Dr. Cyril Stanley Smith of the Massachusetts Institute of Technology has carefully studied the metallurgy of arms and armor, see, for example, *A History of Metallography*, University of Chicago Press, Chicago, 1960; and "Sectional Textures in the Decorative Arts," *Stereology*, Second International Congress for Stereology, 1967, Springer-Verlag, New York.

Figure 4   Detail of Japanese Sword Blade, sixteenth century. The light area at the cutting edge (bottom) is fully martensitic and intensely hard, grading into the softer body (dark) which is insulated during the quench and cools more slowly. Slightly enlarged. (Courtesy, C. S. Smith)

Figure 3   Fifteenth Century German Armor (circa 1475) Showing Structure of Alternating Layers of Steel and Iron. (Courtesy Metropolitan Museum of Art, New York)

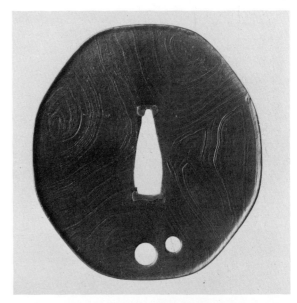

Figure 5   Japanese Sword Guard, XVIII century. Made from laminated steel, deeply etched to develop decorative pattern. (Courtesy, C. S. Smith)

Figure 6   Persian sword, made in Isfahan, XVIII century. (Courtesy, Wallace Collection, London)

fine Japanese blade not only combines several different steels or steels and irons to provide an extremely hard and keen edge with a softer tough body, but, by superb craftsmanship, laminates these materials into an exquisite example of the swordmaker's art. In one example of the many variants of the compound blade, three types of steel are employed. Edge steel is folded 20 times, core steel 8 times, and the three layers of skin steel are folded respectively 16, 6, and 7 times; all hammered out and welded together into a blade containing, theoretically at least, more than 8,000,000 layers. Special heat treatment involving rapid quenching of the edge and slow cooling of the body converts the edge into hard martensite, as shown in

Figure 4. The laminated sword guard or tsuba is often highly decorative as in the "wood grain" pattern shown in Figure 5.

The Persian blade illustrated in Figure 6 displays a laminated structure originating in the segregation of iron carbide in the solidification of the very high carbon steel ingot from which the blade is forged.

Other types of swords, both oriental and occidental, utilized strips of iron and steel welded, twisted, and again welded to make the final blade. The Merovingian sword, Figure 7, was made in Lorraine in the sixth century A.D. The handle of the Malayan kris shown in Figure 8 also displays a distinct laminar structure of textured steel.

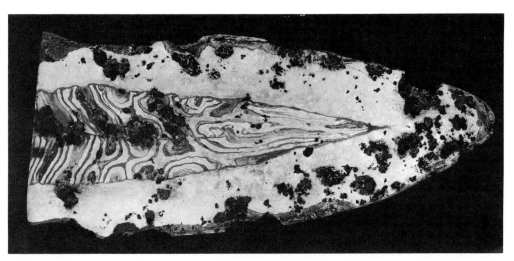

Figure 7   End of Merovingian Sword blade (Lorraine, VI century) repolished and etched. The famed Viking sword was of similar manufacture and had visible texture. (Courtesy, C. S. Smith)

Figure 8   Forged steel handle of Indonesian kris, Majapahit Empire, ca. 1400 A. D. (Courtesy, C. S. Smith)

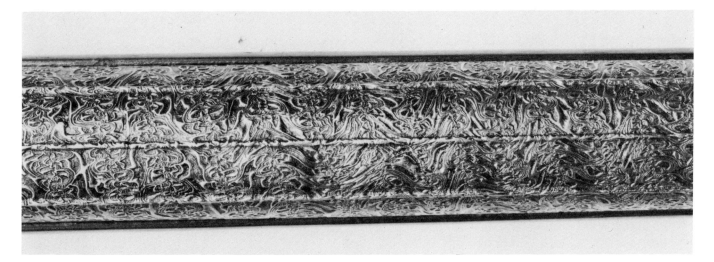

Figure 9   Detail of barrel of Indian Flintlock gun, ca. 1650 A. D. (Courtesy, Wallace Collection, London)

Oriental gun makers understood the art of combining metals as in the seventeenth-century Indian flintlock shown in Figure 9. Different kinds of iron and steel were first combined into strips; these were contorted and twisted into a helix and welded together to form the gun tube.

Modern composite engineering laminates, of course, are not generally made by such painstaking hand processes as are employed in fabricating fine pieces of armor. The basic principles — orientation of structure and strength properties, combinations of hardness, toughness, lightness, strength, durability, and other desirable engineering attributes — are essentially the same. Because no one material is adequate to the task, materials are combined — laminated — to obtain the combined properties or unique new properties necessary to fulfill the requirements of a particular engineering application.

Chapter 1

Mechanics of Composite Structures

Yehuda Stavsky

Nicholas J. Hoff

INTRODUCTION

"Demands on materials imposed by today's advanced technologies have become so diverse and severe that they often cannot be met by simple single-component materials acting alone. It is frequently necessary to combine several materials into a composite to which each constituent not only contributes its share, but whose combined action transcends the sum of the individual properties, and provides new performance unattainable by the constituents acting alone. Space vehicles, heat shields, rocket propellants, deep submergence vessels, buildings, vehicles for water and land transport, aircraft, pressure tanks, and many others impose requirements that are best met, and in many instances met only, by composite materials." These few sentences from a basic survey on composite materials by Dietz[1] explain the great interest in mechanics of heterogeneous systems which arose in the engineering and scientific community during the last quarter of a century.

In a chapter of this size it is almost impossible to cover every part of a field as vast as the mechanics of composites. It is hoped that the chapter may clarify the elastic behavior of composites and thus contribute to improvements in the design of heterogeneous structures. Furthermore, it is believed that the material presented will stimulate and enhance further research and understanding of the many unsolved problems of composite systems.

ACKNOWLEDGMENTS AND APOLOGIES

One co-author, Yehuda Stavsky, appreciates the partial support of the Technion–Israel Institute of Technology for carrying out his work on this chapter. He is also indebted to Mr. I. Smolash and Mr. S. Friedland, students at the Technion, for their efficient help in some of the examples.

No list of references or bibliography can hope to cover the wide field of the mechanics of composites. The references cited give only a very brief indication of the vast literature available. The authors apologize for not citing some important papers on composites.

YEHUDA STAVSKY is Associate Professor, Department of Mechanics, Technion–Israel Institute of Technology, Haifa, Israel.

NICHOLAS J. HOFF is Professor and Head, Department of Aeronautics and Astronautics, Stanford University, Stanford, California.

NOTATION

a, a_{ij}	matrix defined by equation 2.1.32
A, A_{ij}	matrix of elastic areas defined by equation 2.1.14
$A^* A_{ij}^*$	matrix defined by equation 2.1.20
b, b_{ij}	matrix defined by equation 2.1.32
B	beam elastic statical moment defined by equation 4.3.8
B, B_{ij}	matrix of elastic statical moments defined by equation 2.1.14
B^*, B_{ij}^*	matrix defined by equation 2.1.20
c, c_{ij}	matrix defined by equation 2.1.32
C	constant defined by equation 7.1.3
C^*, C_{ij}^*	matrix defined by equation 2.1.20
d	deflection defined by equation 7.1.1
d, d_{ij}	matrix defined by equation 2.1.32
D, D_{ij}	matrix of elastic moments of inertia defined by equation 2.1.14
D^*, D_{ij}^*	matrix defined by equation 2.1.20
e_{ij}^{kl}	elastic compliances, see equation 1.6.11
e_{ij}	elastic compliances, see equation 1.6.12
E_{ij}^{kl}	elastic stiffnesses, see equation 1.6.2
E_{ij}	elastic stiffnesses, see equation 1.6.8
E_i	Young's modulus in direction i for orthotropic sheet
E	Young's modulus for isotropic material
f_b	boundary equation
F	Airy stress function, defined by equation 2.1.18
g_{ij}	components of metric tensor
G_{LT}	shear modulus in L, T axes for orthotropic sheet
G	shear modulus for isotropic material
h	plate or shell thickness
H	horizontal stress resultant, rise of shallow spherical shell
I	the unit matrix
J	Jacobian determinant, defined by equation 1.7.2
k	modulus of compression, constant defined by equations 2.4.4, 3.2.14, and 4.1.46, coefficient of thermal conduction
k_x, k_y, k_{xy}	constant values of curvature changes $\kappa_x, \kappa_y, \kappa_{xy}$

K_i functional operators

l_{ij} direction cosines in equation 1.7.9

l, m direction cosines in equation 1.7.17

L longitudinal direction in orthotropic sheet

L_{ij}, L_i functional operators

m_1, m_2 shear-extension ratios, see equations 1.5.6 and 1.5.11

m_x, m_y, m_{xy} constant values of moments

M_x, M_y, M_{xy}, M_r, M_θ moments

M_{iT} thermal quantities defined by equation 5.1.7

n_x, n_y, n_{xy} constant values of resultants

n normal

N_x, N_y, N_{xy}, N_r, N_θ resultants

N_{iT} thermal quantities defined by equation 5.1.7

p constant defined by equation 7.1.4, hydrostatic pressure, load intensity

p_x, p_y, p_z, p_H, p_V, p_r components of load intensity

P_i constants

P force

Q_x, Q_y, Q transverse shear resultants

r radial coordinate

R_ξ, R_θ principal radii of curvature of shell of revolution

s tangent

t sheet thickness

T transverse direction in orthotropic sheet, temperature change

u, v, w displacement components

x_i, x, y, z Cartesian coordinate axes

α angle, constant defined by equation 6.2.4

α_i coefficients of thermal coefficients

A_i thermal coefficients defined by equation 5.1.5

β change of slope, constant defined by equation 6.2.3

γ_{ij} change of angle between x_i, x_j axes

γ_1, γ_2 constants defined by equation 3.2.21

δ_{ij} Kronecker's delta

$\varepsilon_{ij}, \varepsilon_i$ strain components in equations 1.6.2 and 1.6.9, respectively

$\epsilon_{ij}, \epsilon_i$ strains defined by equation 2.1.2

ζ thickness coordinate in shells

θ angle, circumferential coordinate

$\kappa_x, \kappa_y, \kappa_{xy}$, κ_r, κ_θ curvature changes

λ, μ Lamé coefficients

ν_{ij} Poisson's ratio for orthotropic sheet defined by $-\varepsilon_j/\varepsilon_i$

ν Poisson's ratio for isotropic material

ξ arc coordinate, see equation 3.1.3

σ_i normal stress in i direction

τ_{ij} shear stress, components of stress tensor

ϕ, Φ sloping angles of tangent to meridian curve before and after deformation of shell of revolution

Φ generalized stress function, see equation 2.3.14

ψ constant defined by equation 7.1.2

Ψ stress function, see equation 2.5.15

Ω potential function

$( )_,$ partial differentiation with respect to subscript following the comma

$| \; |$ absolute value, determinant

$[\;]$ matrix

1 STRESS AND STRAIN IN AN ANISOTROPIC LAMINA

1.1 Hooke's law for an orthotropic lamina

Figure 1.1(a) represents a small flat rectangular piece cut out of a lamina. On its horizontal edges is acting a tensile stress, $\sigma_1 = 200$ psi; on its vertical edges a compressive stress, $\sigma_2 = -150$ psi; and on all the four edges a shear stress, $\tau_{12} = 50$ psi. The lamina is assumed to be orthotropic, and the direction of one principal axis of elasticity is indicated in the figure by the hatching which subtends an angle $\alpha = 30°$ with the vertical. If the rectangle were cut out of a wooden board, the hatching would indicate the fibers.

It is of interest to determine the changes caused by the stresses in the length of the edges of the small rectangle and the changes in the angles subtended by the edges. These angles are right angles before the application of the loads. The elastic constants needed for such a calculation are, in general, not available in the case of an orthotropic material for an arbitrary direction of the load, but they are known when the loads are applied in the principal directions of elasticity, or natural axes.

For instance, when a tensile stress σ_L is applied in the direction of the grain as shown in Figure 1.1(c), it causes a tensile strain $\varepsilon_L = \sigma_L/E_L$ in its own direction and a compressive strain $\varepsilon_T = -\nu_{LT}\varepsilon_L$ in the perpendicular (transverse) direction. The symbol E_L is Young's modulus of elasticity for the longitudinal direction, and ν_{LT} is a material constant known as Poisson's ratio. Similarly, a transverse tensile stress σ_T causes a tensile strain $\varepsilon_T = \sigma_T/E_T$ in the transverse direction and a compressive strain $\varepsilon_L = -\nu_{TL}\varepsilon_T$ in the longitudinal direction. Here E_T is Young's modulus of elasticity in the transverse direction, and ν_{TL} is a second Poisson's ratio. When the material is isotropic, $E_L = E_T$ and $\nu_{LT} = \nu_{TL}$, but these equalities do not hold for orthotropic materials. No shear strain γ_{LT} is set up when the normal stresses are applied in the principal directions, and shear stresses τ_{LT} applied parallel to the grain and perpendicularly to it do not give rise to normal strains in the principal directions. The effect of a shear stress is just a change in angle; that is, a shear strain $\gamma_{LT} = \tau_{LT}/G_{LT}$, where G_{LT} is the modulus of elasticity in shear, or modulus of rigidity associated with the longitudinal and transverse directions. The deformations under the simultaneous action of normal and shearing stresses in the L and T directions can be summarized in the equations

$$\varepsilon_L = \sigma_L/E_L - \nu_{TL}\sigma_T/E_T, \tag{1.1.1}$$

$$\varepsilon_T = \sigma_T/E_T - \nu_{LT}\sigma_L/E_L, \tag{1.1.2}$$

$$\gamma_{LT} = \tau_{LT}/G_{LT}. \tag{1.1.3}$$

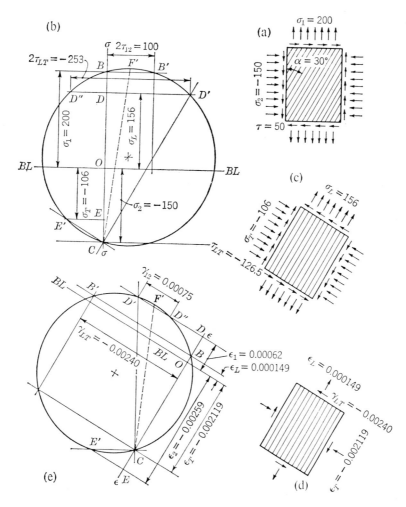

Figure 1.1   Stress and Strain in an Orthotropic Plate

Equations 1.1.1–1.1.3 are a generalization of Hooke's law for orthotropic plates.

Since the laws of elasticity are known only in directions L and T, and not in directions 1 and 2, the deformations caused by the stresses σ_1, σ_2, and τ_{12} cannot be determined directly. First the state of stress in the L, T directions corresponding to the given stresses σ_1, σ_2, and τ_{12} must be found. Then the state of strain in the L, T directions can be calculated from equations 1.1.1–1.1.3. Finally the state of strain in the 1, 2 directions is determined from the calculated strains in the L, T directions.

The calculations just outlined involve a transformation of stress and a transformation of strain. These purely geometric transformations can be carried out by means of a geometric construction involving the use of Mohr's circle.

1.2   Transformation of stress and strain by Mohr's circle

Mohr's circle for stresses

The graphic construction used here was devised by O. C. Mohr[2] and is described in standard textbooks, as, for example, in Timoshenko's *Strength of Materials* (Ref. 3, Part I, p. 38). It was applied by Norris[4] to the solution of the

stress-strain problems of wood and plywood. The technique adopted in this chapter is that suggested by Hoff.[5]

A horizontal base line BL and a vertical normal stress axis σ-σ are drawn, as shown in Figure 1.1(b). Along the vertical axis are laid out distances proportional to the normal stresses, in the present case to $\sigma_1 = 200$ psi upward, in the positive direction, from the base line, and $\sigma_2 = -150$ psi downward, that is, in the negative direction. The point on the σ-σ axis corresponding to the horizontal normal stress is marked by the letter C. From point B corresponding to the vertical normal stress on the σ-σ axis a line is drawn horizontally, that is, parallel to the shear stress indicated on the upper horizontal edge of the rectangle in Figure 1.1(a), and a distance proportional to twice the shear stress is measured in the direction of the shear stress, that is, to the right in the present example. Point B' so obtained, as well as points B and C, are on the stress circle whose center can be easily found as the point of intersection of the perpendicular bisectors of segments BB' and BC.

If the stress in any other direction is required, a parallel to the direction is drawn through point C. The normal stress in the new direction is proportional to the distance from the base line to the intersection of this parallel with the stress circle. Twice the shear stress corresponding to the new direction and to that perpendicular to it is proportional to the horizontal chord through the intersection point.

In the example of Figure 1.1. the principal directions of elasticity subtend 30° and 120° with the vertical. The former is parallel to the fibers indicated in Figure 1.1(a) and is denoted as the longitudinal (L) direction, and the latter is perpendicular to the fibers and is designated as the transverse (T) direction. A parallel to the fibers through C cuts the circle in point D'. Hence the normal stress along the fibers is represented by the segment OD. Similarly, the normal stress in the transverse direction is obtained by drawing a perpendicular to the fibers through C, projecting point E' upon the σ-σ axis, and measuring the distance OE. The shear stress associated with the L, T directions is one-half of the distance $D'D''$ measured with the appropriate scale. It is negative, since the length of the chord representing twice the shear stress decreases from BB' to zero as line CB is rotated to the position CF', changes sign, and increases again in absolute value as the line is rotated further from position CF' to position CD'.

Strain components in natural axes

The values scaled off from Figure 1.1(b) are

$$\sigma_L = 156 \text{ psi}, \quad \sigma_T = -106 \text{ psi}, \quad \tau_{LT} = -126.5 \text{ psi}. \quad (1.2.1)$$

These values are indicated in Figure 1.1(c).

The strain components in the principal elastic axes can be calculated from equations 1.1.1 for given elastic moduli of the material. If the plate is a plank of flat-sawn spruce, the elastic moduli are*

$$E_L = 1,430,000 \text{ psi}, \quad E_T = 51,400 \text{ psi}, \quad G_{LT} = 52,800 \text{ psi},$$
$$\nu_{LT} = 0.5390, \quad \nu_{TL} = 0.0194. \quad (1.2.2)$$

Substitution of these values of the elastic constants and the values of the stresses just determined in equations 1.1.1–1.1.3 yields

$$\varepsilon_L = 0.000149, \quad \varepsilon_T = -0.002119, \quad \gamma_{LT} = -0.002400. \quad (1.2.3)$$

Mohr's circle for strains

The state of strain in the 1, 2 directions can be obtained from that in the L, T directions by means of Mohr's circle. The strains are indicated in Figure 1.1(d), and the circle is constructed in Figure 1.1(e). First, lines are drawn parallel to the known normal strains. Line ε-ε is parallel to the longitudinal strain ε_L prevailing in the direction of the fibers, and the base line BL is parallel to the transverse strain ε_T which is perpendicular to the fibers. The line $\varepsilon_L = 0.000149$ is laid out along the ε-ε axis upward (in the positive direction) from point O, the intersection of the ε-ε axis with the base line, and $\varepsilon_T = -0.002119$ is laid out downward (in the negative direction). The latter point is marked C. The shear strain $\gamma_{LT} = -0.002400$ causes a leftward shift of the upper edge of the rectangle shown in Figure 1.1(d), and consequently γ_{LT} is laid out to the left from point B. The new point B' as well as points B and C lie on the strain circle whose center is

---

* These data are taken from Table 2-3 and Table 2-5 of Reference 6. In agreement with the procedure recommended, the value of the modulus in the table was increased by 10 per cent to correct for shear deflections.

easily found as the intersection of the perpendicular bisectors of segments BB' and BC.

The strain circle contains all the data related to the state of strain of the rectangle under the action of the stresses indicated in Figure 1.1(a). Of particular interest are the normal strains in directions 1 and 2. The former can be determined by passing a vertical (parallel to direction 1) through C. The distance of its intersection D' from the base line, that is, distance DO, is proportional to ε_1. Similarly, the strain ε_2 in direction 2 is proportional to the distance EO. The shear strain γ_{12} associated with directions 1, 2 is proportional to the chord $D'D''$ through D' drawn parallel to the base line.

From Figure 1.1(e) the following numerical values can be scaled off:

$$\varepsilon_1 = 0.00062, \quad \varepsilon_2 = -0.00259, \quad \gamma_{12} = 0.00075. \quad (1.2.4)$$

The shear strain is positive since the negative value of -0.002400 associated with the L, T directions decreases to zero as line CB is rotated into position CF' where the chord parallel to the base line has zero length. From position CF' on the length of the chord it increases positively as position CD' is approached.

The numerical values just given constitute the solution of the strain problem of the orthotropic plate under the action of the stresses indicated in Figure 1.1(a).

Accuracy of the graphic construction

The accuracy of the graphic construction depends on the scale of the drawings, the reliability of the instruments used, and the care exercised by the draftsman-computer. With average instruments, ordinary care, and diagrams drawn on $8\frac{1}{2} \times 11$-in. sheets the error in the results is about 1 per cent of the maximum stress or strain represented.

1.3 Particular stress circles

In Figure 1.2(a) a particular type of stress circle is shown, namely, the one that corresponds to a small rectangular plate loaded only by a tensile stress $\sigma_1 = 100$ psi [Fig. 1.2(b)]. This stress is laid out vertically upward from the base line, and the end point of the segment is marked B. No shear stress is acting along the horizontal edges of the rectangle so that points B' and B coincide. Since $\sigma_2 = 0$, point C lies on the base line. Consequently the center of the stress circle is the point of the σ-σ axis lying midway between B and C.

The stresses corresponding to any arbitrary direction can be found as in the preceding example. For instance, the normal stress for $\alpha = 45°$ is obtained by drawing line CD' and scaling off CD. Twice the shearing stresses τ_{LT} is proportional to the horizontal chord through D' which, in this case, happens to be a diameter. The fact that τ_{LT} is negative can be understood if a similar but slightly different state of stress is investigated.

In Figure 1.2(c) the stress circle is presented for the loading indicated in Figure 1.2(d) which comprises a small positive shear stress $\tau_{12} = 10$ psi in addition to the tensile stress $\sigma_1 = 100$ psi. The circle is constructed in accordance with the rules given earlier. The positive shear stress corresponding to the vertical direction CB decreases to zero as line CB is rotated to position CF' and increases again, but negatively, as the line rotates further toward position CD'. When the initial shear stress corresponding to the horizontal direction is smaller, the change in sign from positive to negative occurs earlier. When the shear stress is zero for the 1, 2 directions, the shear stress is negative for any direction of line CD' between the vertical and the horizontal swept during a clockwise rotation.

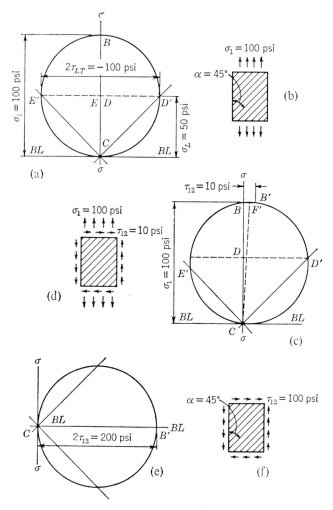

(a)

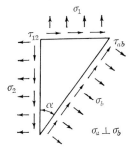

(b)

(d)

(c)

(e)   (f)

Figure 1.2   Particular Stress Circles

Figure 1.2(e) is a stress circle that represents the state of stress when there are neither vertical nor horizontal normal stresses, and the only loading of the small rectangle shown in Figure 1.2(f) is the positive shear stress $\tau_{12} = 100$ psi. The diagrams are self-explanatory, but it may be mentioned that in the case of a negative shear stress point B' would lie to the left of point C.

1.4   Analytic transformation of stress and strain in orthotropic plates

The graphic constructions presented in Sections 1.2 and 1.3 represent the following transformation formulas for stresses and strains. These equations are derived from the equilibrium of a prismatic element of a plate (see, for example, Ref. 7, p. 13).

With the sign convention of Figure 1.3 the relationships among the stresses are

$$\sigma_a = \sigma_1 \cos^2 \alpha + \sigma_2 \sin^2 \alpha + \tau_{12} \sin 2\alpha,$$

$$\sigma_b = \sigma_1 \sin^2 \alpha + \sigma_2 \cos^2 \alpha - \tau_{12} \sin 2\alpha, \qquad (1.4.1)$$

$$\tau_{ab} = \tau_{12} \cos 2\alpha + \tfrac{1}{2}(\sigma_2 - \sigma_1) \sin 2\alpha.$$

Similar equations govern the transformation of strain:

$$\varepsilon_a = \varepsilon_1 \cos^2 \alpha + \varepsilon_2 \sin^2 \alpha + \tfrac{1}{2}\gamma_{12} \sin 2\alpha,$$

$$\varepsilon_b = \varepsilon_1 \sin^2 \alpha + \varepsilon_2 \cos^2 \alpha - \tfrac{1}{2}\gamma_{12} \sin 2\alpha, \qquad (1.4.2)$$

$$\gamma_{ab} = \gamma_{12} \cos 2\alpha + (\varepsilon_2 - \varepsilon_1) \sin 2\alpha.$$

The analytic expressions for the stress-strain relations of an orthotropic plate referred to the natural axes are given by equations 1.1.1–1.1.3. The stress-strain relations in other coordinate systems are more involved since normal stresses give

Figure 1.3   Stresses Acting on a Prism

rise to shear strains as well as to normal strains, and shear stresses cause normal strains, in addition to shear strains (see Sec. 1.6)

When the stresses are given in two mutually perpendicular directions 1, 2 which do not coincide with the principal directions of elasticity L, T, the first step in getting the strain components in the 1, 2 coordinate axes is the transformation of the given stress components to the natural axes L, T by using equations 1.4.1. Equations 1.1.1–1.1.3 can then be used to get the strain components in the natural axes. Finally the strains are transformed, with the aid of appropriate equations of the type 1.4.2, to the 1, 2 directions. The calculations indicated were carried out for pure tension and pure shear and the transformation equations were derived for the elastic moduli of an orthotropic plate, as outlined in the following section.

1.5   Transformation of elastic moduli of orthotropic plates

Case of pure tension

In Figure 1.4(a) the longitudinal or L direction is shown by the hatched lines in the rectangle. L and T are the natural

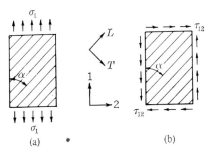

(a)   (b)

Figure 1.4   Notation and Sign Convention for Analytic Expressions for Strain

axes of the orthotropic plate shown. The loading consists of tensile stresses σ_1 subtending an angle α with the grain. The normal strain ε_1 in the direction of the stress can be calculated from the formula

$$\varepsilon_1 = \sigma_1/E_1, \tag{1.5.1}$$

where the modulus of elasticity E_1 in direction 1 can be obtained from the transformation equation

$$\frac{E_L}{E_1} = \cos^4\alpha + \frac{E_L}{E_T}\sin^4\alpha + \tfrac{1}{4}\left(\frac{E_L}{G_{LT}} - 2\nu_{LT}\right)\sin^2 2\alpha. \tag{1.5.2}$$

Young's moduli of elasticity in the longitudinal (with-the-grain) and transverse (across-the-grain) directions are represented by E_L and E_T, G_{LT} is the shear modulus corresponding to these natural axes, and ν_{LT} is Poisson's ratio associated with a contraction in the transverse direction caused by a tensile stress in the longitudinal direction.

The transverse contraction in direction 2 due to the application of σ_1 is governed by the equation

$$\varepsilon_2 = -\nu_{12}\varepsilon_1 = -\nu_{12}\,\sigma_1/E_1, \tag{1.5.3}$$

where Poisson's ratio ν_{12} is given by the transformation formula

$$\nu_{12} = \frac{E_1}{E_L}\left[\nu_{LT} - \tfrac{1}{4}\left(1 + 2\nu_{LT} + \frac{E_L}{E_T} - \frac{E_L}{G_{LT}}\right)\sin^2 2\alpha\right]. \tag{1.5.4}$$

The shear strain γ_{12}, which is the change of the right angle between the 1, 2 axes due to σ_1, is

$$\gamma_{12} = -m_1\sigma_1/E_L, \tag{1.5.5}$$

where the shear-extension ratio m_1 is

$$m_1 = \sin 2\alpha\left[\nu_{LT} + \frac{E_L}{E_T} - \frac{E_L}{2G_{LT}}\right.$$
$$\left. - \cos^2\alpha\left(1 + 2\nu_{LT} + \frac{E_L}{E_T} - \frac{E_L}{G_{LT}}\right)\right]. \tag{1.5.6}$$

In Figure 1.5 the elastic moduli expressed in equations 1.5.2, 1.5.4 and 1.5.6 are plotted for the case when the material is spruce having the properties given by equations 1.2.2. Other orthotropic materials exhibit a similar behavior, but the numerical values may differ considerably.

Case of pure shear

When the loading consists only of shearing stresses acting along edges situated relative to the natural axes L, T as shown in Figure 1.4(b), the shear strain γ_{12} can be calculated from the equation

$$\gamma_{12} = \tau_{12}/G_{12}, \tag{1.5.7}$$

where the modulus of shear G_{12} associated with directions 1, 2 is defined by the equation

$$\frac{G_{LT}}{G_{12}} = \frac{G_{LT}}{E_L}\left[\left(1 + 2\nu_{LT} + \frac{E_L}{E_T}\right)\right.$$
$$\left. - \left(1 + 2\nu_{LT} + \frac{E_L}{E_T} - \frac{E_L}{G_{LT}}\right)\cos^2 2\alpha\right]. \tag{1.5.8}$$

At the same time the applied shear stress causes a normal strain ε_1 in the 1 direction,

$$\varepsilon_1 = -m_1\tau_{12}/E_L, \tag{1.5.9}$$

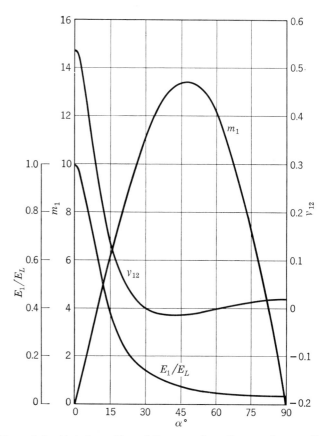

Figure 1.5   Variation with α of E_1, ν_{12} and m_1 (for notation see Fig. 1.4). When α is negative, E_1/E_L and ν_{12} remain unchanged, and m_1 becomes negative

where m_1 is the shear-extension constant given in equation 1.5.6, and a normal strain ε_2 in the 2 direction,

$$\varepsilon_2 = -m_2\tau_{12}/E_L, \tag{1.5.10}$$

where the second shear-extension constant m_2 can be calculated from the equation

$$m_2 = \sin 2\alpha\left[\nu_{LT} + \frac{E_L}{E_T} - \frac{E_L}{2G_{LT}}\right.$$
$$\left. - \sin^2\alpha\left(1 + 2\nu_{LT} + \frac{E_L}{E_T} - \frac{E_L}{G_{LT}}\right)\right]. \tag{1.5.11}$$

It may be noted that the multiplier of τ_{12} in equation 1.5.9 is identical with the multiplier of σ_1 in equation 1.5.5. This is in agreement with Maxwell's reciprocal theorem as given, for example, by Hoff (Ref. 8, p. 373), according to which the extension in the 1 direction caused by a unit shear stress acting along the 1, 2 directions must be the same as the angular change in the 1, 2 directions caused by a unit normal stress in the 1 direction.

In Figure 1.6 the elastic constants expressed in equations

1.5.8–1.5.11 are plotted for the case when the material of the plate is spruce having the properties stated by 1.2.2. Other orthotropic materials exhibit a similar behavior, but the numerical values may differ considerably.

Connections between the elastic constants

As was shown by equations 1.1.1 the normal strain in direction T caused by a normal stress σ_L in direction L is $-\nu_{LT}\sigma_L/E_L$, and the normal strain in direction L caused by a normal stress σ_T in direction T is $-\nu_{TL}\sigma_T/E_T$. According

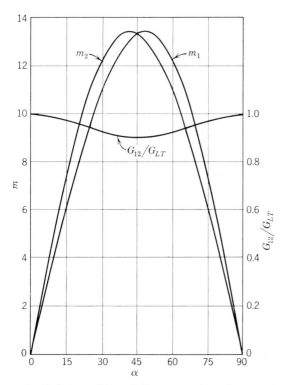

Figure 1.6 Variation with α of G_{12}, m_1, and m_2 (for notation see Fig. 1.4). When α is negative, G_{12}/G_{LT} remains unchanged and m_1 and m_2 change sign

to Maxwell's reciprocal theorem (Hoff, Ref. 8, p. 373) the two strains must be equal if the two stresses are of equal magnitude and sense. Consequently,

$$\nu_{LT}/\nu_{TL} = E_L/E_T. \qquad (1.5.12)$$

This equation is useful in the evaluation of tests carried out for determining the elastic constants. It can be stated in a more general form,

$$\nu_{ij}/\nu_{ji} = E_i/E_j, \qquad (1.5.13)$$

where i and j refer to any two mutually perpendicular directions.

The expression for the modulus of elasticity E_2 may be obtained from equation 1.5.2 by introducing for α the angle $\alpha + 90°$; hence

$$\frac{E_L}{E_2} = \sin^4 \alpha + \frac{E_L}{E_T}\cos^4 \alpha + \frac{1}{4}\left(\frac{E_L}{G_{LT}} - 2\nu_{LT}\right)\sin^2 2\alpha. \quad (1.5.14)$$

It is interesting to note that for $\alpha = 45°$ the shear modulus

G_{12} is independent of G_{LT}:

$$G_{12} = \frac{E_L E_T}{E_L + E_T + 2E_T \nu_{LT}}. \qquad (1.5.15)$$

For isotropic materials $E_L = E_T = E$, $G_{12} = G$, and $\nu_{LT} = \nu$ and equation 1.5.15 reduces to the known relation

$$G = \frac{E}{2(1 + \nu)}. \qquad (1.5.16)$$

Calculation of the example solved graphically in Figure 1.1

The numerical example shown in Figure 1.1 was solved with the aid of Mohr's circle. The graphic construction can be replaced by calculation if use is made of equations 1.5.1–1.5.14. The effects of the three stresses σ_1, σ_2, and τ_{12} are treated separately. The strain-stress relations in the 1, 2 coordinate axes are as follows:

$$\varepsilon_1 = \frac{\sigma_1}{E_1} - \nu_{21}\frac{\sigma_2}{E_2} - m_1\frac{\tau_{12}}{E_L}, \qquad (1.5.17)$$

$$\varepsilon_2 = -\nu_{12}\frac{\sigma_1}{E_1} + \frac{\sigma_2}{E_2} - m_2\frac{\tau_{12}}{E_L}, \qquad (1.5.18)$$

$$\gamma_{12} = -m_1\frac{\sigma_1}{E_L} - m_2\frac{\sigma_2}{E_L} + \frac{\tau_{12}}{G_{12}}. \qquad (1.5.19)$$

The elastic moduli in the 1, 2 axes are

$E_1 = 199{,}239$ psi, $E_2 = 69{,}459$ psi, $\nu_{12} = 0.001541$,

$\nu_{21} = 0.000537$, $m_1 = 11.004$, $m_2 = 12.223$, $G_{12} = 48{,}981$ psi.

$$(1.5.20)$$

(a) $\sigma_1 = 200$ psi:
  The strain components become

  $\varepsilon_1^a = 0.0010038$, $\varepsilon_2^a = -0.0000016$, $\gamma_{12}^a = -0.0015391$.

$$(1.5.21)$$

(b) $\sigma_2 = -150$ psi:
  The strain components take the values

  $\varepsilon_1^b = 0.0000012$, $\varepsilon_2^b = -0.0021596$, $\gamma_{12}^b = 0.0012822$.

$$(1.5.22)$$

(c) $\tau_{12} = 50$ psi:
  Using the τ_{12} terms in equations 1.5.17–1.5.19 and the suitable constants of 1.5.20 one finds for the strain components

  $\varepsilon_1^c = -0.0003848$, $\varepsilon_2^c = -0.0004274$, $\gamma_{12}^c = 0.0010208$.

$$(1.5.23)$$

The combined action of σ_1, σ_2, and τ_{12} as indicated in Figure 1.1 results in the following strains which are the sum of the corresponding quantities given in equations 1.5.21–1.5.23:

$$\varepsilon_1 = 0.0006202, \qquad (1.5.24)$$

$$\varepsilon_2 = -0.0025885, \qquad (1.5.25)$$

$$\gamma_{12} = 0.0007639. \qquad (1.5.26)$$

It may be noted that because of the low value of Poisson's ratios (ν_{12} and ν_{21}) in this example, ε_1^b and ε_2^a can be neglected as compared to ε_1 and ε_2. Furthermore the maximum deviation between the strain values obtained graphically and those computed with the aid of the diagrams is less than 1 per cent of the maximum strain.

1.6 The generalized Hooke's law for anisotropic media

In equations 1.1.1–1.1.3 Hooke's law was expressed for an orthotropic layer referred to its natural axes. When another reference system is used to describe the stress-strain relations for the same orthotropic layer, equations 1.5.17–1.5.19 are obtained showing that the normal stresses depend on the shear strains as well as on the normal strains. A natural generalization of Hooke's law for the linear elastic medium is expressed by the statement (Ref. 9, p. 97):

Each of the components of stress at any point of the medium is a linear function of the components of strain at the point.

Thus

$$\tau_{11} = E_{11}^{11}\varepsilon_{11} + E_{11}^{22}\varepsilon_{22} + E_{11}^{33}\varepsilon_{33} + E_{11}^{23}\varepsilon_{23} + E_{11}^{32}\varepsilon_{32}$$
$$\vdots \qquad + E_{11}^{31}\varepsilon_{31} + E_{11}^{13}\varepsilon_{13} + E_{11}^{21}\varepsilon_{21} + E_{11}^{12}\varepsilon_{12},$$
$$\tau_{12} = E_{12}^{11}\varepsilon_{11} + E_{12}^{22}\varepsilon_{22} + E_{12}^{33}\varepsilon_{33} + E_{12}^{23}\varepsilon_{23} + E_{12}^{32}\varepsilon_{32}$$
$$+ E_{12}^{31}\varepsilon_{31} + E_{12}^{13}\varepsilon_{13} + E_{12}^{21}\varepsilon_{21} + E_{12}^{12}\varepsilon_{12},$$

$$(1.6.1)$$

or in tensor notation

$$\tau_{ij} = E_{ij}^{\alpha\beta}\varepsilon_{\alpha\beta} \qquad (i, j, \alpha, \beta = 1, 2, 3). \qquad (1.6.2)$$

The two second-rank stress and strain tensors are related through *elastic stiffnesses* E_{kl}^{ij} which according to the quotient theorem (see, for example, Ref. 10, p. 30) constitute the components of a tensor of rank four.

If couple stresses (see, for example, Refs. 11 and 12) and body moments[13] are assumed to vanish, then the stress tensor τ is symmetric; consequently

$$E_{ij}^{kl} = E_{ji}^{kl}, \qquad (1.6.3)$$

and the number of 81 stiffnesses appearing in equation 1.6.2 is reduced to 54.

Furthermore, one can assume without loss of generality (see, for example, Ref. 14, p. 59) that

$$E_{ij}^{kl} = E_{ij}^{lk}, \qquad (1.6.4)$$

and so the number of the independent elastic stiffnesses is reduced to 36 at most.

In Section 1.8 it will be shown that the stiffnesses are coefficients in an expansion of the elastic potential W in terms of components of strain. It can be shown from thermodynamical consideration that W exists, at least, for the reversible isothermal or adiabatic processes. The existence of the strain-energy function is ensured by the symmetry relation,

$$E_{ij}^{rs} = E_{rs}^{ij}, \qquad (1.6.5)$$

which holds for rectangular coordinates, and the number of the stiffnesses for the most general linear anisotropic body is 21. It is noted that the generalization of the symmetry relations (1.6.5) for general coordinates is given by equations 1.8.4 (see also Ref. 15, p. 155).

For a heterogeneous medium the E's are functions of position and possibly of temperature or time. It is convenient, especially in applications, to avoid dealing with double sums and so the following notation is introduced (see, for example, Ref. 16, p. 134):

$$\tau_{11} = \tau_1, \ \tau_{22} = \tau_2, \ \tau_{33} = \tau_3, \ \tau_{23} = \tau_4, \ \tau_{31} = \tau_5, \ \tau_{12} = \tau_6; \qquad (1.6.6)$$

$$\varepsilon_{11} = \varepsilon_1, \ \varepsilon_{22} = \varepsilon_2, \ \varepsilon_{33} = \varepsilon_3, \ \varepsilon_{23} = \varepsilon_4, \ \varepsilon_{31} = \varepsilon_5, \ \varepsilon_{12} = \varepsilon_6; \qquad (1.6.7)$$

$$\begin{aligned}
E_{11}^{11} &= E_{11}, \ E_{11}^{22} = E_{12}, \ E_{11}^{33} = E_{13}, \ E_{11}^{23} = E_{14}, \ E_{11}^{31} = E_{15}, \ E_{11}^{12} = E_{16}, \\
E_{22}^{22} &= E_{22}, \ E_{22}^{33} = E_{23}, \ E_{22}^{23} = E_{24}, \ E_{22}^{31} = E_{25}, \ E_{22}^{12} = E_{26}, \\
E_{33}^{33} &= E_{33}, \ E_{33}^{23} = E_{34}, \ E_{33}^{31} = E_{35}, \ E_{33}^{12} = E_{36}, \\
E_{23}^{23} &= E_{44}, \ E_{23}^{31} = E_{45}, \ E_{23}^{12} = E_{46}, \\
E_{31}^{31} &= E_{55}, \ E_{31}^{12} = E_{56}, \\
E_{12}^{12} &= E_{66}.
\end{aligned} \right\} \qquad (1.6.8)$$

Instead of equations 1.6.2 we write

$$\begin{bmatrix} \tau_1 \\ \tau_2 \\ \tau_3 \\ \tau_4 \\ \tau_5 \\ \tau_6 \end{bmatrix} \begin{bmatrix} E_{11} & E_{12} & E_{13} & E_{14} & E_{15} & E_{16} \\ & E_{22} & E_{23} & E_{24} & E_{25} & E_{26} \\ & & E_{33} & E_{34} & E_{35} & E_{36} \\ & & & E_{44} & E_{45} & E_{46} \\ & & & & E_{55} & E_{56} \\ & & & & & E_{66} \end{bmatrix} \begin{bmatrix} \varepsilon_1 \\ \varepsilon_2 \\ \varepsilon_3 \\ \varepsilon_4 \\ \varepsilon_5 \\ \varepsilon_6 \end{bmatrix}, \qquad (1.6.9)$$

or more compactly

$$\tau_i = E_{i\alpha}\varepsilon_\alpha \qquad (i, \alpha = 1, 2, \ldots, 6). \qquad (1.6.10)$$

Since the relation between τ_{ij} and ε_{ij} must be reversible, we have

$$\varepsilon_{ij} = e_{ij}^{\alpha\beta}\tau_{\alpha\beta} \qquad (i, j, \alpha, \beta = 1, 2, 3), \qquad (1.6.11)$$

where the e's are named *elastic compliances* of the linear elastic anisotropic medium and like the E's constitute the components of a fourth-rank tensor.

In order to use the contracted notation 1.6.6 and 1.6.7 in the strain-stress relations of the form

$$\varepsilon_i = e_{i\alpha}\tau_\alpha \qquad (i, \alpha = 1, 2, \ldots, 6), \qquad (1.6.12)$$

the following relations must hold (see, for example, Ref. 17, p. 8):

$$e_{11}^{11} = e_{11}, \; e_{11}^{22} = e_{12}, \; e_{11}^{33} = e_{13}, \; e_{11}^{23} = \tfrac{1}{2}e_{14}, \; e_{11}^{31} = \tfrac{1}{2}e_{15}, \; e_{11}^{12} = \tfrac{1}{2}e_{16},$$

$$e_{22}^{22} = e_{22}, \; e_{22}^{33} = e_{23}, \; e_{22}^{23} = \tfrac{1}{2}e_{24}, \; e_{22}^{31} = \tfrac{1}{2}e_{25}, \; e_{22}^{12} = \tfrac{1}{2}e_{26},$$

$$e_{33}^{33} = e_{33}, \; e_{33}^{23} = \tfrac{1}{2}e_{34}, \; e_{33}^{31} = \tfrac{1}{2}e_{35}, \; e_{33}^{12} = \tfrac{1}{2}e_{36},$$

$$e_{23}^{23} = \tfrac{1}{4}e_{44}, \; e_{23}^{31} = \tfrac{1}{2}e_{45}, \; e_{23}^{12} = \tfrac{1}{2}e_{46},$$

$$e_{31}^{31} = \tfrac{1}{4}e_{55}, \; e_{31}^{12} = \tfrac{1}{2}e_{56},$$

$$e_{12}^{12} = \tfrac{1}{4}e_{66}.$$

$$(1.6.13)$$

1.7 Transformation of stresses, strains, stiffnesses, and compliances

Let Hooke's law, equations 1.6.2 or 1.6.11, be given in any coordinate system x_1, x_2, x_3. If a single-valued reversible transformation of coordinates is introduced

$$\bar{x}_i = \bar{x}_i(x_1, x_2, x_3) \qquad (i = 1, 2, 3), \qquad (1.7.1)$$

with a nonvanishing Jacobian determinant

$$J \equiv \left| \frac{\partial \bar{x}_i}{\partial x_j} \right| \neq 0, \qquad (1.7.2)$$

the inverse transformation may be written

$$x_i = x_i(\bar{x}_1, \bar{x}_2, \bar{x}_3) \qquad (i = 1, 2, 3), \qquad (1.7.3)$$

Hooke's law may be written in the new system of coordinates $\bar{x}$ as follows:

$$\bar{\tau}_{ij} = \bar{E}_{ij}^{\alpha\beta} \, \bar{\varepsilon}_{\alpha\beta}, \qquad \bar{\varepsilon}_{ij} = \bar{e}_{ij}^{\alpha\beta} \, \bar{\tau}_{\alpha\beta}. \qquad (1.7.4)$$

Our object is to relate the components of the stress, strain, stiffness, and compliance tensors in the $\bar{x}$ system to the corresponding components in the original x system.

The stress and strain components transform according to the law of transformation of a second-rank tensor

$$\bar{\tau}_{ij} = \frac{\partial x_\alpha}{\partial \bar{x}_i} \frac{\partial x_\beta}{\partial \bar{x}_j} \tau_{\alpha\beta}, \qquad \bar{\varepsilon}_{ij} = \frac{\partial x_\alpha}{\partial \bar{x}_i} \frac{\partial x_\beta}{\partial \bar{x}_j} \varepsilon_{\alpha\beta}. \qquad (1.7.5)$$

The E's and the e's transform according to the transformation law of a tensor of rank four:

$$\bar{E}_{ij}^{mn} = \frac{\partial \bar{x}_m}{\partial x_\gamma} \frac{\partial \bar{x}_n}{\partial x_\delta} \frac{\partial x_\alpha}{\partial \bar{x}_i} \frac{\partial x_\beta}{\partial \bar{x}_j} E_{\alpha\beta}^{\gamma\delta}, \qquad (1.7.6)$$

$$\bar{e}_{ij}^{mn} = \frac{\partial \bar{x}_m}{\partial x_\gamma} \frac{\partial \bar{x}_n}{\partial x_\delta} \frac{\partial x_\alpha}{\partial \bar{x}_i} \frac{\partial x_\beta}{\partial \bar{x}_j} e_{\alpha\beta}^{\gamma\delta}, \qquad (1.7.7)$$

where all the indices may take the values 1, 2, 3.

As a special case of frequent use we indicate the transformation of orthogonal Cartesian systems x and $\bar{x}$. Equations 1.7.1 and 1.7.3 become

$$\bar{x}_i = \frac{\partial \bar{x}_i}{\partial x_\alpha} x_\alpha, \qquad x_i = \frac{\partial x_i}{\partial \bar{x}_\alpha} \bar{x}_\alpha \qquad (i, \alpha = 1, 2, 3). \quad (1.7.8)$$

Each of the $\partial \bar{x}_i / \partial x_j$ is the direction cosine of the angle between the $\bar{x}_i$ and the x_j axis; consequently

$$\frac{\partial \bar{x}_i}{\partial x_j} = \frac{\partial x_j}{\partial \bar{x}_i} \equiv l_{ij} \qquad \text{(say)}. \qquad (1.7.9)$$

Equations 1.7.8 can be summarized in a table which reads, for example,

$$\bar{x}_1 = l_{11} x_1 + l_{12} x_2 + l_{13} x_3,$$
$$x_1 = l_{11} \bar{x}_1 + l_{21} \bar{x}_2 + l_{31} \bar{x}_3 \qquad (1.7.10)$$

(see Table 1.1).

Table 1.1

	x_1	x_2	x_3
$\bar{x}_1$	l_{11}	l_{12}	l_{13}
$\bar{x}_2$	l_{21}	l_{22}	l_{23}
$\bar{x}_3$	l_{31}	l_{32}	l_{33}

The transformation law of equations 1.7.4, 1.7.6, and 1.7.7 reads

$$\bar{\tau}_{ij} = l_{i\alpha} l_{j\beta} \tau_{\alpha\beta}, \qquad \bar{\varepsilon}_{ij} = l_{i\alpha} l_{j\beta} \varepsilon_{\alpha\beta}, \qquad (1.7.11)$$

$$\bar{E}_{ij}^{mn} = l_{m\gamma} l_{n\delta} l_{i\alpha} l_{j\beta} E_{\alpha\beta}^{\gamma\delta}, \qquad (1.7.12)$$

$$\bar{e}_{ij}^{mn} = l_{m\gamma} l_{n\delta} l_{i\alpha} l_{j\beta} e_{\alpha\beta}^{\gamma\delta}, \qquad (1.7.13)$$

$$(m, n, i, j, \alpha, \beta, \gamma, \delta = 1, 2, 3).$$

Note that for an orthogonal transformation

$$l_{\alpha m} l_{\alpha n} = \delta_{mn} \qquad (\alpha, m, n = 1, 2, 3). \qquad (1.7.14)$$

The Kronecker delta, δ_{mn}, is defined to have the value one if $m = n$ and zero if $m \neq n$. Accordingly, there are three equations of the type $m = n$,

$$l_{11}^2 + l_{21}^2 + l_{31}^2 = 1, \qquad (1.7.15)$$

and three equations of the type $m \neq n$,

$$l_{11} l_{12} + l_{21} l_{22} + l_{31} l_{32} = 0. \qquad (1.7.16)$$

Further, it can be easily verified (see, for example, Ref. 18, p. 28) that $|l_{ij}| = \pm 1$; the case when the determinant is $+1$ corresponds to the transformation of rotation, whereas -1 corresponds to the transformation of reflection or a reflection followed by a rotation.

As an example we consider the case where the $\bar{x}$ system is obtained from the x-coordinate axes by rotation about the x_3 axis through an angle θ in counterclockwise direction (see Fig. 1.7). For this transformation, Table 1.1 has the values

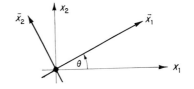

Figure 1.7

Table 1.2

	x_1	x_2	x_3
$\bar{x}_1$	l_{11}	l_{12}	0
$\bar{x}_2$	l_{21}	l_{22}	0
$\bar{x}_3$	0	0	1

shown in Table 1.2, where

$$l_{11} = \cos\theta = l, \qquad l_{12} = \sin\theta = m,$$
$$l_{21} = -\sin\theta = -m, \quad l_{22} = \cos\theta = l. \qquad (1.7.17)$$

According to the transformation law of equations 1.7.11, 1.7.12, and 1.7.13, and in view of Table 1.2, one finds

$$\bar{\tau}_{11} = l_{11}^2\,\tau_{11} + 2l_{11}\,l_{12}\,\tau_{12} + l_{12}^2\,\tau_{22},$$
$$\bar{\varepsilon}_{11} = l_{11}^2\,\varepsilon_{11} + 2l_{11}\,l_{12}\,\varepsilon_{12} + l_{12}^2\,\varepsilon_{22}, \qquad (1.7.18)$$

$$\bar{E}_{11}^{11} = l_{11}^4\,E^{11} + 2l_{11}^2\,l_{12}^2\,E^{22} + 4l_{11}^3\,l_{12}\,E^{12}$$
$$+ l_{12}^4\,E^{22} + 4l_{11}\,l_{12}^3\,E^{12} + 4l_{11}^2\,l_{12}^2\,E^{12}; \qquad (1.7.19)$$

or, in the contracted notation of equations 1.6.8 and 1.7.17, equation 1.7.19 reads

$$\bar{E}_{11} = l^4 E_{11} + 2l^2 m^2 E_{12} + 4l^3 m E_{16}$$
$$+ m^4 E_{22} + 4lm^3 E_{26} + 4l^2 m^2 E_{66}. \qquad (1.7.20)$$

Similarly the transformation equation for the elastic compliance $\bar{e}_{11}^{11}$ is

$$\bar{e}_{11}^{11} = l_{11}^4\,e^{11} + 2l_{11}^2\,l_{12}^2\,e^{22} + 4l_{11}^3\,l_{12}\,e^{12}$$
$$+ l_{12}^4\,e^{22} + 4l_{11}\,l_{12}^3\,e^{12} + 4l_{11}^2\,l_{12}^2\,e^{12}; \qquad (1.7.21)$$

or, in the contracted notation of equations 1.6.13 and 1.7.17, equation 1.7.21 reads

$$\bar{e}_{11} = l^4 e_{11} + 2l^2 m^2 e_{12} + 2l^3 m e_{16} + m^4 e_{22} + 2lm^3 e_{26} + l^2 m^2 e_{66}. \qquad (1.7.22)$$

The 21 equations for the transformed stiffnesses and compliances, using the contracted notation, are summarized in a form similar to that given by Hearmon (Ref. 17, p. 12):

$$\begin{bmatrix} \bar{E}_{13}(\bar{e}_{13}) \\ \bar{E}_{23}(\bar{e}_{23}) \\ 2\bar{E}_{36}(\bar{e}_{36}) \end{bmatrix} = \begin{bmatrix} l^2 & m^2 & lm \\ m^2 & l^2 & -lm \\ -2lm & 2lm & l^2 - m^2 \end{bmatrix} \begin{bmatrix} E_{13}(e_{13}) \\ E_{23}(e_{23}) \\ 2E_{36}(e_{36}) \end{bmatrix}, \qquad (1.7.25)$$

$$\begin{bmatrix} \bar{E}_{44}(\bar{e}_{44}) \\ \bar{E}_{45}(\bar{e}_{45}) \\ \bar{E}_{55}(\bar{e}_{55}) \end{bmatrix} = \begin{bmatrix} l^2 & -2lm & m^2 \\ lm & l^2 - m^2 & -lm \\ m^2 & 2lm & l^2 \end{bmatrix} \begin{bmatrix} E_{44}(e_{44}) \\ E_{45}(e_{45}) \\ E_{55}(e_{55}) \end{bmatrix}, \qquad (1.7.26)$$

$$\begin{bmatrix} \bar{E}_{34}(\bar{e}_{34}) \\ \bar{E}_{35}(\bar{e}_{35}) \end{bmatrix} = \begin{bmatrix} l & -m \\ m & l \end{bmatrix} \begin{bmatrix} E_{34}(e_{34}) \\ E_{35}(e_{35}) \end{bmatrix}, \qquad (1.7.27)$$

$$\bar{E}_{33} = E_{33}, \quad \bar{e}_{33} = e_{33}. \qquad (1.7.28)$$

1.8 The strain-energy function

The volume density of strain energy or the elastic potential W is taken in linear elasticity as

$$W = \tfrac{1}{2}\tau_{\alpha\beta}\,\varepsilon^{\alpha\beta} = \tfrac{1}{2}\tau^{\alpha\beta}\varepsilon_{\alpha\beta} \qquad (\alpha, \beta = 1, 2, 3), \qquad (1.8.1)$$

which holds for general coordinates.

In view of Hooke's law (equations 1.6.2 or 1.6.11), the strain-energy density function may be expressed as a homogeneous quadratic form in the strains or the stresses which is invariant under all transformations of coordinates:

$$W = \tfrac{1}{2}E_{\alpha\beta}^{\gamma\delta}\,\varepsilon^{\alpha\beta}\,\varepsilon_{\gamma\delta} = \tfrac{1}{2}E_{\alpha\beta\gamma\delta}\,\varepsilon^{\alpha\beta}\,\varepsilon^{\gamma\delta} = \tfrac{1}{2}E^{\alpha\beta\gamma\delta}\,\varepsilon_{\alpha\beta}\,\varepsilon_{\gamma\delta}. \qquad (1.8.2)$$

Similarly

$$W = \tfrac{1}{2}e_{\alpha\beta}^{\gamma\delta}\,\tau^{\alpha\beta}\,\tau_{\gamma\delta} = \tfrac{1}{2}e_{\alpha\beta\gamma\delta}\,\tau^{\alpha\beta}\,\tau^{\gamma\delta} = \tfrac{1}{2}e^{\alpha\beta\gamma\delta}\,\tau_{\alpha\beta}\,\tau_{\gamma\delta}. \qquad (1.8.3)$$

The following symmetry properties of E and e are observed:

$$E^{ijrs} = E^{jirs} = E^{ijsr} = E^{rsij}, \qquad (1.8.4)$$

and the same relations hold for E_{ijrs}, e^{ijrs}, and e_{ijrs}. The associated tensors E_{rs}^{ij} (and e_{rs}^{ij}) possess the following symmetric properties:

$$E_{rs}^{ij} = g_{r\alpha}\,g_{s\beta}E^{ij\alpha\beta} = E_{rs}^{ji} = E_{sr}^{ij} = E_{sr}^{ji}, \qquad (1.8.5)$$

where the g's are the covariant components of the metric tensor.

If rectangular coordinates are used, 1.8.5 is supplemented by

$$E_{rs}^{ij} = E_{ij}^{rs}, \quad e_{rs}^{ij} = e_{ij}^{rs}, \qquad (1.8.6)$$

as was indicated in equation 1.6.5.

$$\begin{bmatrix} \bar{E}_{11}(\bar{e}_{11}) \\ \bar{E}_{12}(\bar{e}_{12}) \\ 2\bar{E}_{16}(\bar{e}_{16}) \\ \bar{E}_{22}(\bar{e}_{22}) \\ 2\bar{E}_{26}(\bar{e}_{26}) \\ 4\bar{E}_{66}(\bar{e}_{66}) \end{bmatrix} = \begin{bmatrix} l^4 & 2l^2 m^2 & 2l^3 m & m^4 & 2lm^3 & l^2 m^2 \\ l^2 m^2 & l^4 + m^4 & lm^3 - l^3 m & l^2 m^2 & l^3 m - lm^3 & -l^2 m^2 \\ -2l^3 m & 2(l^3 m - lm^3) & l^4 - 3l^2 m^2 & 2lm^3 & 3l^2 m^2 - m^4 & l^3 m - lm^3 \\ m^4 & 2l^2 m^2 & -2lm^3 & l^4 & -2l^3 m & l^2 m^2 \\ -2lm^3 & 2(lm^3 - l^3 m) & 3l^2 m^2 - m^4 & 2l^3 m & l^4 - 3l^2 m^2 & lm^3 - l^3 m \\ 4l^2 m^2 & -8l^2 m^2 & 4(lm^3 - l^3 m) & 4l^2 m^2 & 4(l^3 m - lm^3) & (l^2 - m^2)^2 \end{bmatrix} \begin{bmatrix} E_{11}(e_{11}) \\ E_{12}(e_{12}) \\ 2E_{16}(e_{16}) \\ E_{22}(e_{22}) \\ 2E_{26}(e_{26}) \\ 4E_{66}(e_{66}) \end{bmatrix}, \qquad (1.7.23)$$

$$\begin{bmatrix} \bar{E}_{14}(\bar{e}_{14}) \\ \bar{E}_{15}(\bar{e}_{15}) \\ \bar{E}_{24}(\bar{e}_{24}) \\ \bar{E}_{25}(\bar{e}_{25}) \\ 2\bar{E}_{46}(\bar{e}_{46}) \\ 2\bar{E}_{56}(\bar{e}_{56}) \end{bmatrix} = \begin{bmatrix} l^3 & -l^2 m & lm^2 & -m^3 & l^2 m & -lm^2 \\ l^2 m & l^3 & m^3 & lm^2 & lm^2 & l^2 m \\ lm^2 & -m^3 & l^3 & -l^2 m & -l^2 m & lm^2 \\ m^3 & lm^2 & l^2 m & l^3 & -lm^2 & -l^2 m \\ -2l^2 m & 2lm^2 & 2l^2 m & -2lm^2 & l^3 - lm^2 & m^3 - l^2 m \\ -2lm^2 & -2l^2 m & 2lm^2 & 2l^2 m & l^2 m - m^3 & l^3 - lm^2 \end{bmatrix} \begin{bmatrix} E_{14}(e_{14}) \\ E_{15}(e_{15}) \\ E_{24}(e_{24}) \\ E_{25}(e_{25}) \\ 2E_{46}(e_{46}) \\ 2E_{56}(e_{56}) \end{bmatrix}, \qquad (1.7.24)$$

The strain-energy density function (1.8.1) then becomes

$$W = \tfrac{1}{2}\tau_{\alpha\beta}\varepsilon_{\alpha\beta}$$
$$= \tfrac{1}{2}(\tau_{11}\varepsilon_{11} + \tau_{22}\varepsilon_{22} + \tau_{33}\varepsilon_{33} + 2\tau_{23}\varepsilon_{23} + 2\tau_{31}\varepsilon_{31}$$
$$+ 2\tau_{12}\varepsilon_{12}), \tag{1.8.7}$$

and equations 1.8.2 and 1.8.3 read

$$W = \tfrac{1}{2}E_{\alpha\beta}^{\gamma\delta}\varepsilon_{\alpha\beta}\varepsilon_{\gamma\delta} = \tfrac{1}{2}e_{\alpha\beta}^{\gamma\delta}\tau_{\alpha\beta}\tau_{\gamma\delta} \qquad (\alpha, \beta, \gamma, \delta = 1, 2, 3). \tag{1.8.8}$$

If we use the contracted notations introduced in 1.6.8 and 1.6.13, W takes the form

$$W = \tfrac{1}{2}\tau_\alpha\varepsilon_\alpha = \tfrac{1}{2}E_{\alpha\beta}\varepsilon_\alpha\varepsilon_\beta = \tfrac{1}{2}e_{\alpha\beta}\tau_\alpha\tau_\beta \qquad (\alpha, \beta = 1, 2, \ldots, 6). \tag{1.8.9}$$

As one can always assume without loss of generality that the homogeneous quadratic form W is symmetric,

$$\frac{\partial W}{\partial \varepsilon_i} = \tfrac{1}{2}(E_{i\alpha} + E_{\alpha i})\varepsilon_\alpha = E_{i\alpha}\varepsilon_\alpha = \tau_i \qquad (i, \alpha = 1, 2, \ldots, 6), \tag{1.8.10}$$

where $E_{ij} = E_{ji}$ which is the contracted form of 1.8.6.

The strain energy per unit volume is written out in full as a quadratic form in strains:

$$
\begin{aligned}
2W = \; & E_{11}\varepsilon_1^2 + 2E_{12}\varepsilon_1\varepsilon_2 + 2E_{13}\varepsilon_1\varepsilon_3 + 2E_{14}\varepsilon_1\varepsilon_4 + 2E_{15}\varepsilon_1\varepsilon_5 + 2E_{16}\varepsilon_1\varepsilon_6 \\
& + E_{22}\varepsilon_2^2 + 2E_{23}\varepsilon_2\varepsilon_3 + 2E_{24}\varepsilon_2\varepsilon_4 + 2E_{25}\varepsilon_2\varepsilon_5 + 2E_{26}\varepsilon_2\varepsilon_6 \\
& + E_{33}\varepsilon_3^2 + 2E_{34}\varepsilon_3\varepsilon_4 + 2E_{35}\varepsilon_3\varepsilon_5 + 2E_{36}\varepsilon_3\varepsilon_6 \\
& + E_{44}\varepsilon_4^2 + 2E_{45}\varepsilon_4\varepsilon_5 + 2E_{46}\varepsilon_4\varepsilon_6 \\
& + E_{55}\varepsilon_5^2 + 2E_{56}\varepsilon_5\varepsilon_6 \\
& + E_{66}\varepsilon_6^2. \tag{1.8.11}
\end{aligned}
$$

Similarly

$$\frac{\partial W}{\partial \tau_i} = \tfrac{1}{2}(e_{i\alpha} + e_{\alpha i})\tau_\alpha = e_{i\alpha}\tau_\alpha = \varepsilon_i \qquad (i, \alpha = 1, 2, \ldots, 6), \tag{1.8.12}$$

and equation 1.8.11 in terms of stresses is obtained according to 1.8.9, replacing E_{ij} by e_{ij} and ε_i by τ_i.

As was noted by Kirchhoff (Ref. 19, Ch. 27) and stated by Love (Ref. 9, p. 99) the elastic potential must be a positive-definite form to secure the stability of the material. This means that W is greater than zero for all real nonzero values of the strain or stress components. The strain energy W vanishes if, and only if, all ε_i or τ_i vanish.

The necessary and sufficient conditions for W to be a positive-definite form are

$$E_{11} > 0, \quad \begin{vmatrix} E_{11} & E_{12} \\ E_{12} & E_{22} \end{vmatrix} > 0, \ldots, \quad \begin{vmatrix} E_{11} & \cdots & E_{16} \\ E_{16} & \cdots & E_{66} \end{vmatrix} > 0, \tag{1.8.13}$$

as was shown, for example, by Ferrar (Ref. 20, p. 138). The same inequalities hold, of course, for the elastic compliances. Note that relations 1.8.13 imply that the diagonal elements are positive.

An invariant of elastic compliances

Let an element of an anisotropic material be subjected to hydrostatic pressure p; then the only nonvanishing stresses are

$$\tau_{11} = \tau_{22} = \tau_{33} = -p, \qquad \tau_{23} = \tau_{31} = \tau_{12} = 0. \tag{1.8.14}$$

The strain energy W (1.8.9) takes the form

$$W = p^2/2k, \tag{1.8.15}$$

where

$$1/k = e_{11} + e_{22} + e_{33} + 2(e_{12} + e_{13} + e_{23}) \tag{1.8.16}$$

is an invariant of the tensor of elastic compliances. In view of equations 1.6.12 and 1.8.14, the first invariant of the strain tensor, which gives the cubical compression, is related to k by

$$\varepsilon_v = \varepsilon_1 + \varepsilon_2 + \varepsilon_3 = -p/k. \tag{1.8.17}$$

The invariant quantity k represents the ratio of the compressive stress to the cubical compression and is called the *modulus of compression*. Since for all stable materials hydrostatic pressure decreases the volume, it is obvious that k is positive.

Some other specialized invariants of elastic compliances are shown in Lekhnitskii's text (Ref. 21, p. 43) taken from Chentsov's work.[22] A further insight into the question of invariants in anisotropic media seems to be pending.

1.9 Elastic symmetry

The elastic symmetry of a body is defined as the *invariance* of the stiffnesses E under a specified transformation of coordinates.

Symmetry with respect to a plane: aeolotropy (13 coefficients)

This symmetry is defined by the *invariance* of the E's under the change of axes

$$\bar{x}_1 = x_1, \qquad \bar{x}_2 = x_2, \qquad \bar{x}_3 = -x_3. \tag{1.9.1}$$

The direction cosines of this transformation of reflection are given in Table 1.3.

Table 1.3

	x_1	x_2	x_3
$\bar{x}_1$	1	0	0
$\bar{x}_2$	0	1	0
$\bar{x}_3$	0	0	−1

In view of equation 1.7.12, one finds that

$$E_{ij}^{3k} = E_{3k}^{ij} = E_{3i}^{33} = E_{33}^{3i} \qquad (i, j, k = 1, 2). \tag{1.9.2}$$

The matrix of stiffnesses for a material with one plane of elastic symmetry (the $x_3 = 0$ plane) takes the following form, when the notation of 1.6.8 is used:

$$[E] = \begin{bmatrix} E_{11} & E_{12} & E_{13} & 0 & 0 & E_{16} \\ & E_{22} & E_{23} & 0 & 0 & E_{26} \\ & & E_{33} & 0 & 0 & E_{36} \\ & & & E_{44} & E_{45} & 0 \\ & & & & E_{55} & 0 \\ & & & & & E_{66} \end{bmatrix}, \quad E_{ij} = E_{ji}. \quad (1.9.3)$$

The coordinate axes x or $\bar{x}$ are denoted as the *natural axes* or *principal axes* of elastic symmetry of an aeolotropic medium characterized by 13 independent elastic coefficients. Note that in any coordinate system obtained from the natural axes by equation 1.7.3 all 21 stiffnesses or compliances are generally

$$[E] = \begin{bmatrix} E_{11} & E_{12} & E_{12} & 0 & 0 & 0 \\ & E_{11} & E_{12} & 0 & 0 & 0 \\ & & E_{11} & 0 & 0 & 0 \\ & & & \frac{1}{2}(E_{11} - E_{12}) & 0 & 0 \\ & & & & \frac{1}{2}(E_{11} - E_{12}) & 0 \\ & & & & & \frac{1}{2}(E_{11} - E_{12}) \end{bmatrix}, \quad E_{ij} = E_{ji}. \quad (1.9.9)$$

necessary; according to equations 1.7.6 and 1.7.7, however they are given functions of the 13 independent coefficients.

Symmetry with respect to two orthogonal planes: orthotropy (9 coefficients)

A body is said to be elastically symmetric with respect to two perpendicular planes $x_3 = 0$ and $x_1 = 0$ if the elastic coefficients must remain invariant under transformation equations 1.9.1 and the following change of axes:

$$\bar{x}_1 = -x_1, \quad \bar{x}_2 = x_2, \quad \bar{x}_3 = x_3. \quad (1.9.4)$$

Then, in view of 1.7.12, the matrix 1.9.3 simplifies to

$$[E] = \begin{bmatrix} E_{11} & E_{12} & E_{13} & 0 & 0 & 0 \\ & E_{22} & E_{23} & 0 & 0 & 0 \\ & & E_{33} & 0 & 0 & 0 \\ & & & E_{44} & 0 & 0 \\ & & & & E_{55} & 0 \\ & & & & & E_{66} \end{bmatrix}, \quad E_{ij} = E_{ji}. \quad (1.9.5)$$

It is noted that elastic symmetry with respect to two orthogonal planes (say, $x_3 = 0$, $x_1 = 0$) implies elastic symmetry with respect to a third plane ($x_2 = 0$).

Transverse isotropy (5 coefficients)

A specialized orthotropic body is defined by the invariance of the elastic coefficients of 1.9.5 under a transformation of rotation in the $x_3 = 0$ plane:

$$\bar{x}_i = l_{i\alpha} x_\alpha, \ \bar{x}_3 = x_3 \quad (i, \alpha = 1, 2). \quad (1.9.6)$$

In view of 1.7.12 the following relations must hold for the stiffnesses in 1.9.5:

$$E_{11} = E_{22}, E_{13} = E_{23}, E_{44} = E_{55}, E_{66} = \tfrac{1}{2}(E_{11} - E_{12}). \quad (1.9.7)$$

The matrix 1.9.5 takes the form

$$[E] = \begin{bmatrix} E_{11} & E_{12} & E_{13} & 0 & 0 & 0 \\ & E_{11} & E_{13} & 0 & 0 & 0 \\ & & E_{33} & 0 & 0 & 0 \\ & & & E_{44} & 0 & 0 \\ & & & & E_{44} & 0 \\ & & & & & \frac{1}{2}(E_{11} - E_{12}) \end{bmatrix}, \quad E_{ij} = E_{ji}. \quad (1.9.8)$$

Isotropy (2 coefficients)

For an isotropic medium the elastic properties are independent of the orientation of coordinate axes. The number of independent elastic coefficients can be reduced to two, namely

The alternative notation is frequently used:

$$E_{11} = \lambda + 2\mu, E_{12} = \lambda, \tfrac{1}{2}(E_{11} - E_{12}) = \mu, \quad (1.9.10)$$

where λ and μ are the Lamé coefficients.

2 THEORY OF COMPOSITE PLATES

2.1 Theory of certain types of heterogeneous aeolotropic plates

The class of plates with which this section is concerned includes as an important special case plates consisting of two identical orthotropic sheets of thickness $h/2$ which are so laminated that the natural axes enclose an angle $+\theta$ with the x, y axes in one sheet ($0 < z \leq h/2$), and an angle $-\theta$ in the other sheet ($-h/2 \leq z < 0$).

In view of equations 1.6.9 and 1.9.3 the following stress-strain relations, written in matrix notation, hold for the orthotropic sheets with reference to the natural axes x, y (Fig. 1.8):

$$\begin{bmatrix} \tau'_{xx} \\ \tau'_{yy} \\ \tau'_{xy} \end{bmatrix} = \begin{bmatrix} E'_{xx} & E'_{xy} & 0 \\ E'_{xy} & E'_{yy} & 0 \\ 0 & 0 & E'_{ss} \end{bmatrix} \begin{bmatrix} \varepsilon'_{xx} \\ \varepsilon'_{yy} \\ 2\varepsilon'_{xy} \end{bmatrix}. \quad (2.1.1)$$

If we introduce the notation

$$\varepsilon_{xx} = \epsilon_x, \ \varepsilon_{yy} = \epsilon_y, \ 2\varepsilon_{xy} = \epsilon_{xy}, \quad (2.1.2)$$

equations 2.1.1 can be written more compactly in the form

$$[\tau'] = [E'][\epsilon']. \quad (2.1.3)$$

If z is a third coordinate normal to the plane of undeflected plate, measured from the interface of the two layers, and

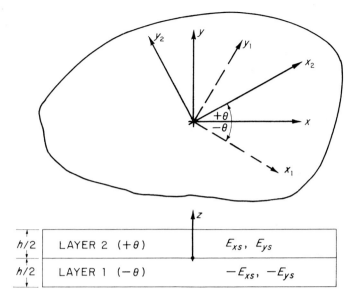

Figure 1.8   View and Section of a 2-Layer Plate with Indication of the Natural Axes

transformation equations 1.7.23 are used, the following system of stress-strain relations is obtained for the composite plate:

$$\begin{bmatrix} \tau_x \\ \tau_y \\ \tau_{xy} \end{bmatrix} = \begin{bmatrix} E_{xx} & E_{xy} & \mathrm{sgn}\,(z)E_{xs} \\ E_{xy} & E_{yy} & \mathrm{sgn}\,(z)E_{ys} \\ \mathrm{sgn}\,(z)E_{xs} & \mathrm{sgn}\,(z)E_{ys} & E_{ss} \end{bmatrix} \begin{bmatrix} \epsilon_x \\ \epsilon_y \\ \epsilon_{xy} \end{bmatrix}, \quad (2.1.4)$$

in which E_{ij} are constants and $\mathrm{sgn}\,(z)$ is $+1$ when $z>0$ and -1 when $z<0$. Note that the repeated index is omitted, as is usually done in the engineering literature (for example, Ref. 23, p. 98).

It was shown by Reissner and Stavsky[24] that for plates of this type there occurs a coupling phenomenon between in-plate stretching and transverse bending which does not occur in the linear theory of homogeneous plates and which may be of practical importance (see Refs. 25 and 26).

The problem was first discussed by Smith[27] and is also reported by Lekhnitskii (Ref. 28, pp. 270–271), without consideration of the coupling between stretching and bending.

Formulation of problem

Consider a 2-layer plate element as shown in Figure 1.9. Defining stress resultants, stress couples, mid-surface strains, and bending curvatures as is usual in plate theory, we have the following equilibrium relations and strain-displacement relations of linear plate theory based on the Euler-Bernoulli hypothesis:

$$N_{x,x} + N_{xy,y} + p_x = 0, \qquad N_{xy,x} + N_{y,y} + p_y = 0, \quad (2.1.5)$$

$$M_{x,x} + M_{xy,y} - Q_x = 0, \qquad M_{xy,x} + M_{y,y} - Q_y = 0, \quad (2.1.6)$$

$$Q_{x,x} + Q_{y,y} + p_z = 0, \quad (2.1.7)$$

$$\epsilon_x = \epsilon_x^0 + z\kappa_x, \qquad \epsilon_y = \epsilon_y^0 + z\kappa_y, \qquad \epsilon_{xy} = \epsilon_{xy}^0 + z\kappa_{xy}, \quad (2.1.8)$$

$$\epsilon_x^0 = u_{,x}, \qquad \epsilon_y^0 = v_{,y}, \qquad \epsilon_{xy}^0 = 2\varepsilon_{xy}^0 = u_{,y} + v_{,x}, \quad (2.1.9)$$

$$\kappa_x = -w_{,xx}, \qquad \kappa_y = -w_{,yy}, \qquad \kappa_{xy} = -2w_{,xy}. \quad (2.1.10)$$

The components of strain according to equations 2.1.8 define components of stress through the following stress-strain relations:

$$\begin{bmatrix} \tau_x \\ \tau_y \\ \tau_{xy} \end{bmatrix} = \begin{bmatrix} E_{xx} & E_{xy} & E_{xs} \\ E_{xy} & E_{yy} & E_{ys} \\ E_{xs} & E_{ys} & E_{ss} \end{bmatrix} \begin{bmatrix} \epsilon_x \\ \epsilon_y \\ \epsilon_{xy} \end{bmatrix}. \quad (2.1.11)$$

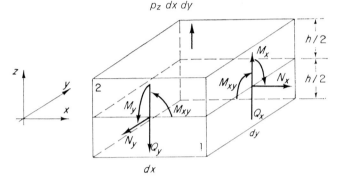

Figure 1.9   Plate Element with Stress Resultants, Stress Couples, and External Loading

It is assumed that E_{xx}, E_{xy}, E_{yy}, and E_{ss} are *even* functions of z whereas E_{xs} and E_{ys} are *odd* functions of z.

In order to obtain a complete system of plate equations, equations 2.1.11 are introduced into the definitions

$$N_x = \int_{-h/2}^{+h/2} \tau_x \, dz, \qquad M_x = \int_{-h/2}^{+h/2} \tau_x \, z \, dz, \quad (2.1.12)$$

and so on, for resultants and couples. The following system of plate stress-strain relations is obtained:

$$\begin{bmatrix} N_x \\ N_y \\ N_{xy} \\ M_x \\ M_y \\ M_{xy} \end{bmatrix} = \begin{bmatrix} A_{xx} & A_{xy} & 0 & 0 & 0 & B_{xs} \\ A_{xy} & A_{yy} & 0 & 0 & 0 & B_{ys} \\ 0 & 0 & A_{ss} & B_{xs} & B_{ys} & 0 \\ 0 & 0 & B_{xs} & D_{xx} & D_{xy} & 0 \\ 0 & 0 & B_{ys} & D_{xy} & D_{yy} & 0 \\ B_{xs} & B_{ys} & 0 & 0 & 0 & D_{ss} \end{bmatrix} \begin{bmatrix} \epsilon_x^0 \\ \epsilon_y^0 \\ \epsilon_{xy}^0 \\ \kappa_x \\ \kappa_y \\ \kappa_{xy} \end{bmatrix}. \quad (2.1.13)$$

The elastic areas A_{ij}, the elastic statical moments B_{ij}, and the elastic moments of inertia D_{ij} are given by

$$(A_{ij}, B_{ij}, D_{ij}) = \int_{-h/2}^{+h/2} (1, z, z^2) E_{ij} \, dz \qquad (i, j = x, y, s). \quad (2.1.14)$$

Equations 2.1.13 may also be written with the help of appropriate auxiliary matrices in the form

$$\begin{bmatrix} N \\ M \end{bmatrix} = \begin{bmatrix} A & B \\ B & D \end{bmatrix} \begin{bmatrix} \epsilon^0 \\ \kappa \end{bmatrix}. \quad (2.1.15)$$

Equations 2.1.5 to 2.1.7 together with 2.1.13 represent a system of eleven equations for eleven unknowns N_x, N_y, N_{xy}, Q_x, Q_y, M_x, M_y, M_{xy}, u, v, w. This system may be reduced to three simultaneous equations for u, v, w as is shown by Stavsky.[29] However, in order to see the relations of the present problem to the usual problems of stretching and bending of homogeneous plates, it is preferable to carry out a reduction to two simultaneous equations in terms of w and an Airy stress function F.

Reduction to two simultaneous equations

In order to obtain the required reduction, equations 2.1.6 and 2.1.7 are combined to read

$$M_{x,xx} + 2M_{xy,xy} + M_{y,yy} + p_z = 0, \qquad (2.1.16)$$

and the compatibility relation,

$$\epsilon^0_{x,yy} + \epsilon^0_{y,xx} = \epsilon^0_{xy,xy}, \qquad (2.1.17)$$

is used which is a consequence of equations 2.1.9.

Equilibrium equations 2.1.5 are satisfied by means of an Airy stress function F, where it is assumed that the body force components p_x, p_y are restricted so as to be derivable from a potential Ω as $p_x = -\Omega_{,x}$, $p_y = -\Omega_{,y}$, by setting

$$N_x = F_{,yy} + \Omega, \qquad N_y = F_{,xx} + \Omega, \qquad N_{xy} = -F_{,xy}. \qquad (2.1.18)$$

In order to obtain the final system for w and F the system of stress-strain relations 2.1.15 must be written so that ϵ^0 and M are given as functions of N and κ:

$$\begin{bmatrix} \epsilon^0 \\ M \end{bmatrix} = \begin{bmatrix} A^* & B^* \\ C^* & D^* \end{bmatrix} \begin{bmatrix} N \\ \kappa \end{bmatrix}. \qquad (2.1.19)$$

The starred matrices are given in terms of the unstarred matrices and their inverses as follows:

$$[A^*] = [A^{-1}], \; [B^*] = -[A^{-1}][B],$$
$$[C^*] = [B][A^{-1}] = -[B^*]^T, \; [D^*] = [D] - [B][A^{-1}][B]. \qquad (2.1.20)$$

It is noted that while A^* and D^* are *symmetric* matrices, B^* and C^* are not necessarily so.

Equations 2.1.10 and 2.1.18 are now introduced on the right of 2.1.19 and then the appropriate portions of 2.1.19 are introduced into 2.1.16 and 2.1.17. In this way the following system of simultaneous fourth-order equations is derived:

$$L_1 w - L_3 F = p_z + L_4 \Omega, \qquad (2.1.21)$$

$$L_2 F + L_3 w = L_5 \Omega. \qquad (2.1.22)$$

The operators L_i are as follows:

$$L_1 = D^*_{xx}(\;)_{,xxxx} + 2(D^*_{xy} + 2D^*_{ss})(\;)_{,xxyy} + D^*_{yy}(\;)_{,yyyy}, \qquad (2.1.23)$$

$$L_2 = A^*_{yy}(\;)_{,xxxx} + 2(2A^*_{xy} + A^*_{ss})(\;)_{,xxyy} + A^*_{xx}(\;)_{,yyyy}, \qquad (2.1.24)$$

$$L_3 = (B^*_{sx} - 2B^*_{ys})(\;)_{,xxxy} + (B^*_{sy} - 2B^*_{xs})(\;)_{,xyyy}, \qquad (2.1.25)$$

$$L_4 = -2(B^*_{xs} + B^*_{ys})(\;)_{,xy}, \qquad (2.1.26)$$

$$L_5 = -(A^*_{xy} + A^*_{yy})(\;)_{,xx} - (A^*_{xx} + A^*_{xy})(\;)_{,yy}. \qquad (2.1.27)$$

Equations 2.1.21 through 2.1.27, established by Reissner and Stavsky,[24] are remarkable for two specific properties:

1. There is a coupling between transverse bending and in-plane stretching effects, which is governed by the L_3 operator, the structure of which is connected with the B matrix in stress-strain relations 2.1.13.

2. The operator L_1 involves the elements of the matrix D^* rather than the elements of the matrix D.

Associated with differential equations 2.1.21 and 2.1.22 are four boundary conditions, specified along the boundary $f_b(x, y) = 0$ (Fig. 1.10), which take the following form:

$$\begin{aligned} u_n &= \bar{u}_n & &\text{or} & N_n &= \bar{N}_n, \\ u_s &= \bar{u}_s & &\text{or} & N_{ns} &= \bar{N}_{ns}, \\ w &= \bar{w} & &\text{or} & Q_n + M_{ns,s} &= \bar{Q}_n + \bar{M}_{ns,s}, \\ w_{,n} &= \bar{w}_{,n} & &\text{or} & M_n &= \bar{M}_n. \end{aligned} \qquad (2.1.28)$$

Note that the coupling of w and F enters not only into the system of equations 2.1.21 and 2.1.22 but into the boundary conditions as well.

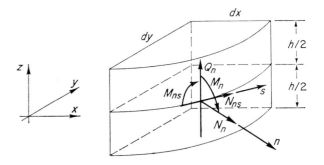

Figure 1.10   Infinitesimal Element of 2-Layer Plate at the Boundary Showing Orientation of Edge Resultants and Couples

A cross-elasticity effect

Expressions for stresses in terms of stress resultants and stress couples can be obtained if stress-strain relations 2.1.11 are appropriately transformed. Combination of 2.1.11 and 2.1.8 leads to a formula which in matrix notation reads

$$[\tau] = [E][\epsilon^0] + z[E][\kappa]. \qquad (2.1.29)$$

Next, equations 2.1.13 are inverted in the form

$$\begin{bmatrix} \epsilon^0_x \\ \epsilon^0_y \\ \epsilon^0_{xy} \\ \kappa_x \\ \kappa_y \\ \kappa_{xy} \end{bmatrix} = \begin{bmatrix} a_{xx} & a_{xy} & 0 & 0 & 0 & b_{xs} \\ a_{yx} & a_{yy} & 0 & 0 & 0 & b_{ys} \\ 0 & 0 & a_{ss} & b_{sx} & b_{sy} & 0 \\ 0 & 0 & c_{xs} & d_{xx} & d_{xy} & 0 \\ 0 & 0 & c_{ys} & d_{yx} & d_{yy} & 0 \\ c_{sx} & c_{sy} & 0 & 0 & 0 & d_{ss} \end{bmatrix} \begin{bmatrix} N_x \\ N_y \\ N_{xy} \\ M_x \\ M_y \\ M_{xy} \end{bmatrix}; \qquad (2.1.30)$$

or more compactly

$$\begin{bmatrix} \epsilon^0 \\ \kappa \end{bmatrix} = \begin{bmatrix} a & b \\ c & d \end{bmatrix} \begin{bmatrix} N \\ M \end{bmatrix}, \qquad (2.1.31)$$

in which a, b, c, d are 3×3 matrices related to the matrices A^*, B^*, D^* by the following expressions:

$$[a] = [A^*] + [B^*][D^{*-1}][B^*]^T, \quad [b] = [B^*][D^{*-1}],$$
$$[c] = [b]^T, \quad [d] = [D^{*-1}], \tag{2.1.32}$$

where the starred matrices are given in terms of the unstarred matrices by equations 2.1.20.

Introduction of 2.1.31 into 2.1.29 leads to a matrix relation of the form

$$[\tau] = [E]\{([a] + z[c])[N] + ([b] + z[d])[M]\}. \tag{2.1.33}$$

Equation 2.1.33, first shown by Reissner and Stavsky,[24] indicates the remarkable fact that, in general, τ_x depends not only on N_x and M_x but also on N_y, M_y, N_{xy}, M_{xy} and that an analogous "cross-elasticity" effect is encountered in the expressions for τ_y and τ_{xy}. Note that this effect still exists for symmetrically layered plates where b and c vanish, as shown by Stavsky.[30]

2.2 Some particular cases derived from theory of Section 2.1

Uniform distributions of stress resultants and couples

Insight, in a relatively simple manner, into some of the consequences of elastic heterogeneity of the kind considered in Section 2.1 may be obtained by determining the state of stress and deformation associated with resultants N and couples M which are independent of x, y.

In view of equation 2.1.18 we have, when N_x, N_y, N_{xy} have constant values n_x, n_y, n_{xy} and there are no body forces p_x, p_y,

$$F = \tfrac{1}{2}n_x y^2 + \tfrac{1}{2}n_y x^2 - n_{xy} xy. \tag{2.2.1}$$

In order that M_x, M_y, M_{xy} at the same time have constant values m_x, m_y, m_{xy} it is necessary, in view of 2.1.19, that κ_x, κ_y, κ_{xy} have constant values k_x, k_y, k_{xy}. Accordingly, the deflection function w must be of the form

$$w = \tfrac{1}{2}k_x x^2 + \tfrac{1}{2}k_y y^2 + k_{xy} xy, \tag{2.2.2}$$

and differential equations 2.1.21 and 2.1.22, with $p_z = \Omega = 0$, are automatically satisfied.

Consideration of stress-strain relations 2.1.30 indicates that in-plane forces are associated with in-plane deformations *and* transverse deflections of the plate. Similarly, bending and twisting couples are associated with transverse deflections *and* in-plane deformations of the heterogeneous plate. Specifically, when $m_x = m_y = m_{xy} = 0$, then

$$[\epsilon^0] = [a][n], \quad [k] = [c][n], \tag{2.2.3}$$

and when $n_x = n_y = n_{xy} = 0$, then

$$[\epsilon^0] = [b][m], \quad [k] = [d][m]. \tag{2.2.4}$$

Corresponding formulas for stresses are

$$[\tau] = [E]([a] + z[c])[N], \quad [\tau] = [E]([b] + z[d])[M]. \tag{2.2.5}$$

Numerical examples

To obtain a quantitative insight into the results that apply to problems with uniform distributions, we consider an orthotropic material with

$$E'_{xx} = 5E'_{yy} = \tfrac{10}{3}E'_{ss}, \quad E'_{xy} = 0, \tag{2.2.6}$$

in stress-strain relations 2.1.1. For $\theta = 45°$ equations 2.1.4, in view of 1.7.23, assume the form

$$\begin{bmatrix} \tau_x \\ \tau_y \\ \tau_{xy} \end{bmatrix} = E'_{xx} \begin{bmatrix} \tfrac{3}{5} & 0 & \tfrac{1}{5}\operatorname{sgn}(z) \\ 0 & \tfrac{3}{5} & \tfrac{1}{5}\operatorname{sgn}(z) \\ \tfrac{1}{5}\operatorname{sgn}(z) & \tfrac{1}{5}\operatorname{sgn}(z) & \tfrac{1}{10} \end{bmatrix} \begin{bmatrix} \epsilon_x \\ \epsilon_y \\ \epsilon_{xy} \end{bmatrix}. \tag{2.2.7}$$

We now consider a rectangular plate subject to (i) a uniform axial force N_x and (ii) a pure bending moment M_x. According to Reissner and Stavsky[24] and Stavsky[31] the following results are obtained for the case of a uniform axial force N_x:

$$w = \frac{3}{2}\frac{N_x}{E_{xx}h^2}xy, \tag{2.2.8}$$

$$\begin{bmatrix} \epsilon_x \\ \epsilon_y \\ \epsilon_{xy} \end{bmatrix} = \frac{N_x}{E_{xx}h}\begin{bmatrix} \tfrac{5}{4} \\ \tfrac{1}{4} \\ -\tfrac{3}{2}z/(h/2) \end{bmatrix}; \tag{2.2.9}$$

$$\begin{bmatrix} \tau_x \\ \tau_y \\ \tau_{xy} \end{bmatrix} = \frac{N_x}{h}\begin{bmatrix} \tfrac{5}{4} - |z|/h \\ \tfrac{1}{4} - |z|/h \\ \tfrac{1}{2}z/|z| - \tfrac{3}{4}z/(h/2) \end{bmatrix}. \tag{2.2.10}$$

It seems worth while to give the corresponding results when coupling between stretching and bending is neglected, as in the approach of Smith,[27] indicated by a superscript S:

$$w^S = 0; \quad \epsilon_x^S = \frac{N_x}{E_{xx}h}, \quad \epsilon_y^S = \epsilon_{xy}^S = 0;$$
$$\tau_x^S = \frac{N_x}{h}, \quad \tau_y^S = 0, \quad \tau_{xy}^S = \frac{N_x}{3h}\frac{z}{|z|}. \tag{2.2.11}$$

A graphical representation of these results is given in Figure 1.11.

Analogous results for the case of a uniform bending moment M_x are

$$w = -\frac{6M_x}{E_{xx}h^3}(\tfrac{5}{4}x^2 + \tfrac{1}{4}y^2), \tag{2.2.12}$$

$$\begin{bmatrix} \epsilon_x \\ \epsilon_y \\ \epsilon_{xy} \end{bmatrix} = \frac{3M_x}{E_{xx}h^2}\begin{bmatrix} \tfrac{5}{4}z/(h/2) \\ \tfrac{1}{4}z/(h/2) \\ \tfrac{1}{2} \end{bmatrix}, \tag{2.2.13}$$

$$\begin{bmatrix} \tau_x \\ \tau_y \\ \tau_{xy} \end{bmatrix} = \frac{6M_x}{h^2}\begin{bmatrix} \tfrac{1}{6}z/|z| + \tfrac{5}{4}z/(h/2) \\ \tfrac{1}{6}z/|z| + \tfrac{1}{4}z/(h/2) \\ \tfrac{1}{4} + |z|/h \end{bmatrix}. \tag{2.2.14}$$

We list again the corresponding results according to Smith[27]:

$$w^S = -\frac{6M_x}{E_{xx}h^3}x^2; \quad \epsilon_x^S = \frac{6M_x}{E_{xx}h^2}\frac{z}{h/2}, \quad \epsilon_y^S = \epsilon_{xy}^S = 0;$$
$$\tau_x^S = \frac{6M_x}{h^2}\frac{z}{h/2}, \quad \tau_y^S = 0, \quad \tau_{xy}^S = \frac{6M_x}{5h^2}\frac{z}{h/2}. \tag{2.2.15}$$

It is noted that in case (i) the axial tension deforms the flat plate into a hyperbolic paraboloid surface. The coupling of

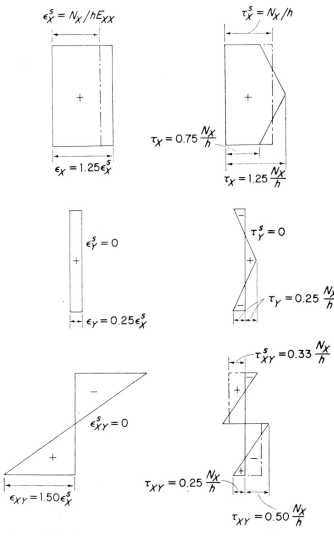

Figure 1.11   Stress and Strain Components in a 2-Layer Plate Subjected to Uniform Axial Tension N_x, When $E_{xx} = 5E_{yy} = \frac{10}{3} E_{ss}$ and $\theta = 45°$

stretching and bending is clearly indicated by equations 2.2.9 and the cross elasticity effect is shown in equations 2.2.10.

In case (ii), equation 2.2.12 yields larger deflections than 2.2.15 does, and the strain and stress components in 2.2.13 and 2.2.14 are entirely different from those in 2.2.15.

For any other values of θ in transformation equations 1.7.23 we would get all the terms in the E matrix in 2.1.11, but qualitatively the previous results are still valid.

Sinusoidal deflection of an infinite plate due to sinusoidal transverse load distributions

We now assume that the two-layer aeolotropic plate is subjected to a transverse load distribution of the form $p_z = p_0 \sin \alpha x \sin \beta y$, where p_0, α, β are constants. We also assume that $p_x = p_y = 0$, so that $\Omega = 0$ in 2.1.21 and 2.1.22. The system of equations 2.1.21 and 2.1.22 then has solutions of the form

$$w = W_0 \sin \alpha x \sin \beta y, \qquad F = F_0 \cos \alpha x \cos \beta y, \quad (2.2.16)$$

where the constants W_0 and F_0 are given by the following expressions:

$$\frac{W_0}{p_0} = \left\{ \alpha^4 D_{xx}^* + 2\alpha^2\beta^2(D_{xy}^* + 2D_{ss}^*) + \beta^4 D_{yy}^* \right.$$
$$\left. + \frac{[\alpha^3\beta(B_{sx}^* - 2B_{ys}^*) + \alpha\beta^3(B_{sy}^* - 2B_{xs}^*)]^2}{\alpha^4 A_{yy}^* + 2\alpha^2\beta^2(2A_{xy}^* + A_{ss}^*) + \beta^4 A_{xx}^*} \right\}^{-1}, \quad (2.2.17)$$

$$\frac{F_0}{W_0} = \frac{\alpha^3(B_{sx}^* + 2B_{ys}^*) + \alpha\beta^3(B_{sy}^* - 2B_{xs}^*)}{\alpha^4 A_{yy}^* + 2\alpha^2\beta^2(2A_{xy}^* + A_{ss}^*) + \beta^4 A_{xx}^*}. \quad (2.2.18)$$

With W_0 and F_0 we find stress resultants from

$$N_x = -\beta^2 F_0 \cos \alpha x \cos \beta y, \quad N_y = \alpha^2\beta^{-2}N_x,$$
$$N_{xy} = -\alpha\beta F_0 \sin \alpha x \sin \beta y. \quad (2.2.19)$$

In order to determine stress couples we first calculate

$$\kappa_x = \alpha^2 W_0 \sin \alpha x \sin \beta y, \quad \kappa_y = \beta^2\alpha^{-2}\kappa_x,$$
$$\kappa_{xy} = -2\alpha\beta W_0 \cos \alpha x \cos \beta y, \quad (2.2.20)$$

and then substitute the result into the matrix formula

$$\begin{bmatrix} M_x \\ M_y \\ M_{xy} \end{bmatrix} = \begin{bmatrix} 0 & 0 & C_{xs}^* & D_{xx}^* & D_{xy}^* & 0 \\ 0 & 0 & C_{yx}^* & D_{yx}^* & D_{yy}^* & 0 \\ C_{sx}^* & C_{sy}^* & 0 & 0 & 0 & D_{ss}^* \end{bmatrix} \begin{bmatrix} N_x \\ N_y \\ N_{xy} \\ \kappa_x \\ \kappa_y \\ \kappa_{xy} \end{bmatrix}.$$
$$(2.2.21)$$

We note that the presence of the numerous zeros in the coefficient matrix in 2.2.21 is due to the symmetry properties of the material which we have assumed in so far as the stiffness matrix in stress-strain relation 2.1.11 is concerned. If some or all of these properties are not assumed, then none of the coefficients in $[C^*]$ and $[D^*]$ necessarily vanishes. For detailed numerical results the reader is referred to the works of Reissner and Stavsky[24] and Stavsky and Roy[32,33] from which it becomes apparent that considerable differences exist between the results of the present theory and the results of theories that do not consider all the cross-coupling terms associated with the particular kind of nonisotropy and heterogeneity which has been assumed.

2.3   General theory of composite aeolotropic plates

Coupling of bending and stretching was shown in Section 2.1 to characterize the behavior of a special class of heterogeneous aeolotropic plates as defined by stress-strain relations 2.1.11 with E_{xs}, E_{ys} that are *odd* functions of z, whereas the other stiffnesses are *even* functions of z.

In what follows a general heterogeneous aeolotropic plate is considered for which no specific z dependence is assumed for the stiffnesses in stress-strain relations 2.1.11. Two basic types of composite systems are considered: (i) *laminated* and (ii) *continuously heterogeneous* systems. A *laminated* plate or shell is assumed to be composed of homogeneous sheets; consequently the stiffnesses are piecewise continuous functions of z, whereas in a heterogeneous plate they are continuous functions of the thickness coordinate z.

It will be shown that the behavior of the two types of composite plates is almost identical and that a coupling phenomenon occurs between in-plane forces and transverse bending (see Ref. 34).

Formulation of problem

Let us consider a laminated plate composed of n homogeneous aeolotropic sheets (Fig. 1.12). Taking the positive

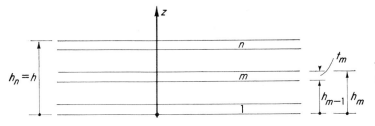

Figure 1.12   Section of Laminated Plate

direction of the thickness coordinate z upward, measured from the bottom face of layer 1, then layer m, for example, is enclosed by its lower and upper faces $h_{m-1} \leq z \leq h_m$, its thickness is t_m, and the total plate thickness is $h_n = h$.

Following Lekhnitskii,[35] we can write stress-strain relations 2.1.11 in the form

$$\begin{bmatrix} \tau_x^m \\ \tau_y^m \\ \tau_{xy}^m \end{bmatrix} = \begin{bmatrix} E_{xx}^m & E_{xy}^m & E_{xs}^m \\ E_{xy}^m & E_{yy}^m & E_{ys}^m \\ E_{xs}^m & E_{ys}^m & E_{ss}^m \end{bmatrix} \begin{bmatrix} \epsilon_x \\ \epsilon_y \\ \epsilon_{xy} \end{bmatrix}, \qquad (2.3.1)$$

where the super index m designates the mth homogeneous sheet of the laminated plate.

If a heterogeneous plate is considered, then 2.1.11 holds and the z dependence of E_{ij} is continuous.

In order to obtain the governing equations of composite plates we define stress resultants, stress couples, reference surface strains (at $z = 0$), and bending curvatures as in equations 2.1.5–2.1.10. Defining resultants and couples by

$$(N_x, N_y, N_{xy}) = \int_0^h (\tau_x, \tau_y, \tau_{xy})\, dz, \qquad (2.3.2)$$

$$(M_x, M_y, M_{xy}) = \int_0^h (\tau_x, \tau_y, \tau_{xy})z\, dz, \qquad (2.3.3)$$

and introducing for the stress components equations 2.1.11 or 2.3.1, we obtain the following plate stress-strain relations:

$$\begin{bmatrix} N_x \\ N_y \\ N_{xy} \\ M_x \\ M_y \\ M_{xy} \end{bmatrix} = \begin{bmatrix} A_{xx} & A_{xy} & A_{xs} & B_{xx} & B_{xy} & B_{xs} \\ A_{xy} & A_{yy} & A_{ys} & B_{xy} & B_{yy} & B_{ys} \\ A_{xs} & A_{ys} & A_{ss} & B_{xs} & B_{ys} & B_{ss} \\ B_{xx} & B_{xy} & B_{xs} & D_{xx} & D_{xy} & D_{xs} \\ B_{xy} & B_{yy} & B_{ys} & D_{xy} & D_{yy} & D_{ys} \\ B_{xs} & B_{ys} & B_{ss} & D_{xs} & D_{ys} & D_{ss} \end{bmatrix} \begin{bmatrix} \epsilon_x^0 \\ \epsilon_y^0 \\ \epsilon_{xy}^0 \\ \kappa_x \\ \kappa_y \\ \kappa_{xy} \end{bmatrix}, \qquad (2.3.4)$$

which may be more compactly written in the form of equation 2.1.15.

The elastic areas, the elastic statical moments, and the elastic moments of inertia for continuously heterogeneous plates are given, respectively, by

$$(A_{ij}, B_{ij}, D_{ij}) = \int_0^h (1, z, z^2)E_{ij}\, dz \qquad (i, j = x, y, s), \quad (2.3.5)$$

whereas for laminated plates these relations take the form

$$(A_{ij}, B_{ij}, D_{ij}) = \sum_{m=1}^n \int_{h_{m-1}}^{h_m} (1, z, z^2)E_{ij}^m\, dz \qquad (i, j = x, y, s). \quad (2.3.6)$$

If we follow the same method that appears in Section 2.1, the governing plate equations are again the same as 2.1.21 and 2.1.22:

$$L_1 w - L_3 F = p_z + L_4 \Omega, \qquad (2.3.7)$$

$$L_2 F + L_3 w = L_5 \Omega. \qquad (2.3.8)$$

However, the operators L_i include some more terms:

$$L_1 = D_{xx}^*(\ )_{,xxxx} + 4D_{xs}^*(\ )_{,xxxy} + 2(D_{xy}^* + 2D_{ss}^*)(\ )_{,xxyy} \\ + 4D_{ys}^*(\ )_{,xyyy} + D_{yy}^*(\ )_{,yyyy}, \qquad (2.3.9)$$

$$L_2 = A_{yy}^*(\ )_{,xxxx} - 2A_{ys}^*(\ )_{,xxxy} + (2A_{xy}^* + A_{ss}^*)(\ )_{,xxyy} \\ - 2A_{xs}^*(\ )_{,xyyy} + A_{xx}^*(\ )_{,yyyy}, \qquad (2.3.10)$$

$$-L_3 = B_{yx}^*(\ )_{,xxxx} \\ + (2B_{ys}^* - B_{sx}^*)(\ )_{,xxxy} + (B_{xx}^* + B_{yy}^* - 2B_{ss}^*)(\ )_{,xxyy} \\ + (2B_{xs}^* - B_{sy}^*)(\ )_{,xyyy} + B_{xy}^*(\ )_{,yyyy}, \qquad (2.3.11)$$

$$-L_4 = (B_{xx}^* + B_{yx}^*)(\ )_{,xx} + 2(B_{xs}^* + B_{ys}^*)(\ )_{,xy} \\ + (B_{xy}^* + B_{yy}^*)(\ )_{,yy}, \qquad (2.3.12)$$

$$-L_5 = (A_{xy}^* + A_{yy}^*)(\ )_{,xx} - (A_{sx}^* + A_{sy}^*)(\ )_{,xy} \\ + (A_{xx}^* + A_{xy}^*)(\ )_{,yy}. \qquad (2.3.13)$$

The A^*, B^*, D^* matrices are defined in terms of unstarred matrices 2.3.5 through equations 2.1.20.

Equations 2.3.7–2.3.13 are remarkable for two specific properties:

1. There is a coupling between transverse bending and in-plane stretching which is governed by the L_3 operator connected with the elastic statical moments B in plate stress-strain relations 2.3.4. In generalization of results 2.1.13 for the particular class of heterogeneous plates treated in Section 2.1 we regularly obtain now all the elements in the B matrix, and consequently the full form of the L_3 operator exists. This means, for example, that for the general heterogeneous plate, w and F remain coupled even for one-dimensional problems.

2. When transferred to the middle surface of the plate, the operator L_1 involves the elements of the matrix D^*, rather than the elements of the matrix D.

The eighth-order system 2.3.7 and 2.3.8 is associated with four boundary conditions of the form given by equations 2.1.28. Once the system of equations for w and F is solved, subject to appropriate boundary conditions, all other unknowns can be determined directly. The stress resultants are given by equations 2.1.18, the curvatures by equations 2.1.10, the strains at the reference plane and the stress couples by equations 2.1.19, and the transverse shear resultants by equations

2.1.6. Finally the stress components that show the cross-elasticity phenomenon are obtained from equations 2.1.11, or conveniently from 2.1.33.

Generalized stress function Φ

When body forces are not considered it is convenient in some problems to transform the system of two simultaneous differential equations for w and F (equations 2.3.7 and 2.3.8) into a single equation of eighth order in terms of a generalized stress function $\Phi = \Phi(x, y)$. Let

$$w = L_2\Phi, \quad F = -L_3\Phi; \quad (2.3.14)$$

then 2.3.8 is identically satisfied (for $\Omega = 0$), whereas equation 2.3.7 becomes

$$L\Phi = p_z, \quad (2.3.15)$$

where

$$L = L_1L_2 + L_3^2, \quad (2.3.16)$$

which can be expressed as follows:

$$L = P_1\frac{\partial^8}{\partial x^8} + P_3\frac{\partial^8}{\partial x^7\,\partial y} + P_5\frac{\partial^8}{\partial x^6\,\partial y^2} + P_7\frac{\partial^8}{\partial x^5\,\partial y^3}$$

$$+ P_9\frac{\partial^8}{\partial x^4\,\partial y^4} + P_8\frac{\partial^8}{\partial x^3\,\partial y^5} + P_6\frac{\partial^8}{\partial x^2\,\partial y^6} + P_4\frac{\partial^8}{\partial x\,\partial y^7}$$

$$+ P_2\frac{\partial^8}{\partial y^8}, \quad (2.3.17)$$

$$P_1 = A_{yy}^* D_{xx}^* + (B_{yx}^*)^2, \qquad P_2 = A_{xx}^* D_{yy}^* + (B_{xy}^*)^2, \quad (2.3.18)$$

$$P_3 = -2A_{ys}^* D_{xx}^* + 4A_{yy}^* D_{xs}^* + 2B_{yx}^*(2B_{ys}^* - B_{sx}^*), \quad (2.3.19)$$

$$P_4 = -2A_{xs}^* D_{yy}^* + 4A_{xx}^* D_{ys}^* + 2B_{xy}^*(2B_{xs}^* - B_{sy}^*), \quad (2.3.20)$$

$$P_5 = (2A_{xy}^* + A_{ss}^*)D_{xx}^* - 8A_{ys}^* D_{xs}^* + 2A_{yy}^*(D_{xy}^* + 2D_{ss}^*)$$
$$+ 2B_{yx}^*(B_{xx}^* + B_{yy}^* - 2B_{ss}^*) + (2B_{ys}^* - B_{sx}^*)^2 \quad (2.3.21)$$

$$P_6 = (2A_{xy}^* + A_{ss}^*)D_{yy}^* - 8A_{xs}^* D_{ys}^* + 2A_{xx}^*(D_{xy}^* + 2D_{ss}^*)$$
$$+ 2B_{xy}^*(B_{xx}^* + B_{yy}^* - 2B_{ss}^*) + (2B_{xs}^* - B_{sy}^*)^2, \quad (2.3.22)$$

$$P_7 = -2A_{xs}^* D_{xx}^* + 4(2A_{xy}^* + A_{ss}^*)D_{xs}^* - 4A_{ys}^*(D_{xy}^* + 2D_{ss}^*)$$
$$+ 4A_{yy}^* D_{ys}^* + 2B_{yx}^*(2B_{xs}^* - B_{sy}^*)$$
$$+ 2(2B_{ys}^* - B_{sx}^*)(B_{xx}^* + B_{yy}^* - 2B_{ss}^*), \quad (2.3.23)$$

$$P_8 = -2A_{ys}^* D_{yy}^* + 4(2A_{xy}^* + A_{ss}^*)D_{ys}^* - 4A_{xs}^*(D_{xy}^* + 2D_{ss}^*)$$
$$+ 4A_{xx}^* D_{xs}^* + 2B_{xy}^*(2B_{ys}^* - B_{sx}^*) + 2(2B_{xs}^* - B_{sy}^*)$$
$$\times (B_{xx}^* + B_{yy}^* - 2B_{ss}^*), \quad (2.3.24)$$

$$P_9 = A_{xx}^* D_{xx}^* - 8A_{xs}^* D_{xs}^* + 2(2A_{xy}^* + A_{ss}^*)(D_{xy}^* + 2D_{ss}^*)$$
$$- 8A_{ys}^* D_{ys}^* + A_{yy}^* B_{yy}^* + 2B_{xy}^* B_{yx}^* + 2(2B_{ys}^* - B_{sx}^*)$$
$$\times (2B_{xs}^* - B_{sy}^*) + (B_{xx}^* + B_{yy}^* - 2B_{ss}^*)^2. \quad (2.3.25)$$

Similarly, boundary conditions 2.1.28 can be expressed in terms of Φ and so the problem is completely defined.

The expressions of stresses in terms of resultants and couples are again given by equations 2.1.33; however, for the general heterogeneous plates zero elements do not necessarily appear in the a, b, c, and d matrices.

2.4   Some particular cases of composite aeolotropic plates
Cylindrical bending of long rectangular plates

Insight into some of the consequences of the general plate heterogeneity considered in Section 2.3 may be obtained by determining the state of stress and deformation in long rectangular plates subject to cylindrical bending.

Consider a composite plate with stress-strain relations 2.1.11 or 2.3.1. Let the y axis coincide with the left longitudinal edge of the plate, and let l be its free span (see Fig. 1.13). The edges

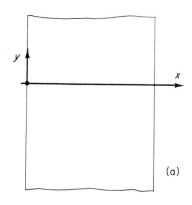

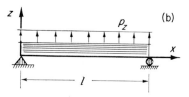

Figure 1.13   View and Section of Long Rectangular Composite Plate Loaded Uniformly

$x = 0, l$ are assumed to be simply supported; the edge $x = l$ is free to move in the x direction while the other one is fixed in place.

If we suppose that the plate is also supported at $y = \pm\infty$, it is clear that the behavior of the composite structure under the load $p_z = p_z(x)$ is independent of y and the problem is one dimensional. This simplification permits an immediate integration of equation 2.3.15, which now takes the form

$$P_1\frac{d^8\Phi}{dx^8} = p_z, \quad (2.4.1)$$

in which P_1 is given by the first equation in 2.3.18. The solution of equation 2.4.1 is subject to the following boundary conditions:

$$\text{at } x = 0, l: \quad N_x = N_{xy} = w = M_x = 0. \quad (2.4.2)$$

For $p_z = \text{const}$ one finds by integration of 2.4.1

$$\Phi = \frac{p_z}{P_1}\frac{x^8}{56 \times 720} + C_1\frac{x^7}{7 \times 720} + C_2\frac{x^6}{720} + C_3\frac{x^5}{120} + C_4\frac{x^4}{24}$$

$$+ C_5\frac{x^3}{6} + C_6\frac{x^2}{2} + C_7 x + C_8. \quad (2.4.3)$$

In view of equations 2.3.14, 2.1.18, 2.1.10, 2.1.19, 2.1.6, and 2.4.2 the following results are obtained:

$$w = \frac{A^*_{yy}}{24P_1} p_z l^4 \left[\left(\frac{x}{l}\right)^4 - 2\left(\frac{x}{l}\right)^3 + \left(\frac{x}{l}\right) \right] = A^*_{yy} k,$$

$$F = -B^*_{yx} k, \quad (2.4.4)$$

$$N_x = 0, \ N_y = B^*_{yx} \frac{p_z}{2P_1} x(l-x), \ N_{xy} = 0, \quad (2.4.5)$$

$$M_x = \frac{p_z}{2} x(l-x), \ M_y = [B^*_{yx} B^*_{yy} + A^*_{yy} D^*_{xy}] \frac{M_x}{P_1},$$

$$M_{xy} = [B^*_{yx} B^*_{ys} + A^*_{yy} D^*_{xs}] \frac{M_x}{P_1}, \quad (2.4.6)$$

$$\epsilon_x = [A^*_{xy} B^*_{yx} + (z - B^*_{xx}) A^*_{yy}] \frac{M_x}{P_1}, \ \epsilon_y = 0,$$

$$\epsilon_{xy} = [A^*_{ys} B^*_{yx} - A^*_{yy} B^*_{sx}] \frac{M_x}{P_1}, \quad (2.4.7)$$

$$Q_x = \frac{p_z}{2}\left(1 - 2\frac{x}{l}\right), \ Q_y = [B^*_{yx} B^*_{ys} + A^*_{yy} D^*_{xs}] \frac{M_x}{P_1}. \quad (2.4.8)$$

The stresses can be readily obtained from 2.3.1 for a given laminated plate by inserting the values of strain components given by equations 2.4.7. The transverse shear components τ_{xz}, τ_{yz} can be obtained by using the macroscopic equilibrium equations, which for the one-dimensional problem with absence of body forces are of the form

$$\tau_{x,x} + \tau_{xz,z} = 0, \ \tau_{xy,x} + \tau_{yz,z} = 0. \quad (2.4.9)$$

For the detailed expressions for all stress components see Stavsky.[34] Note that in order to maintain the deformed shape in the first equation in 2.4.4 a cross resultant force N_y is developed in the plate which does not have any counterpart in homogeneous aeolotropic plate theory.

Uniform distributions of stress resultants and couples

As shown in Section 2.2 we find equations 2.2.1 and 2.2.2 for F and w, respectively, and equations 2.2.3 and 2.2.4 for the strains at the reference plane and curvatures when in-plane forces or couples are applied.

The cross-elasticity effect (equation 2.2.5) also holds; however, now the a, b, c, and d matrices generally include all elements.

The solution for pure bending, twisting, and stretching of *skewed* heterogeneous aeolotropic plates is shown in a note by Stavsky.[36]

Numerical Example: To obtain a quantitative insight into the results of uniform distributions of resultants and couples, consider a continuously heterogeneous plate with the following stiffness matrix:

$$[E] = E^0_{xx} \begin{bmatrix} 1.0(1 + 0.6 \cos \pi z/h) & 0.2(1 + 0.6 \cos \pi z/h) & 0.002(1 + 100 \cos \pi z/h) \\ 0.2(1 + 0.6 \cos \pi z/h) & 1.0(1 + 0.6 \cos \pi z/h) & 0.002(1 + 100 \cos \pi z/h) \\ 0.002(1 + 100 \cos \pi z/h) & 0.002(1 + 100 \cos \pi z/h) & 0.4(1 + 0.6 \cos \pi z/h) \end{bmatrix}. \quad (2.4.10)$$

For the case of a uniform bending moment M_x we find

$$w = -\frac{6M_x}{E^0_{xx} h^3} (1.38525x^2 - 0.13431y^2 - 0.48820xy). \quad (2.4.11)$$

The stress and strain distributions are shown in Figure 1.14.

Note that the stress and strain components deviate considerably, quantitatively and qualitatively, from corresponding results for homogeneous plates. The occurrence of cross stresses is clearly exemplified and it seems that they have to be considered in the design of laminated structures.

Composite plates without coupling of bending and stretching

Consider a symmetrically laminated plate; namely, let the plate described in Figure 1.12 be the upper half of a composite plate made up of $2n$ layers and of total thickness h. For such a plate it is convenient to take the reference plane $z = 0$ to coincide with the plate plane of symmetry. It is obvious from equations 2.3.5 that the B matrix vanishes, and, consequently, in view of 2.1.20b, 2.3.11, and 2.3.12,

$$L_3 = 0, \ L_4 = 0. \quad (2.4.12)$$

The system of equations 2.3.7 and 2.3.8 is uncoupled and reads

$$L_1 w = p_z, \ L_2 F = L_5 \Omega, \quad (2.4.13)$$

in which L_1 is given in terms of unstarred D_{ij}, defined by

$$D_{ij} = 2 \sum_{m=1}^{n} \int_{h_{m-1}}^{h_m} E^m_{ij} z^2 \, dz = \frac{2}{3} \sum_{m=1}^{n} E^m_{ij}(h^3_m - h^3_{m-1})$$

$$(i, j = x, y, s), \quad (2.4.14)$$

and L_2 and L_5 remained unchanged as in equations 2.3.10 and 2.3.13, respectively, except for the definition of the elastic areas (equation 2.3.6) which now becomes

$$A_{ij} = 2 \sum_{m=1}^{n} \int_{h_{m-1}}^{h_m} E^m_{ij} \, dz = 2 \sum_{m=1}^{n} E^m_{ij}(h_m - h_{m-1})$$

$$(i, j = x, y, s). \quad (2.4.15)$$

Note that for a continuously heterogeneous plate with E_{ij} that are *even* functions of z no summation is necessary in equations 2.4.14 and 2.4.15, and the integral limits are $z = 0$ and $z = h/2$.

The bending system (first equation in 2.4.13) and the stretching system (second equation in 2.4.13) are completely analogous to respective equations for homogeneous plates except for the value of the constants that appear in L_1, L_2, and L_5. Known results of homogeneous aeolotropic plates can be fully utilized for the analysis of symmetrically composed plates, as far as the solution for w and F goes. This was first shown for the bending case by Lekhnitskii[35] and for the stretching case by Stavsky.[34] Note that the stress distribution in a composite plate is completely different from that in a homogeneous plate, in view of the cross-elasticity effect (2.1.33) which is only simplified now in the sense that

$$[\tau] = [E][a][N] \quad \text{or} \quad [\tau] = z[E][d][M], \quad (2.4.16)$$

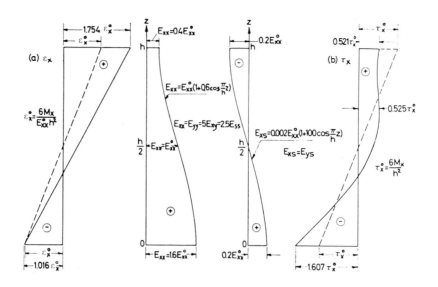

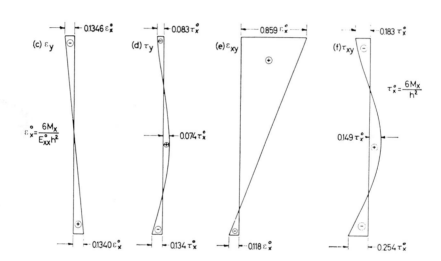

Figure 1.14   Strain and Stress Distributions in a Continuously Heterogeneous
Plate Subject to Pure Bending Moment M_x

where

$$[a] = [A^{-1}], \quad [d] = [D^{-1}]. \tag{2.4.17}$$

For a further discussion and examples see Sections 4.1 and 4.2.

Composite plates with no cross-elasticity effect

As an example of composite plates for which w and F are uncoupled and the cross-elasticity effect (2.1.33) disappears, consider a multilayer aeolotropic plate with an identical z dependence of the stiffnesses E_{ij}:

$$E_{ij}^m = c^m E_{ij}^0 \qquad (i, j = x, y, s; \, m = 1, 2 \ldots, n), \tag{2.4.18}$$

where E_{ij}^m and c^m vary from one layer to another and the E_{ij}^0 are constants.

For a continuously heterogeneous plate we have, instead of equation 2.4.18,

$$E_{ij}(z) = c(z) E_{ij}^0 \qquad (i, j = x, y, s). \tag{2.4.19}$$

Introduction of equations 2.4.18 or 2.4.19 into 2.3.6 or 2.3.5 and 2.1.20 gives

$$[B] = z_0[A], \quad [A^*] = [A^{-1}], \quad [B^*] = -z_0[I],$$
$$[D^*] = [D] - z_0^2[A], \tag{2.4.20}$$

where

$$z_0 = \frac{\frac{1}{2} \sum_{m=1}^{n} c^m (h_m^2 - h_{m-1}^2)}{\sum_{m=1}^{n} c^m (h_m - h_{m-1})} \quad \text{or} \quad z_0 = \frac{\int_0^h c(z) z \, dz}{\int_0^h c(z) \, dz}. \tag{2.4.21}$$

for laminated or continuously heterogeneous plates, respectively, and I is the unit matrix.

As B^* is a diagonal matrix with equal elements, the coupling operator L_3 given by equation 2.3.11 vanishes identically, and the separate equations for w and F (2.4.13) also hold for this class of heterogeneous plates. The operators L_1, L_2, and L_5 are given by equations 2.3.9, 2.3.10, and 2.3.13 if equation 2.4.20 is taken into consideration. A similar situation was shown by Pister and Dong[37] to exist in a plate composed of isotropic layers with constant Poisson's ratio.

The quantity z_0 is found to designate the distance of the neutral plane from the arbitrarily chosen reference plane at the lower face of the plate. If the coordinate axes are transformed to this neutral plane (Fig. 1.15) one finds that $B = 0$; consequently equations 2.1.31 become

$$\begin{bmatrix} \epsilon^0 \\ \kappa \end{bmatrix} = \begin{bmatrix} a & 0 \\ 0 & d \end{bmatrix}\begin{bmatrix} N \\ M \end{bmatrix}, \qquad (2.4.22)$$

where

$$[a] = [A^{-1}], \qquad d = [D^{*-1}], \qquad (2.4.23)$$

which can be obtained from equations 2.4.20.

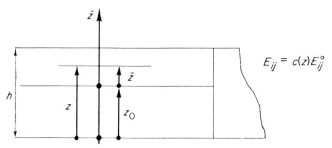

Figure 1.15 Section of Heterogeneous Plate with Same z-Dependence of E_{ij}

The stress components in terms of resultants and couples are given by the following relations obtained from 2.1.33 by specialization:

$$[\tau] = c\left(\frac{[N]}{A^0} + \bar{z}\frac{[M]}{D^0}\right), \quad A^0 = \int_{-z_0}^{h-z_0} c\, d\bar{z}, \quad D^0 = \int_{-z_0}^{h-z_0} c\bar{z}^2\, d\bar{z}. \qquad (2.4.24)$$

These relations clearly show that the cross-elasticity effect disappears. Therefore, for example,

$$\tau_x = c\left(\frac{N_x}{A^0} + \bar{z}\frac{M_x}{D^0}\right), \qquad (2.4.25)$$

where

$$N_x = \int_{-z_0}^{h-z_0} \tau_x\, d\bar{z}, \quad M_x = \int_{-z_0}^{h-z_0} \tau_x \bar{z}\, d\bar{z}. \qquad (2.4.26)$$

The far-reaching consequences of the disappearance of the cross-elasticity effect shows up in a plate theory that considers transverse shear and normal stress deformations. For a detailed discussion see the works of Stavsky.[31,38]

Numerical Example: Consider an aeolotropic heterogeneous rectangular plate with the following E matrix:

$$E_{ij} = E_{ij}^0(1 + \cos \pi z/h), \qquad (2.4.27)$$

and

$$[E^0] = E_{xx}^0\begin{bmatrix} 1 & 0.2 & 0.2 \\ 0.2 & 1 & 0.2 \\ 0.2 & 0.2 & 0.3 \end{bmatrix}. \qquad (2.4.28)$$

For the case of a pure bending moment M_x the deflection pattern is given by

$$w = -\frac{6M_x}{E_{xx}^0 h^3}(2.288326x^2 - 0.176025y^2 - 1.408201xy), \qquad (2.4.29)$$

and the strain and stress distributions are shown in Figure 1.16.

2.5 Symmetric deformations of composite circular orthotropic plates

The linear theory of axisymmetric deformations of heterogeneous circular elastic thin plates with cylindrical orthotropy is established by specialization of the theory of composite shells of revolution as given by Reissner[39] and Stavsky.[40]

Strain-displacement relations

The radial and circumferential strain components are written in the form

$$\varepsilon_r = \varepsilon_{r0} + z\kappa_r, \qquad \varepsilon_\theta = \varepsilon_{\theta 0} + z\kappa_\theta, \qquad (2.5.1)$$

with

$$\varepsilon_{r0} = du/dr \equiv u', \qquad \varepsilon_{\theta 0} = u/r, \qquad (2.5.2)$$

$$\kappa_r = \beta', \qquad \kappa_\theta = \beta/r, \qquad w' = -\beta. \qquad (2.5.3)$$

where u is radial displacement of the reference surface ($z = 0$) and w is its transverse displacement.

Stress-strain relations

The plate material is taken to be rotationally orthotropic and heterogeneous in accordance with Hooke's law:

$$\tau_r = E_{rr}\varepsilon_r + E_{r\theta}\varepsilon_\theta, \qquad (2.5.4)$$

$$\tau_\theta = E_{\theta r}\varepsilon_r + E_{\theta\theta}\varepsilon_\theta, \qquad (2.5.5)$$

$$\tau_{r\theta} = 0 \text{ by symmetry of deformations}, \qquad (2.5.6)$$

$$E_{\theta r} = E_{r\theta}, \quad E_{ij} = E_{ij}(z) \qquad (i, j = r, \theta). \qquad (2.5.7)$$

Stress resultants and stress couples are defined by (see Fig. 1.17)

$$(N_r, N_\theta) = \int_{-h_1}^{+h_2} (\tau_r, \tau_\theta)\, dz,$$

$$(M_r, M_\theta) = \int_{-h_1}^{+h_2} (\tau_r, \tau_\theta)z\, dz. \qquad (2.5.8)$$

The only transverse shear stress resultant that occurs in the symmetric plate problem is defined by

$$Q = \int_{-h_1}^{+h_2} \tau_{rz}\, dz. \qquad (2.5.9)$$

The plate stress-strain relations are obtained by introduction of equations 2.5.4, and 2.5.5 into 2.5.8:

$$\begin{bmatrix} N_r \\ N_\theta \\ \hline M_r \\ M_\theta \end{bmatrix} = \left[\begin{array}{cc|cc} A_{rr} & A_{r\theta} & B_{rr} & B_{r\theta} \\ A_{r\theta} & A_{\theta\theta} & B_{r\theta} & B_{\theta\theta} \\ \hline B_{rr} & B_{r\theta} & D_{rr} & D_{r\theta} \\ B_{r\theta} & B_{\theta\theta} & D_{r\theta} & D_{\theta\theta} \end{array}\right]\begin{bmatrix} \varepsilon_{r0} \\ \varepsilon_{\theta 0} \\ \kappa_r \\ \kappa_\theta \end{bmatrix}, \qquad (2.5.10)$$

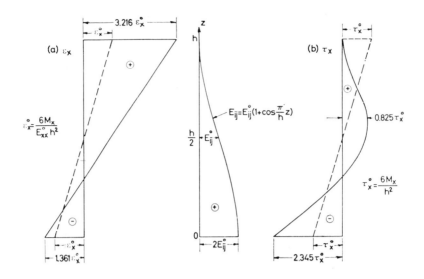

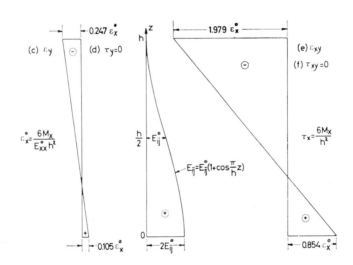

Figure 1.16 Strain and Stress Components in a Heterogeneous Plate with
Same z-Variation of E_{ij}, Subject to Pure Bending Moment M_x

where

$$(A_{ij}, B_{ij}, D_{ij}) = \int_{-h_1}^{+h_2} (1, z, z^2) E_{ij}\, dz \qquad (i, j = r, \theta).$$
$$(2.5.11)$$

Equilibrium equations

Force and moment equilibrium conditions of the undeformed plate element (Fig. 1.17) are

$$(rN_r)' + rp_r - N_\theta = 0, \qquad (2.5.12)$$

$$(rQ)' + rp_z = 0, \qquad (2.5.13)$$

$$(rM_r)' - M_\theta - rQ = 0. \qquad (2.5.14)$$

These three equilibrium equations together with the four plate
stress-strain relations 2.5.10 constitute a system of seven
equations for seven unknowns, N_r, N_θ, Q, M_r, M_θ and, u, β.

Reduction to two simultaneous equations

The introduction of a resultant stress function Ψ in the
form

$$\Psi = rN_r \qquad (2.5.15)$$

enables us to satisfy identically equation 2.5.12 by taking

$$N_\theta = \Psi' + rp_r, \qquad (2.5.16)$$

whereas the transverse shear resultant is obtained by direct
integration of equation 2.5.13:

$$rQ = -\int rp_z\, dr. \qquad (2.5.17)$$

The third equilibrium equation (2.5.14) together with a
compatibility relation which is written

$$(r\varepsilon_{\theta 0})' = \varepsilon_{r0} \qquad (2.5.18)$$

are transformed to two simultaneous equations for β and Ψ:

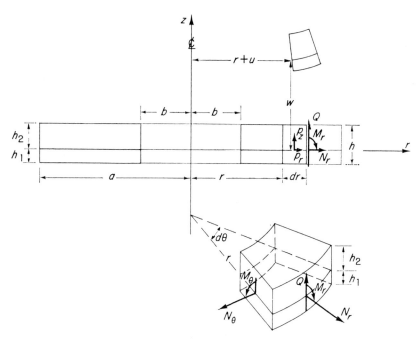

Figure 1.17   Circular Ring Plate with Stress Resultants, Couples, and Loading Intensity Shown on Plate Element

$$L_{11}\beta + L_{12}\Psi = L_{13}(p_r, p_z), \qquad (2.5.19)$$

$$L_{21}\beta + L_{22}\Psi = L_{23}(p_r), \qquad (2.5.20)$$

where

$$L_{11} = D_{rr}^* r^2(\ )'' + D_{rr}^* r(\ )' - D_{\theta\theta}^*(\ ), \qquad (2.5.21)$$

$$-L_{12} = B_{\theta r}^* r^2(\ )'' + [B_{\theta r}^* + B_{rr}^* - B_{\theta\theta}^*]r(\ )' - B_{r\theta}^*(\ ), \qquad (2.5.22)$$

$$L_{13}(p_r, p_z) = r^2 Q + B_{\theta r}^* r(r^2 p_r)' - B_{\theta\theta}^*(r^2 p_r), \qquad (2.5.23)$$

$$L_{21} = B_{\theta r}^* r^2(\ )'' + [B_{\theta r}^* - B_{rr}^* + B_{\theta\theta}^*]r(\ )' - B_{r\theta}^*(\ ), \qquad (2.5.24)$$

$$L_{22} = A_{\theta\theta}^* r^2(\ )'' + A_{\theta\theta}^* r(\ )' - A_{rr}^*(\ ), \qquad (2.5.25)$$

$$L_{23}(p_r) = -A_{\theta\theta}^* r(r^2 p_r)' + A_{r\theta}^*(r^2 p_r). \qquad (2.5.26)$$

To the system of equations 2.5.19 and 2.5.20 for β and Ψ the following boundary conditions are adjoined:

For a ring plate at $r = a, b$:

$$u = \bar{u} \quad \text{or} \quad N_r = \bar{N}_r, \qquad (2.5.27)$$

$$w = \bar{w} \quad \text{or} \quad Q = \bar{Q}, \qquad (2.5.28)$$

$$\beta = \bar{\beta} \quad \text{or} \quad M_r = \bar{M}_r. \qquad (2.5.29)$$

For a complete plate $b = 0$ and the conditions at the inner boundary are replaced by the regularity requirement for β and Ψ.

The coupling operators L_{12} and L_{21} vanish only when the plate is symmetrically composed or when all elastic moduli have the same z dependence. Then solutions of homogeneous orthotropic plates can be almost directly utilized for the determination of β and Ψ. In all other cases of plate heterogeneity the coupled system 2.5.19 and 2.5.20 must be considered

and we indicate some examples for the case of isotropic composite circular plates.

2.6   Examples of composite circular isotropic plates

For isotropic heterogeneous plates Hooke's law (equations 2.5.4 and 2.5.5) takes the form

$$\begin{bmatrix} \tau_r \\ \tau_\theta \end{bmatrix} = \frac{E}{1 - \nu^2} \begin{bmatrix} 1 & \nu \\ \nu & 1 \end{bmatrix} \begin{bmatrix} \varepsilon_r \\ \varepsilon_\theta \end{bmatrix} \qquad (2.6.1)$$

and

$$E = E(z), \quad \nu = \nu(z). \qquad (2.6.2)$$

The plate stress-strain relations (2.5.10) are simplified as follows:

$$\begin{bmatrix} N_r \\ N_\theta \\ M_r \\ M_\theta \end{bmatrix} = \begin{bmatrix} A_1 & A_2 & B_1 & B_2 \\ A_2 & A_1 & B_2 & B_1 \\ B_1 & B_2 & D_1 & D_2 \\ B_2 & B_1 & D_2 & D_1 \end{bmatrix} \begin{bmatrix} \varepsilon_{r0} \\ \varepsilon_{\theta 0} \\ \kappa_r \\ \kappa_\theta \end{bmatrix}, \qquad (2.6.3)$$

where

$$A_1 = A_{rr} = A_{\theta\theta} = \int_{-h_1}^{h_2} \frac{E}{1 - \nu^2} \, dz,$$

$$A_2 = A_{r\theta} = A_{\theta r} = \int_{-h_1}^{h_2} \frac{\nu E}{1 - \nu^2} \, dz, \qquad (2.6.4)$$

$$B_1 = B_{rr} = B_{\theta\theta} = \int_{-h_1}^{h_2} \frac{E}{1 - \nu^2} \, z \, dz,$$

$$B_2 = B_{r\theta} = B_{\theta r} = \int_{-h_1}^{h_2} \frac{\nu E}{1 - \nu^2} \, z \, dz, \qquad (2.6.5)$$

$$D_1 = D_{rr} = D_{\theta\theta} = \int_{-h_1}^{h_2} \frac{E}{1-\nu^2} z^2 \, dz,$$

$$D_2 = D_{r\theta} = D_{\theta r} = \int_{-h_1}^{h_2} \frac{\nu E}{1-\nu^2} z^2 \, dz. \quad (2.6.6)$$

The governing system of equations, 2.5.19 and 2.5.20, becomes

$$D_1^* L\beta - B_2^* L\Psi = L_{13}(p_r, p_z), \quad (2.6.7)$$

$$B_2^* L\beta + A_1^* L\Psi = L_{23}(p_r), \quad (2.6.8)$$

where

$$L = r^2(\ )'' + r(\ )' - (\ ) \quad (2.6.9)$$

and

$$\begin{aligned}L_{13}(p_r, p_z) &= r^2 Q + B_2^* r(r^2 p_r)' - B_1^*(r^2 p_r), \\ L_{23}(p_r) &= -A_1^* r(r^2 p_r)' + A_2^*(r^2 p_r).\end{aligned} \quad (2.6.10)$$

It is noted that similar equations in terms of the radial displacement u and the slope β were given by Grigolyuk (Ref. 41, p. 90) for a two-layer isotropic circular plate.

The starred quantities are given in terms of the basic quantities 2.6.4–2.6.6 as follows:

$$A_1^* = \frac{1}{\Delta} A_1, \quad B_1^* = \frac{1}{\Delta}(A_2 B_2 - A_1 B_1),$$

$$B_2^* = \frac{1}{\Delta}(A_2 B_1 - A_1 B_2), \quad (2.6.11)$$

$$D_1^* = D_1 - \frac{1}{\Delta}[A_1(B_1^2 + B_2^2) - 2A_2 B_1 B_2], \quad \Delta = A_1^2 - A_2^2. \quad (2.6.12)$$

Taking $p_r = 0$ we rewrite equations 2.6.7, 2.6.8 in the form

$$\left\{\frac{1}{r}\left[r\left(\beta - \frac{B_2^*}{D_1^*}\Psi\right)\right]_{,r}\right\}_{,r} = \frac{Q}{D_1^*}, \quad (2.6.13)$$

$$\left\{\frac{1}{r}\left[r\left(\frac{B_2^*}{A_1^*}\beta + \Psi\right)\right]_{,r}\right\}_{,r} = 0, \quad (2.6.14)$$

which can be directly integrated to give

$$\beta - \frac{B_2^*}{D_1^*}\Psi = \frac{1}{r}\int\left[r\int\frac{Q}{D_1^*}\,dr\right]dr + \bar{C}_1 r + \frac{\bar{C}_2}{r}, \quad (2.6.15)$$

$$\frac{B_2^*}{A_1^*}\beta + \Psi = \bar{C}_3 r + \frac{\bar{C}_4}{r}, \quad (2.6.16)$$

and finally

$$\beta = \frac{A_1}{A_1 D_1 - B_1^2}\left\{\frac{1}{r}\int\left[r\int Q\,dr\right]dr\right\} + C_1 r + \frac{C_2}{r}, \quad (2.6.17)$$

$$\Psi = \frac{A_1 B_2 - A_2 B_1}{A_1 D_1 - B_1^2}\left\{\frac{1}{r}\int\left[r\int Q\,dr\right]dr\right\} + C_3 r + \frac{C_4}{r}. \quad (2.6.18)$$

Note that in going over from equations 2.6.15 and 2.6.16 to 2.6.17 and 2.6.18 the starred coefficients were expressed in terms of the original unstarred quantities (equations 2.6.4–2.6.6).

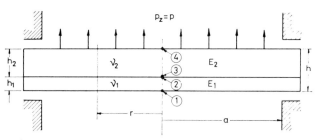

Figure 1.18 Cross-Section of Clamped Two-Layer Circular Plate Uniformly Loaded

Uniformly loaded clamped circular plates

Consider a clamped two-layer circular plate uniformly loaded by $p_z = p$, as shown in Figure 1.18, and subjected to the following boundary conditions at $r = a$:

$$N_r = w = \beta = 0, \quad (2.6.19)$$

whereas at $r = 0$, β and Ψ must be regular. The transverse shear resultant (2.5.17) becomes

$$Q = -\tfrac{1}{2}pr, \quad (2.6.20)$$

whereas 2.6.17 and 2.6.18 become, in view of the boundary conditions,

$$\begin{aligned}\beta &= \frac{A_1}{A_1 D_1 - B_1^2} \cdot \frac{pa^3}{16}\left[\frac{r}{a} - \left(\frac{r}{a}\right)^3\right], \\ \Psi &= \frac{A_1 B_2 - A_2 B_1}{A_1 D_1 - B_1^2} \cdot \frac{pa^3}{16}\left[\frac{r}{a} - \left(\frac{r}{a}\right)^3\right].\end{aligned} \quad (2.6.21)$$

The radial and transverse displacement components are

$$\begin{aligned}u &= -\frac{B_1}{A_1 D_1 - B_1^2} \cdot \frac{pa^3}{16}\left[\frac{r}{a} - \left(\frac{r}{a}\right)^3\right], \\ w &= \frac{A_1}{A_1 D_1 - B_1^2} \cdot \frac{pa^4}{64}\left[1 - 2\left(\frac{r}{a}\right)^2 + \left(\frac{r}{a}\right)^4\right].\end{aligned} \quad (2.6.22)$$

The stress resultants and couples are obtained from 2.6.3 in view of 2.5.2, 2.5.3, 2.6.21, and 2.6.22:

$$N_r = \frac{A_1 B_2 - A_2 B_1}{A_1 D_1 - B_1^2} \cdot \frac{pa^2}{16}\left[1 - \left(\frac{r}{a}\right)^2\right], \quad (2.6.23)$$

$$N_\theta = \frac{A_1 B_2 - A_2 B_1}{A_1 D_1 - B_1^2} \cdot \frac{pa^2}{16}\left[1 - 3\left(\frac{r}{a}\right)^2\right], \quad (2.6.24)$$

$$M_r = \frac{pa^2}{16}\left[1 - 3\left(\frac{r}{a}\right)^2\right] + \frac{A_1 D_2 - B_1 B_2}{A_1 D_1 - B_1^2}\frac{pa^2}{16}\left[1 - \left(\frac{r}{a}\right)^2\right], \quad (2.6.25)$$

$$M_\theta = \frac{pa^2}{16}\left[1 - \left(\frac{r}{a}\right)^2\right] + \frac{A_1 D_2 - B_1 B_2}{A_1 D_1 - B_1^2}\frac{pa^2}{16}\left[1 - 3\left(\frac{r}{a}\right)^2\right]. \quad (2.6.26)$$

The stress components are derived from relations 2.6.1 in view of 2.5.1–2.5.3 and 2.6.21 or alternatively by equations of the type indicated by 2.1.33.

Numerical Example: The results 2.6.21–2.6.26 obtained for a uniformly loaded clamped circular plate are evaluated for the following case (Fig. 1.19):

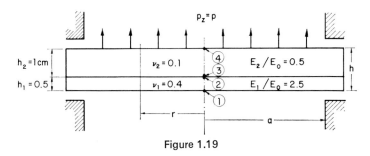

Figure 1.19

$$-h_1 \leqslant z < 0, \quad E_1/E_0 = 2.5, \quad \nu_1 = 0.4, \quad h_1/h = \tfrac{1}{3},$$
$$0 < z \leqslant h_2, \quad E_2/E_0 = 0.5, \quad \nu_2 = 0.1, h_2/h = \tfrac{2}{3},$$
$$E_0 = 10^6 \, \text{kg/cm}^2, \quad h = 1.5 \text{ cm}. \qquad (2.6.27)$$

The numerical results for the stress resultants and couples are shown in Figure 1.20. It is noted that N_r and N_θ are not developed at all in a homogeneous plate under the same boundary conditions (2.6.19).

The couples M_r and M_θ differ somewhat from the corresponding values for a circular homogeneous plate and are about 20 per cent higher at the center. In Figure 1.21 the r dependence of the radial and circumferential stress components is shown for four characteristic locations across the plate

thickness. The dotted curves show the stress distribution for the same two-layer plate shown in Figure 1.18, but with a constant Poisson's ratio of $\nu_1 = \nu_2 = \nu = 0.3$. Except for the quantitative differences between the dotted and full curves (Fig. 1.21) it is pointed out that in the case of the plate with constant Poisson's ratio there exists a neutral plane at $z_0/h = 0.3095$ where all strains and stresses vanish. One can say that the coupling of β and Ψ stems from the fact that no such neutral plane exists in a heterogeneous plate of the type shown in Figure 1.18.

2.7 Finite deflections of composite plates

Aeolotropic heterogeneous plates

We briefly indicate in which way the linear plate theory with which we have been concerned in Sections 2.1 and 2.3 is modified if finite transverse deflections are taken into account.

Equilibrium equations 2.1.5 and 2.1.6 are valid as before, but equation 2.1.7 is replaced by

$$[Q_x + N_x w_{,x} + N_{xy} w_{,y}]_{,x}$$
$$+ [Q_y + N_{xy} w_{,x} + N_y w_{,y}]_{,y} + p_z = 0. \quad (2.7.1)$$

Strain-displacement relations 2.1.9 are replaced by the following nonlinear equations:

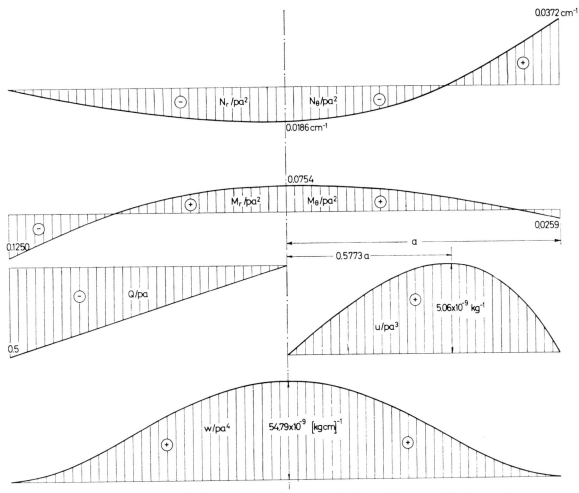

Figure 1.20   Diagrams of Moments, Stress Resultants, Radial, and Transverse Displacements

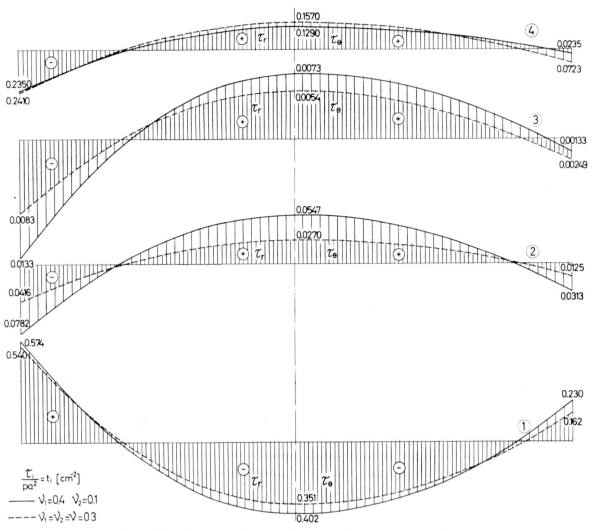

Figure 1.21   Diagrams of Stresses Across the Plate and Along the Radius

$$\epsilon_x^0 = u_{,x} + \tfrac{1}{2} w_{,x}^2, \quad \epsilon_y^0 = v_{,y} + \tfrac{1}{2} w_{,y}^2,$$

$$\epsilon_{xy}^0 = u_{,y} + v_{,x} + w_{,x} w_{,y}, \tag{2.7.2}$$

whereas equations 2.1.8 and 2.1.10 remain unchanged.

Plate stress-strain relations 2.3.4 remain valid and together with equations 2.1.5, 2.1.6, and 2.7.1 constitute a system of eleven equations for eleven unknowns N_x, N_y, N_{xy}, Q_x, Q_y, M_x, M_y, M_{xy}, u, v, w. This system may be reduced to three simultaneous nonlinear equations for the displacement components u, v, w if we use a method applied by Stavsky[29] for a specialized heterogeneous plate. Here we give a reduction of the system of equations to two simultaneous equations in terms of w and an Airy stress function F, as done by Reissner and Stavsky[24] and Stavsky,[42] by using equation 2.7.1 and the modified compatibility relation

$$\epsilon_{x,yy}^0 + \epsilon_{y,xx}^0 - \epsilon_{xy,xy}^0 = w_{,xy}^2 - w_{,xx} w_{,yy}, \tag{2.7.3}$$

which is a consequence of equation 2.7.2.

The resulting nonlinear simultaneous equations for w and F are

$$L_1 w - L_3 F - K_0(F, w) - K_1(\Omega, w) = p_z + L_4 \Omega, \tag{2.7.4}$$

$$L_2 F + L_3 W + \tfrac{1}{2} K_0(w, w) = L_5 \Omega, \tag{2.7.5}$$

where the functional operators L_1, ..., L_5 are given by expressions 2.3.9–2.3.13, respectively, and the operators K_i have the following form:

$$K_0(F, w) = F_{,yy} w_{,xx} - 2F_{,xy} w_{,xy} + F_{,xx} w_{,yy}, \tag{2.7.6}$$

$$\tfrac{1}{2} K_0(w, w) = w_{,xx} w_{,yy} - w_{,xy}^2, \tag{2.7.7}$$

$$K_1(\Omega, w) = \Omega(w_{,xx} + w_{,yy}) + \Omega_{,x} w_{,x} + \Omega_{,y} w_{,y}. \tag{2.7.8}$$

The system 2.7.4 and 2.7.5 combines the linear coupling of bending and stretching characteristic of general composite plates[34] with the von Kármán classical nonlinear coupling effect expressed by the K operators, which is seen to be the same as for homogeneous plates.

Circular orthotropic plates undergoing symmetric deformations

In order to extend the linear theory of Section 2.5 to account for finite deflections some modifications should be introduced.

The radial strain component takes the nonlinear form

$$\varepsilon_{r0} = u' + \tfrac{1}{2} \beta^2, \tag{2.7.9}$$

whereas the other strain-displacement relations, 2.5.1–2.5.3, remain unchanged.

Equilibrium equations 2.5.12 and 2.5.13 are, respectively, replaced by

$$(rH)' + rp_r - N_\theta = 0, \qquad (2.7.10)$$

$$(rV)' + rp_z = 0, \qquad (2.7.11)$$

and the third equation (2.5.14), as well as plate stress-strain relations 2.5.10, are still valid.

The reduction of the system of seven equations (2.5.10, 2.7.10, 2.7.11, and 2.5.14) for 3 stress resultants, 2 couples, and 2 deformation variables to two simultaneous equations is achieved by introducing the stress function Ψ in the form

$$\Psi = rH. \qquad (2.7.12)$$

Then 2.7.10 is identically satisfied by taking

$$N_\theta = \Psi'' + rp_r, \qquad (2.7.13)$$

whereas direct integration of 2.7.11 gives

$$rV = -\int rp_z \, dr. \qquad (2.7.14)$$

For the relations between $N_r(N_\xi)$, Q, H, and V see equations 3.1.21 and 3.1.22.

The moment equilibrium equation (2.5.14), together with the compatibility relation

$$(r\varepsilon_{\theta 0})' - \varepsilon_{r 0} = -\tfrac{1}{2}\beta^2 \qquad (2.7.15)$$

that modifies 2.5.18, are expressed in terms of β and Ψ to give the governing nonlinear system of simultaneous equations:

$$L_{11}\beta + L_{12}\Psi + L_{13}(\beta, \Psi) = L_{14}(\beta, p_r, p_z), \qquad (2.7.16)$$

$$L_{21}\beta + L_{22}\Psi = L_{24}(\beta, p_r, p_z). \qquad (2.7.17)$$

The functional operators L_{11}, L_{12}, L_{22} are given, respectively, by equations 2.5.21, 2.5.22, and 2.5.25; as for the other operators, we have

$$L_{13}(\beta, \Psi) = -r\beta\Psi, \qquad (2.7.18)$$

$$L_{14}(\beta, p_r, p_z) = -rB_{rr}^* \beta(rV)' + [r + B_{r\theta}^*\beta - rB_{rr}^*(\beta' - 1)](rV) \\ - B_{\theta\theta}^* r^2 p_r + rB_{\theta r}^*(r^2 p_r)', \qquad (2.7.19)$$

$$L_{21}\beta = B_{\theta r}^* r^2 \beta'' + [B_{\theta r}^* - B_{rr}^* + B_{\theta\theta}^*]r\beta' - B_{r\theta}^*\beta + \tfrac{1}{2}r\beta^2, \qquad (2.7.20)$$

$$L_{24}(\beta, p_r, p_z) = -A_{rr}^* \beta(rV) + A_{\theta r}^* r(\beta rV)' - A_{\theta\theta}^* r(r^2 p_r)' \\ + A_{r\theta}^*(r^2 p_r). \qquad (2.7.21)$$

Associated with the system 2.7.16 and 2.7.17 are boundary conditions of the form 2.5.22–2.5.29 for a ring plate. In case the composed circular plate is of a special type for which the matrix B^* vanishes then $L_{12} = L_{21} = 0$.

3 THEORY OF COMPOSITE SHELLS

3.1 Symmetric finite deformations of composite shells of revolution

Introduction

This section is concerned with the general theory of rotationally orthotropic shells of revolution heterogeneous in the thickness and meridional directions and subject to symmetric

deformations. Such heterogeneity may be achieved, for example, by composing a multilayer shell (see reports by Dietz *et al.*[43] and Tsai[25]) which makes it possible to adapt the material properties to a desired design rather than the design to the material. Hence the analysis of composite shells gained much interest in recent years as one can see from Ambartsumyan's[44] survey. Most of the work on heterogeneous shells was toward the development of linear theories, including Ambartsumyan's monograph[45] on anisotropic shells. In what follows we introduce the general nonlinear theory for non-shallow heterogeneous shells of revolution as given by Stavsky.[40]

Geometry of shell

Consider a laminated shell bounded by two coaxial surfaces of revolution and loaded in such a manner that it remains a shell of revolution. The geometry of the shell is defined by a "reference surface of revolution" and by a coordinate ζ measuring the distance of any material point from this reference surface ($\zeta = 0$). Note that the reference surface does not necessarily coincide with the middle surface of the shell. A similar situation exists in heterogeneous plates (Sec. 2.3), where the reference plane was located at the bottom face of the plate rather than at its middle plane.

The parametric equations of the undeformed reference surface of the shell (see Fig. 1.22) are given in polar coordinates by

$$r = r(\xi), \; z = z(\xi). \qquad (3.1.1)$$

The bounding surfaces of the shell are

$$\zeta = -h_1(\xi), \; \zeta = h_2(\xi), \qquad (3.1.2)$$

and the total shell thickness h is the sum of h_1 and h_2. Any material point of the undeformed shell is designated by ξ, θ, and ζ which constitute a system of orthogonal curvilinear coordinates in space. The meridional differential arc length ds_0 on the undeformed reference surface is given in terms of r and z by

$$(ds_0)^2 = (dr)^2 + (dz)^2 = (\alpha \, d\xi)^2. \qquad (3.1.3)$$

Then

$$\alpha^2 = (r')^2 + (z')^2, \qquad (3.1.4)$$

where primes indicate differentiation with respect to the parameter ξ. Other useful geometrical relations are the following:

$$r' = \alpha \cos \phi, \; z' = \alpha \sin \phi, \qquad (3.1.5)$$

$$\tan \phi = z'/r', \; \tan \Phi = (z + w)'/(r + u)', \qquad (3.1.6)$$

where ϕ and Φ denote the sloping angle of the tangent to a meridian curve before and after deformation of shell, whereas u and w are the radial and axial displacement components, respectively. The radius of curvature of the curve generating the undeformed reference surface is given by

$$R_\xi = \alpha/\phi', \qquad (3.1.7)$$

and the other principal radius of curvature is

$$R_\theta = r/\sin \phi. \qquad (3.1.8)$$

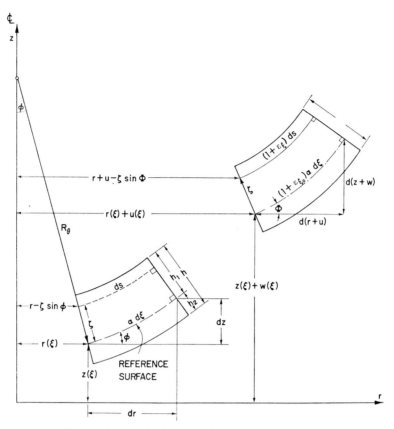

Figure 1.22 Shell Element Before and After Deformation

Strain-displacement relations

Assuming that the Euler-Bernoulli hypothesis holds for the heterogeneous shells under consideration and following Reissner[46] we write the expressions for the meridional strain ε_ξ and the circumferential strain ε_θ in the form

$$\varepsilon_\xi = \frac{\varepsilon_{\xi 0} + \zeta \kappa_\xi}{1 - \zeta/R_\xi}, \quad \varepsilon_\theta = \frac{\varepsilon_{\theta 0} + \zeta \kappa_\theta}{1 - \zeta/R_\theta}. \qquad (3.1.9)$$

The reference surface strains $\varepsilon_{\xi 0}$ and $\varepsilon_{\theta 0}$ are given by

$$\varepsilon_{\xi 0} = \left(1 + \frac{u'}{r'}\right) \frac{\cos \phi}{\cos \Phi} - 1, \qquad (3.1.10)$$

or alternatively

$$\varepsilon_{\xi 0} = \cos(\Phi - \phi) - 1 + \frac{u'}{\alpha} \cos \Phi + \frac{w'}{\alpha} \sin \Phi, \quad (3.1.11)$$

and

$$\varepsilon_{\theta 0} = u/r. \qquad (3.1.12)$$

The curvature changes κ_ξ, κ_θ of the reference surface are of the form

$$\kappa_\xi = -\frac{1}{\alpha}(\Phi' - \phi'), \quad \kappa_\theta = -\frac{1}{r}(\sin \Phi - \sin \phi), \quad (3.1.13)$$

whereas the radial and axial displacement components are given, respectively, by the following expressions:

$$u = r\varepsilon_{\theta 0}, \quad w = \int \alpha[(1 + \varepsilon_{\xi 0}) \sin \Phi - \sin \phi] \, d\xi. \quad (3.1.14)$$

Stress-strain relations

The shell material is taken to be rotationally orthotropic and Hooke's law takes the form

$$\begin{bmatrix} \tau_\xi \\ \tau_\theta \end{bmatrix} = \begin{bmatrix} E_{\xi\xi} & E_{\xi\theta} \\ E_{\xi\theta} & E_{\theta\theta} \end{bmatrix} \begin{bmatrix} \varepsilon_\xi \\ \varepsilon_\theta \end{bmatrix}. \qquad (3.1.15)$$

In view of the shell heterogeneity in the meridional and thickness directions it is noted that

$$E_{ij} = E_{ij}(\xi, \zeta) \qquad (i, j = \xi, \theta). \qquad (3.1.16)$$

In order to establish a two-dimensional shell theory we introduce stress resultants and couples (see Fig. 1.23) as follows:

$$N_\xi(1 + \varepsilon_{\theta 0})r \, d\theta = \int_{-h_1}^{h_2} \tau_\xi(1 + \varepsilon_\theta)r(1 - \zeta/R_\theta) \, d\theta \, d\zeta, \quad (3.1.17)$$

which, after neglecting $\varepsilon_{\theta 0}$, ε_θ, and ζ/R_θ as compared to unity, becomes

$$N_\xi = \int_{-h_1}^{+h_2} \tau_\xi \, d\zeta. \qquad (3.1.18)$$

Similarly

$$N_\theta = \int_{-h_1}^{h_2} \tau_\theta \, d\zeta, \quad (M_\xi, M_\theta) = \int_{-h_1}^{h_2} (\tau_\xi, \tau_\theta)\zeta \, d\zeta. \quad (3.1.19)$$

The only transverse shear stress resultant that occurs in a symmetric shell problem is defined by

$$Q = \int_{-h_1}^{h_2} \tau_{\xi\zeta} \, d\zeta. \qquad (3.1.20)$$

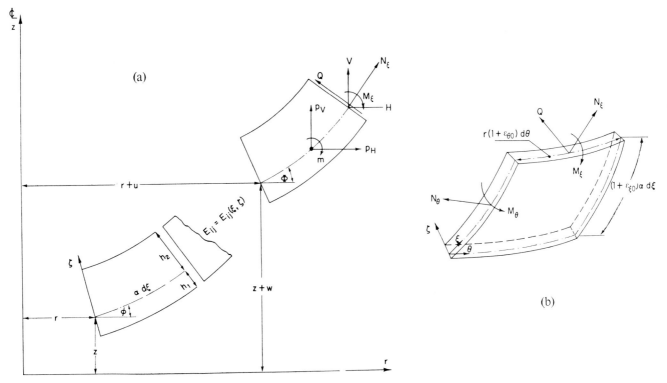

Figure 1.23   Shell Element with Applied Loads, Stress Resultants, and Couples

If H and V denote the radial and axial components, respectively, of N_ξ and Q, the following relations hold for the deformed shell:

$$N_\xi = H \cos \Phi + V \sin \Phi, \qquad (3.1.21)$$

$$Q = -H \sin \Phi + V \cos \Phi. \qquad (3.1.22)$$

Introducing equations 3.1.15 into definitions 3.1.18 and 3.1.19 and neglecting ζ/R_ξ and ζ/R_θ in 3.1.9 as compared to unity, we obtain the following stress-strain relations:

$$
\begin{bmatrix} N_\xi \\ N_\theta \\ M_\xi \\ M_\theta \end{bmatrix}
=
\begin{bmatrix}
A_{\xi\xi} & A_{\xi\theta} & B_{\xi\xi} & B_{\xi\theta} \\
A_{\xi\theta} & A_{\theta\theta} & B_{\xi\theta} & B_{\theta\theta} \\
B_{\xi\xi} & B_{\xi\theta} & D_{\xi\xi} & D_{\xi\theta} \\
B_{\xi\theta} & B_{\theta\theta} & D_{\xi\theta} & D_{\theta\theta}
\end{bmatrix}
\begin{bmatrix} \varepsilon_{\xi 0} \\ \varepsilon_{\theta 0} \\ \kappa_\xi \\ \kappa_\theta \end{bmatrix} ; \qquad (3.1.23)
$$

or more compactly,

$$
\begin{bmatrix} N \\ M \end{bmatrix}
=
\begin{bmatrix} A & B \\ B & D \end{bmatrix}
\begin{bmatrix} \varepsilon_0 \\ \kappa \end{bmatrix}. \qquad (3.1.24)
$$

The elastic areas, the elastic statical moments and the elastic moments of inertia are given, respectively, by

$$(A_{ij}, B_{ij}, D_{ij}) = \int_{-h_1}^{h_2} (1, \zeta, \zeta^2) E_{ij}\, d\zeta \qquad (i, j = \xi, \theta). \qquad (3.1.25)$$

Such quantities appear also in the theory of heterogeneous plates (see equations 2.3.5).

Equilibrium of shell element

Force and moment equilibrium equations of the deformed shell element (Fig. 1.23) are of the form

$$(rH)' + r\alpha p_H - \alpha N_\theta = 0, \qquad (3.1.26)$$

$$(rV)' + r\alpha p_V = 0, \qquad (3.1.27)$$

$$(rM_\xi)' - \alpha M_\theta \cos \Phi - r\alpha Q + r\alpha m = 0, \qquad (3.1.28)$$

where the strains $\varepsilon_{\xi 0}$, $\varepsilon_{\theta 0}$ were assumed small compared to unity. The components of surface load intensity in radial and axial directions are denoted by p_H and p_V, respectively, whereas the surface or body couple per unit area is m.

In view of equations 3.1.10, 3.1.12, 3.1.13, 3.1.21, and 3.1.22, the three equilibrium equations, 3.1.26–3.1.28, together with the heterogeneous shell stress-strain relations, 3.1.23, constitute a system of seven equations for three stress resultants N_ξ, N_θ, Q (or H, V, N_θ), two couples M_ξ, M_θ and two deformation variables u, Φ.

Compatibility relation for strains

Noting that

$$u' = (r\varepsilon_{\theta 0})', \qquad (3.1.29)$$

and introducing this relation into equation 3.1.10, we obtain the following compatibility relation:

$$(r\varepsilon_{\theta 0})' - r'\varepsilon_{\xi 0} = \alpha(1 + \varepsilon_{\xi 0})(\cos \Phi - \cos \phi), \qquad (3.1.30)$$

which is valid for large strains.

Reduction to two simultaneous equations

The reduction of the system of seven equations, 3.1.23, 3.1.26—3.1.28, to two simultaneous equations follows along lines similar to Reissner's work.[47]

Let us introduce a stress resultant function Ψ defined by

$$\Psi = rH. \qquad (3.1.31)$$

The shell equations are now expressed in terms of the stress function Ψ and the meridian angle Φ of the deformed shell. The expressions for stress resultants become

$$rV = -\int r\alpha p_V \, d\xi, \tag{3.1.32}$$

$$rN_\xi = \Psi' \cos\Phi + rV \sin\Phi, \tag{3.1.33}$$

$$rQ = -\Psi' \sin\Phi + rV \cos\Phi, \tag{3.1.34}$$

$$\alpha N_\theta = \Psi'' + r\alpha p_H. \tag{3.1.35}$$

Stress-strain relations 3.1.23 are transformed to the form

$$
\begin{bmatrix} \varepsilon_{\xi 0} \\ \varepsilon_{\theta 0} \\ M_\xi \\ M_\theta \end{bmatrix} =
\begin{bmatrix}
A^*_{\xi\xi} & A^*_{\xi\theta} & B^*_{\xi\xi} & B^*_{\xi\theta} \\
A^*_{\theta\xi} & A^*_{\theta\theta} & B^*_{\theta\xi} & B^*_{\theta\theta} \\
\hline
C^*_{\xi\xi} & C^*_{\xi\theta} & D^*_{\xi\xi} & D^*_{\xi\theta} \\
C^*_{\theta\xi} & C^*_{\theta\theta} & D^*_{\theta\xi} & D^*_{\theta\theta}
\end{bmatrix}
\begin{bmatrix} N_\xi \\ N_\theta \\ \kappa_\xi \\ \kappa_\theta \end{bmatrix}; \tag{3.1.36}
$$

or, more compactly,

$$
\begin{bmatrix} \varepsilon_0 \\ M \end{bmatrix} =
\begin{bmatrix} A^* & B^* \\ C^* & D^* \end{bmatrix}
\begin{bmatrix} N \\ \kappa \end{bmatrix}, \tag{3.1.37}
$$

which have the same form as equations 2.1.19. The starred matrices are given in terms of the unstarred matrices and in terms of their inverses by equations 2.1.20. It is again noted that A^* and D^* are *symmetric* matrices, whereas B^* and C^* are not necessarily so.

The third equilibrium equation (3.1.28) and the compatibility relation (3.1.30) are expressed, correspondingly, by the following two simultaneous equations in terms of Φ and Ψ for $E_{ij} = E_{ij}(\zeta)$:

$$L_{11}\Phi + L_{12}\Psi + L_{13}(\Phi, \Psi) = L_{14}(\Phi, p_H, p_V, m), \tag{3.1.38}$$

$$L_{21}\Phi + L_{22}\Psi + L_{23}(\Phi, \Psi) = L_{24}(\Phi, p_H, p_V). \tag{3.1.39}$$

The functional operators L_{ij} are nondimensional and are of the form

$$
\begin{aligned}
L_{11}\Phi = {}& r^2(\Phi'' - \phi'') + rr'(\Phi' - \phi') + (r\alpha D^*_{\xi\theta}/D^*_{\xi\xi}) \\
& \times (\cos\Phi - \cos\phi)\phi' - (\alpha^2 D^*_{\theta\theta}/D^*_{\xi\xi})\cos\Phi(\sin\Phi - \sin\phi),
\end{aligned} \tag{3.1.40}
$$

$$L_{12}\Psi = -(r^2 C^*_{\xi\theta}/D^*_{\xi\xi})\Psi'' - (rr' C^*_{\xi\theta}/D^*_{\xi\xi})\Psi', \tag{3.1.41}$$

$$
\begin{aligned}
L_{13}(\Phi, \Psi) = {}& -[r\alpha(C^*_{\xi\xi} - C^*_{\theta\theta})/D^*_{\xi\xi}]\Psi'' \cos\Phi + (r\alpha/D^*_{\xi\xi}) \\
& \times [C^*_{\xi\xi}\Phi' \sin\Phi + (\alpha C^*_{\theta\xi}/r) \cos^2\Phi - \alpha \sin\Phi)\Psi', \tag{3.1.42}
\end{aligned}
$$

$$
\begin{aligned}
L_{14}(\Phi, p_H, p_V, m) = {}& (r\alpha C^*_{\xi\theta}/D^*_{\xi\xi})(r^2 p_H)' - [(\alpha^2 C^*_{\theta\theta}/D^*_{\xi\xi}) \cos\Phi] \\
& \times (r^2 p_H) + [(r\alpha C^*_{\xi\xi}/D^*_{\xi\xi}) \sin\Phi](rV)' \\
& + (r\alpha/D^*_{\xi\xi})[C^*_{\xi\xi}\Phi' \cos\Phi - (\alpha C^*_{\theta\xi}/2r) \sin 2\Phi - \alpha \cos\Phi] \\
& \times (rV) + (r^2\alpha^2/D^*_{\xi\xi})m, \tag{3.1.43}
\end{aligned}
$$

$$
\begin{aligned}
L_{21}\Phi = {}& -(r/\alpha)B^*_{\theta\xi}(\Phi'' - \phi'') + (B^*_{\xi\xi}\cos\Phi - B^*_{\theta\xi}r'/\alpha)(\Phi' - \phi') \\
& - B^*_{\theta\theta}\phi'(\cos\Phi - \cos\phi) + (\alpha/r)B^*_{\xi\theta} \\
& \times \cos\Phi(\sin\Phi - \sin\phi) - \alpha(\cos\Phi - \cos\phi), \tag{3.1.44}
\end{aligned}
$$

$$L_{22}\Psi = (r/\alpha)A^*_{\theta\theta}\Psi'' + (A^*_{\theta\theta}r'/\alpha)\Psi', \tag{3.1.45}$$

$$L_{23}(\Phi, \Psi) = -[A^*_{\theta\xi}\Phi' \sin\Phi - (\alpha/r)A^*_{\xi\xi} \cos^2\Phi]\Psi', \tag{3.1.46}$$

$$
\begin{aligned}
L_{24}(\Phi, p_H, p_V) = {}& -A^*_{\theta\theta}(r^2 p_H)' + [(\alpha/r)A^*_{\xi\theta} \cos\Phi](r^2 p_H) \\
& - A^*_{\theta\xi} \sin\Phi(rV)' + [(\alpha/2r)A^*_{\xi\xi} \sin 2\Phi - A^*_{\theta\xi}\phi' \cos\Phi](rV). \tag{3.1.47}
\end{aligned}
$$

The governing shell equations (3.1.38 and 3.1.39) in terms of Φ and Ψ are valid for finite deformations and are useful especially for a stability analysis. Homogeneous shell theory is obtained as a special case of the present shell theory by deleting all B^* and C^* terms in equations 3.1.40–3.1.44.

It is noted that in view of Reissner's discussion,[49] compatibility relation 3.1.30 can be modified by neglecting $\varepsilon_{\xi 0}$ as compared to unity. This results in a simplified equation 3.1.39, where all Φ terms in the coefficients of Ψ', p_H, p_V become terms with ϕ, as indicated by Stavsky.[40]

Small finite deflections theory

In many applications it is sufficient to consider rather small finite changes of the tangent angle ϕ. If we define

$$\beta = \phi - \Phi \tag{3.1.48}$$

and retain terms up to the second degree in β and Ψ' in governing equations 3.1.38 and 3.1.39, they take the following form:

$$L_{11}\beta + L_{12}\Psi + L_{13}(\beta, \Psi) = L_{14}(\beta, p_H, p_V, m), \tag{3.1.49}$$

$$L_{21}\beta + L_{22}\Psi + L_{23}(\beta, \Psi) = L_{24}(\beta, p_H, p_V). \tag{3.1.50}$$

The functional operators in equilibrium equation 3.1.49 are given by

$$
\begin{aligned}
L_{11}\beta = {}& r^2\beta'' + rr'\beta' + [rr''(D^*_{\xi\theta}/D^*_{\xi\xi}) - (r')^2(D^*_{\theta\theta}/D^*_{\xi\xi})]\beta \\
& + \tfrac{1}{2}[rz''(D^*_{\theta\theta}/D^*_{\xi\xi}) - 3r'z'(D^*_{\theta\theta}/D^*_{\xi\xi})]\beta^2, \tag{3.1.51}
\end{aligned}
$$

$$
\begin{aligned}
L_{12}\Psi = {}& (r^2 C^*_{\xi\theta}/D^*_{\xi\xi})\Psi'' + [rr'(C^*_{\xi\xi} - C^*_{\theta\theta})/D^*_{\xi\xi} + rr' C^*_{\xi\theta}/D^*_{\xi\xi}]\Psi' \\
& + [rr'' C^*_{\xi\xi}/D^*_{\xi\xi} + (r\alpha^2/D^*_{\xi\xi})\sin\phi - (r')^2 C^*_{\theta\xi}/D^*_{\xi\xi}]\Psi, \tag{3.1.52}
\end{aligned}
$$

$$
\begin{aligned}
L_{13}(\beta, \Psi) = {}& [rz'(C^*_{\xi\xi} - C^*_{\theta\theta})/D^*_{\xi\xi}]\beta\Psi' + [rz'' C^*_{\xi\xi}/D^*_{\xi\xi} \\
& - (r\alpha^2/D^*_{\xi\xi})\cos\phi - 2r'z'(C^*_{\theta\xi}/D^*_{\xi\xi})]\beta\Psi + rz'(C^*_{\xi\xi}/D^*_{\xi\xi})\beta'\Psi, \tag{3.1.53}
\end{aligned}
$$

$$
\begin{aligned}
L_{14}(\beta, p_H, p_V, m) = {}& -(r\alpha C^*_{\theta\theta}/D^*_{\xi\xi})(r^2 p_H)' + [(\alpha^2 C^*_{\theta\theta}/D^*_{\xi\xi}) \\
& \times (\cos\phi + \beta \sin\phi)](r^2 p_H) + [r\alpha(C^*_{\xi\xi}/D^*_{\xi\xi})(\beta \cos\phi - \sin\phi)] \\
& \times (rV)' + [(r\alpha^2/D^*_{\xi\xi})(\cos\phi + \beta \sin\phi) + (\alpha^2 C^*_{\theta\xi}/D^*_{\xi\xi}) \\
& \times (\tfrac{1}{2}\sin 2\phi - \beta \cos 2\phi) - (r\alpha C^*_{\xi\xi}/D^*_{\xi\xi})(\beta \cos\phi - \sin\phi)'] \\
& \times (rV) - (r^2\alpha^2/D^*_{\xi\xi})m. \tag{3.1.54}
\end{aligned}
$$

The functional operators in compatibility equation 3.1.50 are of the form

$$
\begin{aligned}
L_{21}\beta = {}& (r/\alpha)B^*_{\theta\xi}\beta'' + (r'/\alpha)(B^*_{\theta\theta} - B^*_{\xi\xi} + B^*_{\theta\xi})\beta' \\
& + [-\alpha \sin\phi + (r''/\alpha)B^*_{\theta\theta} - (r')^2 B^*_{\xi\theta}/r\alpha]\beta \\
& + (z'/\alpha)(B^*_{\theta\theta} - B^*_{\xi\xi})\beta\beta' + \tfrac{1}{2}[\alpha \cos\phi + (z''/\alpha)B^*_{\theta\theta} \\
& - 3(r'z'/r\alpha)B^*_{\theta\xi}]\beta^2, \tag{3.1.55}
\end{aligned}
$$

$$
\begin{aligned}
L_{22}\Psi = {}& (r/\alpha)A^*_{\theta\theta}\Psi'' + (r'/\alpha)A^*_{\theta\theta}\Psi' \\
& + [(r''/\alpha)A^*_{\theta\xi} - (r')^2 A^*_{\xi\xi}/r\alpha]\Psi, \tag{3.1.56}
\end{aligned}
$$

$$L_{23}(\beta, \Psi) = [(z''/\alpha)A^*_{\theta\xi} - 2(r'z'/r\alpha)A^*_{\xi\xi}]\beta\Psi + (z'/\alpha)A^*_{\theta\xi}\beta'\Psi, \tag{3.1.57}$$

$$L_{24}(\beta, p_H, p_V) = -A_{\theta\theta}^*(r^2 p_H)' + [(r' + z'\beta)A_{\xi\theta}^*/r](r^2 p_H)$$

$$- [(z'/\alpha)A_{\theta\xi}^*](rV)' + [(r'z'/r\alpha)A_{\xi\xi}^* - (z''/\alpha)A_{\theta\xi}^*](rV)$$

$$+ (r'/\alpha)A_{\theta\xi}^*(\beta rV)' + [(r''/\alpha)A_{\theta\xi}^*$$

$$+ ((z')^2 - (r')^2)A_{\xi\xi}^*/r\alpha]\beta(rV). \tag{3.1.58}$$

The two simultaneous second-order nonlinear ordinary differential equations 3.1.49 and 3.1.50 for rotationally symmetric deformations of composite shells of revolution are characterized by the appearance of the B^* and C^* terms that introduce a stronger coupling of β and Ψ than encountered in homogeneous shells. The heterogeneous shell equations include as special cases the homogeneous shell equations as given, for example, by Reissner.[47–49] The solution of the equations of the nonlinear theory of shells of revolution seems to be an open question in most cases. Of special interest would be an analysis of the edge effect in heterogeneous anisotropic shells and the range of validity of the various simplified theories, such as linear-bending and nonlinear-membrane theory.

Finally we give the expressions for resultants as obtained from equations 3.1.32–3.1.35, retaining terms up to the second degree in β and Ψ:

$$rV = -\int r\alpha p_V \, d\xi, \tag{3.1.59}$$

$$rN_\xi = \Psi \cos\phi + rV \sin\phi + \beta(\Psi \sin\phi - rV \cos\phi), \tag{3.1.60}$$

$$rQ = -\Psi \sin\phi + rV \cos\phi + \beta(\Psi \cos\phi + rV \sin\phi), \tag{3.1.61}$$

$$\alpha N_\theta = \Psi' + r\alpha p_H. \tag{3.1.62}$$

The expressions of the curvature changes (3.1.13) become

$$\kappa_\xi = \beta'/\alpha, \quad \kappa_\theta = (\beta \cos\phi + \tfrac{1}{2}\beta^2 \sin\phi)/r, \tag{3.1.63}$$

whereas the strains and couples are given by equations 3.1.36, in view of the modified equations 3.1.60, 3.1.62, and 3.1.63.

Linear theory

The linear small-deflection theory of heterogeneous shells of revolution is obtained from equations 3.1.49 and 3.1.50 by omitting all nonlinear terms. In the following sections some examples will be shown for the solution of the linear equations for cylindrical and spherical shells.

3.2 Symmetric deformations of composite circular cylindrical shells

Small finite deflections of composite circular cylindrical shells

In order to obtain the equations for small finite deflections of heterogeneous orthotropic cylindrical shells we set in equations 3.1.49 and 3.1.50

$$r = a, \ z = a\xi, \ \alpha = a, \ \sin\phi = 1, \ \cos\phi = 0, \ \Phi = \tfrac{1}{2}\pi - \beta, \tag{3.2.1}$$

so that the governing equations become

$$D_{\xi\xi}^*\beta'' - B_{\theta\xi}^*\Psi'' - (B_{\xi\xi}^* - B_{\theta\theta}^*)\beta\Psi'' + B_{\xi\xi}^*\beta'\Psi' + a\Psi$$

$$= B_{\theta\xi}^* a^2 p_H' - B_{\theta\theta}^* \beta a^2 p_H + B_{\xi\xi}^* aV' + (a - B_{\xi\theta}^*)\beta aV - a^2 m, \tag{3.2.2}$$

$$B_{\theta\xi}^*\beta'' - (B_{\xi\xi}^* - B_{\theta\theta}^*)\beta'\beta - a\beta + A_{\theta\theta}^*\Psi'' + A_{\theta\xi}^*\beta'\Psi$$

$$= -A_{\theta\theta}^* a^2 p_H + A_{\xi\theta}^*\beta a^2 p_H - A_{\theta\xi}^* aV' + A_{\xi\xi}^*\beta aV. \tag{3.2.3}$$

The equations for homogeneous orthotropic cylindrical shells are obtained as a special case of equations 3.2.2 and 3.2.3 after taking the reference surface halfway between the outer and inner surfaces of the shell so that all B^* terms vanish. It is noted that, whereas in a homogeneous shell these equations are linear in the variables β and Ψ except for the nonlinear term $A_{\theta\xi}^*\beta'\Psi$ in 3.2.3 which turns out to be negligible, for the heterogeneous shell the nonlinearity of 3.2.2 and 3.2.3 cannot be avoided.

Equations 3.1.59–3.1.62 for resultants become

$$V = -a \int p_V \, d\xi, \tag{3.2.4}$$

$$aN_\xi = \beta\Psi + aV, \tag{3.2.5}$$

$$aQ = -\Psi + \beta aV, \tag{3.2.6}$$

$$aN_\theta = \Psi' + a^2 p_H. \tag{3.2.7}$$

The curvature changes (3.1.63) become

$$\kappa_\xi = \beta'/a, \quad \kappa_\theta = \beta^2/2r. \tag{3.2.8}$$

The displacements are obtained from 3.1.11 and 3.1.12 and have the form

$$u = a\varepsilon_{\theta 0}, \quad w = a \int \varepsilon_{\xi 0} \, d\xi, \tag{3.2.9}$$

whereas the strains are given in terms of β and Ψ through equations 3.1.36.

Linear theory of composite circular cylindrical shells

The linearized theory follows from the foregoing equations (3.2.2 and 3.2.3) by referring the differential equations of equilibrium to the undeformed shell and by omitting all nonlinear terms in the expressions for the strains. The resulting governing linear equations are as follows:

$$D_{\xi\xi}^*\beta'' - B_{\theta\xi}^*\Psi'' + a\Psi = B_{\theta\xi}^* a^2 p_H' + B_{\xi\xi}^* aV' - a^2 m, \tag{3.2.10}$$

$$B_{\theta\xi}^*\beta'' - a\beta + A_{\theta\theta}^*\Psi'' = -A_{\theta\theta}^* a^2 p_H - A_{\theta\xi}^* aV'. \tag{3.2.11}$$

This system of two simultaneous equations can be transformed to the following two single equations for β or Ψ:

$$\beta^{\text{iv}} - 2B_{\theta\xi}^*\left(\frac{a}{k}\right)\beta'' + \frac{a^2}{k}\beta = \frac{1}{k}[A_{\theta\theta}^* a^3 p_H' + (A_{\theta\theta}^* B_{\xi\xi}^*$$

$$- A_{\theta\xi}^* B_{\theta\xi}^*)aV''' + A_{\theta\xi}^* a^2 V' - A_{\theta\theta}^* a^2 m''], \tag{3.2.12}$$

$$\Psi^{\text{iv}} - 2B_{\theta\xi}^*\left(\frac{a}{k}\right)\Psi'' + \frac{a^2}{k}\Psi = -a^2 p_H''' + B_{\theta\xi}^*\frac{a^3}{k}p_H'$$

$$- aV''' + B_{\xi\xi}^*\frac{a^2}{k}V' - \frac{a^3}{k}m, \tag{3.2.13}$$

where

$$k = A_{\theta\theta}^* D_{\xi\xi}^* + (B_{\theta\xi}^*)^2. \tag{3.2.14}$$

For the case where p_H and p_V are constant and m vanishes, the right-hand sides of equations 3.2.12 and 3.2.13 vanish so the equations for determining β or Ψ are identical:

$$L\beta = 0, \quad L\Psi = 0, \tag{3.2.15}$$

where

$$L = (\ )^{iv} + 2c(\ )'' + d^2(\ ), \quad c = -B_{\theta\xi}^* a/k, \quad d^2 = a^2/k. \quad (3.2.16)$$

The boundary conditions associated with the shell equations may take the following form along the edge $r = r_b$:

$$
\begin{aligned}
u &= \bar{u} &\text{or}& &H &= \bar{H}, \\
w &= \bar{w} &\text{or}& &V &= \bar{V}, \\
\beta &= \bar{\beta} &\text{or}& &M_\xi &= \bar{M}_\xi.
\end{aligned}
\quad (3.2.17)
$$

Composite cylindrical shell acted upon by edge bending moments

Consider a semi-infinite shell acted upon by bending moments M_0 at the edge $\xi = 0$ (Fig. 1.24). Since for this

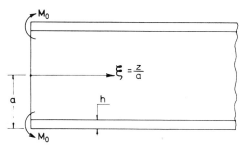

Figure 1.24   Semi-infinite Cylindrical Shell with Applied Edge Moments

problem $p_H = p_V = 0$, differential equations 3.2.15 hold with the following boundary conditions:

$$H = 0, M_\xi = M_0 \quad \text{at} \quad \xi = 0, \quad (3.2.18)$$

together with the condition that stresses and displacements vanish for $\xi = \infty$.

Satisfying the regularity conditions for $\xi = \infty$ we find the following expressions for β and Ψ:

$$\beta = B_1 e^{-\gamma_1 \xi} \cos \gamma_2 \xi + B_2 e^{-\gamma_1 \xi} \sin \gamma_2 \xi, \quad (3.2.19)$$

$$\Psi = C_1 e^{-\gamma_1 \xi} \cos \gamma_2 \xi + C_2 e^{-\gamma_1 \xi} \sin \gamma_2 \xi; \quad (3.2.20)$$

the constants γ_1 and γ_2 are given by

$$2\gamma_1^2 = d - c, \quad 2\gamma_2^2 = d + c, \quad (3.2.21)$$

where c and d are defined by 3.2.16.

Since β and Ψ must satisfy the homogeneous part of equations 3.2.10 and 3.2.11 and the boundary conditions 3.2.18 at the edge of the shell, we obtain the final expressions 3.2.19 and 3.2.20 in the form

$$
\begin{aligned}
\beta = \frac{M_0 a}{D_{\xi\xi}^*} e^{-\gamma_1 \xi} \Bigg\{ &-\frac{2\gamma_1}{\gamma_1^2 + \gamma_2^2} \cos \gamma_2 \xi + \frac{1}{\gamma_2} \bigg[\frac{B_{\theta\xi}^*}{a} (\gamma_1^2 + \gamma_2^2) \\
&- \left(\frac{\gamma_1^2 - \gamma_2^2}{\gamma_1^2 + \gamma_2^2} \right) \bigg] \sin \gamma_2 \xi \Bigg\}, \quad (3.2.22)
\end{aligned}
$$

$$\Psi = M_0 \frac{\gamma_1^2 + \gamma_2^2}{\gamma_2} e^{-\gamma_1 \xi} \sin \gamma_2 \xi. \quad (3.2.23)$$

Having the fundamental variables β and Ψ as given for the considered boundary-value problem by equations 3.2.22 and 3.2.23 one can easily obtain all other quantities of interest.

$$V = 0, N_\xi = 0, Q = -\Psi/a, N_\theta = \Psi''/a, \quad (3.2.24)$$

$$\kappa_\xi = \beta'/a, \quad \kappa_\theta = 0. \quad (3.2.25)$$

If we substitute relations 3.2.24 and 3.2.25 into 3.1.36, the strains and stress couples can be expressed in terms of β and Ψ as follows:

$$\varepsilon_{\xi 0} = A_{\xi\theta}^* \Psi'/a + B_{\xi\xi}^* \beta'/a, \quad \varepsilon_{\theta 0} = A_{\theta\theta}^* \Psi'/a + B_{\theta\xi}^* \beta'/a, \quad (3.2.26)$$

$$M_\xi = -B_{\theta\xi}^* \Psi'/a + D_{\xi\xi}^* \beta'/a, \quad M_\theta = -B_{\theta\theta}^* \Psi'/a + D_{\theta\xi}^* \beta'/a. \quad (3.2.27)$$

Specifically, the final form of M_ξ is

$$M_\xi = M_0 e^{-\gamma_1 \xi} \left(\cos \gamma_2 \xi + \frac{\gamma_1}{\gamma_2} \sin \gamma_2 \xi \right). \quad (3.2.28)$$

The stress distribution across the thickness of the shell and along its generator is obtained in terms of stress resultants and couples by using equations 2.1.33.

The results obtained include as a special case the class of heterogeneous shells for which B^* vanishes when we set

$$2\gamma_1^2 = 2\gamma_2^2 = d, \quad d^2 = a^2 (A_{\theta\theta}^* D_{\xi\xi}^*)^{-1} \quad (3.2.29)$$

in equations 3.2.22–3.2.28.

Numerical Examples: To obtain a quantitative insight into the results of the previous section we consider the following heterogeneous orthotropic cylindrical shells:

Case 1:

$$[E] = E_0 \begin{bmatrix} 1 + 0.6 \cos \dfrac{\pi}{h} \zeta & 0.2 \\[2ex] 0.2 & 5\left(1 - 0.6 \cos \dfrac{\pi}{h} \zeta\right) \end{bmatrix},$$

Case 2:

$$[E] = E_0 \begin{bmatrix} 5\left(1 - 0.6 \cos \dfrac{\pi}{h} \zeta\right) & 0.2 \\[2ex] 0.2 & 1 + 0.6 \cos \dfrac{\pi}{h} \zeta \end{bmatrix}. \quad (3.2.30)$$

In Case 1 the circumferential direction is stronger than the axial direction; in Case 2 the converse is true. The results for deformations and stresses are represented graphically in Figures 1.25 and 1.26, and the influence of the different elastic properties in axial and circumferential directions is indicated. Numerical results are given for the following values of a, h, and E_0:

$$a/h = 100, a = 150 \text{ cm}, E_0 = 10^6 \text{ kg/cm}^2. \quad (3.2.31)$$

The nondimensional parameters γ_1 and γ_2 assume the following values:

Case 1: $\gamma_1 = 20.1900, \gamma_2 = 21.0450, \gamma_2/\gamma_1 = 1.0430; \quad (3.2.32)$

Case 2: $\gamma_1 = 9.4140, \gamma_2 = 9.0285, \gamma_2/\gamma_1 = 0.9591. \quad (3.2.33)$

Note that the ratio γ_2/γ_1 is quite close in both cases, whereas the values of γ for Case 1 are more than twice the corresponding values for Case 2.

Figure 1.25 shows that the penetration of the edge effect in Case 2 is deeper than in Case 1. This is associated with a reduction of the peak values of N_θ, Q, and M_θ.

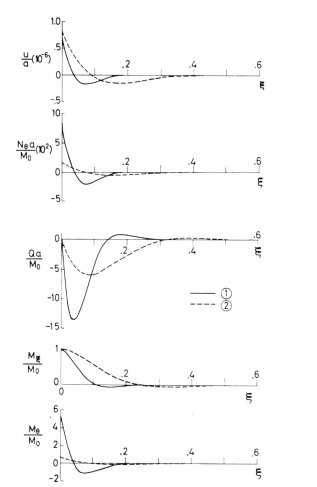

Figure 1.25   Distribution of Radial Displacements, Resultants, and Couples Along Shell Generator

The axial and circumferential stress distribution at the shell edge (Fig. 1.26) seems to be controllable, to some extent, by the thickness variation of the elastic moduli of the composite shell.

The results are expressed in terms of τ_ξ^0, which is the maximum stress that arises in a homogeneous shell subject to the same edge moment. It is seen that in the considered heterogeneous shell the stresses may well exceed the corresponding values for a homogeneous shell. However, there is quite a substantial part of the shell where the contrary is true. Further results seem to be necessary toward an efficient design.

3.3 Symmetric deformations of composite spherical shells

Linear theory of composite spherical shells

The governing equations for a spherical shell of radius a are obtained by setting

$$r = a \sin \xi, \ z = -a \cos \xi, \ \alpha = a, \ \phi = \xi, \ \Phi = \xi - \beta. \quad (3.3.1)$$

The reference surface strains (3.1.10 and 3.1.12) and the curvature changes (3.1.13) become

$$\varepsilon_{\xi 0} = \frac{\cos \xi}{\cos \Phi} + \frac{u'}{a \cos \Phi} - 1, \ \varepsilon_{\theta 0} = \frac{u}{a \sin \xi}, \quad (3.3.2)$$

$$a\kappa_\xi = 1 - \Phi' = \beta', \ a\kappa_\theta = \left(1 - \frac{\sin \Phi}{\sin \xi}\right) = \beta \cot \xi + O(\beta^2). \quad (3.3.3)$$

If we specialize equations 3.1.49 and 3.1.50 to the spherical shell and omit all nonlinear terms, the following fourth-order system for β and Ψ results:

$$D_{\xi\xi}^*(\beta'' + \cot \xi \ \beta') - (D_{\xi\theta}^* + D_{\theta\theta}^* \cot^2 \xi)\beta + C_{\xi\theta}^*\Psi''$$
$$+ (C_{\xi\xi}^* - C_{\theta\theta}^* + C_{\xi\theta}^*) \cot \xi \Psi' + (a - C_{\xi\xi}^* - C_{\theta\xi}^* \cot^2 \xi)\Psi$$
$$= -\frac{C_{\xi\theta}^*}{\sin \xi}(r^2 p_H)' + C_{\theta\theta}^* \frac{\cot \xi}{\sin \xi}(r^2 p_H) - C_{\xi\xi}^*(rV)'$$
$$+ (a + C_{\theta\xi}^* + C_{\xi\xi}^*) \cot \xi (rV) - a^2 m, \quad (3.3.4)$$

$$B_{\theta\xi}^*\beta'' + (B_{\theta\theta}^* - B_{\xi\xi}^* + B_{\theta\xi}^*) \cot \xi \beta' - (a + B_{\theta\theta}^* + B_{\xi\xi}^* \cot^2 \xi)\beta$$
$$+ A_{\theta\theta}^*\Psi'' + A_{\theta\theta}^* \cot \xi \Psi' - (A_{\theta\xi}^* + A_{\xi\xi}^* \cot^2 \xi)\Psi$$
$$= -\frac{A_{\theta\theta}^*}{\sin \xi}(r^2 p_H)' + A_{\xi\theta}^* \frac{\cot \xi}{\sin \xi}(r^2 p_H) - A_{\theta\xi}^*(rV)'$$
$$+ (A_{\xi\xi}^* - A_{\theta\xi}^*) \cot \xi (rV). \quad (3.3.5)$$

In terms of solutions of these two simultaneous equations one finds the following expressions for stress resultants and couples in accordance with the linearized equations 3.1.60 and 3.1.61:

$$aN_\xi = \Psi \cot \xi + rV, aQ = -\Psi' + (rV) \cot \xi,$$
$$aN_\theta = \Psi' + arp_H. \quad (3.3.6)$$

The reference surface strains and the couples are given in terms of the basic variables β and Ψ through relations 3.1.36. The boundary conditions at the edge $r = r_b$ of the spherical shell are given by equations 3.2.17, whereas at $r = 0$ the following symmetry conditions hold:

$$u = 0, \ v = 0, \ \beta = 0. \quad (3.3.7)$$

Approximate analysis of edge loads for composite spherical shells

In order to get some insight into the behavior of heterogeneous isotropic spherical shells we assume that ξ is not too small (say larger than 30°) and that the effect of the edge loads is restricted to a narrow edge zone. Then the derivatives of β and Ψ are large in comparison with the original functions, and one may simplify equations 3.3.4 and 3.3.5 as follows, taking $p_H = p_V = m = 0$:

$$D_{\xi\xi}^*\beta'' - B_{\theta\xi}^*\Psi'' + a\Psi = 0, \quad (3.3.8)$$

$$B_{\theta\theta}^*\beta'' - a\beta + A_{\theta\theta}^*\Psi'' = 0. \quad (3.3.9)$$

This simplification, also suggested by Grigolyuk[50] for a two-layer shell, is equivalent to replacing the vicinity of the spherical shell edge by a tangent conical shell and applying to it the equations developed for the cylindrical shell. The admissibility of such an approximate analysis for a homogeneous shell is well known (see, for example, Ref. 23, p. 549); as for the heterogeneous shell some further studies are necessary regarding the existence of a "narrow edge zone," especially for anisotropic shells.

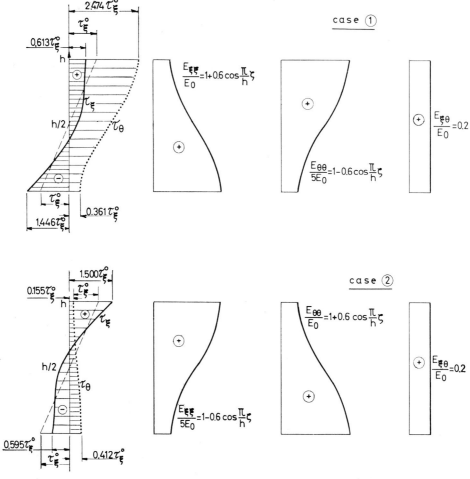

Figure 1.26   Thickness Variation of Elastic Moduli, Axial, and Circumferential Stresses at the Shell Edge

As shown in Section 3.2 this system of two simultaneous equations can be transformed to the following two single equations for β or Ψ:

$$L\beta = 0, \quad L\Psi = 0, \tag{3.3.10}$$

where

$$L = (\ )^{iv} + 2c(\ )'' + d^2(\ ), \quad c = -B^*_{\theta\xi}\,a/k, \quad d^2 = a^2/k, \tag{3.3.11}$$

and k is defined by 3.2.14. Recall that primes indicate here differentiation with respect to ϕ.

3.4 Symmetric deformations of composite shallow shells of revolution

This section is concerned with the problem of symmetrical bending of thin elastic shells of revolution for shells that may be considered shallow before and after deformation.

A shell will be considered to remain shallow as long as

$$\sin\Phi \sim \Phi, \quad 1 - \cos\Phi \sim \tfrac{1}{2}\Phi^2; \tag{3.4.1}$$

where $\cos\Phi$ occurs by itself it is approximated by 1.

We set

$$r = \xi, \quad z = z(\xi) = z(r), \tag{3.4.2}$$

$$r' = 1, \quad z' = z'(r). \tag{3.4.3}$$

We also have

$$r' \sim \alpha, \quad z' \sim \phi. \tag{3.4.4}$$

Geometry of deformation

The reference surface strains (3.1.10–3.1.12) and the curvature changes (3.1.13) become

$$\varepsilon_{r0} = u' - \phi\beta + \tfrac{1}{2}\beta^2, \quad \varepsilon_{\theta0} = u/r, \tag{3.4.5}$$

$$\kappa_r = \beta', \quad \kappa_\theta = \beta/r. \tag{3.4.6}$$

Axial displacement relation 3.1.14 becomes

$$w' = -\beta. \tag{3.4.7}$$

Stress-strain relations

Equations 3.1.23 hold, whereas expressions 3.1.21 and 3.1.22 for N_r and Q in terms of horizontal and axial resultants become

$$N_r = H + V\phi \sim H, \tag{3.4.8}$$

$$Q = -H\Phi + V. \tag{3.4.9}$$

Statics of shell element

The following equilibrium differential equations are obtained from 3.1.26–3.1.28 by setting for α and $\cos\phi$ the value 1:

$$(rH)' + rp_H - N_\theta = 0, \tag{3.4.10}$$

$$(rV)' + rp_V = 0, \tag{3.4.11}$$

$$(rM_r)' - M_\theta - rQ + rm = 0. \tag{3.4.12}$$

Compatibility relation

Relation 3.1.30 now takes the form

$$(r\varepsilon_{\theta 0})' - \varepsilon_{r0} = (\phi - \tfrac{1}{2}\beta)\beta, \tag{3.4.13}$$

which follows also from 3.4.5.

Reduction to two simultaneous equations

The stress resultants are written in terms of β, Ψ, p_H, and p_V as follows:

$$rV = -\int rp_V\, dr, \tag{3.4.14}$$

$$N_r = \Psi/r + V\Phi \sim \Psi/r, \tag{3.4.15}$$

$$Q = -\Phi\Psi/r + V, \tag{3.4.16}$$

$$N_\theta = \Psi' + rp_H. \tag{3.4.17}$$

The reference surface strains and the bending moments are expressed in terms of stress resultants and curvature changes through equations 3.1.36.

The third equilibrium equation (3.4.12) and the compatibility relation (3.4.13) are expressed by the following two simultaneous equations for a deformation variable β and a stress variable Ψ:

$$D_{rr}^* \beta'' + D_{rr}^* \frac{\beta'}{r} - D_{\theta\theta}^* \frac{\beta}{r^2} - \left[B_{\theta r}^* \Psi'' + (B_{rr}^* - B_{\theta\theta}^* + B_{\theta r}^*) \frac{\Psi'}{r} \right.$$
$$\left. - B_{r\theta}^* \frac{\Psi}{r^2} \right] + (\phi - \beta)\frac{\Psi}{r} = B_{\theta r}^* \frac{1}{r}(r^2 p_H)' - B_{\theta\theta}^* p_H - m + V, \tag{3.4.18}$$

$$B_{\theta r}^* \beta'' + (B_{\theta\theta}^* - B_{rr}^* + B_{\theta r}^*)\frac{\beta'}{r} - B_{r\theta}^* \frac{\beta}{r^2} + A_{\theta\theta}^* \Psi'' + A_{\theta\theta}^* \frac{\Psi'}{r}$$
$$- A_{rr}^* \frac{\Psi}{r^2} - \frac{1}{r}(\phi - \tfrac{1}{2}\beta)\beta = -\frac{1}{r} A_{\theta\theta}^* (r^2 p_H)' + A_{r\theta}^* p_H. \tag{3.4.19}$$

These equations are essentially identical with those developed by Reissner[39] for a homogeneous shell except that the middle surface is not taken as the reference.

The linear theory of shallow shells of revolution with polar orthotropy is obtained from equations 3.4.18 and 3.4.19 by deleting $\beta\Psi'$ and β^2 terms.

Composite shallow spherical shell

Consider a shell with a surface equation

$$z = H - r^2/2R, \tag{3.4.20}$$

and meridian slope

$$z' = \phi = -r/R, \tag{3.4.21}$$

in the interval $0 \le r \le a$. Base radius a, rise H, and curvature radius R of the shallow spherical shell are related as follows:

$$a^2 = 2HR. \tag{3.4.22}$$

The assumption of shallowness requires that we limit the discussion to shells for which $|\phi_{\max}| \lesssim \frac{1}{3}$. According to 3.4.21 and 3.4.22 this means that a spherical shell may be considered shallow as long as

$$H/a \lesssim \tfrac{1}{6}. \tag{3.4.23}$$

If we assume that $p_H = 0$, $m = 0$, and p_V is constant, governing equations 3.4.18 and 3.4.19 become

$$D_{rr}^* \beta'' + D_{rr}^* \frac{\beta'}{r} - D_{\theta\theta}^* \frac{\beta}{r^2}$$
$$- \left[B_{\theta r}^* \Psi'' + (B_{rr}^* - B_{\theta\theta}^* + B_{\theta r}^*)\frac{\Psi'}{r} - B_{r\theta}^* \frac{\Psi}{r^2} \right] - \frac{\Psi}{R} - \frac{1}{r}\beta\Psi$$
$$= -\tfrac{1}{2}pr, \tag{3.4.24}$$

$$B_{\theta r}^* \beta'' + (B_{\theta\theta}^* - B_{rr}^* + B_{\theta r}^*)\frac{\beta'}{r} - B_{r\theta}^* \frac{\beta}{r^2} + \frac{\beta}{R} + \frac{1}{2}\frac{\beta^2}{r}$$
$$+ A_{\theta\theta}^* \Psi'' + A_{\theta\theta}^* \frac{\Psi'}{r} - A_{rr}^* \frac{\Psi}{r^2} = 0. \tag{3.4.25}$$

Linear theory of heterogeneous isotropic shallow spherical pseudomembranes

Let the pseudomembrane state be defined by the vanishing of the following bending stiffnesses:

$$D_{rr}^* = D_{\theta\theta}^* = 0. \tag{3.4.26}$$

Then the linear pseudomembrane equations become

$$B_{\theta r}^* L\Psi + \Psi/R = \tfrac{1}{2}pr, \tag{3.4.27}$$

$$B_{\theta r}^* L\beta + \beta/R + A^* L\Psi = 0, \tag{3.4.28}$$

where

$$A_{rr}^* = A_{\theta\theta}^* = A^*, \quad L = (\ )'' + \frac{1}{r}(\ )' - \frac{1}{r^2}(\ ). \tag{3.4.29}$$

Note that in contrast to the case of homogeneous membranes for which $B_{\theta r}^*$ vanishes and the order of the system 3.4.27 and 3.4.28 reduces to two, for the heterogeneous membrane its order is four, as of the original system 3.4.24 and 3.4.25.

Certain types of heterogeneous orthotropic shallow spherical membranes

If the considered shell is symmetrically composed with respect to its middle surface, or if all elastic moduli have the same thickness variation, one finds (for example, see Stavsky[34]) that B^* vanishes. The linear membrane equations take the following form by specialization of 3.4.24 and 3.4.25:

$$\Psi = \tfrac{1}{2}pRr, \quad \beta = \tfrac{1}{2}p\frac{R}{r}(A_{rr}^* - A_{\theta\theta}^*). \tag{3.4.30}$$

Except for the A^* coefficients these equations coincide with the result obtained for homogeneous orthotropic membranes by Reissner,[39] whose interesting observations equally apply in our case. Inasmuch as β becomes infinite as r approaches zero, unless $A_{rr}^* = A_{\theta\theta}^*$, we conclude that in contrast to the behavior of isotropic shells linear membrane theory may not always be adequate to describe the state of stress in the interior of a polar orthotropic shell of revolution subject to continuous surface-load distributions. In order to establish the nature of the actual solution in the interior of the shell for a given boundary-value problem it is necessary to consider the complete differential equations 3.4.24 and 3.4.25.

Certain types of heterogeneous orthotropic pseudomembrane shells

Consider such composed orthotropic shells for which B^* vanishes and define a pseudomembrane shell by the equation

$$D_{rr}^* = 0. \tag{3.4.31}$$

The linear pseudomembrane shell equations are derived from 3.4.24 and 3.4.25 and take the form

$$-D_{\theta\theta}^* \frac{\beta}{r^2} = \frac{\Psi}{R} - \tfrac{1}{2}pr, \tag{3.4.32}$$

$$A_{\theta\theta}^*\left(\Psi'' + \frac{\Psi'}{r}\right) - \left(A_{rr}^* + \frac{r^4}{R^2 D_{\theta\theta}^*}\right)\frac{\Psi}{r^2} = -\frac{pr^3}{2RD_{\theta\theta}^*}. \tag{3.4.33}$$

These equations were shown by Reissner[39] to hold for homogeneous orthotropic shells and we apply here his solution for the pseudomembrane shell.

The homogeneous part of equation 3.4.33 has the following solution in terms of modified Bessel functions:

$$\Psi_h = c_1 I_q(r^2/L^2) + c_2 K_q(r^2/L^2), \tag{3.4.34}$$

where

$$q = \tfrac{1}{2}\sqrt{A_{rr}^*/A_{\theta\theta}^*}, \quad L^2 = 2R\sqrt{A_{\theta\theta}^* D_{\theta\theta}^*}. \tag{3.4.35}$$

A particular solution of the complete equation 3.4.33 which is regular at the origin is as follows:

$$\Psi_p = \tfrac{1}{2}pRL\left[\int_0^{r^2/L^2} \eta^{3/2} I_q(\eta)\, d\eta\, K_q(r^2/L^2) \right.$$
$$\left. + \int_{r^2/L^2}^{a^2/L^2} \eta^{3/2} K_q(\eta)\, d\eta\, I_q(r^2/L^2)\right]. \tag{3.4.36}$$

The solution 3.4.34–3.4.36 reduces to elementary form when $q = \tfrac{1}{2}, \tfrac{3}{2}, \tfrac{5}{2}, \ldots$. The simplest case is given when $q = \tfrac{1}{2}$, that is, when $A_{rr}^* = A_{\theta\theta}^*$. For this case the particular solution Ψ_p becomes $\tfrac{1}{2}pRr$, which is the same as for the case $D_{\theta\theta}^* = 0$.

The next simplest case corresponds to $q = \tfrac{3}{2}$, with $A_{rr}^* = 9A_{\theta\theta}^*$, where the particular solution becomes an elementary function, and the general solution of 3.4.33 takes the form

$$q = \tfrac{3}{2}: \quad \Psi = \tfrac{1}{2}pRL\left[\frac{r}{L} - 2\frac{L^3}{r^3} + \hat{c}_1 \frac{L}{r}\left(1 + \frac{L^2}{r^2}\right)e^{-r^2/L^2}\right.$$
$$\left. + \hat{c}_2 \frac{L}{r}\left(1 - \frac{L^2}{r^2}\right)e^{r^2/L^2}\right]. \tag{3.4.37}$$

The behavior of the functions I_q and K_q for small r is given by the following approximations:

$$I_q(\eta) \sim \frac{1}{2^q q!}\eta^q, \quad K_q \sim 2^{q-1}(q-1)!\,\eta^{-q} \quad (q \neq 0). \tag{3.4.38}$$

Consequently the particular solution Ψ_p in 3.4.36 behaves

like r^5 for small values of r, whereas the regular part of homogeneous solution 3.4.34 behaves like r^{2q}.

Accordingly, the stress resultants N_r and N_θ associated with the homogeneous solution behave like r^{2q-1} for small values of r. This means that when $A_{rr}^* > A_{\theta\theta}^*$, we have vanishing N_r and N_θ at the origin; when $A_{rr}^* < A_{\theta\theta}^*$, N_r and N_θ become infinite at the origin.

Numerical Example of a Pseudomembrane Shell: In order to get an insight into the nature of the stress distribution in the special class of composed orthotropic shells for which B^* vanishes, we consider a pseudomembrane shell and assume the following boundary conditions which lead to linear membrane theory for an isotropic shell:

$$\Psi(0) = 0, \quad \Psi(a)\tfrac{1}{2} = pRa. \tag{3.4.39}$$

The first of these conditions means that $c_2 = 0$ in equation 3.4.34. The second condition determines c_1 in the form

$$c_1 I_q(a^2/L^2) + \tfrac{1}{2}pRL \int_0^{a^2/L^2} \eta^{3/2} I_q(\eta)\, d\eta\, K_q(a^2/L^2) = \tfrac{1}{2}pRa. \tag{3.4.40}$$

For the case $q = \tfrac{3}{2}$, c_1 is obtained from 3.4.37, and values of the stress resultants N_r and N_θ are shown in Figures 1.27

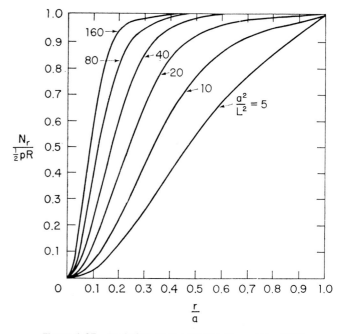

Figure 1.27   Variation of N_r with the Parameter a^2/L^2

and 1.28. The nature of the distribution of N_r and N_θ depends on the value of the dimensionless parameter

$$\frac{a^2}{L^2} = \frac{a^2}{2R\sqrt{A_{\theta\theta}^* D_{\theta\theta}^*}}; \tag{3.4.41}$$

as it increases the stress resultants approach the values that they would have according to linear membrane theory for isotropic shells except in the neighborhood of $r = 0$.

While N_r approaches the value $\tfrac{1}{2}pR$ from below, N_θ approaches this value from above with a localized overshoot of about 35 per cent irrespective of a^2/L^2.

The possible deviation of N_θ from the classical "membrane solution" in the interior of the shell shows that the behavior of the orthotropic homogeneous or heterogeneous shell may be entirely different from the corresponding isotropic shell. This situation seems to be of fundamental importance in analyzing anisotropic shells. Further extended studies are necessary to understand the boundary-layer penetration problem that arises in view of the given example.

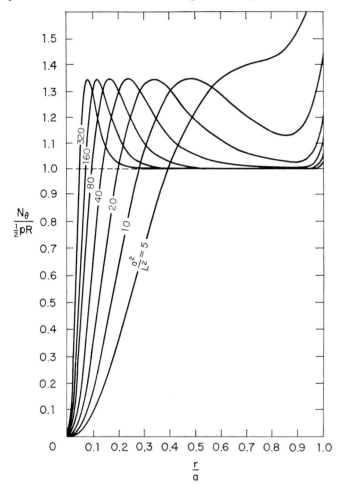

Figure 1.28   Variation of N_θ with the Parameter a^2/L^2

4   ANALYSIS OF SYMMETRICALLY LAMINATED BEAMS, PLATES, AND SHELLS

4.1   Extension and in-plane shear of symmetrically laminated plates

When laminae are connected to form laminated plates it is common practice to match them in order to obtain a balanced construction. This means that instead of bonding together — for instance, two sheets a and b of thickness t_a and t_b, respectively — a sheet of the type a of thickness $t_a/2$ is cemented to each face of sheet b as shown in Figure 1.29. Such a balanced laminate does not warp when the temperature or humidity of the air changes or when loads are applied in the middle plane of the laminate. It is permissible, therefore, to investigate the stresses and strains in this plane alone.

It was shown in Section 1.5 that the application of a tensile stress to a lamina in the 1 direction causes normal strains in

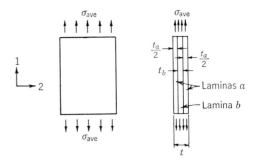

Figure 1.29   Symmetrically Laminated Plate

directions 1 and 2 *and* a change in the angle subtended by the edges since the directions 1 and 2 are not the natural axes of the orthotropic material. The applied tensile load can be distributed between the laminae in proportion to the tensile rigidity E_1t, of each so that the strain ε_1 is the same for each lamina. However, the transverse normal strain ε_2 and the shear strain γ_{12} would then, as a rule, be different in the two laminae. This cannot happen if the laminae are bonded together. Consequently, in order to equalize the strain, normal stresses σ_2 and shear stresses τ_{12} must be set up between the laminae.

The strains in lamina a are according to equations 1.5.17–1.5.19 as follows:

$$\varepsilon_{1a}^0 = \frac{\sigma_{1a}}{E_{1a}} - \nu_{21a}\frac{\sigma_{2a}}{E_{2a}} - m_{1a}\frac{\tau_{12a}}{E_{La}}, \qquad (4.1.1)$$

$$\varepsilon_{2a}^0 = -\nu_{12a}\frac{\sigma_{1a}}{E_{1a}} + \frac{\sigma_{2a}}{E_{2a}} - m_{2a}\frac{\tau_{12a}}{E_{La}}, \qquad (4.1.2)$$

$$\gamma_{12a}^0 = -m_{1a}\frac{\sigma_{1a}}{E_{La}} - m_{2a}\frac{\sigma_{2a}}{E_{La}} + \frac{\tau_{12a}}{G_{12a}}. \qquad (4.1.3)$$

If the subscript is changed from a to b, the equations represent the strain-stress relations for lamina b. The super index 0 designates independence of the thickness coordinate.

Because of the type of loading assumed, the normal stresses σ_{1a} and σ_{1b} must be equivalent to the average applied normal stress σ_1, while the normal stresses σ_{2a} and σ_{2b} as well as the shear stresses τ_{12a} and τ_{12b} must be in equilibrium along the edges:

$$\sigma_{1a}t_a + \sigma_{1b}t_b = \sigma_1 t = n_1, \qquad (4.1.4)$$

$$\sigma_{2a}t_a + \sigma_{2b}t_b = n_2 = 0, \qquad (4.1.5)$$

$$\tau_{12a}t_a + \tau_{12b}t_b = n_{12} = 0. \qquad (4.1.6)$$

The requirements of consistent deformations are

$$\varepsilon_{1a}^0 = \varepsilon_{1b}^0 = \varepsilon_1^0, \qquad (4.1.7)$$

$$\varepsilon_{2a}^0 = \varepsilon_{2b}^0 = \varepsilon_2^0, \qquad (4.1.8)$$

$$\gamma_{12a}^0 = \gamma_{12b}^0 = \gamma_{12}^0. \qquad (4.1.9)$$

Substitution and algebraic manipulations allow us to write the following three simultaneous equations:

$$A_{11}\sigma_{1a} + A_{12}\sigma_{2a} + A_{13}\tau_{12a} = (\sigma_1/E_{1b})(t/t_a t_b), \qquad (4.1.10)$$

$$A_{21}\sigma_{1a} + A_{22}\sigma_{2a} + A_{23}\tau_{12a} = -(\nu_{12b}\sigma_1/E_{1b})(t/t_a t_b),$$

$$(4.1.11)$$

$$A_{31}\sigma_{1a} + A_{32}\sigma_{2a} + A_{33}\tau_{12a} = -(m_{1b}\sigma_1/E_{Lb})(t/t_a t_b),$$

$$(4.1.12)$$

where

$$A_{11} = \frac{1}{E_{1a}t_a} + \frac{1}{E_{1b}t_b},\quad A_{22} = \frac{1}{E_{2a}t_a} + \frac{1}{E_{2b}t_b},$$

$$A_{33} = \frac{1}{G_{12a}t_a} + \frac{1}{G_{12b}t_b}, \quad (4.1.13)$$

$$A_{12} = A_{21} = -\frac{\nu_{12a}}{E_{1a}t_a} - \frac{\nu_{12b}}{E_{1b}t_b}, \quad (4.1.14)$$

$$A_{13} = A_{31} = -\frac{m_{1a}}{E_{La}t_a} - \frac{m_{1b}}{E_{Lb}t_b}, \quad (4.1.15)$$

$$A_{23} = A_{32} = -\frac{m_{2a}}{E_{La}t_a} - \frac{m_{2b}}{E_{Lb}t_b}. \quad (4.1.16)$$

If numerical values are substituted for the coefficients A, it is not difficult to solve equations 4.1.10–4.1.12 for the unknown stresses acting in lamina a. Solutions in a closed algebraic form are, however, cumbersome.

The general case of uniform extension n_1 and n_2 together with a uniform shear force n_{12} leads to the following equations:

$$A_{11}\sigma_{1a} + A_{12}\sigma_{2a} + A_{13}\tau_{12a}$$
$$= (\sigma_1/E_{1b} - \nu_{21b}\sigma_2/E_{2b} - m_{1b}\tau_{12}/E_{Lb})(t/t_a t_b), \quad (4.1.17)$$
$$A_{21}\sigma_{1a} + A_{22}\sigma_{2a} + A_{23}\tau_{12a}$$
$$= (-\nu_{12b}\sigma_1/E_{1b} + \sigma_2/E_{2b} - m_{2b}\tau_{12}/E_{Lb})(t/t_a t_b), \quad (4.1.18)$$
$$A_{31}\sigma_{1a} + A_{32}\sigma_{2a} + A_{33}\tau_{12a}$$
$$= (-m_{1b}\sigma_1/E_{Lb} - m_{2b}\sigma_2/E_{Lb} + \tau_{12}/G_{12b})(t/t_a t_b), \quad (4.1.19)$$

where the constant coefficients A are those given in equations 4.1.13–4.1.16.

Numerical Example: The application of the theory to actual problems will now be shown in connection with the laminate sketched in Figure 1.30. It consists of five layers a totaling

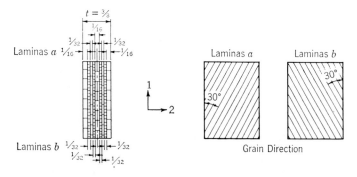

Figure 1.30   Symmetrically Laminated Plate Subject to Axial Tension n_1

$t_a = \frac{1}{4}$ in. having the grain, or L direction, subtend an angle $\alpha = 30°$ with the vertical, and four layers b totaling $t_b = \frac{1}{8}$ in. having the grain, or L direction, subtend an angle $\alpha = -30°$ with the vertical. The loading consists of a uniformly distributed tensile force n_1 per unit length having an average intensity of $\sigma_1 = 1000$ psi.

The orthotropic material is spruce with moduli given by equations 1.2.2, where L and T designate the natural axes of elastic symmetry.

Making use of the appropriate transformation equations of Section 1.5 one finds the following moduli in the 1, 2 axes for layers a and b:

$$(E_1/E_L)_a = (E_1/E_L)_b = 0.139328,$$
$$(E_2/E_L)_a = (E_2/E_L)_b = 0.048572, \quad (4.1.20)$$
$$\nu_{12a} = \nu_{12b} = 0.001541,\ \nu_{21a} = \nu_{21b} = 0.000537, \quad (4.1.21)$$
$$m_{1a} = -m_{1b} = 11.004,\ m_{2a} = -m_{2b} = 12.223, \quad (4.1.22)$$
$$(G_{12}/G_{LT})_a = (G_{12}/G_{LT})_b = 0.927667. \quad (4.1.23)$$

The coefficients A defined by equations 4.1.13–4.1.16 assume the following numerical values:

$$A_{11} = 6.02292 \times 10^{-5},$$
$$A_{22} = 17.27649 \times 10^{-5}, \quad (4.1.24)$$
$$A_{33} = 24.49938 \times 10^{-5},$$

$$A_{12} = -0.928 \times 10^{-7},$$
$$A_{13} = 3.07810 \times 10^{-5}, \quad (4.1.25)$$
$$A_{23} = 3.41915 \times 10^{-5}.$$

The solution of equations 4.1.10–4.1.12 can be obtained, for instance, by the method of determinants with the results

$$\sigma_{1a} = 858.5\text{ psi},\ \sigma_{2a} = -54.8\text{ psi},\ \tau_{12a} = 276.7\text{ psi}. \quad (4.1.26)$$

The stresses in layers b can be calculated with the aid of equations 4.1.4–4.1.6:

$$\sigma_{1b} = 1283.0\text{ psi},\ \sigma_{2b} = 109.7\text{ psi},\ \tau_{12b} = -553.4\text{ psi}. \quad (4.1.27)$$

Next the strains are calculated from equations 4.1.1–4.1.3:

$$\varepsilon_1^0 = 2.1799 \times 10^{-3},\ \varepsilon_2^0 = -3.1615 \times 10^{-3},$$
$$\gamma_{12}^0 = -0.4882 \times 10^{-3}. \quad (4.1.28)$$

The applied average stress $\sigma_1 = 1000$ psi divided by the strain ε_1 yields the apparent modulus, $E_1 = 458,700$ psi. The strain ε_2 divided by the strain ε_1 and multiplied by -1 is the apparent Poisson's ratio $\nu_{12} = 1.450$.

The numerical example shows that the state of stress and strain in a symmetrically laminated plate is quite complex. They cannot be calculated by short-cut methods.

It is worth noting that in this example plates b carry 128 per cent of the average stress in the direction of the applied tension and are subjected in addition to shearing stresses amounting to 55 per cent of the applied average tensile stress.

Poisson's ratio as calculated by equation 1.5.4 may become negative for some values of α. In the case of the spruce veneer here considered, negative values do occur in the range between 30°31′ and 59°28′. The effect of a negative ν_{12} may be seen by recalculating the foregoing numerical example for the case of $\alpha = 45°$. Then

$$A_{11} = A_{22} = 11.508 \times 10^{-5},\ A_{33} = 25.098 \times 10^{-5}, \quad (4.1.24')$$
$$A_{12} = 13.84 \times 10^{-7},\ A_{13} = A_{23} = 3.752 \times 10^{-5}, \quad (4.1.25')$$
$$\sigma_{1a} = 893.4\text{ psi},\ \sigma_{2a} = -106.6\text{ psi},\ \tau_{12a} = 330.9\text{ psi}, \quad (4.1.26')$$
$$\sigma_{1b} = 1213.3\text{ psi},\ \sigma_{2b} = 213.3\text{ psi},\ \tau_{12b} = -661.8\text{ psi}, \quad (4.1.27')$$

$$\epsilon_1^0 = 5.4474 \times 10^{-3}, \ \epsilon_2^0 = -4.0223 \times 10^{-3},$$
$$\gamma_{12}^0 = -0.4593 \times 10^{-3}. \quad (4.1.28')$$

Note that the shear stress τ_{12b} is over 66 per cent of the applied average tensile stress as compared to 55 per cent obtained for the case $\alpha = 30°$.

Results by the theory of composite plates

The analysis of symmetrically laminated plates composed of layers of type a and b was shown by elementary considerations for the case of constant in-plane forces.

The results for a multilayer symmetrically laminated plate subject to constant edge forces are obtained from equations 2.2.3a and 2.4.16a as follows:

$$[\epsilon^0] = [A^{-1}][n], \ [\tau] = [E][A^{-1}][n], \quad (4.1.29)$$

where the A terms designate the elastic areas and are defined by equations 2.4.15.

For the numerical example considered, the strain and stress components (4.1.29) have the following values:

$$\begin{bmatrix} \epsilon_1^0 \\ \epsilon_2^0 \\ \epsilon_{12}^0 \end{bmatrix} = \begin{bmatrix} 0.5813 \times 10^{-5} & -0.8431 \times 10^{-5} & -0.1302 \times 10^{-5} \\ -0.8431 \times 10^{-5} & 2.9050 \times 10^{-5} & -0.1446 \times 10^{-5} \\ -0.1302 \times 10^{-5} & -0.1446 \times 10^{-5} & 1.0362 \times 10^{-5} \end{bmatrix} \begin{bmatrix} n_1 \\ 0 \\ 0 \end{bmatrix},$$
$$(4.1.30)$$

$$\begin{bmatrix} \tau_1 \\ \tau_2 \\ \tau_{12} \end{bmatrix}_a = \begin{bmatrix} 2.2893 & -0.4191 & 3.0030 \\ -0.1462 & 2.5042 & 1.1639 \\ 0.7379 & 0.8197 & 2.1269 \end{bmatrix} \begin{bmatrix} n_1 \\ 0 \\ 0 \end{bmatrix}, \quad (4.1.31)$$

$$\begin{bmatrix} \tau_1 \\ \tau_2 \\ \tau_{12} \end{bmatrix}_b = \begin{bmatrix} 3.4213 & 0.8383 & -6.0067 \\ 0.2925 & 2.9916 & -2.3279 \\ -1.4758 & -1.6393 & 3.7462 \end{bmatrix} \begin{bmatrix} n_1 \\ 0 \\ 0 \end{bmatrix}. \quad (4.1.32)$$

Setting $n_1 = 375$ lb/in. one gets

$$\begin{bmatrix} \epsilon_1^0 \\ \epsilon_2^0 \\ \epsilon_{12}^0 \end{bmatrix} = 10^{-3} \begin{bmatrix} 2.1799 \\ -3.1615 \\ -0.4882 \end{bmatrix}, \quad \begin{bmatrix} \tau_1 \\ \tau_2 \\ \tau_{12} \end{bmatrix}_a = \begin{bmatrix} 858.5 \\ -54.8 \\ 276.7 \end{bmatrix} \text{psi},$$

$$\begin{bmatrix} \tau_1 \\ \tau_2 \\ \tau_{12} \end{bmatrix}_b = \begin{bmatrix} 1283.0 \\ 109.7 \\ -553.4 \end{bmatrix} \text{psi}, \quad (4.1.33)$$

exactly as given previously in 4.1.26–4.1.28.

Formulas for symmetrically laminated orthotropic plates

When each of the laminae possesses elastic symmetry and the laminae are glued together so that their principal axes of orthotropy are parallel, the resulting composite plate is also orthotropic. If the axes 1, 2 are taken to coincide with the principal axes of the layers then the following constants in equations 4.1.1–4.1.3, 4.1.10–4.1.12 vanish:

$$m_{1a} = m_{1b} = m_{2a} = m_{2b} = 0, \quad (4.1.34)$$
$$A_{13} = A_{31} = A_{23} = A_{32} = 0. \quad (4.1.35)$$

Consequently the stresses in layers a become

$$\begin{bmatrix} \sigma_1 \\ \sigma_2 \end{bmatrix}_a = \frac{\sigma_1 t}{E_{1b} t_a t_b} \frac{1}{A_{11} A_{22} - A_{12}^2} \begin{bmatrix} A_{22} + \nu_{12b} A_{12} \\ -\nu_{12b} A_{11} - A_{12} \end{bmatrix},$$
$$(4.1.36)$$

and the strains are

$$\begin{bmatrix} \epsilon_1^0 \\ \epsilon_2^0 \end{bmatrix} = \frac{\sigma_1 t}{E_{1a} t_a E_{1b} t_b} \frac{1}{A_{11} A_{22} - A_{12}^2}$$
$$\times \begin{bmatrix} A_{22} - \left(\dfrac{\nu_{12a}^2}{E_{1a} t_a} + \dfrac{\nu_{12b}^2}{E_{1b} t_b} \right) \\ -\nu_{12a} \nu_{12b} A_{12} - \left(\dfrac{\nu_{12a}}{E_{2b} t_b} + \dfrac{\nu_{12b}}{E_{2a} t_a} \right) \end{bmatrix}. \quad (4.1.37)$$

Note that $\tau_{12} = \gamma_{12}^0 = 0$ in the orthotropic composite plate unless a shearing force $n_{12} = \tau_{12} t$ is applied. Then one finds from 4.1.19 the shear stress and shear strain that are developed:

$$\tau_{12a} = \tau_{12} \frac{t}{t_a} \frac{G_{12a} t_a}{G_{12a} t_a + G_{12b} t_b} = G_{12a} \gamma_{12a}. \quad (4.1.38)$$

The apparent modulus E_1, when σ_1 is applied, takes the form

$$E_1 = \frac{\sigma_1}{\epsilon_1^0} = \frac{E_{1a} t_a E_{1b} t_b}{t} \frac{A_{11} A_{22} - A_{12}^2}{A_{22} - \left(\dfrac{\nu_{12a}^2}{E_{1a} t_a} + \dfrac{\nu_{12b}^2}{E_{1b} t_b} \right)}, \quad (4.1.39)$$

where the apparent Poisson's ratio ν_{12} is given by

$$\nu_{12} = -\frac{\epsilon_2^0}{\epsilon_1^0} = \frac{\nu_{12a} \nu_{12b} A_{12} + \left(\dfrac{\nu_{12a}}{E_{2b} t_b} + \dfrac{\nu_{12b}}{E_{2a} t_a} \right)}{A_{22} - \left(\dfrac{\nu_{12a}^2}{E_{1a} t_a} + \dfrac{\nu_{12b}^2}{E_{1b} t_b} \right)}. \quad (4.1.40)$$

Note that values of E_2 and ν_{21} can be obtained from equations 4.1.39 and 4.1.40 upon interchanging the subscripts 1 and 2.

Finally, the effective shearing modulus becomes

$$G_{12} = \tau_{12}/\gamma_{12} = (1/t)(G_{12a} t_a + G_{12b} t_b). \quad (4.1.41)$$

Cross-laminated plates

When the laminate is composed of orthotropic layers with the principal elastic axes alternatively parallel and perpendicular to the reference layer, a cross-laminated plate is obtained. For the special case of a plate composed of identical layers the following relations hold:

$$E_{1b} = E_{2a}, \ E_{2b} = E_{1a}, \ \nu_{12b} = \nu_{21a}, \ \nu_{21b} = \nu_{12a}. \quad (4.1.42)$$

The effective moduli (4.1.39–4.1.41) become

$$E_1 = E_{1a} \frac{t_a}{t} \frac{(1 + k t_b/t_a) - \nu_{12a}^2 k^2 t^2/[t_a(k t_a + t_b)]}{1 - \nu_{12a}^2 k}, \quad (4.1.43)$$

$$\nu_{12} = \nu_{12a} k t/(k t_a + t_b), \quad (4.1.44)$$

$$G_{12} = G_{12a} = G_{12b}, \quad (4.1.45)$$

where

$$k = E_{2a}/E_{1a}. \quad (4.1.46)$$

For the special case $t_a = t_b = t/2$ equations 4.1.43 and 4.1.44 simplify to the following:

$$E_1 = \frac{E_{1a}}{2} \frac{(1 + k)^2 - 4\nu_{12a}^2 k^2}{(1 + k)(1 - \nu_{12a}^2 k)}, \quad (4.1.47)$$

$$\nu_{12} = \nu_{12a} 2k/(k+1). \qquad (4.1.48)$$

The variation of E_1 and ν_{12} as a function of k is shown in Figure 1.31 for three values of ν_{12a} : 0, 0.5, 1. It may be seen that the assumption $\nu_{12a} = 0$ is a good enough approximation in engineering calculations except when ν_{12a} is close to unity.

Practical calculation of the properties of cross laminated plates

When the plate is cross laminated, equation 4.1.43 is simplified and becomes

$$E_1 t = E_{1a} t_a + E_{1b} t_b. \qquad (4.1.49)$$

The value of E_2 is obtained by taking 2 instead of 1. An even more simplified calculation omits those layers that have the grain perpendicular to the direction in which the apparent modulus is sought.

The results obtainable by the various methods are now compared by means of an example. It is assumed that the plywood is composed of the same layers as those shown in Figure 1.30 but that the fibers of the adjacent layers are arranged perpendicularly to one another.

Equation 4.1.49 yields

$$E_1 = \tfrac{1}{3}(2 \times 1{,}430{,}000 + 1 \times 51{,}400) = 970{,}500 \text{ psi},$$
$$E_2 = \tfrac{1}{3}(1 \times 1{,}430{,}000 + 2 \times 51{,}400) = 510{,}900 \text{ psi}, \qquad (4.1.50)$$

where 1 and 2 indicate the direction of the fibers in layers a and b, respectively.

If the transverse layers are neglected, the results are

$$E_1 = 953{,}300 \text{ psi}, \quad E_2 = 476{,}700 \text{ psi}. \qquad (4.1.51)$$

The values obtained from equation 4.1.43 are

$$E_1 = 978{,}000 \text{ psi}, \quad E_2 = 515{,}000 \text{ psi}. \qquad (4.1.52)$$

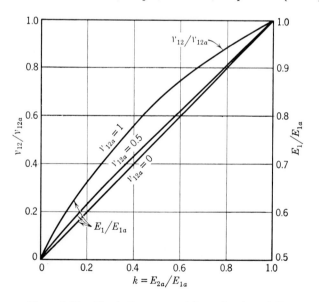

Figure 1.31   Elastic Constants of Cross-Laminated Plates

It can be concluded, therefore, that the differences between the results obtained by the exact formulas and those computed from the simple connection given in 4.1.49 are small enough to permit the use of equation 4.1.49 in practical calculations.

Poisson's ratio is, in accordance with equation 4.1.44,

$$\nu_{12} = 0.539 \, \frac{0.036 \times 3}{0.036 \times 2 + 1} = 0.0543. \qquad (4.1.53)$$

Once the apparent moduli and Poisson's ratios of a laminated sheet are determined for the natural axes, they can be calculated for all the other directions according to the transformation equations of Section 1.5. This was done for cross-laminated spruce plywood for the case when $t_a = t_b$. Equations 4.1.47, 4.1.48, and 4.1.45 yield

$$E_1 = E_2 = 747{,}000 \text{ psi}, \quad \nu_{12} = \nu_{21} = 0.0375,$$
$$G_{12} = G_{21} = 52{,}800 \text{ psi}. \qquad (4.1.54)$$

With these values the apparent elastic constants were determined for any angle α subtended by the grain with the principal directions. The results are plotted in Figures 1.32

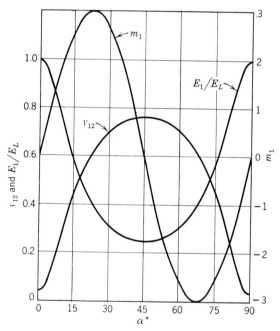

Figure 1.32   Variation with α of E_1, ν_{12}, and m_1 for a Cross-Laminated Spruce

and 1.33. Comparison with Figures 1.5 and 1.6 shows that cross lamination reduces materially the variation in the value of the apparent Young's modulus. For the single spruce veneer the maximum is 1,430,000 psi and the minimum 51,400 psi, while for the plywood the corresponding values are 747,000 psi and 186,000 psi. The shearing rigidity is increased by cross lamination for every angle α, and the maximum value, amounting to 6.76 times that found for the single veneer, is reached when $\alpha = 45°$.

4.2   Pure bending and torsion of symmetrically laminated plates

Symmetrically laminated plates that are subject to uniform edge bending moments m_x, m_y and twisting couples m_{xy} can be analyzed by using equations 2.2.4b and 2.4.16b as follows:

$$[\epsilon] = z[D^{-1}][m], \quad [\tau] = z[E][D^{-1}][m], \qquad (4.2.1)$$

where the D terms designate the elastic moments of inertia defined by equations 2.4.14.

Numerical Example: Consider the laminated plate of Figure 1.30 and designate the thickness coordinate by z, measured

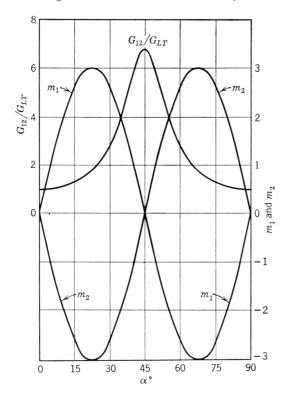

Figure 1.33 Variation with α of G_{12}, m_1, and m_2 for a Cross-Laminated Spruce

from the midsurface of the plate. The strains (first equation in 4.2.1) have the following values for the numerical example at hand:

$$
\begin{bmatrix} \epsilon_1 \\ \epsilon_2 \\ \epsilon_{12} \end{bmatrix} = z \begin{bmatrix} 0.5386 \times 10^{-3} & -0.6722 \times 10^{-3} & -0.2528 \times 10^{-3} \\ -0.6722 \times 10^{-3} & 2.5314 \times 10^{-3} & -0.2808 \times 10^{-3} \\ -0.2528 \times 10^{-3} & -0.2808 \times 10^{-3} & 1.1317 \times 10^{-3} \end{bmatrix} \begin{bmatrix} m_1 \\ m_2 \\ m_{12} \end{bmatrix}.
$$
(4.2.2)

The stresses (second equation in 4.2.1) in laminae a and b are given, respectively, by the following matrix equations:

$$
\begin{bmatrix} \tau_1 \\ \tau_2 \\ \tau_{12} \end{bmatrix}_a = z \begin{bmatrix} 182.7832 & -49.7330 & 200.4490 \\ -17.3516 & 208.2815 & 77.6842 \\ 49.2489 & 54.7055 & 163.5092 \end{bmatrix} \begin{bmatrix} m_1 \\ m_2 \\ m_{12} \end{bmatrix},
$$
(4.2.3a)

$$
\begin{bmatrix} \tau_1 \\ \tau_2 \\ \tau_{12} \end{bmatrix}_b = z \begin{bmatrix} 402.5747 & 194.4105 & -783.5732 \\ 67.8288 & 302.8995 & -303.6746 \\ -192.5184 & -213.8486 & 477.9186 \end{bmatrix} \begin{bmatrix} m_1 \\ m_2 \\ m_{12} \end{bmatrix}.
$$
(4.2.3b)

Case of pure bending moment $m_1 = 10$ lb-in./in. setting

$$m_1 = 10 \text{ lb-in./in.}, \quad m_2 = 0, \quad m_{12} = 0, \quad (4.2.4)$$

in equations 4.2.2 we find the following *continuous* z dependence of the strains:

$$
\begin{bmatrix} \epsilon_1 \\ \epsilon_2 \\ \epsilon_{12} \end{bmatrix} = z \begin{bmatrix} -0.6722 \times 10^{-2} \\ 2.5314 \times 10^{-2} \\ -0.2808 \times 10^{-2} \end{bmatrix}.
$$
(4.2.5)

From equations 4.2.3 the *discontinuous* z dependence of the stresses is obtained and is given in Table 1.4. For the other

Table 1.4

$\dfrac{z}{t/2}$	lamina	τ_1	τ_2	τ_{12}
$\tfrac{1}{6}$	a	57.1 psi	−5.4 psi	15.4 psi
	b	125.8	21.2	−60.2
$\tfrac{2}{6}$	a	114.2	−10.8	30.8
	b	251.6	42.4	−120.3
$\tfrac{3}{6}$	a	171.3	−16.3	46.2
	b	377.4	63.6	−180.5
$\tfrac{4}{6}$	a	228.4	−21.7	61.6
	b	503.2	84.8	−240.6
1	a	342.6	−32.5	92.3

half of the plate thickness z changes sign and so do all the stresses as shown in Figure 1.34.

Case of twisting couples $m_{12} = 10$ lb-in./in. setting

$$m_1 = 0, \quad m_2 = 0, \quad m_{12} = 10 \text{ lb-in./in.} \quad (4.2.6)$$

in equations 4.2.2 we obtain the following *continuous* strains:

$$
\begin{bmatrix} \epsilon_1 \\ \epsilon_2 \\ \epsilon_{12} \end{bmatrix} = z \begin{bmatrix} -0.2528 \times 10^{-2} \\ -0.2808 \times 10^{-2} \\ 1.1317 \times 10^{-2} \end{bmatrix}.
$$
(4.2.7)

The *discontinuous* stress distribution follows from equations 4.2.3 and is given in Table 1.5. For negative z values, cor-

Table 1.5

$\dfrac{z}{t/2}$	lamina	τ_1	τ_2	τ_{12}
$\tfrac{1}{6}$	a	62.6 psi	24.3 psi	51.1 psi
	b	−244.9	−94.9	149.3
$\tfrac{2}{6}$	a	125.3	48.6	102.2
	b	−489.7	−189.8	298.7
$\tfrac{3}{6}$	a	187.9	72.8	153.3
	b	−734.6	−284.7	448.0
$\tfrac{4}{6}$	a	250.6	97.1	204.4
	b	−979.5	−379.6	597.4
1	a	375.8	145.7	306.6

responding to the other half of the plate thickness, all stresses change sign.

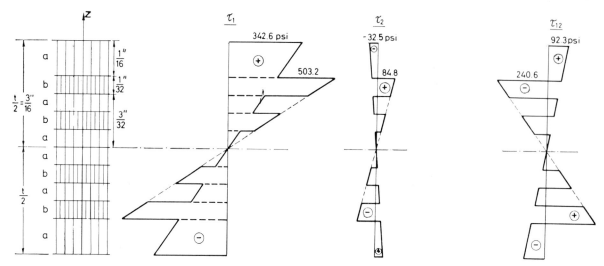

Figure 1.34   Section of Symmetrically Laminated Plate and Stress Distribution for $m_1 = 10$ lb-in./in.

4.3 Pure bending and transverse shear of symmetrically laminated beams

Pure bending

In a theory of homogeneous beams subject to a pure bending moment M_x the only nonvanishing stress component is σ_x (see, for example, Ref. 14, p. 100). As shown in Section 2.4 bending and stretching become uncoupled for symmetrically laminated plates, or beams, whereas the cross-elasticity effect (2.1.33) simplifies to 2.4.16. This effect disappears whenever *all* elastic moduli have the same thickness dependence, as shown by equations 2.4.24. In particular we consider those beams that are also symmetrically laminated, as is common in structural applications (Fig. 1.35).

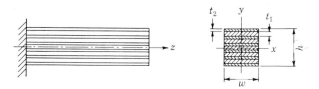

Figure 1.35   Symmetrically Laminated Beam

A beam composed of isotropic layers of different thicknesses t_m and different Young's moduli E^m, but with *constant* Poisson's ratio ν, is a simple example for which the cross-elasticity phenomenon disappears, and consequently the application of a couple M_x gives rise only to σ_x. This would of course be the stress field if the composite isotropic beam with constant ν were also symmetrically layered. Specialization of equations 2.4.22 and 2.4.24 gives

$$\epsilon_x = \frac{M_x}{D} z, \qquad (4.3.1)$$

$$\sigma_x^m = \frac{M_x}{D} E^m z, \qquad (4.3.2)$$

where

$$D = \tfrac{2}{3} b \sum_{m=1}^{n} E^m (h_m^3 - h_{m-1}^3), \qquad (4.3.3)$$

and b is the width of the beam. The curvature of the composed beam is given by

$$\kappa_x = \frac{M_x}{D} = -\frac{d^2 z}{dx^2}, \qquad (4.3.4)$$

from which the deflection of the beam can be readily determined.

In case the beam is composed of orthotropic layers with x, y, and z axes (Fig. 1.35) as their principal axes of elasticity, the previous equations hold precisely whenever *all* elastic moduli have an identical z dependence. The Young's modulus in the x direction for the mth orthotropic layer is then designated by E^m.

It may be mentioned that the maximum normal stress in a laminate does not necessarily occur in the fibers farthest from the neutral axis. Not z but $E^m z$ must be a maximum in equation 4.3.2 for the normal stress to be a maximum, and this may be true in a lamina closer to the neutral axis if its modulus is larger than that of the extreme lamina.

When the restriction on Poisson's ratio is removed and ν varies from one layer to the other, it is clear from the following equation that ϵ_y is discontinuous on the dividing surfaces between the layers:

$$\epsilon_y = -\nu \epsilon_x = -\nu \frac{M_x}{D} z. \qquad (4.3.5)$$

The problem that arises in trying to remove the discontinuity in ϵ_y is discussed in Chapter 25 of Muskhelishvili's monograph[51] and will not be pursued further here. Since for isotropic layers ν generally varies between zero and one-half it is not expected that the stress field (4.3.2) is substantially altered.

Transverse shear

When a transverse shear force V_x is acting (Fig. 1.35), the transverse shear stress is obtained from equilibrium considerations in the form

$$\tau_{xz} = V_x B / D b, \qquad (4.3.6)$$

where τ_{xz} is the shear stress along any horizontal line distant z from the neutral axis y and B is the weighted static moment with respect to y of the portion of the cross section lying above the horizontal line. The maximum shear stress always occurs at the neutral axis since B has its maximum value there. However, the danger of failure by shearing may be greater in another lamina if its load-carrying capacity in shear is smaller than that of the lamina at the neutral axis.

Note that when E is a continuous function of z and the beam is symmetrically composed the expression 4.3.3 for D becomes

$$D = \int_{-h/2}^{h/2} E z^2 \, dA, \qquad (4.3.7)$$

whereas B is expressed by

$$B = \int_{z}^{h/2} E \xi \, dA \qquad (z < \xi < h/2). \qquad (4.3.8)$$

It is clear from equation 4.3.4 how the deflection of a laminated beam can be calculated by considering only the effect of the bending rigidity. When the material has a small shearing rigidity the deflection due to shear should be added to the deflection due to the bending moment. A more detailed investigation of the deflections is needed in the case of sandwich-type beams, the core of which has very little resistance to shear. This problem is discussed in Section 7.1.

4.4 Symmetrically laminated shells of revolution

It was already indicated in Section 2.4 that for symmetrically laminated plates the elastic-statical moments, with respect to the symmetry plane, disappear:

$$B_{ij} = 0. \qquad (4.4.1)$$

Consequently one finds, in view of equations 2.1.20, the following simplified relations:

$$B_{ij}^* = 0, \quad C_{ij}^* = 0, \quad D_{ij}^* = D_{ij}, \qquad (4.4.2)$$

and the general plate equations 2.3.7 and 2.3.8 reduce to 2.4.13 which are completely analogous to the respective equations for homogeneous plates. A similar situation exists in symmetrically laminated shells when the reference surface ($\zeta = 0$) coincides with the surface of elastic symmetry. In view of relations 4.4.2, governing equations 3.1.38–3.1.47 for heterogeneous shells of revolution are considerably simplified to the form of analogous equations for homogeneous shells, except for the values of the constants that appear in the operators L. Known results for Φ and Ψ from homogeneous shell theory can be almost directly utilized for the corresponding symmetrically layered shell problem. Note that relations 4.4.2 hold for the class of shells with elastic moduli that vary with ζ according to the same law, and the analogy to homogeneous shell equations applies equally well. This analogy was exemplified for shallow shells of revolution by equations 3.4.30–3.4.40. For further examples of application the reader may wish to consult Ambartsumyan's book[45] on anisotropic shells.

Finally, it is pointed out that the stress distribution in a composite shell is entirely different from that in a homogeneous shell. For a symmetrically layered shell, equation 2.1.33 simplifies as follows:

$$[\tau] = [E]\{[a][N] + z[d][M]\}, \qquad (4.4.3)$$

whereas for shells with the same ζ variation of all moduli equations 2.4.24 hold.

5 THERMAL STRESSES IN COMPOSITE PLATES AND SHELLS

5.1 Thermoelasticity of heterogeneous aeolotropic plates

The interest in thermoelasticity started as early as 1835, shortly after the formulation of elasticity theory itself, when Duhamel published his memoir on thermal stresses. Quite a number of solutions are available for homogeneous systems, such as those given in the texts of Melan and Parkus,[52] Parkus,[53] and Nowacki.[54] In recent years, the development of atmospheric-reentry vehicles and nuclear reactors made the analysis of thermal effects, especially in composite systems, a problem of major interest.

A general quasi-static thermoelastic theory as developed by Stavsky[55] is introduced in this section for thin heterogeneous aeolotropic plates. It is assumed that the temperature distribution within the heterogeneous structure is determined by the theory of heat conduction based on Fourier's law. For detailed examples the reader may wish to consult, for example, the texts by Schneider (Ref. 56, Ch. 2) and Gebhart (Ref. 57, Ch. 11). Further we restrict the discussion to the case where the temperature distribution throughout the plate is independent of its deformation (see, for example, Boley and Weiner, Ref. 58, Ch. 2).

Thermoelastic strain-stress relations

The following strain-stress relations may be written for an aeolotropic thermoelestic plate material:

$$\begin{bmatrix} \epsilon_x \\ \epsilon_y \\ \epsilon_{xy} \end{bmatrix} = \begin{bmatrix} e_{xx} & e_{xy} & e_{xs} & \alpha_x \\ e_{xy} & e_{yy} & e_{ys} & \alpha_y \\ e_{xs} & e_{ys} & e_{ss} & \alpha_s \end{bmatrix} \begin{bmatrix} \tau_x \\ \tau_y \\ \tau_{xy} \\ T \end{bmatrix}, \qquad (5.1.1)$$

where the coordinates x, y are taken in the undeflected bottom face ($z = 0$) of the plate. Since the plate is heterogeneous in the thickness direction the elastic compliance moduli e_{ij} and the coefficients of thermal expansion α_i are given functions of z. The change in temperature from the initial stress-free state is denoted by T and is a function of the space coordinates. It is assumed here that both the elastic and thermal properties are independent of time and temperature. Conveniently we rewrite equations 5.1.1 in the form

$$\begin{bmatrix} \epsilon_x - \alpha_x T \\ \epsilon_y - \alpha_y T \\ \epsilon_{xy} - \alpha_s T \end{bmatrix} = \begin{bmatrix} e_{xx} & e_{xy} & e_{xs} \\ e_{xy} & e_{yy} & e_{ys} \\ e_{xs} & e_{ys} & e_{ss} \end{bmatrix} \begin{bmatrix} \tau_x \\ \tau_y \\ \tau_{xy} \end{bmatrix}, \qquad (5.1.2)$$

which upon inversion become

$$\begin{bmatrix} \tau_x - \mathsf{A}_x T \\ \tau_y - \mathsf{A}_y T \\ \tau_{xy} - \mathsf{A}_s T \end{bmatrix} = \begin{bmatrix} E_{xx} & E_{xy} & E_{xs} \\ E_{xy} & E_{yy} & E_{ys} \\ E_{xs} & E_{ys} & E_{ss} \end{bmatrix} \begin{bmatrix} \epsilon_x \\ \epsilon_y \\ \epsilon_{xy} \end{bmatrix}, \qquad (5.1.3)$$

where

$$\begin{bmatrix} -A_x \\ -A_y \\ -A_s \end{bmatrix} = \begin{bmatrix} E_{xx} & E_{xy} & E_{xs} \\ E_{xy} & E_{yy} & E_{ys} \\ E_{xs} & E_{ys} & E_{ss} \end{bmatrix} \begin{bmatrix} \alpha_x \\ \alpha_y \\ \alpha_s \end{bmatrix}. \tag{5.1.4}$$

The Duhamel-Neumann law for the stress-strain relations for an aeolotropic plate takes the following compact form, in view of 5.1.3:

$$[\tau] = [E][\epsilon] + [A]T. \tag{5.1.5}$$

Plate stress-strain relations

The reference surface strains (at $z = 0$), and the bending curvatures are defined, as for the isothermal plate theory, by equations 2.1.8–2.1.10. Consequently the stress resultants and couples defined by 2.3.2 and 2.3.3 take the following form, in view of 5.1.5:

$$\begin{bmatrix} N_x - N_{xT} \\ N_y - N_{yT} \\ N_{xy} - N_{sT} \\ M_x - M_{xT} \\ M_y - M_{yT} \\ M_{xy} - M_{sT} \end{bmatrix} = \begin{bmatrix} A_{xx} & A_{xy} & A_{xs} & B_{xx} & B_{xy} & B_{xs} \\ A_{xy} & A_{yy} & A_{ys} & B_{xy} & B_{yy} & B_{ys} \\ A_{xs} & A_{ys} & A_{ss} & B_{xs} & B_{ys} & B_{ss} \\ B_{xx} & B_{xy} & B_{xs} & D_{xx} & D_{xy} & D_{xs} \\ B_{xy} & B_{yy} & B_{ys} & D_{xy} & D_{yy} & D_{ys} \\ B_{xs} & B_{ys} & B_{ss} & D_{xs} & D_{ys} & D_{ss} \end{bmatrix} \begin{bmatrix} \epsilon_x^0 \\ \epsilon_y^0 \\ \epsilon_{xy}^0 \\ \kappa_x \\ \kappa_y \\ \kappa_{xy} \end{bmatrix}, \tag{5.1.6}$$

where the thermal quantities N_{xT}, M_{xT}, and so forth are given by

$$(N_{iT}, M_{iT}) = \int_0^h (1, z) A_i T \, dz \qquad (i = x, y, s), \tag{5.1.7}$$

and the A, B, D terms are given by equations 2.3.5 for the heterogeneous plate, or by 2.3.6 for the layered plate.

Plate equations

Following the same procedure of Section 2.1 we formulate the plate thermoelastic theory in terms of the transverse deflection w and the Airy stress function F which must satisfy an eighth-order system of differential equations of the form

$$L_1 w - L_3 F = p_z + L_{Bx} N_{xT} + L_{By} N_{yT} + L_{Bs} N_{sT}$$
$$+ M_{xT,xx} + 2M_{sT,xy} + M_{yT,yy}, \tag{5.1.8}$$

$$L_2 F + L_3 w = L_{Ax} N_{xT} + L_{Ay} N_{yT} + L_{As} N_{sT}. \tag{5.1.9}$$

The functional operators L_1, L_2, L_3 are given by expressions 2.3.9–2.3.11, whereas the other operators are of the following form:

$$L_{Ai} = A_{yi}^*(\ )_{,xx} - A_{si}^*(\ )_{,xy} + A_{xi}^*(\ )_{,yy} \qquad (i = x, y, s), \tag{5.1.10}$$

$$L_{Bi} = B_{ix}^*(\ )_{,xx} + 2B_{is}^*(\ )_{,xy} + B_{iy}^*(\ )_{,yy} \qquad (i = x, y, s), \tag{5.1.11}$$

where the A^* and B^* terms are given by relations 2.1.20. The two simultaneous equations for w and F (5.1.8 and 5.1.9) are remarkable for the thermal coupling indicated by the L_{Bi} operators in addition to the coupling through the L_3 operator which also appears in the isothermal case.

Boundary conditions

Associated with the system of differential equations 5.1.8 and 5.1.9 are four conditions at each boundary point as given by relations 2.1.28. It is noted that, in addition to the coupling of w and F in the boundary conditions for transverse shear or bending moment, the thermal quantities N_{iT}, M_{iT} are also present. In order to see this we first use equation 5.1.6 to express ϵ_x^0, M_x, and so forth in terms of N_x, κ_x, and so forth with the following result:

$$\begin{bmatrix} \epsilon_x^0 \\ \epsilon_y^0 \\ \epsilon_{xy}^0 \\ M_x - M_{xT} \\ M_y - M_{yT} \\ M_{xy} - M_{sT} \end{bmatrix} = \begin{bmatrix} A_{xx}^* & A_{xy}^* & A_{xs}^* & B_{xx}^* & B_{xy}^* & B_{xs}^* \\ A_{yx}^* & A_{yy}^* & A_{ys}^* & B_{yx}^* & B_{yy}^* & B_{ys}^* \\ A_{sx}^* & A_{sy}^* & A_{ss}^* & B_{sx}^* & B_{sy}^* & B_{ss}^* \\ C_{xx}^* & C_{xy}^* & C_{xs}^* & D_{xx}^* & D_{xy}^* & D_{xs}^* \\ C_{yx}^* & C_{yy}^* & C_{ys}^* & D_{yx}^* & D_{yy}^* & D_{ys}^* \\ C_{sx}^* & C_{sy}^* & C_{ss}^* & D_{sx}^* & D_{sy}^* & D_{ss}^* \end{bmatrix} \begin{bmatrix} N_x - N_{xT} \\ N_y - N_{yT} \\ N_{xy} - N_{sT} \\ \kappa_x \\ \kappa_y \\ \kappa_{xy} \end{bmatrix}, \tag{5.1.12}$$

where the A^*, B^*, C^*, D^* terms are given by 2.1.20. Therefore, for example, a boundary condition for M_x would take the form

$$M_x = \bar{M}_x = C_{xx}^*(F_{,yy} - N_{xT}) + C_{xy}^*(F_{,xx} - N_{yT})$$
$$+ C_{xs}^*(-F_{,xy} - N_{sT}) - D_{xx}^* w_{,xx} - D_{xy}^* w_{,yy}$$
$$- 2D_{xs}^* w_{,xy} + M_{xT}, \tag{5.1.13}$$

which clearly shows that in addition to w and F the thermal quantities N_{xT}, N_{yT}, N_{sT}, and M_{xT} enter into the boundary-value problem.

A cross-thermoelastic phenomenon

The cross-elasticity effect that was expressed by 2.1.33 can be extended to the thermoelastic case as was shown by Stavsky.[59] In order to express the stresses in terms of resultants and couples we first invert the system 5.1.6, which in matrix notation takes the form

$$\begin{bmatrix} \epsilon^0 \\ \kappa \end{bmatrix} = \begin{bmatrix} a & b \\ c & d \end{bmatrix} \begin{bmatrix} N^* \\ M^* \end{bmatrix}, \tag{5.1.14}$$

where

$$[N^*] = \begin{bmatrix} N_x - N_{xT} \\ N_y - N_{yT} \\ N_{xy} - N_{sT} \end{bmatrix}, \quad [M^*] = \begin{bmatrix} M_x - M_{xT} \\ M_y - M_{yT} \\ M_{xy} - M_{sT} \end{bmatrix}, \tag{5.1.15}$$

and the matrices a, b, c, d are expressed by 2.1.32 in terms of A^*, B^*, D^*.

In view of relations 2.1.8 the strain components are the sum of

$$[\epsilon] = [\epsilon^0] + z[\kappa], \tag{5.1.16}$$

which in conjunction with 5.1.14 and 5.1.5 gives

$$[\tau] = [E]\{([a] + z[c])[N^*] + ([b] + z[d])[M^*]\} + [A]T. \tag{5.1.17}$$

These expressions clearly show a cross-thermoelasticity effect: namely, *each* stress component is a linear function of *all* stress resultants and couples as well as of *all* the thermal quantities N_{iT}, M_{iT}.

5.2 Thermoelasticity of heterogeneous shells of revolution

Formulation of problem

The general theory of Section 3.1 for symmetric finite deformations of composite shells of revolution is extended to include thermal effects, following Stavsky.[40] The geometry of the shell is defined by equations 3.1.1–3.1.8 and the strain-displacement relations 3.1.9–3.1.14 are valid for the thermoelastic case. Taking the shell material to be rotationally orthotropic, mechanically and thermally, the following stress-strain relations hold:

$$
\begin{bmatrix} \tau_\xi \\ \tau_\theta \end{bmatrix} = \begin{bmatrix} E_{\xi\xi} & E_{\xi\theta} & \mathsf{A}_\xi \\ E_{\xi\theta} & E_{\theta\theta} & \mathsf{A}_\theta \end{bmatrix} \begin{bmatrix} \varepsilon_\xi \\ \varepsilon_\theta \\ T \end{bmatrix}. \qquad (5.2.1)
$$

In view of the shell heterogeneity it is noted that

$$
E_{ij} = E_{ij}(\zeta),\ \mathsf{A}_i = \mathsf{A}_i(\zeta),\ T = T(\xi, \zeta) \qquad (i, j = \xi, \theta), \quad (5.2.2)
$$

and $\tau_{\xi\theta}$ vanishes as the discussion is restricted to symmetric deformations.

The strains are expressed in terms of the reference surface strains and the curvature changes as follows:

$$
\varepsilon_\xi = \varepsilon_{\xi 0} + \zeta \kappa_\xi,\ \varepsilon_\theta = \varepsilon_{\theta 0} + \zeta \kappa_\theta. \qquad (5.2.3)
$$

Introduction of thermal stress-strain relations 5.2.1 into the definitions of shell resultants and couples (3.1.18 and 3.1.19) and consideration of 5.2.3 give

$$
\begin{bmatrix} N_\xi - N_{\xi T} \\ N_\theta - N_{\theta T} \\ M_\xi - M_{\xi T} \\ M_\theta - M_{\theta T} \end{bmatrix} = \begin{bmatrix} A_{\xi\xi} & A_{\xi\theta} & B_{\xi\xi} & B_{\xi\theta} \\ A_{\xi\theta} & A_{\theta\theta} & B_{\xi\theta} & B_{\theta\theta} \\ \hline B_{\xi\xi} & B_{\xi\theta} & D_{\xi\xi} & D_{\xi\theta} \\ B_{\xi\theta} & B_{\theta\theta} & D_{\xi\theta} & D_{\theta\theta} \end{bmatrix} \begin{bmatrix} \varepsilon_{\xi 0} \\ \varepsilon_{\theta 0} \\ \kappa_\xi \\ \kappa_\theta \end{bmatrix}; \quad (5.2.4)
$$

or, more compactly,

$$
\begin{bmatrix} N^* \\ M^* \end{bmatrix} = \begin{bmatrix} A & B \\ \hline B & D \end{bmatrix} \begin{bmatrix} \varepsilon_0 \\ \kappa \end{bmatrix}, \qquad (5.2.5)
$$

where

$$
N_i^* = N_i - N_{iT},\ M_i^* = M_i - M_{iT} \qquad (i = \xi, \theta). \quad (5.2.6)
$$

The coefficients A_{ij}, B_{ij}, D_{ij} are given by expressions 3.1.25 and the thermal quantities N_{iT}, M_{iT} are defined by

$$
(N_{iT}, M_{iT}) = \int_{-h_1}^{+h_2} (1, \zeta) \mathsf{A}_i\, T d\zeta \qquad (i = \xi, \theta), \quad (5.2.7)
$$

where h_1 and h_2 denote the bounding surfaces of the shell (Fig. 1.22).

Using equilibrium equations 3.1.26–3.1.28 together with compatibility relation 3.1.30 and following the method shown in Section 3.1 for formulating the shell theory in terms of Φ and Ψ, one finds that governing equations 3.1.38 and 3.1.39 assume the form

$$
L_{11}\Phi + L_{12}\Psi + L_{13}(\Phi, \Psi) = L_{14}(\Phi, p_H, p_V, m) + L_{15}(\Phi, T),
$$
$$
(5.2.8)
$$
$$
L_{21}\Phi + L_{22}\Psi + L_{23}(\Phi, \Psi) = L_{24}(\Phi, p_H, p_V) + L_{25}(\Phi, T).
$$
$$
(5.2.9)
$$

The thermal functional operators L_{15}, L_{25} are given by the following expressions:

$$
L_{15}(\Phi, T) = (r\alpha/D_{\xi\xi}^*)[-C_{\xi\xi}^*(rN_{\xi T})' - C_{\xi\theta}^*(rN_{\theta T})' + (rM_{\xi T})'
$$
$$
+ \alpha \cos \Phi\, (C_{\theta\xi}^* N_{\xi T} + C_{\theta\theta}^* N_{\theta T} - M_{\theta T})], \quad (5.2.10)
$$
$$
L_{25}(\Phi, T) = A_{\theta\xi}^*(rN_{\xi T})' + A_{\theta\theta}^*(rN_{\theta T})'
$$
$$
- \alpha \cos \Phi\, (A_{\xi\xi}^* N_{\xi T} + A_{\xi\theta}^* N_{\theta T}), \quad (5.2.11)
$$

whereas the other operators are given by equations 3.1.40–3.1.47. Primes indicate differentiation with respect to the parameter ξ.

It is noted that for vanishing p_H, p_V, m equations 5.2.8 and 5.2.9 remain nonhomogeneous in contrast to the isothermic case where equations 3.1.38 and 3.1.39 become homogeneous.

Small-finite-deflections theory

In a small-finite-deflections shell theory the following approximation is made:

$$
\cos \Phi = \cos (\phi - \beta) = \cos \phi + \beta \sin \phi - \tfrac{1}{2}\beta^2 \cos \phi, \quad (5.2.12)
$$
$$
\sin \Phi = \sin (\phi - \beta) = \sin \phi - \beta \cos \phi - \tfrac{1}{2}\beta^2 \sin \phi. \quad (5.2.13)
$$

If we retain in equations 5.2.8 and 5.2.9 terms up to the second degree in β and Ψ, the governing system of equations becomes

$$
L_{11}\beta + L_{12}\Psi + L_{13}(\beta, \Psi) = L_{14}(\beta, p_H, p_V, m) + L_{15}(\beta, T),
$$
$$
(5.2.14)
$$
$$
L_{21}\beta + L_{22}\Psi + L_{23}(\beta, \Psi) = L_{24}(\beta, p_H, p_V) + L_{25}(\beta, T).
$$
$$
(5.2.15)
$$

The thermal functional operators L_{15}, L_{25} are given by

$$
L_{15}(\beta, T) = (r\alpha/D_{\xi\xi}^*)[C_{\xi\xi}^*(rN_{\xi T})' + C_{\xi\theta}^*(rN_{\theta T})' - (rM_{\xi T})'
$$
$$
- \alpha(\cos \phi + \beta \sin \phi)(C_{\theta\xi}^* N_{\xi T} + C_{\theta\theta}^* N_{\theta T} - M_{\theta T})],
$$
$$
(5.2.16)
$$
$$
L_{25}(\beta, T) = A_{\theta\xi}^*(rN_{\xi T})' + A_{\theta\theta}^*(rN_{\theta T})'
$$
$$
- \alpha(\cos \phi + \beta \sin \phi)(A_{\xi\xi}^* N_{\xi T} + A_{\xi\theta}^* N_{\theta T}), \quad (5.2.17)
$$

whereas the other functional operators are given by equations 3.1.51–3.1.58. The expressions for resultants and curvature changes, in terms of β and Ψ, are again given by equations 3.1.59–3.1.63.

5.3 Thermal stresses in composite cylindrical shells

Linear theory of composite circular cylindrical shells

In order to obtain the linear thermoelastic equations for circular cylindrical shells the geometrical relations 3.2.1 are introduced into equations 5.2.14 and 5.2.15, omitting nonlinear terms in β and Ψ, with the result

$$
D_{\xi\xi}^*\beta'' - B_{\theta\xi}^*\Psi'' + a\Psi = B_{\xi\xi}^* a^2 p_H' + B_{\theta\xi}^* aV' - a^2 m
$$
$$
- a(B_{\xi\xi}^* N_{\xi T}' + B_{\theta\xi}^* N_{\theta T}' + M_{\xi T}'), \quad (5.3.1)
$$
$$
B_{\theta\xi}^*\beta'' - a\beta + A_{\theta\theta}^*\Psi'' = -A_{\theta\theta}^* a^2 p_H' - A_{\theta\theta}^* aV'
$$
$$
+ a(A_{\theta\xi}^* N_{\xi T}' + A_{\theta\theta}^* N_{\theta T}'). \quad (5.3.2)
$$

Equations 3.2.4–3.2.7 for stress resultants become

$$V = -a \int p_V \, d\xi = N_\xi, \qquad (5.3.3)$$

$$aQ = -\Psi, \qquad (5.3.4)$$

$$aN_\theta = \Psi'' + a^2 p_H. \qquad (5.3.5)$$

The curvature changes (3.2.8) become

$$\kappa_\xi = \beta'/a, \quad \kappa_\theta = 0, \qquad (5.3.6)$$

whereas the displacements are expressed in terms of the strains by equations 3.2.9.

The two simultaneous equations 5.3.1 and 5.3.2 are transformed into the following single equations for β and Ψ:

$$\beta^{iv} - 2B_{\theta\xi}^* \frac{a}{k} \beta'' + \frac{a^2}{k} \beta = \frac{1}{k}(A_{\theta\theta}^* at_1'' + B_{\theta\xi}^* at_2'' - a^2 t_2), \qquad (5.3.7)$$

$$\Psi^{iv} - 2B_{\theta\xi}^* \frac{a}{k} \Psi'' + \frac{a^2}{k} \Psi = \frac{1}{k}(-B_{\theta\xi}^* at_1'' + a^2 t_1 + D_{\xi\xi}^* at_2''), \qquad (5.3.8)$$

where the thermal quantities t_1 and t_2 are of the form

$$t_1 = -B_{\xi\xi}^* N_{\xi T}' - B_{\theta\xi}^* N_{\theta T}' - M_{\xi T}', \qquad (5.3.9)$$

$$t_2 = A_{\theta\xi}^* N_{\xi T}' + A_{\theta\theta}^* N_{\theta T}'. \qquad (5.3.10)$$

The mechanical load is assumed to vanish and the value of k is given by equation 3.2.14. Note that primes indicate differentiation with respect to the nondimensional coordinate

$$\xi = z/a. \qquad (5.3.11)$$

It is indicated that equations 5.3.7 and 5.3.8 are nonhomogeneous, whereas the corresponding composite shell equations (3.2.15) for the isothermic case are homogeneous. Boundary conditions 3.2.17 remain valid for the thermoelastic case; however, some of them involve the thermal quantities N_{iT}, M_{iT}. Having the solution for β and Ψ, we can readily obtain the stress resultants and couples, and the stress distribution is given by equation 5.1.17.

Semi-infinite two-layer cylindrical shell

Consider a semi-infinite two-layer isotropic cylindrical shell subject to a given thermal field (see Fig. 1.36a):

$$T(\xi, \zeta) = T_0 + (T_e^0 - T_i^0)e^{-g\xi}f(\zeta), \qquad (5.3.12)$$

where T_0 denotes the temperature at the reference $\zeta = 0$ and is given by

$$T_0 = T_e^0 - (T_e^0 - T_i^0)e^{-g\xi} \frac{k_e \ln(r_e/r_0)}{k_i \ln(r_0/r_i) + k_e \ln(r_e/r_0)}, \qquad (5.3.13)$$

and the thickness variation of the temperature is determined by

$$f(\zeta) = \frac{k_j \ln(1 - \zeta/r_0)}{k_i \ln(r_0/r_i) + k_e \ln(r_e/r_0)} \qquad (j = i, e). \qquad (5.3.14)$$

The external and internal temperatures at $\xi = 0$, denoted by T_e^0, T_i^0, respectively, and the factor g are given. The coefficients of thermal conductivity of the external and internal layers are k_e, k_i, respectively.

Associated with differential equation 5.3.7 or 5.3.8 are the following boundary conditions:

$$u = 0, \beta = 0 \qquad \text{at} \quad \xi = 0, \qquad (5.3.15)$$

together with the requirement that stresses and displacements are bounded at infinity.

The complete solution of equations 5.3.7 and 5.3.8 satisfying regularity conditions at $\xi = \infty$ is given by the following formulas for β and Ψ:

$$\beta = B_1 e^{-\gamma_1 \xi} \cos \gamma_2 \xi + B_2 e^{-\gamma_1 \xi} \sin \gamma_2 \xi + \beta_p, \qquad (5.3.16)$$

$$\Psi = C_1 e^{-\gamma_1 \xi} \cos \gamma_2 \xi + C_2 e^{-\gamma_1 \xi} \sin \gamma_2 \xi + \Psi_p, \qquad (5.3.17)$$

where β_p, Ψ_p designate the particular solutions; γ_1 and γ_2 are defined by equations 3.2.21; and c and d are given by 3.2.16. The difference between β and Ψ for the thermoelastic case (5.3.16 and 5.3.17) and for the isothermic case (3.2.19 and 3.2.20) lies in the addition of the particular solutions. For detailed determination of the basic variables β and Ψ, subject to conditions 5.3.15, and of stress resultants, stress couples, and stress distribution, the reader may wish to consult a report by Stavsky and Smolash.[60]

Numerical Examples: To get some insight into the behavior of two-layer cylindrical shells subject to a thermal field as specified by equations 5.3.12 and 5.3.13 we consider the following numerical values:

$$T_e^0 = 200°C, T_i^0 = 100°C, g = 150, \qquad (5.3.18)$$

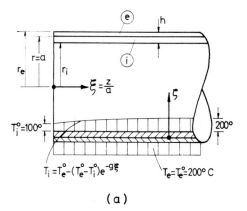

(a)

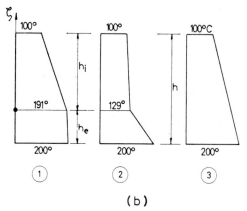

(b)

Figure 1.36   Semi-infinite Two-Layer Cylindrical Shell with Temperature Variation

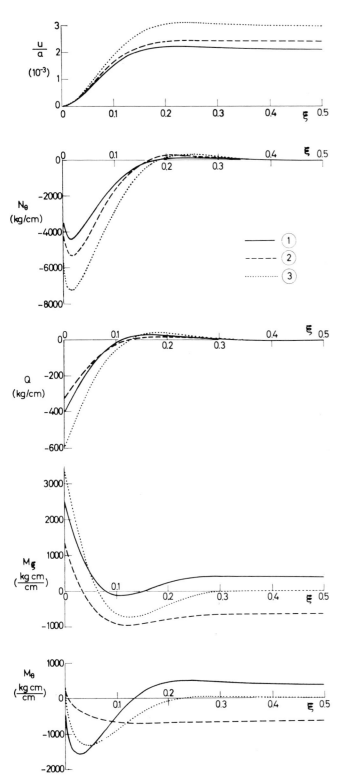

Figure 1.37 Distribution of Radial Displacement, Resultants, and Couples Along Shell Generator

$$r_0 = a = 150 \text{ cm}, \ r_e - r_0 = h_e = 0.5 \text{ cm}, \ r_0 - r_i = h_i = 1.0 \text{ cm}. \tag{5.3.19}$$

The thermoelastic properties of the composite isotropic shell were assumed for two cases:

Case	E_e/E_0	E_i/E_0	ν_e	ν_i	α_e/α_0	α_i/α_0	k_i/k_e
1	0.5	2.5	0.1	0.4	2	1	5
2	2.5	0.5	0.4	0.1	1	2	0.2

$$\tag{5.3.20}$$

with

$$E_0 = 10^6 \text{ kg/cm}^2, \ \alpha_0 = 10^{-5}/°\text{C}. \tag{5.3.21}$$

For comparison we take as Case 3 an isotropic homogeneous shell with the same total thickness as the composite shells, $h = 1.5$ cm, subject to the same external and internal temperatures (5.3.18), and with

$$E/E_0 = 1.7, \ \nu = 0.33, \ \alpha/\alpha_0 = 1.5. \tag{5.3.22}$$

The detailed temperature distribution across the shell thickness at the edge $\xi = 0$ for these three cases is shown in Figure 1.36(b). Figure 1.37 shows that the results for Cases 1 and 2 are considerably different quantitatively and qualitatively. For example, the axial moment in Case 1 is almost twice the corresponding value in Case 2. Furthermore, M_ξ approaches a positive asymptotic value for Case 1, whereas for Case 2 it approaches a negative value. Comparison with a homogeneous shell (Case 3) shows that the forces and couples that arise due to the thermal field are quite sensitive to a change in the values of the thermoelastic parameters.

Figure 1.38 shows that the stress distribution in the axial direction is considerably controlled by the heterogeneity of the shell. It is noted that the highest stress appears in the axial direction for Cases 1 and 3 and in the circumferential direction for Case 2.

6 BUCKLING OF COMPOSITE COLUMNS, PLATES, AND SHELLS

6.1 Buckling of composite columns

Figure 1.39 shows a symmetrically laminated column, supported on knife edges and compressed in a testing machine. The load P is applied along the axis of the column which is assumed to be perfectly straight. It has been known since the investigations of Euler,[61] Engesser,[62] and von Kármán[63] that under such conditions a perfect column remains straight until a critical value P_{cr} of the load is reached. When the critical value is exceeded by the slightest amount, the column becomes unstable and buckles as shown in Figure 1.39(b).

If the compressive stress in every lamina is less than the proportional limit of the material when the column is under the action of P_{cr}, the following formula can be used for the calculation of the critical load:

$$P_{cr} = \pi^2 D/L'^2, \tag{6.1.1}$$

where D is the bending rigidity of the composite isotropic column as computed from expression 4.3.3 and L' is the so-called reduced length of the column which is equal to the distance between two consecutive inflection points of the

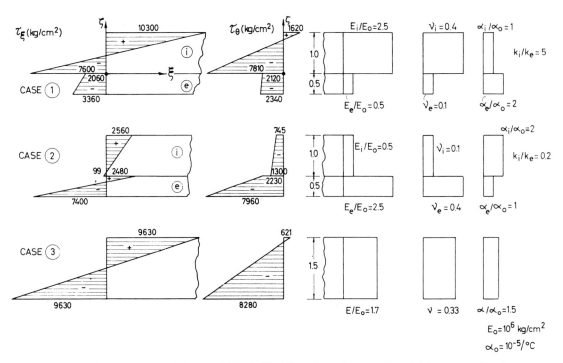

Figure 1.38   Thickness Variation of Elastic Moduli, Axial and Circumferential Stresses at $\xi = 0$

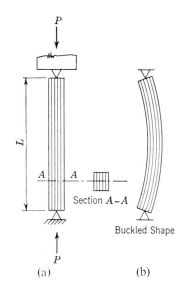

Figure 1.39   Laminated Column

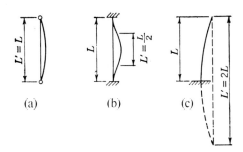

Figure 1.40   Effective Length of a Column

deflected shape of the axis of the column. Figure 1.40 illustrates that $L' = L$ when the ends of the column are simply supported, $L' = L/2$ when they are rigidly clamped, and $L' = 2L$ when one end is clamped and the other free.

If the proportional limit is exceeded in a lamina when the column is under the action of the critical load P_{cr}, Young's modulus ceases to be the factor of proportionality between stress and strain for that lamina. When the column buckles, the various laminas are stretched or compressed, in addition to being bent, depending on their location in the composite. The increases in the compressive stress are proportional to the instantaneous value of the modulus, the so-called *tangent modulus* E_t, while the decrements are proportional to the original modulus E because only the elastic portion of the deformation can be recovered. Consequently, the resistance of each lamina to bending at the moment of buckling is proportional to some average modulus, the value of which lies between E and E_t and depends on the location of the lamina. The buckling load can then be calculated from the formula

$$P_{cr} = \pi^2 D_r / L'^2, \qquad (6.1.2)$$

where D_r is the reduced bending rigidity which can be obtained from expression 4.3.3 if in the summation E^m is replaced by the appropriate value E_r^m of the reduced modulus for each layer. The values of the reduced moduli can be determined on the basis of considerations as presented, for instance, by Timoshenko and Gere (Ref. 64, Sec. 3.3), but the algebraic transformations become quite lengthy when there are many layers of different materials.

When the shearing rigidity of the laminate is small, the buckling load is reduced below the value predicted by equations 6.1.1 or 6.1.2. In such a case, part of the lateral deflections shown in Figure 1.39(b) is caused by bending and the rest by shear. The buckling load P_s could be calculated from a modified Engesser formula (see Ref. 64, Sec. 2.17) when relation 4.3.6 is considered. When the shearing rigidity of the central layers of the laminate is considerably smaller than that of the external layers, as is generally true for sandwich construction, the buckling load should be calculated as suggested in Hoff's text (Ref. 8, Art. 3.3).

6.2 Buckling of symmetrically composed plates

When a plate is placed in a testing machine and compressed in such a way that more than the two opposite loaded edges are supported, the buckled shape is not cylindrical but has double curvature. In Figure 1.41(b), (c) the two unloaded

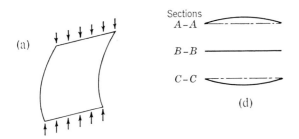

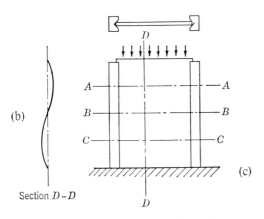

Figure 1.41   The Buckling of Plates

edges are in V grooves and, consequently, must remain straight when the compressive load is applied. The two loaded edges remain straight because of the friction between plate and testing machine. In the case illustrated, the buckled shape comprises two bulges which may be seen from the horizontal sections drawn in Figure 1.41(d) and the vertical section D-D indicated in Figure 1.41(b).

The buckling problem of symmetrically composed plates can be represented by a partial differential equation of the form

$$D_{xx}\frac{\partial^4 w}{\partial x^4} + 2(D_{xy}+2D_{ss})\frac{\partial^4 w}{\partial x^2 \partial y^2} + D_{yy}\frac{\partial^4 w}{\partial y^4}$$
$$= N_x\frac{\partial^2 w}{\partial x^2} + 2N_{xy}\frac{\partial^2 w}{\partial x\, \partial y} + N_y\frac{\partial^2 w}{\partial y^2}, \quad (6.2.1)$$

where the D's are defined by 2.4.14. Note that N_x and N_y are negative when they represent compression. Equation 6.2.1 has been solved for a few combinations of loading and boundary conditions. The solution for the case of buckling of simply supported rectangular plates uniformly compressed in one

direction is given in Timoshenko and Gere (Ref. 64, p. 404). Smith[65] calculated the critical stress of compressed orthotropic plates under the following boundary conditions:

Case 1. Loaded edges clamped, unloaded edges simply supported.
Case 2. Loaded edges simply supported, unloaded edges clamped.
Case 3. All edges rigidly clamped.

The results for all these cases can be represented in the form

$$\sigma_{cr} = k\frac{\pi^2}{hb^2}\sqrt{D_{xx}D_{yy}}, \quad (6.2.2)$$

where h is the total thickness of the plate, b the length of the loaded edge, and k a constant plotted in Figures 1.42 and 1.43. The figures are reproduced from a publication by Dale and Smith.[66] The factor k is plotted against the quantity

$$\beta = (a/b)\sqrt[4]{D_{yy}/D_{xx}}, \quad (6.2.3)$$

where a is the length of the unloaded edge. A family of curves is presented for each set of edge conditions. The parameter α of the family is

$$\alpha = \frac{D_{xy} + 2D_{ss}}{\sqrt{D_{xx}D_{yy}}}. \quad (6.2.4)$$

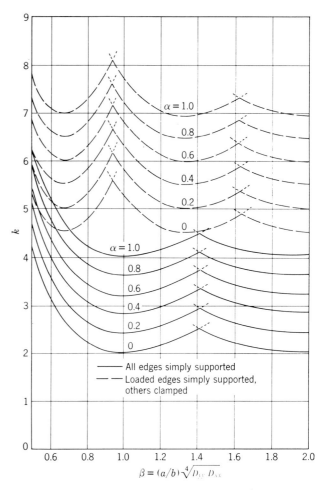

Figure 1.42   Coefficients of Plate Buckling in Compression

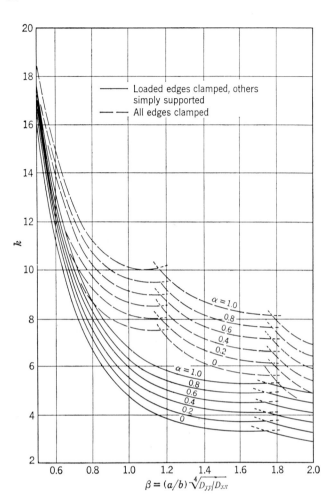

Figure 1.43   Coefficients of Plate Buckling in Compression

The buckling of orthotropic plates under shear loads was investigated by Bergmann and H. Reissner.[67] Equation 6.2.1 was integrated under the following conditions: N_x and N_y are equal to zero, D_{xx} is small or zero, and the plate is infinitely long in the x direction. At H. Reissner's suggestion, this work was continued by Seydel[68-70] who found approximate solutions for both infinite and finite rectangular plates for the case when all D's have arbitrary values. The results of these investigations related to infinite plates were presented by Bergmann and H. Reissner[67] in the following form:

$$\tau_{cr} = \frac{\sqrt{2D_{yy}(D_{xy}+2D_{ss})}}{(b/2)^2 h}\left[8.3 + 1.525\frac{D_{xx}D_{yy}}{(D_{xy}+2D_{ss})^2}\right.$$
$$\left. - 0.493\frac{D_{xx}^2 D_{yy}^2}{(D_{xy}+2D_{ss})^4}\right], \quad (6.2.5)$$

when

$$D_{xx}D_{yy} < (D_{xy}+2D_{ss})^2 \quad (6.2.6)$$

and

$$\tau_{cr} = \frac{\sqrt[4]{D_{xx}D_{yy}^3}}{(b/2)^2 h}\left[8.125 + 5.64\sqrt{\frac{(D_{xy}+2D_{ss})^2}{D_{xx}D_{yy}}}\right.$$
$$\left. - 0.6\frac{(D_{xy}+2D_{ss})^2}{D_{xx}D_{yy}}\right], \quad (6.2.7)$$

when

$$D_{xx}D_{yy} > (D_{xy}+2D_{ss})^2. \quad (6.2.8)$$

The formulas suggested for finite rectangular plates by Seydel[70] are

$$\tau_{cr} = C\,\frac{\sqrt[4]{D_{xx}D_{yy}^3}}{(b/2)^2 h}, \quad (6.2.9)$$

provided the plate parameter θ is not smaller than unity:

$$\theta = \frac{\sqrt{D_{xx}D_{yy}}}{D_{xy}+2D_{ss}} \geq 1. \quad (6.2.10)$$

The value of the numerical factor C must be taken from Figure 1.44. Only the curves corresponding to $\theta = 1$, 2 and infinity, and the points on the C axis for $\theta = 1.41$, 3, 5, and 10 were given by Seydel.[70] The curves corresponding to the latter four values of θ were drawn in by eyesight in agreement with Seydel's proposal.

In Figure 1.44, C is plotted against the effective side ratio $1/\beta$ defined as

$$1/\beta = \frac{b}{a}\,\sqrt[4]{D_{xx}/D_{yy}}. \quad (6.2.11)$$

When β^{-1} is found to be greater than unity, a and b as well as the subscripts x and y should be interchanged in equations 6.2.9–6.2.11 in order to permit the use of Figure 1.44.

An extended theoretical and experimental investigation was carried on at the Forest Products Laboratory in Madison, Wisconsin, for establishing the buckling stresses in compression, shear, or combinations of compression and shear of flat rectangular plywood panels under various edge conditions. The results of the theoretical work, carried out by the energy method, were published in a series of reports of which we mention, for example, the work of March.[71] Many graphs from these reports are reproduced in ANC-18.[6]

The stability of aeolotropic composite plates requires further studies. The same holds for orthotropic composite plates with unsymmetrical arrangement of layers.

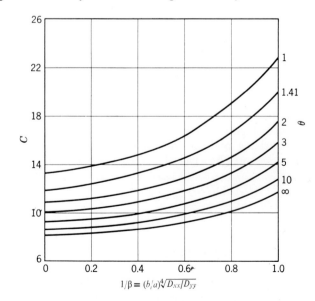

Figure 1.44   Coefficients of Plate Buckling in Shear

6.3 Buckling of composite cylindrical shells

The buckling problem of thin elastic shells seems to be one of the most challenging problems of structural mechanics. The engineering community has shown great interest over the last century in the investigation of stability of shells. Numerous texts on buckling of thin structures were published of which we mention the one by Timoshenko and Gere[64] and the one by Vol'mir.[72] Several review articles appeared recently by Fung and Sechler,[73] Hoff, [74,75] Nash,[76] Flügge,[77] and Langhaar[78] making reference to hundreds of papers, most of which treat the buckling of homogeneous isotropic shells.

A linear theory of buckling of homogeneous orthotropic shells was derived by Flügge[79] who stated the buckling condition as the requirement that a determinant vanishes. Some further studies by Russian authors are reported by Ambartsumyan.[44] An analysis of the postbuckling behavior of long thin-walled plywood cylinders was carried out by March, [80] and more recently by Thielemann[81] and Thielemann and Esslinger,[82] along lines similar to those suggested by von Kármán and Tsien.[83] The orthotropic shell theory was applied to the analysis of reinforced shells when both the shell and stiffeners buckle simultaneously (see Hoff's 1967 review, Ref. 75).

The stability of composite shell structures is still an open field for future research. As indicated by Ambartsumyan[44] in his survey on anisotropic layered shells, "... at present we have not any investigation of static stability of any shell in the general case of anisotropy."

It is worth noting the pioneering work of Grigolyuk on stability of bimetallic cylindrical and conical shells as published in a series of papers (Refs. 41, 50, 84–88).

Some studies on the stability of heterogeneous aeolotropic cylindrical shells appeared in 1963 by Cheng and Ho;[89,90] a simplified analysis was given by Tasi[91] for the case of orthotropy. An attempt to get an insight into the effect of heterogeneity on the stability of composite orthotropic cylindrical shells was undertaken by Stavsky and Friedland.[92]

7 ELEMENTS OF THE MECHANICS OF SANDWICH STRUCTURES

7.1 Bending of sandwich beams

Characteristics of sandwich construction

Sandwich construction is characterized by the use of two thin outer layers of strong material, denoted as faces, between which a thick layer of very light weight and comparatively weak core is sandwiched. The obvious advantage of this construction is the large moment of inertia of the cross section achieved by spacing far apart the main carrying elements, namely, the faces. The weight of the structure is small because of the low density of the core. Typical materials used for faces are plywood, papreg, high-pressure laminates, aluminum alloy, stainless steel, glass-fiber textile, reinforced plastic, titanium, and heat-resistant steel; for the core, expanded synthetic plastics, balsa wood, and built-up grids are among the many materials available. Faces and core are cemented together by means of synthetic glues.

During World War II sandwich structural elements attained some importance in aeronautical engineering where rigidity and light weight are the primary requirements. The suitability of sandwich-type elements for major structural components of aircraft was generally acknowledged after the structural details of the Mosquito fighter bomber were made public.

Analysis of sandwich beams

Early investigations disclosed that the bending rigidity and the buckling load of sandwich structural elements are considerably smaller than may be anticipated on the basis of the conventional formulas of strength of materials. The reason is the small shearing rigidity of the relatively thick core. Consequently the theory of sandwich structures must take into account transverse shear deformations which are usually neglected in homogeneous systems. The theory of sandwich construction has a very extensive literature which has been summarized recently in a book by Plantema.[93] In this section we give the results of the analysis of the deflection of a cantilever sandwich beam (Fig. 1.45) as given by Hoff and

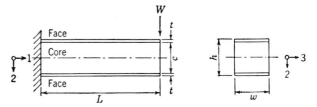

Figure 1.45 Sandwich Beam

Mautner.[94] The differential equation of the problem was obtained by the variational method from the essential portions of the strain energy stored in the beam during deformations, namely, the strain energy of extension and bending of the faces and the strain energy of shear of the core. The end deflection d of the cantilever is expressed in the form (for the symbols see Fig. 1.45)

$$d = \psi W L^3 / 3EI, \tag{7.1.1}$$

where

$$\psi = 1 + CEI_1 / 2EI_f \tag{7.1.2}$$

and

$$C = \frac{3}{(pL)^2}\left(1 - \frac{\tanh pL}{pL}\right). \tag{7.1.3}$$

Moreover,

$$p^2 = Gcw / EI^*, \tag{7.1.4}$$

$$EI^* = EI_1(2EI_f) / (EI_1 + 2EI_f). \tag{7.1.5}$$

The symbols expressing rigidities have the following meaning:

$$EI_1 = \tfrac{1}{2}tw(c+t)^2 E_f \tag{7.1.6}$$

is the bending rigidity of the two faces when each face is thought of as being concentrated along its own horizontal centroidal axis;

$$2EI_f = \tfrac{1}{6}t^3 w E_f \tag{7.1.7}$$

is the bending rigidity of the two faces when they are thought of as not being connected to each other by the core;

$$EI = EI_1 + 2EI_f, \tag{7.1.8}$$

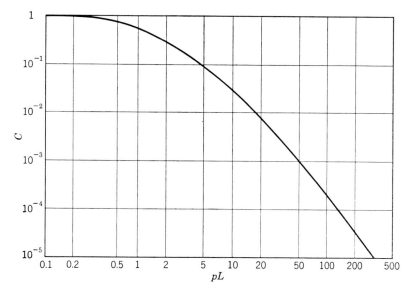

Figure 1.46   Factor C in Formula for End Deflection of Sandwich Beam

is the total bending rigidity of the sandwich beam if the bending rigidity of the core is neglected. The value of C is plotted against the parameter pL in Figure 1.46.

When $\psi = 1$, the deflections of the sandwich beam are in agreement with the values given by the ordinary bending formula. Hence the second term in the expression for ψ represents the shearing deformations. The importance of the latter may be judged from a numerical example shown in Hoff's text (Ref. 8, p. 192):

"When $L = 10$ in., $c = 1$ in., $t = 0.032$ in., $w = 1$ in., $E = 10.5 \times 10^6$ psi, and $G = 4000$ psi, one obtains

$$EI_1 = 178,500 \text{ lb-in}^2., \qquad 2EI_f = 57 \text{ lb-in}^2.,$$
$$Gcw = 4000 \text{ lb}, \qquad p^2 = 70 \text{ in.}^{-2}$$
$$pL = 84, \qquad C = 4.2 \times 10^{-4},$$

Consequently
$$\psi = 1 + 1.32 = 2.32,$$

and the actual end deflection is 132 per cent higher than that corresponding to a core with infinite shearing rigidity."

Equation 7.1.3 can be further simplified for certain values of the nondimensional parameter pL:

1. *When $pl < 0.1$.* The hyperbolic tangent can be replaced by the first two terms of its Taylor expansion so one obtains

$$C = 1, \quad \psi = EI/2EI_f, \tag{7.1.9}$$
and
$$d = WL^3/3(2EI_f). \tag{7.1.10}$$

Consequently, when the shearing rigidity of the core is very small, the sandwich beam deflects like a beam consisting only of the two faces.

2. *When $pL > 100$.* In this case $\tanh pL$ is almost equal to unity, and so one finds

$$C = 3/(pL)^2, \tag{7.1.11}$$
and
$$d = \frac{WL^3}{3EI} + \frac{WL}{Gbw}\left(\frac{EI_1}{EI}\right)^2. \tag{7.1.12}$$

Since in most cases EI_1 differs but little from EI, the total deflection of a sandwich beam with $pL > 100$ can be calculated as the sum of the bending deflection and the shearing deflection.

3. *When $pL \to \infty$.* When pL is very large,

$$d = WL^3/3EI. \tag{7.1.13}$$

The deflection is the simple bending deflection of the Euler-Bernoulli theory when the core is infinitely rigid in shear.

The normal and shear stresses in sandwich beams can be computed from the equations developed in Section 4.3.

7.2   Bending and stretching of sandwich-type plates and shells

Sandwich-type plates and shells have been extensively studied since the beginning of the Second World War. Out of the wealth of information accumulated so far we only mention the monograph of Plantema[93] summarizing the present state of knowledge in the fields of bending and buckling of sandwich plates and shells. This book contains quite a large list of references which will be of great help for further research.

The questions of optimum design, thermal effects, vibrations and related stability problems for sandwich-type systems deserve further exploration.

What characterizes the analysis of sandwich thin structures is the consideration of transverse shear deformation and — to some extent — the effect of normal stress deformation. On the other hand, for multilayer structures coupling of bending stretching is the governing factor in analysis. As sandwiches need not be symmetrically layered, and as composites might include some weak sheets it seems worth while to look into the problem of establishing a unified theory for both main types of heterogeneous systems.

REFERENCES

1. A. G. H. DIETZ, "Composite Materials," 1965 Edgar Marburg Lecture, American Society for Testing and Materials, Philadelphia, Pa., 1965.

2. O. C. MOHR, "Über die Darstellung des Spannungszustandes und des Deformationszustandes eines Körperelementes und über die Anwendung derselben in der Festigkeitslehre," *Civilingenieur* **28**, cols. 113–156 (1882).

3. S. TIMOSHENKO, *Strength of Materials*, Second Edition, D. Van Nostrand Company, Inc., New York, Part I, 1940; Part 2, 1941.

4. C. B. NORRIS, "The Application of Mohr's Stress and Strain Circles to Wood and Plywood," U.S. Dept. Agr. Forest Serv., Forest Prod. Lab. Rept. 1317, Forest Products Laboratory, Madison, Wis., 1943.

5. N. J. HOFF, "A Graphic Resolution of Strain," *J. Appl. Mech.* **12**, A211–A216 (1945).

6. Army-Navy-Civil Committee on Aircraft Design Criteria, *Design of Wood Aircraft Structures*, Bull. ANC-18, U.S. Government Printing Office, Washington, D.C., 1944.

7. S. TIMOSHENKO and J. N. GOODIER, *Theory of Elasticity*, Second Edition, McGraw-Hill Book Company, Inc., New York, 1951.

8. N. J. HOFF, *The Analysis of Structures*, John Wiley & Sons, Inc., New York, 1956.

9. A. E. H. LOVE, *A Treatise on the Mathematical Theory of Elasticity*, Fourth Edition, Dover Publications, Inc., New York, 1944. (American printing of Cambridge University Press 1927 Edition.)

10. R. ADLER, M. BAZIN, and M. SHIFFER, *Introduction to General Relativity*, McGraw-Hill Book Company, Inc., New York, 1965.

11. R. D. MINDLIN, and H. F. TIERSTEN, "Effects of Couple-stresses in Linear Elasticity," *Arch. Ratl. Mech. Anal.* **11**, 415–448 (1962).

12. R. D. MINDLIN, "Influence of Couple-stresses on Stress Concentrations," *Exptl. Mech.* **3**, 1–7 (1963).

13. E. REISSNER, "Note on the Theorem of the Symmetry of the Stress Tensor," *J. Math. Phys.* **23**, 192–194 (1944).

14. I. S. SOKOLNIKOFF, *Mathematical Theory of Elasticity*, Second Edition, McGraw-Hill Book Company, Inc., New York, 1956.

15. A. E. GREEN, and W. ZERNA, *Theoretical Elasticity*, Oxford University Press, London, Corrected First Edition, 1960.

16. J. F. NYE, *Physical Properties of Crystals*, Corrected First Edition, Oxford University Press, London, 1960.

17. R. F. S. HEARMON, *An Introduction to Applied Anisotropic Elasticity*, Oxford University Press, New York, 1961.

18. I. S. SOKOLNIKOFF, *Tensor Analysis*, John Wiley & Sons, Inc., New York, 1960.

19. G. R. KIRCHHOFF, *Vorlesungen über Mechanik*, Vierte Auflage, Druck und Verlag von B. G. Teubner, Leipzig, 1897.

20. W. L. FERRAR, *Algebra*, Oxford University Press, London, 1941.

21. S. G. LEKHNITSKII, *Theory of Elasticity of an Anisotropic Elastic Body*, Holden-Day, Inc., San Francisco, 1963. (Translation of Russian 1950 Edition.)

22. N. G. CHENTSOV, "Investigations on Plywood as an Orthotropic Plate," Tekhnicheskie Zametki Tsentr. Aero-Gidr. Inst. No. 91, 1936 (Technical Notes, Central Aero and Hydrodynamic Institute).

23. S. TIMOSHENKO and S. WOINOWSKY-KRIEGER, *Theory of Plates and Shells*, Second Edition, McGraw-Hill Book Company, Inc., New York, 1959.

24. E. REISSNER and Y. STAVSKY, "Bending and Stretching of Certain Types of Heterogeneous Aeolotropic Elastic Plates," *J. Appl. Mech* **28**, 402–408 (1961).

25. S. W. TSAI, "Structural Behavior of Composite Materials," NASA CR-71, National Aeronautics and Space Administration, Washington, D.C., July 1965.

26. V. D. AZZI and S. W. TSAI, "Elastic Moduli of Laminated Anisotropic Composites," *Exptl. Mech.* **5**, 177–185 (1965).

27. C. B. SMITH, "Some New Types of Orthotropic Plates Laminated of Orthotropic Material," *J. Appl. Mech.* **20**, 286–288 (1953).

28. S. G. LEKHNITSKII, *Anisotropic Plates*, Second Edition, Gosud. Izdat. Tekh.-Teor. Lit., Moskva, 1957 (in Russian).

29. Y. STAVSKY, "Finite Deformations of a Class of Aeolotropic Plates with Material Heterogeneity," *Israel J. Technol.* **1**, 69–74 (1963).

30. Y. STAVSKY, "On a Cross-Elasticity Phenomenon in Symmetrically Non-Homogeneous Plates," *J. Aerospace Sci.* **26**, 607 (1959).

31. Y. STAVSKY, "On the Theory of Heterogeneous Anisotropic Plates," Doctor's Thesis, M.I.T., Cambridge, Mass., 1959.

32. Y. STAVSKY and J. R. ROY, "Some Equilibrium and Stability Solutions for Laminated Aeolotropic Plates," M.I.T. Report, ARDC Cont. No. AF33(616)-6280, May 15, 1960. Also in *Investigation of Mechanics of Reinforced Plastics*, WADD-TR-60-746, Part II, 67–116, 184–204, Mar. 1962.

33. Y. STAVSKY and J. R. ROY, "Some Numerical Results for Symmetrical, Two-layer Aeolotropic Plates, Antisymmetric for C_{16}, C_{26}," M.I.T. Report, ARDC Cont. No. AF33(616)-6280, Mar. 15, 1960.

34. Y. STAVSKY, "Bending and Stretching of Laminated Aeolotropic Plates," *Proc. Am. Soc. Civil Engrs: (J. Eng. Mech. Div.)* **87**, EM6, 31–56, 1961. Also *Trans. Am. Soc. Civil Engrs.* **127**, Part I, 1194–1219 (1962).

35. S. G. LEKHNITSKII, "Bending of Non-Homogeneous Anisotropic Thin Plates of Symmetrical Construction," *Prikl. Mat. Mekh.* **5**, 71–91 (1941).

36. Y. STAVSKY, "Pure Bending, Twisting and Stretching of Skewed Heterogeneous Plates," *A.I.A.A.J.* **1**, 221–222 (1963).

37. K. S. PISTER and S. B. DONG, "Elastic Bending of Layered Plates," *Proc. Am. Soc. Civil Engrs. (J. Eng. Mech. Div.)* **85**, EM4, 1–10 (1959).

38. Y. STAVSKY, "On the Theory of Symmetrically Heterogeneous Plates Having the Same Thickness Variation of the Elastic Moduli," *Topics in Applied Mechanics*, Memorial Volume to the late Professor Edwin Schwerin, D. Abir, F. Ollendorff and M. Reiner, Editors, Elsevier Publishing Company, Amsterdam, 1965, pp. 105–116.

39. E. REISSNER, "Symmetric Bending of Shallow Shells of Revolution," *J. Math. Mech.* **7**, 121–140 (1958).

40. Y. STAVSKY, "Non-linear Theory for Axisymmetric Deformations of Heterogeneous Shells of Revolution," *Contributions to Mechanics*, Markus Reiner 80th Anniversary Volume, D. Abir, Editor, Pergamon Press, London, 1969, pp. 181–194.

41. E. I. GRIGOLYUK, "Thin Bimetallic Plates and Shells," *Inzhen. Sbornik* **17**, 69–120 (1953).

42. Y. STAVSKY, "On the General Theory of Heterogeneous Aeolotropic Plates," *Aeron. Quart.* **15**, 29–38 (1964).

43. A. G. H. DIETZ et al., "Micromechanics of Fibrous Composites," Report MAB-207-M, Materials Advisory Board, Ad Hoc Committee on Micro-Mechanics of Fibrous Composites, National Academy of Sciences–National Research Council, Washington, D.C., May 1965.

44. S. A. AMBARTSUMYAN, "Contributions to the Theory of Anisotropic Layered Shells," *Appl. Mech. Rev.* **15**, 245–249 (1962). A revised version is entitled "Some Current Aspects of the Theory of Anisotropic Layered Shells," *Applied Mechanics Surveys*, H. N. Abramson, H. Liebowitz, J. M. Crowley, and S. Juhasz, Editors, Spartan Books, Inc., Washington, D.C., 1966, pp. 301–314.

45. S. A. AMBARTSUMYAN, *Theory of Anisotropic Shells*, NASA Technical Translation F-118, Washington, D.C., 1964. (Translation of Russian 1961 edition.)

46. E. REISSNER, "On the Theory of Thin Elastic Shells," *Contributions to Applied Mechanics*, H. Reissner Anniversary Volume, J. W. Edwards, Ann Arbor, Mich., 1949, pp. 231–247.

47. E. REISSNER, "On Axisymmetrical Deformations of Thin Shells of Revolution," *Proc. Symp. Appl. Math.* **3**, 27–52 (1950).

48. E. REISSNER, "Rotationally Symmetric Problems in· the Theory of Thin Elastic Shells," *Proceedings of the Third U.S. National Congress of Applied Mechanics*, 1958, pp. 51–69.

49. E. REISSNER, "On the Equations for Finite Symmetrical Deflections of Thin Shells of Revolution," *Progress in Applied Mechanics*, The Prager Anniversary Volume, D. C. Drucker, Editor, Macmillan, London, 1963, pp. 171–178.

50. E. I. GRIGOLYUK, "Equations of Axisymmetric Bimetallic Elastic Shells," *Inzhen. Sbornik* **18**, 89–98 (1954).

51. N. I. MUSKHELISHVILI, *Some Basic Problems of the Mathematical Theory of Elasticity*, P. Noordhoff Ltd., Groningen, Holland, 1953. (Translation of Russian 1949 third edition.)

52. E. MELAN and H. PARKUS, *Wärmespannungen infolge stationärer Temperaturfelder*, Springer-Verlag, Wien, 1953.

53. H. PARKUS, *Instationäre Wärmespannungen*, Springer-Verlag, Wien, 1959.

54. W. NOWACKI, *Thermoelasticity*, Pergamon Press, Oxford, 1962.

55. Y. STAVSKY, "Thermoelasticity of Heterogeneous Aeolotropic Plates," *Proc. Am. Soc. Civil Engrs.* (*J. Eng. Mech. Div.*) **89**, EM2, 89–105 (1963).

56. P. J. SCHNEIDER, *Conduction Heat Transfer*, Addison-Wesley Publishing Co., Inc., Reading, Mass., 1955.

57. B. GEBHART, *Heat Transfer*, McGraw-Hill Book Company, Inc., New York, 1961.

58. A. B. BOLEY and J. H. WEINER, *Theory of Thermal Stresses*, John Wiley & Sons, Inc., New York, 1960.

59. Y. STAVSKY, "Cross-Thermoelastic Phenomenon in Heterogeneous Aeolotropic Plates," *A.I.A.A. J.* **1**, 960–962 (1963).

60. Y. STAVSKY and I. SMOLASH, "Thermoelasticity of Heterogeneous Orthotropic Cylindrical Shells," TDM Report No. 67-7, Technion–Israel Institute of Technology, Haifa, Israel, 1967.

61. L. EULER, *Methodus Inveniendi Lineas Curvas Maximi Minimive Proprietate Gaudentes*, Lausanne, 1744. *Additamentum: De Curvis Elasticis.*

62. F. ENGESSER, "Über Knickfragen," *Schweiz. Bauztg.* **26**, 24 (July 1895).

63. TH. VON KÁRMÁN, "Untersuchungen über Knickfestigkeit," *Forschungsarbeiten, VDI* (Berlin), No. 81, 1910.

64. S. P. TIMOSHENKO and J. M. GERE, *Theory of Elastic Stability*, Second Edition, McGraw-Hill Book Company, Inc., New York, 1961.

65. R. C. T. SMITH, "The Buckling of Flat Plywood Plates in Compression," Australian Council Aeronaut. Rept. ACA-12, Melbourne, Dec. 1944.

66. F. A. DALE and R. C. T. SMITH, "Grid Sandwich Panels in Compression," Australian Council Aeronaut. Rept. ACA-16, Melbourne, Apr. 1945.

67. S. BERGMANN and H. REISSNER, "Neuere Probleme aus der Flugzeugstatik über die Knickung von Wellblechstreifen bei Schubbeanspruchung," *Z. Flugtech. u. Motorluftsch.* **20**, No. 18, 475 (1929); *ibid.* **21**, No. 12, 306 (1930).

68. E. SEYDEL, *Schubknickversuche mit Wellblechtafeln*, Jahrb. DVL, R. Oldenburg, Berlin, 1931, p. 233.

69. E. SEYDEL, "Ausbeul-Schublast rechteckiger Platten-Zahlenbeispiele und Versuchsergebnisse," *Z. Flugtech. u. Motorluftsch.* **24**, No. 3, 78 (1933). English translation as "The Critical Shear Load of Rectangular Plates," NACA Tech. Mem. 705, National Advisory Committee for Aeronautics, Washington, D.C., 1933.

70. E. SEYDEL, "Über das Ausbeulen von rechteckigen, isotropen, oder orthogonalanisotropen Platten bei Schubbeanspruchung," *Ingr.-Arch.* **4**, 169 (1933).

71. H. W. MARCH, "Buckling of Flat Plywood Plates in Compression, Shear or Combined Compression and Shear," U.S. Dept. Agr. Forest Serv., Forest Prod. Lab. Rept. 1316, Forest Products Laboratory, Madison, Wis., Apr. 1942.

72. A. S. VOL'MIR, *Stability of Elastic Systems*, Gosudarstvennoye Izdatel'stvo Fiz.-Mat. Literaturyi, Moskva, 1963.

73. Y. C. FUNG and E. E. SECHLER, "Instability of Thin Elastic Shells," *Structural Mechanics, Proceedings of the First Symposium on Naval Structural Mechanics*, J. N. Goodier and N. J. Hoff, Editors, Pergamon Press, Oxford, 1960, pp. 115–168.

74. N. J. HOFF, "The Perplexing Behavior of Thin Circular Cylindrical Shells in Axial Compression," Second Theodore von Kármán Memorial Lecture, Eighth Israel Ann. Conf. Aviation and Astronautics, February 1966, *Israel J. Technol.* **4**, 1–28 (1966).

75. N. J. HOFF, "Thin Shells in Aerospace Structures," Fourth von Kármán Lecture, *Astronautics & Aeronautics* **5**, 26–45 (Feb. 1967).

76. W. A. NASH, "Instability of Thin Shells," *Applied Mechanics Surveys*, H. N. Abramson, H. Liebowitz, J. M. Crowley, and S. Juhasz, Editors, Spartan Books, Inc., Washington, D.C., 1966, pp. 339–348.

77. W. FLÜGGE, "Recent Trends in the Theories of Plates and Shells," *Applied Mechanics Surveys*, H. N. Abramson, H. Liebowitz, J. M. Crowley and S. Juhasz, Editors, Spartan Books Inc., Washington, D.C., 1966, pp. 291–294.

78. H. L. LANGHAAR, "Theory of Buckling," *Applied Mechanics Surveys*, H. N. Abramson, H. Liebowitz, J. M. Crowley, and S. Juhasz, Editors, Spartan Books, Inc., Washington, D.C., 1966, pp. 317–324.

79. W. FLÜGGE, "Die Stabilität der Kreiszylinderschale," *Ingr.-Arch.* **3**, 463–506 (1932).

80. H. W. MARCH, "Buckling of Long, Thin Plywood Cylinders in Axial Compression," U.S. Dept. Agr. Forest Serv., Forest Prod. Lab. Rept. 1322-A, Forest Products Laboratory, Madison, Wis., Sept. 1943.

81. W. F. THIELEMANN, "New Developments in the Nonlinear Theories of the Buckling of Thin Cylindrical Shells," *Aeronautics and Astronautics*, Proceedings of the Durand Centennial Conference, N. J. Hoff and W. G. Vincenti, Editors, Pergamon Press, Oxford, 1960, pp. 76–119.

82. W. F. THIELEMANN and M. E. ESSLINGER, *Buckling of Thin Elastic Cylindrical Shells*, Technische Hochschule, Braunschweig, 1965.

83. TH. VON KÁRMÁN and H. S. TSIEN, "The Buckling of Thin Cylindrical Shells under Axial Compression," *J. Aeron. Sci.* **8**, 303–312 (1941).

84. E. I. GRIGOLYUK, "On the Strength and Stability of Cylindrical Bimetallic Shells," *Inzhen. Sbornik* **16**, 119–148 (1953).

85. E. I. GRIGOLYUK, "Stability of a Closed Double-Layer Conical Shell Under Action of Uniform Normal Pressure," *Inzhen. Sbornik*, **19**, 73–82 (1954).

86. E. I. GRIGOLYUK, "On the Instability with Large Deflections of a Closed Laminated Conical Shell Subject to Uniform Normal Surface Pressure," *Inzhen. Sbornik* **22**, 111–119 (1955).

87. E. I. GRIGOLYUK, "Elastic Stability of Orthotropic and Laminated Conical and Cylindrical Shells," *Sb. Statei "Raschet Prostranstvennykh Konstruktsii"* **3**, 375–420 (1955).

88. E. I. GRIGOLYUK, "On the Stability Analysis of Bimetallic Cylindrical Shells," *Inzhen. Sbornik* **23**, 28–35 (1956).

89. S. CHENG and B. P. C. HO, "Stability of Heterogeneous Aeolotropic Cylindrical Shells under Combined Loading," *A.I.A.A. J.* **1**, 892–898 (1963).

90. B. P. C. HO and S. CHENG, "Some Problems in Stability of Heterogeneous Aeolotropic Cylindrical Shells under Combined Loading," *A.I.A.A. J.* **1**, 1603–1607 (1963).

91. J. TASI, "Effect of Heterogeneity on the Stability of Composite Cylindrical Shells under Axial Compression," *A.I.A.A. J.* **4**, 1058–1062 (1966)

92. Y. STAVSKY and S. FRIEDLAND "Stability of Heterogeneous Orthotropic Cylindrical Shells in Axial Compression," TDM Report No. 67–9, Technion–Israel Institute of Technology, Haifa, Israel, 1967.
    Y. STAVSKY and S. FRIEDLAND, "Free Edge Buckling of Heterogeneous Cylindrical Shells in Axial Compression," TDM Report No. 68–4, Technion–Israel Institute of Technology, Haifa, Israel, 1968. Also to appear in *Internat. J. Mech. Sci.* **11** (1969).

93. F. J. PLANTEMA, *Sandwich Construction*, Vol. 3 of *Airplane, Missile and Spacecraft Structures*, N. J. Hoff, Editor, John Wiley & Sons, Inc., New York, 1966.

94. N. J. HOFF and S. E. MAUTNER, "Bending and Buckling of Sandwich Beams," *J. Aeron. Sci.* **15**, 707–720 (1948). (Contents originally presented at the Sixth Int. Cong. Appl. Mech., Paris, Sept. 22–29, 1946.)

Chapter 2

Adhesives

Richard F. Blomquist

1 INTRODUCTION

Although adhesive bonding has been practiced for centuries, this important method of joining materials has had its greatest growth only in recent years. The remarkable development of synthetic-resin adhesives has made possible the bonding of a wide variety of materials to provide joints of strength as required over a wide range, and also of a high degree of permanence where necessary. When coupled with important developments in bonding techniques, equipment for handling the various adhesive processes, and improved knowledge of the fundamentals of the bonding processes, joint design, and related technology, the result is an ever growing variety of adhesive-bonded constructions.

Adhesives are now available to provide the necessary adhesion to most currently used materials. They may provide strengths from mere holding actions to levels in excess of the strength of many adherends, permanence ranging from short-term dry conditions to complete outdoor weathering in most climates, as well as to the exotic atmospheres of outer space and the temperatures of supersonic air travel.

On the other side of the picture is the realization that no single adhesive or group of adhesives will begin to meet even the majority of this wide range of applications. In many cases rather complex bonding techniques must be used to provide joints that will satisfy the more severe requirements. Hence, there is still no single all-purpose adhesive, and it is doubtful whether such a material is likely to be developed very soon. However, there is now a fertile area of research and development to produce more versatile adhesive-bonding systems and particularly to develop systems that are more tolerant of a wider range of bonding conditions.

2 BACKGROUND

An adhesive is currently defined as a substance capable of holding materials together by surface attachment. An adhesive is often thought of as a liquid substance that attaches itself to two other substances and then promptly loses its own identity. This is not true. The adhesive layer in a joint is a material in its own right, with its own specific physical, mechanical, and chemical properties — a very definite component in a composite. The adhesive joint or assembly can be likened to a

RICHARD F. BLOMQUIST is at the United States Department of Agriculture Forestry Sciences Laboratory, Athens, Georgia.

Portions of this chapter are taken from Dr. Blomquist's Edgar Marburg Lecture on "Adhesives — Past, Present, and Future," presented to the American Society for Testing and Materials in 1963 (Ref. 1). Permission to reproduce this material was given by ASTM.

chain of five links. The two end links are the substances being bonded, each with its particular cohesive properties. The two adjacent links are the adhesive forces that hold the adhesive to these adherends, the substances being bonded. The middle link is the adhesive film itself, a definite material. Failure of any link in this chain results in failure of the joint.

"Adhesive" is a general term that includes such materials as cement, glue, mucilage, and paste. Although all these terms are loosely used interchangeably, "adhesive" is becoming most widely used and is considered the most acceptable general term for all such bonding agents. The term "adherend" is generally used to refer to the body held to another body by the adhesive. The process of attaching one adherend to another with an adhesive is generally called "bonding," although other terms such as "gluing" or "cementing" are often used in certain branches of the industry. The final assembly of the two adherends and the adhesive is most commonly called the bond, but it is also referred to as a joint.

The term "glue" originally referred to adhesives prepared from animal proteins, such as hides, hoofs, cartilage, and tendons. These glues were widely used in the woodworking industry, and the term "glue" is generally used by the industry to include all adhesives used on wood. The term "paste" refers to certain adhesive compositions that have a characteristic plastic-type consistency with a high order of yield value and are derived by heating mixtures of starch and water and then cooling. A "mucilage" is an adhesive prepared from vegetable gums and water and is mainly used for bonding paper. The term "cement" is commonly used to refer to adhesives based on rubbers and thermoplastic resins dispersed in organic solvents and setting by loss of solvent (this is, of course, quite apart from its meaning in "portland cement"). A complete set of definitions for these and other terms relating to adhesives is given in ASTM Definitions D 907.[2]*

Perhaps the earliest uses of adhesives were for bonding of papyrus or paper with adhesives of plant origin, such as gums or starches, and the bonding of wood with starch or animal proteins. Records of wood bonding go back to the time of the Egyptian pharaohs.[3] Animal glue, made from hoofs and hides of animals, has been used in woodworking, particularly for the assembly of wood furniture, for several hundred years. Veneering of thin wood layers was done on a limited scale by these early Egyptians, but the first significant production of plywood probably took place early in the present century.

* The references listed at the end of this chapter are intended only as a beginning for further study and are by no means an exhaustive survey of the extensive literature available in this rapidly expanding field.

Starch glues were used first for such purposes as bonding hardwood veneers for table tops and piano parts.[4] Soon after, glues based on soybean proteins were developed to bond softwood veneers; this resulted in the establishment of the important softwood plywood industry, using mainly Douglas-fir veneers in the Pacific Northwest.

Vegetable, sodium silicate, and other adhesives of natural origin were used in paper bonding — in fact, some of these uses probably go back to earlier centuries. Examples of such early uses include making and sealing envelopes, making paper bags, attaching paper labels, and attaching stamps. These early glues were not very effective for bonding metals, or such early plastics as celluloid.

The development of synthetic resins and plastics, which began in the 1920's and 1930's, gave a great impetus to the use of adhesives. These new materials offered many advantages over available natural materials, particularly the possibilities of tailor-making new polymers to provide the needed adhesion to other materials, and greater permanence to the bonds when exposed to water, heat, and weathering. The first urea-resin and phenol-resin adhesives were introduced into the United States from Europe in the early 1930's, although the basic resins had been discovered considerably earlier here and abroad.

3 ADVANTAGES AND LIMITATIONS OF ADHESIVE BONDING

To understand more clearly what adhesives can contribute to composites of materials, we should consider the advantages and limitations of adhesive bonding. Adhesive bonding fulfills the need for an easy and rapid joining together of two adherends. In addition, it has several important advantages that are not found with such mechanical fastenings as nails, screws, or bolts, with soldering, or with welding or brazing:

1. Certain materials can be joined that would be impossible or impractical to attach by other means. This advantage is used, for example, in the bonding of paper labels to cans or bottles and attaching wallpaper to walls.

2. Often, materials can be joined more economically and efficiently by adhesive bonding than by other methods. The combining of many small pieces of wood into large glued assemblies is an example of this.

3. Smoother surfaces and contours can be obtained by adhesive bonding of materials, such as brake and clutch facings or aircraft assemblies, in which rivet heads or other projections affect performance.

4. There is a more efficient and uniform transfer of stresses from one member to another than with mechanical fastenings, as illustrated by glued-wood roof trusses, stressed-cover house panels, and helicopter rotor blades. Adhesive bonding has been particularly effective in increasing the fatigue life of assemblies by reducing stress risers, as in rotor blades.

5. Adhesive bonding has made possible the design of new and better composite articles by taking advantage of the best properties of each joining material. This is shown in the design of light but strong sandwich panels, flush doors, metal-faced plywood, adhesive-bonded rubber mounts, and endless combinations of wood, paper, metals, plastics, rubber, and other materials.

6. Corrosion of joints between dissimilar materials is reduced by preventing direct metal-to-metal contact.

7. Better sealing action is provided for gases and liquids in bonded joints than is possible with mechanical fastenings. Examples of this are the bonding of extrusions to seal the edges of sandwich panels and the fabrication of watertight wood boats by adhesive bonding of the components.

In addition, weight can often be reduced considerably by using adhesive-bonded constructions, as in structural metal aircraft assemblies, by permitting the use of thinner gages of metal that are reinforced when necessary, as around cutouts and along edges. The stiff, rigid, but lightweight panels of thin skins with honeycomb cores have been very effective in reducing weight in aircraft constructions.

Adhesive bonding often costs less than conventional mechanical fastenings, as in certain aircraft assemblies. This is often particularly significant as the size of the assembly increases, since the entire assembly may be bonded at one time and thus is cheaper than automatic riveting.

However, cost comparisons of adhesive bonding with other methods of fastenings are based on many interrelated factors, including the cost of the adhesive, related processing and equipment, labor, and other factors that make it impossible to generalize on relative costs of the different methods of attachment.

At the same time, there are certain limitations of adhesive bonding as compared with mechanical fastenings. Generally, relative strengths with adhesive bonding are somewhat more directional than with mechanical fastenings, particularly with metal-bonding structural adhesives. Such adhesives are very good in shear, acceptable in tension, and low in resistance to cleavage or peel. Surfaces to be bonded must be cleaner and more carefully prepared than is usually necessary for mechanical fastenings. Adhesion of a certain chemical type of adhesive is generally better for some adherends than for others. Any given adhesives application, therefore, requires a rather specific adhesive formulation and may also require special bonding conditions, since there is no single adhesive or bonding process that begins to meet the requirements of a truly general-purpose adhesive.

Adhesives generally do not develop their full strength and permanence immediately, as do welded or mechanical joints, but often require appreciable time for the reactions of setting and curing to take place. Production equipment used for adhesive bonding is different from that used for mechanical fastening. Adhesive bonding requires jigs and presses to ensure adequate and intimate contact of the mating adherends while the adhesive solidifies. Presently available adhesives, which are nearly all based on organic compounds, vary considerably in permanence in joints. For example, some have limited heat resistance above 400°F for periods of more than a few hours. However, recent developments indicate that much more thermally stable adhesives can be produced.

Finally, the present lack of reliable nondestructive tests for evaluating the final bond quality in production makes it necessary to practice close quality control over the entire bonding process to be certain of uniform bond quality particularly in structural joints.

4 THE NATURE OF ADHESION

Since adhesion to the substrate is a vital consideration for successful application of adhesives, as well as for paints and coatings, the question of what actually is involved in adhesion has been studied and discussed for many years. In general, no direct experimental techniques are available for studying actual adhesion. Hence the various theories of adhesion have been developed first from observations of actual tests of adhesive joints, in which the joints are made and then broken in a variety of physical tests. Unfortunately in tests of weaker adherends, such as paper or wood, the joints usually fail in the adherends if the bonding is done properly with appropriate adhesives; thus the experimenter learns nothing about the actual strength of the interfacial bond between adhesive and adherend.

Even when stronger adherends, such as metals, are broken in such a test, there is always considerable doubt whether the joint has failed in actual adhesion, or whether failure has occurred cohesively in the adhesive film itself. Limited recent studies utilizing refined microscopy, elipsometry, and radioactive tracer techniques have tended to indicate that at least a very thin layer of adhesive is left on the substrate in such broken joints. Physical chemists and physicists have considered the theoretical aspects of the actual adhesion from adsorption measurements of polymers from solutions onto the selected substrates. They have explained adhesion on the same basis as the forces of attraction between molecules of two different materials and have developed useful hypotheses and computed probable magnitudes of such adhesion. They then related their theoretical computations and hypotheses to the observed results of destructive tests of joints.

The theories of adhesion have been reported in many symposia and in technical literature.[5–17] Such documents will furnish a broader understanding of these various ideas. Perhaps one of the most valuable forums for current interchange of such ideas in the United States has been the series of Gordon Research Conferences on Adhesion, beginning in 1954 and sponsored by the American Society for the Advancement of Science. However, papers from these conferences are not published. Similar forums have been conducted in other countries in recent years.

At present there are at least four different theories of adhesion. These may be classified as: (a) mechanical, (b) adsorption, (c) diffusion, and (d) electrostatic. Each has certain strong points and also limitations. It is quite possible that each theory is at least partially correct for specific combinations of adhesives and adherends.

The oldest theory is often referred to as mechanical adhesion. This hypothesizes that adhesion is due to mechanical interlocking of the solidified adhesive in holes or crevices in the adherend. Such mechanical interlocking is no doubt a factor in some bonding, particularly of porous adherends such as paper or wood. Indeed, penetration of the adhesive into such porous adherends may be important to produce strong joints, if for no other reason than the fact that this tends to reinforce the adherend surface layers, thus reducing premature failures in the adherend. This theory was attractive originally to explain adhesion in porous adhesives such as wood or paper. Even with porous adherends, however, other types of adhesion

are considered to be operating in addition to mechanical interlocking.

The other theories involve what has been referred to as "specific adhesion." The adsorption theory considers adhesion primarily as a surface phenomenon. The liquid adhesive must first spread over the adherend surfaces to wet them and to allow the molecules of the adhesive to come close enough to those of the adherend to permit the intermolecular forces to be established. These forces are thought to be similar in nature to the forces responsible for the cohesive strength of either the cured or hardened adhesive or of the adherend. These forces probably include van der Waals forces, as well as dispersion and induction forces, and the electrovalent, covalent, and coordinate covalent bonds that hold molecules themselves together. DeBruyne is usually credited for first postulating the rule of adhesion. This states, in effect, that polar adhesives bond well to polar adherends and nonpolar adhesives bond well to nonpolar adherends, but that polar adhesives did not bond well to nonpolar adherends, nor nonpolar adhesives to polar adherends. Sharp and Schonhorn[14] have recently pointed out the fallacy of the nonpolar-polar postulate with an experiment involving adhesion between epoxy resins (polar) and polyethylene (nonpolar). Although conventional liquid epoxy resins do not adhere to polyethylene because they do not adequately wet it, good adhesion can be obtained if the polyethylene is melted and applied to the cured solid epoxy resin. This is a matter of surface free-energy relations. The literature contains further explanations of the adsorption theory.[15]

The diffusion theory of adhesion has been closely associated with the work of Voyutskii and associates[16] in their concept of autohesion. Here, adhesion is considered to result from diffusion of chain molecules or their segments and the resultant formation of strong bonds between adhesive and adherend, and is suggested as an explanation of bonds to rubber. Two conditions for good autohesion are that the substance must have good cohesive strength and that there must be sufficient coalescence when the two surfaces meet. For complete coalescence there must be interweaving of the chains of the high polymer, which is characteristic of all parts of its volume. The welding of thermoplastics is thus considered to be due to diffusion.

A fourth theory postulates that adhesion is due to electrostatic forces in which the adhesive joint is somewhat like a condenser,[17] involving an electrical double layer formed at the adhesive-adherend interface. During joint separation a potential difference develops and is discharged when the joint breaks. In some cases, electrical discharge and electron emission have been reported when joints separated. This is presumably due to electron transfers between the adherend and adhesive.

Regardless of which mechanism of adhesion is preferred, theoretical levels of bond quality calculated from thermodynamic data are always considerably larger than strengths observed in actual joint tests. Reinhart[7] has considered this situation and has suggested an explanation shown in Figure 2.1. These reductions in the theoretical strength of the bond are first considered to be due to incomplete wetting of the adherend, thus reducing the number of possible sites for intermolecular forces to be set up at the interface. Further reductions in strength are due to the presence of internal stresses in the

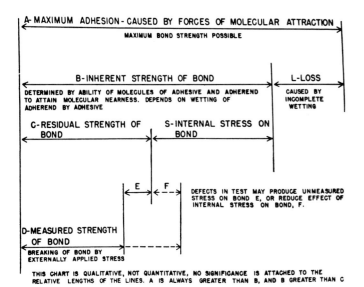

A–MAXIMUM ADHESION–CAUSED BY FORCES OF MOLECULAR ATTRACTION

MAXIMUM BOND STRENGTH POSSIBLE

B–INHERENT STRENGTH OF BOND | L–LOSS

DETERMINED BY ABILITY OF MOLECULES OF ADHESIVE AND ADHEREND TO ATTAIN MOLECULAR NEARNESS. DEPENDS ON WETTING OF ADHEREND BY ADHESIVE | CAUSED BY INCOMPLETE WETTING

C–RESIDUAL STRENGTH OF BOND | S–INTERNAL STRESS ON BOND

E F

DEFECTS IN TEST MAY PRODUCE UNMEASURED STRESS ON BOND E, OR REDUCE EFFECT OF INTERNAL STRESS ON BOND, F.

D–MEASURED STRENGTH OF BOND

BREAKING OF BOND BY EXTERNALLY APPLIED STRESS

THIS CHART IS QUALITATIVE, NOT QUANTITATIVE, NO SIGNIFICANCE IS ATTACHED TO THE RELATIVE LENGTHS OF THE LINES. A IS ALWAYS GREATER THAN B, AND B GREATER THAN C

Figure 2.1   Relations Between the Factors Involved in Adhesion

joints due to contractions in volume of the adhesive in solidifying from the liquid state, to thermal contraction in cooling from the elevated curing temperature when the bond was formed, from voids due to escape of volatile solvents or air from the adhesive during cure and in service, and from strains in the adhesive when curing on irregular adherend surfaces. Observed joint quality may be further reduced by imperfections in the actual test method, due to misalignment, erratic loading rates, and by factors associated with the geometry of the test specimen.

The uncertainties in our understanding of adhesion itself point out the need for further consideration both of adhesion of the adhesive to the adherend, and the improvement of the cohesive properties of the solidified adhesive film in the joint. Related to this is the need for greater attention to the design of joints for effective use of the adhesive; this could involve reducing stress concentrations such as are rather critical in simple overlap joints and the application of new knowledge of fracture mechanics both in the adhesive film and in the adherend in a given joint design. Some of these factors will be considered in the following sections.

5   INTERNAL STRESSES IN JOINTS

As has been pointed out, the presence of internal stresses in the adhesive film can significantly reduce the observed strength of the joint below theoretical values. The problem of internal stresses in adhesive films is therefore worth further consideration. Like the forces of adhesion themselves, the true nature of these internal stresses cannot readily be studied experimentally, and many of our thoughts on this subject are based on intuitive reasoning from actual experience with joints. A typical adhesive must go through a fluid state in order to flow over the adherend surfaces and permit intimate contact with this adherend surface on a molecular scale. However, since such a liquid film has a rather low order of cohesive strength, it is necessary to convert this to a solid state to produce a strong bond.

The process of changing from a liquid to a solid state usually results in some volume change, generally a reduction, which in itself creates internal stresses within the adhesive film. Solvent-based adhesives lose their solvent, often rather slowly, before, during, or after the actual bonding process. The loss of solvent may cause volume changes or voids in the film, which are themselves sources of internal stress concentrations. Adhesives that solidify as a result of a chemical reaction, as do the thermosetting-resin adhesives, also undergo volume changes that result in internal stresses in the bond. The magnitude of the stresses depends on the actual nature of the chemical reaction involved and, to some extent, upon the rate of the reaction. Pressure-sensitive adhesives probably undergo the least change in volume but do not actually change from a liquid to a solid. Thus pressure-sensitive adhesives generally have lower cohesive strength in the final adhesive film than do other types of adhesive systems, such as those based on thermosetting resins.

If the surfaces of the adherends are relatively smooth and flat, shrinkage during bond formation is likely to be rather uniform, which will tend to draw the two surfaces closer together and result in rather thin adhesive films. Rough surfaces, on the other hand, interfere with this drawing together and result in films of variable thickness. This tends to cause voids in the adhesive film and nonuniform shrinkage of the film, thus resulting in more points where internal stress concentrations can develop. Thin films generally do not flow readily during bond formation, whereas thick films are likely to be lower in cohesive strength and give weaker bonds. Therefore, there is probably some optimum adhesive film thickness for each adhesive-adherend combination, but the optimum thickness usually must be established empirically.

Where the temperature of the bond must be raised in bonding, as for certain thermosetting and hot-melt adhesives, thermal expansion of both adherend and adhesive occurs. This relative expansion may be quite different in the two materials for a given temperature change. Cooling such bonds to room temperature after bonding will cause thermal contractions of the same order. These differential dimensional changes in adherend and adhesive layer also result in internal stresses within the bond. Thermally induced stresses in bonds also result because of changes in temperature of the bond during service, such as in aircraft joints alternately on the ground and at high altitudes.

Internal stresses in adhesive bonds can be reduced by several methods. These include modification of the adhesive composition so that the mechanical properties of the final dry adhesive film more nearly approach those of the adherend. This modification may be accomplished by plasticizing a hard and brittle material to make it more flexible so that it tends to deform somewhat in the joint and relieves internal stresses that may exist. The preparation of the adherend surfaces to avoid the need for variable and excessively thick adhesive films will also tend to reduce these internal stresses. Any means of avoiding inclusion of solvent in the adhesive film after bonding will also help. This can be achieved by using high concentrations of adhesives in the solvents or by drying techniques to remove as much solvent as possible before joining and final bond formation. Anything that can be done to match the thermal expansion coefficients of the dry adhesive film and the adherends, such as modification of the adhesive composition or use of special fillers, will also reduce internal stresses in the bonds. This stress reduction will permit the user to develop a higher proportion of the total theoretical strength of the adhesive bond than would otherwise be possible.

These considerations of the internal stresses within the adhesive layer of the bond illustrate one reason why the cohesive properties of the adhesive layer must be considered. The adhesive film is thus a separate material from an engineering standpoint.

6 DESIGN OF JOINTS FOR BONDING

While it may be possible merely to replace rivets, screws, or other mechanical fastenings with an adhesive and achieve acceptable performance, in most cases some changes in joint design are necessary to utilize the most important advantages of the adhesive and to minimize any disadvantages. Most structural adhesives, such as those used in metal bonding for aircraft, are rather rigid, strong in shear, moderately strong in tension, and rather weak in peel or cleavage. Thus, it is normally better to design the joint to be loaded in shear, as in an overlap joint, rather than in tension, as in a butt joint.

Some general principles for good joint design include: (1) the bonded area should be as large as possible, (2) the maximum proportion of the bonded area should contribute to the joint strength, (3) the adhesive should be stressed in the direction of its maximum strength (as in shear, rather than in tension or peel); and (4) stresses in the weakest direction of the adhesive layer should be minimized.

The design problems in metal lap joints to be loaded in shear have probably been studied most completely [18–23]. A simple overlap joint will not be loaded in pure shear when the two ends are pulled apart but will be subjected to bending stresses that deform the adherends and adhesive layer to varying amounts, depending upon the dimensions and stiffness of the two adherends. Such bending introduces peel forces at the ends of the overlap. The strength of such a simple lap joint is directly proportional to the width of the joint. Although this strength is also proportional to the length of the overlap, this strength relationship is by no means linear, and unit loads decrease proportionally as the overlap increases in length. This is caused by high stress concentrations at the end of the lap so that the center of the bonded area may contribute very little to the observed joint strength. Thus, joint strength can be improved by beveling the ends of the lap. Various alternate designs for this lap joint are illustrated in Figure 2.2. Double lap joints and scarf joints will reduce bending and stress concentrations and thus improve joint strength. However, each improved joint design offers certain practical problems in fabrication and economy that must be considered in selecting the final joint design. The thickness of the metal adherend also will influence the shear strength of a lap joint. If a metal is too thin, its yield strength may be lower than the shear strength of the adhesive.

Special joint-design problems are encountered in bonding flexible adherends where peel forces are particularly important. Thus in bonding thin metallic sheets or films to another metal piece, the thin material should be overlapping to place the adhesive in shear; the overlap should be as long as possible

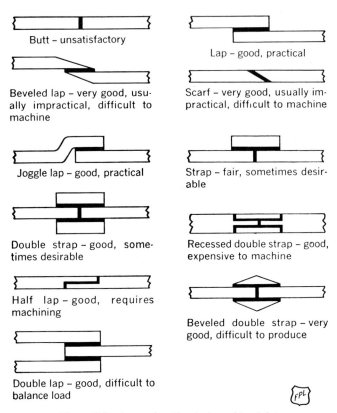

Butt – unsatisfactory

Lap – good, practical

Beveled lap – very good, usually impractical, difficult to machine

Scarf – very good, usually impractical, difficult to machine

Joggle lap – good, practical

Strap – fair, sometimes desirable

Double strap – good, sometimes desirable

Recessed double strap – good, expensive to machine

Half lap – good, requires machining

Beveled double strap – very good, difficult to produce

Double lap – good, difficult to balance load

Figure 2.2   Improving the design of lap joints

in order to place a minimum unit shear stress on the adhesive. Better joint designs for the assembly of corners, stiffeners, and T sections are also possible through careful engineering analysis of the problems involved.

In some applications the joint design and the mechanical properties of the adhesive are chosen to achieve some special performance. For instance, in sealing the flaps on top of a corrugated fiberboard carton for canned or bottled food products, an adhesive is sometimes chosen to produce good shear strength to ensure stiffness and resistance to racking in transit and handling but with rather low tensile strength so that the flaps can be pulled up easily to open the carton for unpacking.

Another solution to the joint design problem is to modify the mechanical properties of the adhesive layer itself so that the particular limitation of the adhesive is minimized. Thus it has been found possible to improve peel resistance of metal-bonding adhesives. Reformulation significantly reduces peel failure, as in bonding thin metal facings to honeycomb core and sandwich panels. Such reformulation has its limitations, however, since it may also change other desirable properties such as resistance to heat distortion or general permanence.

7 CLASSES OF ADHESIVES

Adhesives may be classified by several methods. A common method is based on chemical composition of the main components:

Natural Origin
  1. Starches, dextrins, vegetable gums
  2. Proteins (vegetable and animal):

a. Animal (hides, sinews, bones)
b. Fish (skins)
c. Blood (whole blood or albumin)
d. Casein
e. Soybean meal (also peanut and other vegetable proteins)
3. Other materials
a. Asphalt
b. Shellac
c. Natural (or reclaimed) rubber
d. Sodium silicate, magnesium oxychloride, and other inorganic materials

Synthetic Origin*

1. Thermoplastic resins (cellulose esters and ethers, alkyd and acrylic esters, polyamides, polystyrene, poly-(vinyl alcohol) and derivatives)
2. Thermosetting resins (urea, melamine, phenol, resorcinol, furan, epoxy, unsaturated polyesters)
3. Synthetic rubbers (neoprene, nitrile, polysulfide).

This system works well for the simpler adhesives. However, it does not adequately cover many of the newer formulations in which several different components, all important to the adhesive system and for the actual adhesion, are combined. Thus some of the newer structural metal-bonding adhesives may be combinations of a phenol resin with a synthetic rubber or a thermoplastic resin, or even all three. Other combinations include protein-latex or phenolic resin-blood formulations as well as endless combinations of two or more synthetic resins such as urea and melamine resins or phenolic or resorcinol resins. These may be actual mixtures of the two resins or they may be copolymers of the two resins. Thermoplastic resins may be formulated with other materials to produce certain degrees of thermosetting properties.

Another classification considers the main types of uses. According to this system, adhesives may be classified as (1) structural adhesives, whose primary function is to hold two adherends together and produce high strength in shear, tension, or peel; (2) holding adhesives, which are intended primarily for merely attaching one adherend to another and holding it in place without requiring any significant resistance to external stressing; and (3) sealing adhesives, whose principal function is to close the joint between two adherends to provide a seal against moisture, gases, or vapors without necessarily providing any significant strength.

Obviously, many adhesives applications may require two or even all three functions. Structural adhesives are common in the wood-bonding and metal-bonding fields and fabrication of building components, such as trusses or structural sandwich panels. Holding adhesives have been widely used in labeling and packaging and in attachment of wallpaper and wall, floor, and ceiling tiles. Sealing adhesives are used to close packages for food and drug items and are also used in fabricating containers, such as the wing fuel tanks for airplanes where the adhesive may have to contribute some structural properties as well as sealing.

* For discussions of the chemistry and technology of these individual synthetic resins, see References 24–29.

Adhesives are also classified according to the adherends with which they are commonly used. Thus, there are paper-bonding, wood-bonding, metal-bonding, or plastic-bonding adhesives. Each type may include structural adhesives as well as those used where holding or sealing is more important.

8 ADHESIVE COMPONENTS

The composition of a practical adhesive is often fairly complex, particularly for some of the newer special-purpose adhesives. The principal components of an adhesive include the following.

The *adhesive base* or *binder* is primarily responsible for the adhesion forces that hold the two adherends together. The binder is generally the component from which the name of the adhesive is derived in the classification based on composition. There may be two or more bases in some adhesives, as in a phenolic-poly(vinyl butyral) base adhesive used in structural metal bonding.

Solvents are needed in most adhesives to disperse the binder to a spreadable consistency. In most adhesives for bonding wood and paper the solvent is water. In many adhesives based on synthetic resins, rubbers, and even natural gums, a variety of organic solvents are required to achieve the necessary solubility and provide some minimum percentage of binder solids.

Thinners or diluents are volatile liquids added to an adhesive to modify the consistency or other properties. Such liquids are usually not true solvents for the other components when used alone but are effective in diluting or thinning the composition as might be required for spray application. They provide the desired degree of volatility compatible with the open assembly conditions of use and help avoid blisters and blowups in subsequent hot pressing.

Catalysts are substances that markedly speed up the cure of an adhesive when used in minor amounts as compared to the amounts of the primary reactants. Catalysts must be chosen to increase the rate of a specific chemical reaction involved in the adhesive base as it hardens in the joint. A good example of a catalyst is ammonium chloride, commonly used to speed the curing or crosslinking of urea-formaldehyde wood adhesives. Buffers or retarders such as ammonia or calcium phosphate may also be added to control the acidity and hence the rate of this curing reaction.

Hardeners are substances added to an adhesive to promote or control the curing reaction by taking part in it. That is, a hardener is a chemical reactant in the curing reaction, whereas a catalyst does not react directly but merely controls the rate of reaction. A good example of a hardener is paraformaldehyde, commonly used with resorcinol-formaldehyde wood adhesives.

Fillers are relatively nonadhesive substances added to an adhesive to improve its working properties, permanence, strength, or other qualities. These materials are generally intended to do something useful and not merely reduce cost. Examples are English walnut shell flour or pecan shell flour, which are commonly added to many synthetic-resin wood adhesives to improve their spreading on conventional roll spreaders or to control excessive adhesive penetration into certain porous woods. Metallic and inorganic fillers, such as aluminum powder, alumina, or china clay, are often added to epoxy-resin adhesives as fillers to improve viscosity and spreadability. They can also alter the thermal expansion coefficient of the cured adhesive film to reduce stresses between adherend and film due to different thermal expansion and contraction characteristics.

Extenders are substances generally having some adhesive action. They are added to an adhesive to reduce the amount of primary binder required per unit of joint area and thus reduce the cost of the joint. An example is the wheat flour commonly added to urea-formaldehyde wood adhesives to reduce glueline cost.

Preservatives are agents added to certain adhesives to retard or prevent decomposition by microorganisms, either while the adhesive is being stored or applied or during service of the completed bond. These agents are usually most important in formulations containing carbohydrates or proteins such as flour, starch, or casein proteins that are readily attacked by mold, fungi, or bacteria. Preservatives include inorganic material such as copper or mercury salts and many new organic compounds such as organic mercury compounds and various chlorinated phenols.

Fortifiers, a term not yet generally accepted, are materials that are added to an adhesive binder primarily to improve the permanence of the resultant bond. They should be binders or at least have some distinct good adhesive value. An example is the addition of resorcinol along with paraformaldehyde to a urea-formaldehyde wood glue to improve the resistance of the resultant bond to weathering or exposure to other conditions. Melamine-formaldehyde resins are often added to urea resins at the time of use for the same purpose.

A *carrier* is usually a thin fabric or foil used to support the adhesive composition to provide a dry-film adhesive. An example is the thin paper film used to support the phenol-formaldehyde resin in the conventional phenolic-resin film glues used in the wood industry. Glass cloth and other fabrics are now used to support some of the newer structural metal-bonding adhesives based on phenolic and epoxy resins or phenolic and poly(vinyl butyral) resins. It should be noted, however, that some film adhesives are unsupported and thus have no carrier.

9 SELECTION OF AN ADHESIVE

At the present time there are no all-purpose adhesives that combine all of the desired properties for a wide variety of bonding applications, and it is unlikely that there will be such adhesives in the next few years. Therefore, adhesives must be fitted to the proposed application, and the selection of the adhesive from the wide variety available becomes a confusing problem. From a practical standpoint, the potential user of an adhesive should consider several factors in selecting the most suitable adhesive for this purpose.

9.1 Adherence

Of first importance is the ability of the adhesive to adhere to the adherends. This has been discussed previously.

9.2 General levels of strength

Consideration must be given to various factors of joint design. Some minimum level of shear, tensile, cleavage, peel, fatigue, long-time creep rupture, and other strength properties will be needed. For many structural uses, the possible redesign of a joint should also be considered.

9.3 Working properties

Working properties include the characteristics of the adhesive that influence application (such as mixing, spreading, pressing, curing, speed of curing or rate of strength development, and convenience of cleanup afterwards). In many industrial bonding processes, the working properties of the adhesive are largely dictated by practical considerations. One set of working properties is needed if the adhesive bonding operation must be fitted into a continuous production line, including several other separate operations in sequence. Quite different properties are needed for special field installations that preclude the use of any elaborate apparatus to spread the adhesive, to maintain pressure for any length of time on the joint, or to control bonding temperatures. In the case of many packaging operations carried out on high-speed conveyor lines, rather expensive production line equipment may already be installed and must be used. In such cases, the working properties of a new adhesive must be compatible with the already available equipment. In the application of weather-stripping or sound-absorbing pads or upholstery in automobile body fabrication, the installation must be accomplished instantaneously as the body moves down an assembly line, and each installation must be in its proper sequence. It is impossible under these conditions to apply external bonding pressure and to maintain it for any length of time.

Adhesives used for bonding floor and ceiling tile must tolerate considerable variation in spreading conditions and in time-temperature conditions existing before the tile is pressed into place. The adhesive must then develop sufficient strength almost instantaneously with little or no bonding pressure to hold the part in place, while the adhesive may develop some further strength under ambient room conditions. In these applications, the working properties are often the most important single consideration in the selection of an adhesive, and adhesives without the desired working properties cannot be considered regardless of how cheap, strong, or durable they might be. The working properties of an adhesive are directly related to the equipment requirements for using the adhesive, particularly for industrial bonding. Great progress is being made in the development of special factory equipment to spread, press, and cure adhesives economically and rapidly, in a variety of joints, both in batch operations and in continuous bonding. The greatest developments in equipment for adhesive bonding have probably been in packaging and in softwood plywood manufacturing. Much remains to be done, however, in design and construction of bonding equipment. Such equipment should be designed in close collaboration with the formulator of the adhesive and the adhesive user.

9.4 Permanence

One of the most important properties imparted by many of the modern synthetic-resin adhesives is the high level of permanence they can provide in properly made joints. For such applications as plywood for exterior or marine use or bonded metal sandwich panels in modern aircraft, the permanence of the joint is of prime importance. In such cases, a rather expensive adhesive or one requiring elaborate bonding procedures and special expensive equipment may be necessary to provide the required permanence. On the other hand, some adhesive bonds are intended to be temporary, such as those used to hold parts of shoes together during manufacture before the final stitching or other mechanical fastenings are completed. In such cases, it may actually be necessary to destroy the adhesive bond readily and economically after it has served its purpose. An example is the label adhesives for beverage bottles that must withstand immersion in ice water for some time in use but must then be easily removed in subsequent bottle-washing operations. One of the most serious current problems in selecting adhesives for long-term and severe service applications is the lack of reliable short-term accelerated permanency tests for adhesives. This is particularly true of tests that may be applied with confidence in screening new chemical types of adhesives for which little or no service experience is yet available.

9.5 Cost

Cost is nearly always a factor in the selection of an adhesive. Everything must be considered, including the cost of the adhesive as supplied; the cost of labor, solvents, fillers, and other additives to prepare it for use; the cost of equipment and labor in the bonding process; and the economic loss due to rejects from defective joints.

10 PROCESS BY WHICH ADHESIVES DEVELOP STRENGTH

To understand some of the problems involved in internal stresses in the joint as well as to indicate differences between types of adhesives and their applicability for end-use applications, it is necessary to consider the processes by which liquid adhesives become solids and develop strength in joints.

10.1 Air-drying or solvent-responsive types

Many adhesives are simply solutions or dispersions of the solid components in a suitable solvent-thinner system. The solvent may be water or some organic liquid. In all such cases, the solvent must be lost from the adhesive film in order to produce the gelling or hardening by which final bond strength is achieved. The solvent may be partially lost to the air after being spread on the adherend but before the joint is assembled, or it may be absorbed by certain adherends. Illustrations are the familiar starch and dextrin glues dispersed in water and used for bonding porous material such as wood or paper. Various organic solvents are needed to disperse rubbers and various thermoplastic resin adhesive bases. These types are common in the mastic adhesives of asphalt, rubber, and similar materials used for bonding acoustical or floor tiles and in the fabrication and repair of rubber goods. A variation of this type is the solvent-reactivated adhesive. With these adhesives, the adherend is coated with the adhesive in solution form and

then dried to a tack-free condition for convenience in storage or handling. When ready for use, the dried film is partially liquefied on the surface by application of a suitable solvent or dispersing agent, and then the bond is assembled. This is the situation with gummed paper tapes and labels or postage stamps. Other more complex systems are based on various natural or synthetic resins where organic solvents may be needed for reactivation, as in the conventional tire-tube patching cements.

10.2 Hot-melt or fusible types

Some adhesive bases can be converted from a solid to a liquid state by melting and then be applied to the adherend in the molten form. Hardening of the adhesive on cooling after the joint is assembled then results in conversion to the solid state and the development of the cohesive strength of the adhesive in the bond. An early example of such hot-melt adhesives was the original hot animal glue used by cabinet and furniture makers. The dry animal protein is dispersed in water at room temperature, forming a stiff gel. This gel is then melted by heating in a jacketed container maintained at 140°F to 160°F, at which temperature the liquid form is applied to the adherend. Cooling the adhesive to room temperature then causes rapid solidification and development of bond strength. The water-base animal glues develop some further strength in this case by subsequent loss of water to the wood. Newer adhesives of the hot-melt type are based on various thermoplastic resin components that are converted directly from the solid state to liquid by melting; then they are applied in the molten form to cool in the joint after assembly and pressure is applied. Some materials available for such hot-melt adhesives include the newer cellulose esters, some of the polyvinyl esters and acetals, and certain polyamides. These may be modified with plasticizers to improve the flow properties in the molten state and to reduce the rigidity of the dry film in the joint. Such hot-melt adhesives with essentially 100 per cent solids include those used for modern bookbinding, package sealing, and labeling. The principal advantage of these hot-melt adhesives is very fast development of initial strength by cooling.

10.3 Pressure-sensitive adhesives

Pressure-sensitive adhesives have a high degree of initial tackiness and may be used either as liquid adhesives or in the form of tapes. The pressure-sensitive adhesive tapes are the most common example. Such adhesives have a rather high cohesive strength when manufactured and applied and must usually be applied to both adherend surfaces. The two adhesive-coated adherend surfaces, which have no solvent present at the time the joint is assembled, then merely adhere to each other or may diffuse slightly into each other to form the bond instantaneously. In the case of most pressure-sensitive tapes, the special coating on the fabric or plastic face has been formulated to adhere to an uncoated adherend upon contact and only mild momentary pressure. The compositions of pressure-sensitive tape coatings are rather complex but generally consist of a film-forming elastomeric material, often one of several types of rubber, and added resins or other materials to impart the desired degree of tack, wetting power, and specific adhesion.

10.4 Chemically reactive adhesives

Chemically reactive adhesives undergo a chemical reaction in the actual bonding process. Like all chemical reactions, the rate of the reaction depends on the temperature at which the adhesive film is maintained. Earlier examples included the vulcanization of rubber in adhesives and chemical modifications of protein adhesives with formaldehyde or other tanning agents to produce greater degrees of water or moisture resistance. The principal current application of this type, however, is in the thermosetting resin adhesives, such as the urea, phenolic, resorcinol, epoxy, and melamine resins. In these types the low-molecular-weight polymer is commonly mixed with a catalyst, hardener, or initiator to start the final reaction, known as curing, as it is spread in liquid form on the adherend. A joint is then assembled and brought under pressure, and the final hardening or curing of the resin takes place either at room temperature or at some elevated temperature, depending upon the reactivity of the resin and the desired speed of strength development. This curing reaction is irreversible and, once the polymer has reacted to its cross-linked condition, the strength and durability have been achieved and the joint cannot be separated without degrading the adhesive itself.

11 THE BONDING PROCESS

Modern adhesives technology recognizes that the bonding process with most adhesives is essentially the same in principle, that all phases of this bonding process must be adequately controlled to achieve uniformly satisfactory results. The user must obtain the proper instructions for the specific adhesive and follow them closely to ensure satisfactory results. The various steps in the bonding operation are discussed below.

11.1 Preparation of adherends

In many adhesives applications, such as in the use of pressure-sensitive tapes for wrapping packages, the adherend must be bonded without any sort of previous surface treatment. In such cases, adhesives must be formulated to bond to the adherends in their natural condition. However, for permanent bonds and for high-strength joints, it is generally necessary to prepare the adherend surfaces before applying the adhesive. In the case of bonding wood joints, this includes conditioning the wood to the proper moisture content and surfacing to present a smooth, flat surface that is free of any torn or loose fibers. Metal surfaces generally have to be machined to permit good fit before bonding. Veneers of wood or thinner sheets of metal and plastics are not surfaced before bonding but may require surface cleaning techniques. Particularly in metal bonding for high-strength applications, the metal must be degreased and often further chemically treated to produce a surface that is free of occluded impurities. Such a surface should be chemically acceptable for wetting and adhesion and should reduce possible corrosion failures in or near the metal surface in subsequent service. This may require only solvent degreasing or surface abrasion, but it may also require chemical etching in acid or alkaline solutions specifically selected for the type of metal involved. Recent studies also suggest that metal surface preparation techniques significantly influence the resistance of the resultant bonds to thermal aging.

11.2 Preparation of adhesive

Many adhesives are ready for application when sold. Others may be in powder form to be dispersed in liquids and may also require the addition of specific amounts of hardeners, catalysts, extenders, fillers, or other ingredients. It is very important, therefore, to obtain the proper instructions from the manufacturer for that specific adhesive and to follow these directions closely.

11.3 Application of adhesive

Adhesives may be applied in liquid form to the adherend surfaces by brushing, roller coating, or spraying the liquid adhesives or by roller or knife coating molten adhesives. Tape or dry-form adhesives may be applied by simple insertion into the joint between the adherends. Some adhesives must be applied to both surfaces and some to only one of the surfaces to be bonded. The amount of adhesive applied is often critical, particularly in high-strength adhesives. For highest quality structural joints, the preparation of the adherend and of the adhesive and its application must be correlated so that a uniformly thin layer of adhesive is spread over a flat surface. Variable-thickness adhesive films are likely to cause erratic joint performance because of different amounts of occluded solvent, difficulties in wetting uneven surfaces, and inclusion of air bubbles in the film. Many of the chemically reactive adhesives undergo the hardening reaction during the period between preparing the adhesive and application of pressure on the joint. This reaction depends on the ambient temperature, particularly the temperature of the glue film or glue mix, and these factors must be coordinated with the subsequent bonding operations.

11.4 Assembly or air-drying period

Generally some time must elapse between the time the adhesive is applied to the surface and pressure is applied to the joint. Chemically reactive adhesives continue to increase in viscosity or fluidity during this period, and solvent-type adhesives are continually losing solvent and thus increasing in viscosity. These reactions are also temperature dependent. The conditions during this assembly period must be controlled so that, at the time pressure is applied, the adhesive is at an optimum consistency. It must be neither too stiff to spread out to form a uniform film and wet the opposite surface nor so fluid that it is squeezed out of the joint.

11.5 Bonding pressure and temperature

After the liquid adhesive has reached its optimum consistency on the adherend surface, the assembled joint is brought under pressure. This pressure may be very low and held only momentarily, or it may be rather high and maintained for a period of several hours, depending on the type of adhesive and the temperature. The purpose of pressure is to distribute the adhesive uniformly over the entire joint surface to produce a film of uniform thickness, to expel air bubbles, and to force the two adherends into as close contact as necessary. Pressure must then be maintained until the adhesive film has developed sufficient strength to hold the adherends together and to resist both internal and external stresses that otherwise tend to pull the joint open. The temperature at which the glue line is maintained during bonding depends on the curing system by which the adhesive develops strength in changing from a liquid to a solid. Often heat may be applied by hot platens or high-frequency heating to increase the rate of strength development and shorten the period under pressure.

11.6 Conditioning after bonding

Some adhesives have essentially full strength as soon as they are released from pressure. Others may develop further strength after pressure is removed and during the subsequent conditioning period. With such materials as wood or paper, it is important during this conditioning period to distribute moisture adsorbed in the gluing process throughout the adherends to minimize subsequent dimensional instability of the bonded joint.

12 EQUIPMENT REQUIREMENTS FOR ADHESIVE BONDING

In general, the successful use of adhesives in any application will depend on cooperation and consideration between the adhesive producer or supplier, the user, and the manufacturer of the equipment to be used in the bonding process. Unfortunately in the past these three individuals have tended to work independently and not keep each other adequately informed of the real interdependence of adhesive, the equipment, and the method of application in the user's plant. Better cooperation in the early stages of development of the proposed bonded product can greatly improve the acceptability of the bonding process and reduce the complaints when unsatisfactory performance is obtained.

In some cases, the equipment is already available in the plant of the user and must be used. In this instance the adhesive manufacturer must know of the conditions under which his adhesive must operate so that he may tailor-make, or otherwise select a suitable adhesive formulation with good chance of success. In other cases, no equipment is available and the user is then free to consult with the adhesive manufacturer on the equipment needs best suited to apply and use his adhesive successfully. Many times equipment specialists, knowing the characteristics of the adhesive selected and the type of bonded product to be fabricated, can make real contributions by providing special equipment to apply, press, and cure the joints to best advantage, and perhaps to simplify the materials handling problems for more economical production.

It is beyond the scope of this discussion to describe all the equipment available for mixing, spreading, handling the coated adherends through the assembly and air-drying or precuring stages, or to describe the various types of presses and heating systems for curing various adhesives.

Adhesives can be applied by spraying, roller coating, curtain coating, electrostatic deposition, or by hand brushing or rolling, much as might be used for application of paints and other coatings. Where solvents must be removed before pressing, air drying or infrared and other heating techniques may be used. Application of pressure may be by simple hand pressure, but more typically with mechanical screw or hydraulic pressure. Vacuum-bag pressure systems may be used in some cases where low pressures are adequate, and combinations of vacuum in a bag, in an autoclave where steam or external air pressure can be applied, are also used in bag-moulding or fluid-pressure techniques.

Heat to cure the adhesive can be applied from platens heated with electricity, steam, or hot liquid, by placing the clamped assemblies in heated chambers such as kilns or ovens, or by heating with high-frequency or induction heating. Here the important factor is to raise the temperature of the adhesive film to the desired value as rapidly as possible to produce the necessary curing reaction.

13 TYPICAL STRUCTURAL ADHESIVES AND APPLICATIONS

13.1 Plywood adhesives

The plywood industry is probably the largest current user of adhesives in the structural field. This industry is essentially two separate industries as far as the types and uses of the plywood itself are concerned and also from the standpoint of the adhesives used.

The hardwood plywood industry is the smaller of the two and is situated mainly in the Eastern United States, Southeast, South, and the Upper Midwest around the Great Lakes. Principal species for veneer are birch, gum, yellow poplar, elm, oak, and mahogany, although many other species are used for special panels when available. Most hardwood plywood is used for interior applications in furniture, doors, and architectural paneling where the appearance is more important than strength, and resistance to severe weather or other environmental conditions is not important. A limited amount of mahogany plywood is used for boats and other marine uses. A considerable amount of plywood from lower grade veneers is used for packaging, as in boxes and crates.

The principal adhesive used in hardwood plywood is urea-formaldehyde resin glue. This is usually extended with wheat flour to reduce glueline costs. Such plywood is expected to have only limited resistance to high relative humidity conditions, and is normally intended for use only in protected interior conditions. Container plywood is also made with flour-extended urea resins and may also contain some organic preservative where resistance to microorganism attack is a problem, such as for use under damp storage conditions or for overseas use. Where hardwood plywood is intended for boats or other severe service, melamine-urea resin combinations are used, although a limited amount of phenol-formaldehyde resin is also used.

The softwood plywood industry is many times larger in volume than the hardwood plywood industry. Douglas fir has been the principal species but in recent years other western softwoods (such as white fir, ponderosa pine and western larch) and southern pine have also been used extensively.

Until recently this industry centered largely in the Pacific Northwest and the Intermountain region, but now it has expanded in the Deep South and throughout the Southeast with the growth of the southern pine industry. Softwood plywood is normally used in building constructions, such as wall, floor, and roof sheathing, and in prefabricated building components; in such uses, strength and stiffness are important and the appearance is less critical since much of such plywood is either covered with other materials or painted.[30]

Exterior-type softwood plywood is made with phenol-resin glues. Plywood for interior uses is made with soybean and blood glues, and more recently with specially developed formulations of phenol resins extended with blood or certain lignocellulose products from agricultural processes. Requirements for glue-line quality in interior-type plywood may soon be raised to provide a greater degree of resistance, particularly to high moisture conditions and occasional water soaking. Ultimately this softwood plywood industry may change to an all-exterior glue line.

13.2 Laminated wood products

The development of the laminated wood industry has been rapid during the period following World War II, based largely on wartime technology in laminating durable keels and ribs for wooden naval vessels.[31] Such laminated timbers for exterior or severe service are now glued with resorcinol or phenol-resorcinol resins. Timbers for interior use are normally glued with water-resistant casein glues. There is now a growing interest and development in the production of mass-produced structural wood products such as straight beams. These are being produced with high-frequency techniques for rapid curing of the glues. In one interesting technique the individual boards are first preheated and then a specially formulated reactive phenol-resorcinol resin is continually mixed and applied rapidly to the heated surfaces; the beam is then immediately assembled and pressed with a special continuous press. This results in immediate cure without the necessity of heating to the glue line by conventional external techniques.

13.3 Composites of wood and other materials

Composites of wood, metals, and plastics are important means of producing new products by combining the best properties of each material.[32] Metal-faced plywood has been produced for many years. The original glue most widely used was a casein-latex formulation, cured at room temperature. Later two-stage bonding techniques were developed during World War II in which a special metal-adhesive primer was applied to the metal and cured by baking. The primed metal was then bonded to the wood with a conventional wood adhesive, such as a resorcinol resin, cured at room temperature.

In general, most satisfactory panels of metal-faced plywood, particularly when the metal is on only one face, are produced in room-temperature bonding processes. Hot-setting techniques tend to result in cupped or distorted panels because of unequal shrinkage of the metal and wood when the panel is cooled to room temperature on removal from the hot press. More recently synthetic rubber-base, contact-setting adhesives have found wide acceptance in bonding metal faces to wood.

Such adhesives, normally of a neoprene base, are coated on both mating surfaces, the solvents removed by air drying or heating, and then the two surfaces are joined carefully and bonded under only momentary pressure by means of nip roll presses. This gives a continuous bonding technique, which is also widely used for applying rigid plastic sheets to plywood and other wood-base panels. These panels are used mainly for interior, or other protected-use environments.

Interesting new products are now made by bonding flexible plastic films or sheets to plywood and other wood-base materials. Plastics include polyvinyl chloride, polyester, and polyvinyl fluoride films. Each plastic requires special adhesives to provide the necessary specific adhesion and wetting to the materials involved. Compositions are proprietary. Some of these adhesive systems used in such plastic-overlaid products are applicable to the continuous bonding process with roller pressure. Various resin-treated papers and vulcanized papers are now bonded to wood to provide better surface appearance and improved paint holding power, using modified polyvinyl acetate emulsion adhesive systems with the roll laminating process.[33]

13.4 Sandwich panels

Lightweight, rigid panels with thin and stiff sheets as facings and a variety of honeycomb cores of paper, plastic, or metals have been bonded successfully since World War II with a variety of adhesives similar to those for bonding sheets of metals to themselves. Adhesives include the polyvinyl-butyral, epoxy, phenol-nitrile rubber, and phenol-neoprene combinations, usually pressed in hot presses.[34] Sandwich panels are also being fabricated with metal faces and foamed plastic cores such as polystyrene or cellulose acetate, and wood-base facings are being bonded to paper honeycomb cores with conventional wood adhesives. In bonding to honeycomb cores, one additional property required of the adhesive is the ability to form a good fillet around the end of the honeycomb cell. Such adhesives are generally required to have good resistance to peel in addition to the necessary tensile and shear properties.

13.5 Bonding metal to metal

The bonding of metal to metal with adhesives, in place of welding or brazing, received much attention during World War II in connection with the fabrication of new aircraft structures.[1,34–37] Here the ability of adhesive bonds to distribute stresses over the entire joint areas was of particular importance, as compared to the use of mechanical fastenings. The success of metal helicopter rotor blades, which are subject to high fatigue stresses in service, is generally attributed to successful use of structural metal-bonding adhesives. The original adhesives used in metal-to-metal bonds in aircraft were those hot-setting adhesives based on combinations of phenol resins with either polyvinyl formal, polyvinyl butyral, neoprene, or nitrile rubbers. More recently some of the epoxy resin systems have been used, and other new systems of epoxy and polyamide resins are now important. To avoid difficulties with residual solvents entrapped in the joints between two nonporous adherends cured at temperatures of 250°F or higher, special tape adhesives of the same compositions as the solvent-based systems were developed and are in use. Some such tapes are supported on fabrics of glass fibers, nylon, or cotton, whereas others are unsupported.

A remarkable development in structural metal-bonding adhesives has occurred in recent years to meet the need for supersonic planes and more exotic space vehicles which are subject to temperatures in flight of 300°F or higher, often for prolonged periods. One early example was in the U.S. Air Force's B-58 "Hustler" supersonic bomber, with a speed of 1500 miles per hour. Some of the wing and fuselage surfaces, made of various honeycomb core sandwich panels, were subject to flight temperatures of 260°F.[38] An anticipated target requirement for such adhesives was the ability to retain strength at 325°F for 30,000 hours. These conditions were met by new formulations of phenol and epoxy resins, cured at hot-press temperatures.

Since that time an extensive development program for new high-temperature-resistant polymers for such adhesive applications in space vehicles has resulted in striking new adhesive systems capable of withstanding heat aging at temperatures in excess of 600°F for several hundred hours. One such successful system of adhesives is based on either polybenzimidazole or polyimides. As might be expected, such exotic adhesives are still costly and are subject to certain limitations in operating characteristics, such as inadequate flow, poor solubility for spreading, and very slow cure even at high curing temperatures. However, in spite of such current limitations, the potentials of these new adhesives are impressive and indicative of what can be accomplished by polymer chemists with new organic polymers in service at temperatures formerly considered far beyond the range of stability of conventional organic compounds.

An interesting factor shown in studies of heat aging of adhesive-bonded, metal-to-metal joints is the interrelationship of the metal-adhesive interface. Evidence suggests that in some cases the metal surface may exert undesirable catalytic effects on the adhesive film at high temperatures to accelerate the thermal degradation of some polymer systems. Such effects appear to be specific for certain metals and polymer systems. Much additional basic research is needed in this area to better understand the forces in action and to suggest methods by which they can be minimized to reduce undesirable thermal aging at high service temperatures.

13.6 Bonding plastic materials

As indicated previously, many plastic sheets and films may be successfully bonded. Because of the wide variety of such plastic materials it is impossible to generalize on the adhesive systems and bonding techniques to be used.[39] Phenolic resin-based plastic materials may often be bonded with phenol, resorcinol, and other typical wood-bonding adhesives. The newer epoxy resin adhesives are also important here, although these versatile adhesives are not suitable for all plastic materials. Some plastic materials require special surface preparation for bonding much as in metal bonding.[40] Polyethylene can be modified for bonding by flaming or use of corona discharges. Fluorinated hydrocarbons offer some of the greatest problems in bonding since their surface energies are low; and such surfaces are difficult to wet adequately enough to permit adequate adhesion. Such polymers have been successfully bonded after first treating them with sodium amalgam systems in naphthalene. The growing literature in this field attests to the wide variety of plastic materials being bonded.[39-41]

14 TESTING OF ADHESIVES

The use of adhesives in relatively critical applications requires a good system for the evaluation of new adhesive systems before selection for a new use, and for the routine evaluation during fabrication operations to ensure adequate bonding. Evaluation includes determination of working properties such as viscosity, solids content, pH, working life, curing rates, and storage life. The strength properties in shear, tension, peel, cleavage, fatigue, and under long-time loading conditions must be evaluated. In addition it is often desirable to determine the permanence characteristics of adhesives in actual joints. Fortunately test procedures for most of these properties have been developed and are described, particularly in Standards of the American Society for Testing and Materials.[42] Development of improved test procedures is continuing.

To determine the strength properties of adhesives, it is generally necessary to make joints with the particular adherends of interest and then break these under controlled conditions in some sort of mechanical testing procedure. This is generally acceptable in evaluation of the actual adhesive. However, another important problem is to ensure that bonded joints in production continue to meet prescribed standards of acceptance. This system has the obvious disadvantage of destroying the joint in using the conventional mechanical tests. At present the most suitable means of ensuring uniform bond quality in production is a rigid quality control system of all the bonding conditions known to affect joint quality. This includes control of the preparation of the adherends (such as surface preparation and machining), the mixing and spreading of the adhesive, assembly, and pressure and temperature-time conditions in curing the adhesive in the joint. Such quality control is rather laborious and subject to human failures in execution; naturally it presupposes an accurate knowledge of all factors likely to influence joint quality with the particular adhesive-adherend combination.

A more desirable approach to the problem of ensuring uniformly high joint quality is the use of nondestructive test procedures that can be routinely applied to production-bonded items, using modern statistical sampling techniques. Considerable research and development has been devoted to the development and reliability of such nondestructive procedures for adhesive-bonded joints, with particular attention to structural metal bonding of such great importance in the aircraft and space applications.

The most widely used technique at present is the response to ultrasonic vibration in such joints.[43] At first, such efforts were only successful in distinguishing between joints with some actual adhesion and actual voids. More recent studies tend to indicate some promise in distinguishing between joints of poor and acceptable quality when carefully calibrated for the particular adhesive-adherend combination. Work in this area is continuing, with both refinements in the ultrasonic systems and in the use of other influences such as magnetic effects. Further improvements may be expected to reach the desired goals for reliable nondestructive test procedures that can be used with a higher degree of confidence in the future.

15 ADHESIVE SPECIFICATIONS

The development of specifications for adhesives for specific applications has been under way for many years. Probably the most complete group of published specifications in this field are those developed by the military agencies for their more critical applications, and those of other government agencies for their own specific needs. Often these needs are considerably different than those of commercial firms and may not be directly applicable. Recently the American Society for Testing and Materials has begun the development of adhesive specifications for certain uses. In addition, many industrial firms have developed their own specifications, but most of these are not available for the use of others. Generally it is necessary for the prospective user of adhesives to develop his own specification based upon his particular requirements. These can be built around existing test procedures available in the ASTM Standards[42] and from other sources. The user will have to determine his performance standards of acceptance.

REFERENCES

1. R. F. BLOMQUIST, "Adhesives — Past, Present, and Future," 1963, Edgar Marburg Lecture, American Society for Testing and Materials, Philadelphia, Pa., 1963.
2. American Society for Testing and Materials, "Standard Definitions of Terms Relating to Adhesives," ASTM D 907-64a, American Society for Testing and Materials, Philadelphia, Pa., 1964.
3. T. D. PERRY, *Modern Wood Adhesives*, Pitman Publishing Corporation New York, 1944, p. 14.
4. F. L. DARROW, *The Story of an Ancient Art*, Perkins Glue Company, Lansdale, Pa., 1930.
5. R. HOUWINK and G. SALOMON, *Adhesion and Adhesives*, Vol. I, Elsevier Publishing Company, Amsterdam, 1965.
6. J. J. BIKERMAN, *The Science of Adhesive Joints*, Academic Press Inc., New York, 1960.
7. F. W. REINHART, *J. Chem. Educ.* **31**, 128 (Mar. 1954).
8. J. E. RUTZLER and R. L. SAVAGE, *Adhesion and Adhesives. Fundamentals and Practice*, John Wiley & Sons, Inc., New York, 1954.
9. P. WEISS, *Adhesion and Cohesion*, Elsevier Publishing Company, New York, 1962.
10. S. S. VOYUTSKII, *Autohesion and Adhesion of High Polymers*, Interscience Publishers, New York, 1963.
11. W. A. ZISMAN, *Ind. Eng. Chem.* **55**, 19 (Oct. 1963).
12. F. L. BROWNE and T. R. TRUAX, *Colloid Symposium Monographs* **4**, 258 (1926).
13. F. W. REINHART and I. G. CALLOMON, "Survey of Adhesion and Adhesives," Wright Air Development Center Technical Report 58-450, Wright-Patterson Air Force Base, Ohio, 1959.
14. L. H. SHARPE, H. SCHONHORN and C. J. LYNCH, *Intern. Sci. and Tech.* **28**, 36 (1964).
15. L. H. SHARPE and H. SCHONHORN, Advan. Chem. Ser. **43**, 189 (1964).
16. S. S. VOYUTSKII, *Adhesives Age* **5**, 30 (Apr. 1962).

17. S. V. SKINNER, R. L. SAVAGE and J. E. RUTZLER, *J. Appl. Phys.* **24**, 438 (1953).
18. G. W. KOEHN, "Design Manual on Adhesives," *Machine Design* **26**, 143 (1954).
19. N. K. BENSON, *Appl. Mech. Rev.* **14**, 83 (1961).
20. M. GOLAND and E. REISSNER, *J. Appl. Mech.* **11**, A17 (1944).
21. O. VOLKERSEN, *Luftfahrtforschung* **15**, 41 (1938).
22. E. W. KUENZI, "Determination of Mechanical Properties of Adhesives for Use in Bonded Joints," U.S. Dept. Agr. Forest Serv., Forest Prod. Lab. Rept. 1851, Forest Products Laboratory, Madison, Wis., 1956.
23. DIETER F. K. KUTSCHA, "Mechanics of Adhesive-Bonded Lap-Type Joints: Survey and Review," Technical Documentary Report No. ML-TDR-64-298, Wright-Patterson Air Force Base, Ohio, 1964.
24. C. A. REDFARN and J. BEDFORD, *Experimental Plastics*, Second Edition, Interscience Publishers Inc., New York, 1960.
25. W. R. SORENSON and T. W. CAMPBELL, *Preparative Methods of Polymer Chemistry*, Interscience Publishers, Inc., New York, 1962.
26. F. W. BILLMEYER, JR., *Textbook of Polymer Science*, Interscience Publishers Inc., New York, 1962.
27. T. S. CARSWELL, *Phenoplasts, Their Structure, Properties, and Chemical Technology*, Interscience Publishers, Inc., New York, 1947.
28. R. W. MARTIN, *The Chemistry of Phenolic Resins*, John Wiley & Sons, Inc., New York, 1956.
29. J. K. STILLE, *Introduction to Polymer Chemistry*, John Wiley & Sons, Inc., New York, 1962.
30. N. S. PERKINS, *Plywood — Properties, Design, and Construction*, Douglas-Fir Plywood Association, Tacoma, Wash., 1962.
31. A. D. FREAS and M. L. SELBO, "Fabrication and Design of Glued Laminated Wood Structural Members," U.S. Dept. Agr. Tech. Bull. 1069, U.S. Government Printing Office, Washington, D.C., 1954.
32. T. R. TRUAX and R. F. BLOMQUIST, *Symposium on Adhesives*, American Society for Testing and Materials Special Publication, Philadelphia, Pa., 1946.
33. B. G. HEEBINK, *Forest Prod. J.* **11**, 167 (1961).
34. H. W. EICKNER, "Evaluation of Several Adhesives and Processes for Bonding Sandwich Constructions of Aluminum Facings on Paper Honeycomb Core," NACA Tech. Note 2106, National Advisory Committee for Aeronautics, Washington, D.C., 1950.
35. R. F. BLOMQUIST, *Machine Design* **28**, 99 (May 31, 1956).
36. N. A. DEBRUYNE, *Structural Adhesives for Metal Aircraft*, Fourth Anglo-American Aeronautical Conference, Royal Aeronautical Society, London, 1953.
37. D. K. RIDER, *Prod. Eng.* **35**, 85 (May 25, 1964); *ibid.* **35**, 75 (June 8, 1964).
38. M. J. BODNAR and W. J. POWERS, *Symposium on Adhesives for Structural Applications*, Interscience Publishers, Inc., New York, 1961, p. 9.
39. M. J. BODNAR and W. J. POWERS, *Plastics Technol.* **4**, 721 (1958).
40. M. J. BODNAR and W. J. POWERS, *Symposium on Adhesives for Structural Applications*, Interscience Publishers, Inc., New York, 1961, p. 107.
41. H. A. PERRY, *Adhesive Bonding of Reinforced Plastics*, McGraw-Hill Book Company, Inc., New York, 1959.
42. *ASTM Book of Standards*, American Society for Testing and Materials, Philadelphia, Pa., 1965, Part 16.
43. A. G. H. DIETZ, H. W. BOCKSTRUCK, and G. EPSTEIN, *Symposium on Testing Adhesives for Durability and Permanence*, Spec. Tech. Publ. 138, American Society for Testing and Materials, Philadelphia, Pa., 1952.

Chapter 3

Structural Glued Laminated Timber

Russell P. Wibbens

1 HISTORY

While the art of wood gluing is very old — dating back to the time of the ancient Egyptians — structural gluing as it is thought of today dates from the turn of the twentieth century, when many glued laminated timber bridges and buildings were built in Europe using softwoods and casein adhesives. Structural timber laminating was introduced into the United States in the early 1930's. In 1934 a building using glued laminated timber three-hinged arches was constructed on the grounds of the U.S. Forest Products Laboratory in Madison, Wisconsin. This building is still in use by the Laboratory.

World War II created a great demand for heavy timber construction for military and industrial uses and accelerated the development of the laminating industry. The growth of the industry in the United States since that time is reflected by the statistics of the Bureau of the Census of the U.S. Department of Commerce. The Bureau's Census of Manufactures for the years 1954, 1958, and 1963 list the following quantities of lumber consumed in the fabrication of glued laminated structural elements and framing for bridges, buildings and other structures, including marine laminates.

1954	31,420,000 board feet
1958	47,845,000 board feet
1963	85,937,000 board feet

Another important factor in the growth of the laminating industry was the formation in 1952 of the American Institute of Timber Construction (AITC). The Institute was formed by a group of the nation's leading fabricators of both sawn and glued laminated structural timber as a means of advancing properly engineered, fabricated, and erected structural timber framing. It has established standards, specifications, and design recommendations for engineered timber construction. The laminating industry, acting through AITC, developed and submitted to the U.S. Department of Commerce a proposed commercial standard for structural glued laminated timber. This document gained approval and was promulgated in 1963 as U.S. Commercial Standard CS253-63.[1] It covers the minimum requirements for the production of structural glued laminated timber.

2 INTRODUCTION

The term "structural glued laminated timber," as used in this chapter, refers to an engineered stress-rated product of a timber laminating plant, comprising assemblies of suitably

RUSSELL P. WIBBENS is a structural engineer at the American Institute of Timber Construction, Washington, D.C.

selected and prepared wood laminations securely bonded together with adhesives. The grain of all laminations is approximately parallel longitudinally. The separate laminations may not exceed 2 in. in net thickness. They may be comprised of pieces end-joined to form any length, of pieces placed or glued edge to edge to make wider ones, or of pieces bent to curved form during gluing. Some of the advantages of glued laminated timber are

1. Glued laminated members can be fabricated in almost any length, size, or structural shape. Laminated beams with spans of over 100 ft, and arches and domes with clear spans exceeding 300 ft have been built.

2. Laminating permits better utilization of available lumber supplies, since large members can be fabricated from short lengths and narrow widths of lumber. The individual laminations are selectively placed in the member, permitting better dispersal of knots and other natural characteristics of the lumber, and resulting in improved strength. It is also possible to place higher grades of lumber in areas where high stresses occur, such as in the outer laminations of bending members.

3. Lumber used for laminating must be seasoned prior to gluing to meet the requirements of CS253-63. As a result, the finished member has a high degree of dimensional stability under dry-use conditions.

There are certain limiting factors concerning glued laminated timber which must also be considered. Among these are

1. The cost of the laminating process raises the cost of the finished member above that of a solid sawn member. However, because of the design variations made possible through the use of glued laminated timber, sizes and shapes not available in solid sawn timbers may be produced.

2. Great care and continuous quality control must be provided by the laminator during production in order to meet the requirements of CS253-63. On the other hand, this greater degree of control results in better assurance of a quality product.

3. Large curved laminated members may create shipping difficulties. The laminating industry has offset this limitation to a large extent by designing special connections which permit assembly of the member at the job site, and by the development of special rail cars and truck trailers to handle these members.

Lumber used for structural laminating includes those species whose allowable unit stresses have been developed in accordance with the principles set forth in "Fabrication and Design of Glued Laminated Wood Structural Members,"[2] or by a broadly representative and competent group, including producers and users, such as the American Society for Testing and Materials. Species for which allowable unit stresses for structural glued laminated timber have been so established, and the specifications in which they are set forth, are as follows:

1. Douglas fir and larch: "Standard Specifications for Structural Glued Laminated Douglas Fir (Coast Region) Timber"[3] and "Standards for Structural Glued Laminated Members Assembled with WWPA Grades of Douglas Fir and Larch Lumber."[4]

2. Hardwoods: "Standard, Specifications for the Design and Fabrication of Hardwood Glued Laminated Lumber for Structural, Marine and Vehicular Use."[5]

3. West coast hemlock: "Standard Specifications for Structural Glued Laminated West Coast Hemlock Lumber — Fabrication and Design."[6]

4. Southern pine: "Standard Specifications for Structural Glued Laminated Southern Pine Timber."[7]

5. California redwood: "Standard Specifications for Structural Glued Laminated California Redwood Timber."[8]

Laminating adhesives are of two types — dry-use and wet-use. Dry-use adhesives are those that perform satisfactorily where the moisture content of wood does not exceed 16 per cent for repeated or prolonged periods of service. Casein adhesives, with suitable mold inhibitors, are the standard dry-use adhesives of the structural glued laminated timber industry. Such adhesives have proven their dependability in both Europe and North America, and they are used in large quantities by other wood products manufacturers as well as by the laminating industry. Wet-use adhesives are those that perform satisfactorily for all conditions, including exposure to the weather, marine use, and pressure treatment. Phenol-, resorcinol-, and melamine-base adhesives are the most widely used wet-use adhesives in structural glued laminated members. They will withstand the most severe conditions of exposure. Melamine-urea adhesives are also used, primarily for gluing of end joints in laminations. The use of straight urea adhesives is not permitted by CS253-63.

3 MANUFACTURING AND QUALITY CONTROL

The integrity of glued laminated timber is of primary importance for the protection of life and property whenever the product is used for structural applications. The quality of the product is determined by the quality of the lumber, the suitability of the adhesive and the techniques used in bonding the lumber together with the adhesive.

The manufacture of glued laminated timber differs from many other manufacturing operations in that most of the production is for custom products, that is, the product is manufactured for a specific use. This makes true random sampling of the product difficult, and to assure the proper level of quality CS253-63 requires that the laminator have a quality control system that provides a continuous detailed check of each production process, a visual inspection of the finished production, and physical tests on samples of finished production.

The manufacturing procedures outlined below are followed, in general, by laminating plants in the United States, although individual plants may vary in the detailed procedures. The quality control procedures indicated below meet the requirements of CS253-63 and the "Inspection Manual," AITC 200.[9]

3.1 Lumber preparation

Lumber for laminating is carefully selected for knot size and location, slope of grain, rate of growth, density and other factors that affect the strength and/or appearance of the finished product. The lumber may be based on commercial grades or on special laminating grades. Where it is based on commercial grades, it is regraded where necessary to meet special requirements for laminating.

The finished member must function as a unit; therefore, the lumber must be precisely surfaced so that the faces can be brought into intimate contact without depending upon the adhesive to fill the gap. This requires that individual laminations have a smooth surface and be of uniform thickness across the width and along the length.

The moisture content of the lumber at the time of gluing should approximate that which it will attain in service but may not exceed 16 per cent. Moisture content is measured by means of portable electric meters or by automatic metering devices in the production line.

3.2 Adhesives

Adhesive manufacturers' quality control is of a high caliber; however, as an additional safeguard, the laminator also performs tests for strength and durability on each batch of adhesives received at his plant.

Adhesives must be spread with suitable equipment so as to produce a uniform application of a predetermined amount of adhesive. Weight of adhesive spread can be determined by one of several different methods, but must meet requirements of CS253-63.

3.3 End joints

Since lumber is rarely manufactured in lengths long enough for use in glued laminated structural members, the production of strong, durable end joints is one of the most important steps in the laminating process. Currently, both plain beveled scarf joints and finger joints are used for end-jointing. Finger joints may be cut through the wide face or through the narrow face of the lamination (see Fig. 3.1).

Scarf joints may be preglued and cured prior to assembly of the lamination into the finished member, or they may be glued and cured integrally with the individual laminations. Finger joints are always preglued. The thickness of assembled end joints must be within plus 0.020 in. to minus 0.005 in. to assure full pressure during the joint cure period.

End joints are checked for strength and wood failure by means of bending or tension test equipment. Laminating adhesives are stronger than the wood itself; therefore, a properly made joint, when tested, must meet or exceed bending or tension stress value criteria as specified in CS253-63 and must show more failure in the wood than in the glue line.

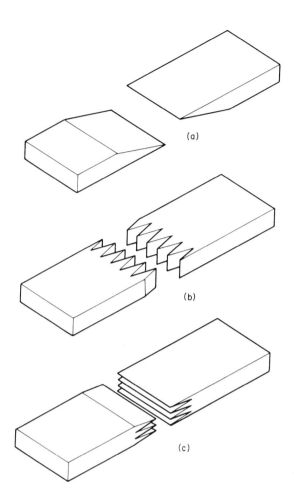

Figure 3.1 Typical End Joints for Structural Glued Laminated Timber. (a) Plain beveled scarf joint; (b) Finger joint—fingers cut through wide face of lamination; (c) Finger joint—fingers cut through narrow face of lamination

3.4 Edge joints

A lamination may consist of two or more pieces of lumber placed edge to edge. The edge joints need not be glued except when occurring in members loaded normal to lamination edges, or in members where torsion is a significant design consideration.

3.5 Face joint assembly and bonding

The individual laminations are passed through equipment that spreads adhesive on one or both faces. As the laminations emerge from the spreader, each is picked up and placed in a clamping form. Prior to this time, the clamping form will have been set to the desired shape of the finished member by means of full-scale lumber or plywood templates which are built in the plant.

As the laminations are placed in the form, quality-control personnel carefully check to determine that proper assembly times are observed. Assembly time consists of open assembly, which is the time elapsed between spreading the adhesive and assembling the spread surfaces into close contact with one

another, and closed assembly, which is the time elapsed from the assembly of the first laminations of the package into intimate contact until final application of pressure or heat or both to the entire package.

Clamping and pressure are necessary to bring the surfaces of the laminations being bonded into intimate contact, to pull the member into shape, to force out excessive adhesive, and to hold the pieces firmly together until the adhesive has developed sufficient strength. The means of pressure application must be such that uniform pressure is applied and maintained throughout the curing period of the adhesive. The entire member is tightened as a unit. In most plants evenly spaced clamps are progressively and uniformly tightened with an air impact wrench. Some plants use hydraulic presses for applying pressure to smaller members. Where clamps are used, bolts are checked with a torque wrench to ensure that the correct pressure has been applied and is being maintained. CS253-63 prohibits the nailing of the structural portion of laminations in lieu of clamping or other positive methods of controlling pressure.

Glue-line cure is important. Some adhesives cure at room temperatures, while others require added heat. Heat may be supplied in a number of ways. Some plants raise the plant temperature overnight, some place space heaters near the curing members, some use special curing chambers, and some place canvas or plastic "tents" over the members and introduce heat.

3.6 Finishing

After the laminated member has been removed from the clamps, it is transported to the plant's finishing area where it is planed and cut to length, fabrication of holes and/or daps for connections is performed, and the member is sealed, stained, and/or painted in accordance with job specifications.

Three finished appearance grades have been established for structural glued laminated members. They are industrial, which is ordinarily suitable for construction in industrial plants, warehouses, garages, and other uses where appearance is not of primary concern; architectural, which is ordinarily suitable for construction where appearance is an important requirement; and premium, which is suitable for uses which demand the finest appearance. The appearance grades apply to the surfaces of glued laminated members. They do not apply to laminating procedures, nor do they modify design stresses, fabrication controls, grades of lumber used or other provisions of standard laminating specifications.

Before leaving the plant, the finished members may be wrapped in water-resistant paper for the protection during transit, storage, and erection. The plant's responsibility also includes proper loading of the products on rail cars or trucks for shipment. It may be necessary to make special routing or transportation arrangements for oversize loads.

3.7 Visual inspection

In order to determine conformance to CS253-63 and the job specifications, all production is visually inspected for dimensions; shape, including camber and cross section; type, quality, and location of edge and end joints; appearance grade;

lumber species and orientation of grades; moisture content; adhesive type; and glue-line thickness.

3.8 Physical tests

Physical tests, consisting of in-line tests and tests on samples of finished production, are run by the laminator to assure a proper day-to-day level of quality. Face, edge, and end-joint bonding and each combination of wood species and adhesive type used by the plant are represented by samples cut either from production members or from special samples made under production conditions.

Quality of face joint bonding is determined by a block shear test. In this test, samples of laminations adjacent to glue lines are cut with the glue lines intact. The forces necessary to shear the blocks along the glue line and the wood failure on the sheared surfaces are measured. The block shear test may also be used to evaluate scarf joint bonding.

For wet-use service an additional test for integrity of glue bonds, known as the cyclic delamination test, is performed. In it, a production sample is saturated with water in an autoclave or similar pressure vessel, a vacuum is drawn, and then pressure is applied. After removal from the autoclave the sample is dried in an oven. The soaking-drying cycle is then repeated twice. At the end of the final drying period, the total length of open glue lines (delamination) is measured and compared to the total length of glue line on the sample.

Physical tests of end joints were previously discussed under that heading.

The plant quality control department maintains careful records of block shear tests, delamination tests, and other laboratory tests to ensure conformance to CS253-63.

3.9 Marking and certification

Provision is made in CS253-63 for marking or certifying structural glued laminated timber members with the identification mark or certificate of a qualified central inspection organization as evidence that the members are in conformance with the Standard and the applicable job specifications. In permitting the use of its marks or certificates, the qualified central inspection organization must determine (a) that the quality control system of the manufacturer of the product, including his quality control and procedures manuals and their application, enable him to manufacture products that conform with CS253-63; and (b) by periodic inspection, that the manufacturer applies identification marks of conformance to CS253-63 and the applicable job specifications only on products which in fact so conform.

The laminating industry, in developing CS253-63 through AITC, added the provisions on marking and certification of products as a means of assuring the purchaser that he is getting structural glued laminated timber of the grade and quality specified. The Institute established the AITC Inspection Bureau, which is a qualified central inspection organization as defined by CS253-63, and permits only qualified laminators to use AITC quality marks and certificates of conformance to CS253-63.

4 DESIGN

Engineering formulas and design criteria applicable to solid wood members are, in general, applicable also to glued laminated members. Timber design information may be found in the following, among other, publications: "Timber Construction Manual",[10] "National Design Specification for Stress-Grade Lumber and Its Fastenings",[11] "Modern Timber Engineering",[12] "Timber Design and Construction Handbook",[13] and "Douglas Fir Use Book".[14]

Most of the design information herein is taken or summarized from the First Edition (1966) of the "Timber Construction Manual," prepared by AITC, and is used with their permission. This publication treats design and related material in depth and contains many time-saving design aids.

The principal differences between the design of glued laminated timber and sawn lumber are due to the use of higher allowable working stresses for laminated material and in the modification of these stresses for various design conditions. The strength values for glued laminated timber are derived from the same basic data as for sawn lumber and the increased stress values for laminated material derive from three basic factors:

1. The increase in strength resulting from drying. The strength properties of wood increase in drying, but in large sawn timbers, these increases are offset by internal stresses developed in drying. These internal stresses are largely eliminated in glued laminated timber because the net thicknesses of individual laminations are 2 in. or less and because the laminated materials are dried to a uniform moisture content.

2. Dispersement of strength-reducing characteristics. The process of laminating incorporates a natural dispersal of the strength-reducing characteristics present in the individual laminations.

3. Placement of laminations. Higher-strength laminations are used in the more highly stressed portions of a laminated member.

4.1 Notations and symbols

The wood products industry has adopted engineering notations and symbols for use in timber design work. These symbols are employed herein.

A area of cross section

b breadth (width) of rectangular member

C coefficient, constant, or factor

C_c curvature factor

C_d depth effect factor

C_f form factor

C_s slenderness factor

c distance from neutral axis to extreme fiber

D diameter

d depth of rectangular member, or least dimension of compression member

E modulus of elasticity

e eccentricity

F_b allowable unit stress for extreme fiber in bending

f_b actual unit stress for extreme fiber in bending

F_c allowable unit stress in compression parallel to grain

F'_c allowable unit stress in compression parallel to grain adjusted for l/d ratio

f_c actual unit stress in compression parallel to grain

$F_{c\perp}$ allowable unit stress in compression perpendicular to grain

$f_{c\perp}$ actual unit stress in compression perpendicular to grain

F_r allowable unit radial stress

f_r actual unit radial stress

F_{rc} allowable unit radial stress in compression

f_{rc} actual unit radial stress in compression

F_{rt} allowable unit radial stress in tension

f_{rt} actual unit radial stress in tension

F_t allowable unit stress in tension parallel to grain

f_t actual unit stress in tension parallel to grain

F_v allowable unit horizontal shear stress

f_v actual unit horizontal shear stress

h rise

I moment of inertia

L span length of beam, or unsupported length of column, in feet

l span length of beam, or unsupported length of column, in inches

M bending moment

m unit bending moment

P total concentrated load, or axial compression load

P/A induced axial load per unit of cross-sectional area

Q statical moment of an area about the neutral axis

R radius of curvature

R_H horizontal reaction

R_V vertical reaction

r radius of gyration

S section modulus

T total axial tension load

t thickness

V total vertical shear

W total uniform load

w uniform load per unit of length

Δ_A allowable deformation or deflection

Δ_a actual deformation or deflection

$\parallel$ parallel

$\perp$ perpendicular

4.2 Allowable unit stresses

Tables 3.1 through 3.3 give maximum allowable unit stress combinations recommended by the regional lumber associations which publish standard specifications for structural glued laminated timber. The tables are based on normal duration of loading and are divided into sections for dry-use and wet-use conditions. Dry-use conditions are those in which the moisture content in service is less than 16 per cent as in most covered structures. Wet-use conditions are those in which the moisture content in service is 16 per cent or more, as it may be in exterior or submerged construction and in some structures housing wet processes or otherwise having constant high relative humidities. Tables 3.1 and 3.2 give recommended maximum allowable stress combinations for horizontally laminated members (load acting normal to wide face of laminations). Table 3.3 gives recommended maximum allowable stress combinations for vertically laminated members (load acting normal to edge of laminations).

Requirements for slope of grain, type and location of end joints, certain manufacturing requirements and other factors as given in the regional lumber association laminating specifications must be met if these recommended maximum allowable stress combinations are to apply. Note that, although maximum allowable stress combinations are given in Tables 3.1 through 3.3, it is likely to be more economical to use lower allowable stress combinations published in the regional lumber association structural laminating specifications for the species (see References 4–8).

Radial tension or compression stress

When a curved laminated member is subjected to a bending moment, a stress is induced in a radial direction at right angles to the grain. The maximum magnitude of this stress, f_r, occurs at the neutral axis of a rectangular section and is given by the formula

$$f_r = 3M/2Rbd.$$

According to presently accepted design criteria, when M is in the direction tending to decrease curvature (increase the radius), this radial stress is tension and is limited to one-third of the value of the allowable unit stress in horizontal shear, F_v. When M is in the direction tending to increase curvature (decrease the radius), the radial stress is compression and is limited to the value of the allowable unit stress in compression perpendicular to the grain, $F_{c\perp}$. Research is presently being conducted in an effort to refine current design criteria for curved structural glued laminated members to resist radial tension stresses.

4.3 Modification of stresses

Allowable unit stress values given in Tables 3.1 through 3.3 are for normal duration of loading. Adjustment of these stresses for other durations of loading and for various service and design conditions are given in the following paragraphs.

Duration of load

Normal load duration means fully stressing a member to its allowable stress by the application of the full maximum normal design load for a period of approximately 10 years (either continuously or cumulatively during the life of the structure). Loads whose application results in stresses less than 90 per cent of this full maximum normal load need not be counted in accumulating the 10-year period. For other durations of load, either continuously or intermittently applied, one of the factors given in Table 3.4 should be applied.

If loads of different durations are applied simultaneously, the size of the member required is determined for the total of all loads applied at the allowable unit stress adjusted by the factor for the load of shortest duration in the combination. In like manner, but neglecting the load of shortest duration, the size of member required to support the remaining loads at the stress adjusted by the factor for the load of next shortest duration is determined. By repeating this procedure for all the

Table 3.1   Recommended Maximum Allowable Unit Stress Combinations for Structural Glued Laminated Softwood Species[a]

Species	Stress Values, in pounds per square inch					
	F_b	F_t	F_c	$F_{c\perp}$	F_v	E^b
Dry Conditions of Use						
Douglas fir (coast region)						
Members stressed principally in bending						
	2600	1600	1500	450	165	1.80
Members stressed principally in axial tension or compression						
	2600	2600	2200	450	165	1.80
Douglas fir and larch						
Members stressed principally in bending						
	2600	1900	1400	410	195	1.82
Members stressed principally in axial tension or compression						
	2400	2400	1800	385	195	1.82
West Coast hemlock	2200	2200	1700	365	140	1.54
Southern pine	2600	2600	2000	385	200	1.80
California redwood						
Members stressed principally in bending						
	2200	2000	2200	325	125	1.30
Members stressed principally in axial tension or compression						
	2200	2200	2200	325	125	1.30
Wet Conditions of Use						
Douglas fir (coast region)						
Members stressed principally in bending						
	2000	1300	1100	305	145	1.60
Members stressed principally in axial tension or compression						
	2000	2000	1600	305	145	1.60
Douglas fir and larch						
Members stressed principally in bending						
	2100	1500	1000	275	175	1.51
Members stressed principally in axial tension or compression						
	1900	1900	1300	260	175	1.51
West coast hemlock	1800	1800	1200	240	120	1.40
Southern pine	2000	2000	1500	260	175	1.60
California redwood						
Members stressed principally in bending						
	1800	1600	1600	215	110	1.20
Members stressed principally in axial tension or compression						
	1800	1800	1600	215	110	1.20

[a] Recommended in regional lumber association standard structural laminating specifications. (See References, 4, 6–8.)
[b] Multiply values by 1,000,000.

remaining loads, the size of member required for the controlling duration of load condition is obtained. However, when the permanently applied load is less than or equals 90 per cent of the total normal load (including the permanently applied load), the normal loading condition will control the size of member required.

Temperature

Some reduction of stresses may be necessary for glued laminated members that are subjected to elevated temperatures for repeated or prolonged periods of time, especially where the high temperature is associated with a high moisture content in the wood. Tests have shown that an increase in strength occurs at low temperatures.

Treatments

The allowable unit stress values given in Tables 3.1 through 3.3 for glued laminated timber also apply to wood that has been treated with a preservative when this treatment is in accordance with American Wood-Preservers' Association standards[15] which limit pressure and temperature. Investigations have indicated that, in general, any weakening of timber as a result of preservative treatment is caused almost entirely by subjecting the wood to temperatures and pressures which are above the AWPA limits, rather than by the preservative used.

Highly acidic salts such as zinc chloride, if present in appreciable concentrations, tend to hydrolyze wood. Fortunately, the concentrations used in wood preservative treatments are sufficiently small that the strength properties other than impact resistance are not greatly affected under normal use conditions. A significant loss in impact strength may occur, however, if higher concentrations are used. As none of the other common salt preservatives is likely to form solutions as highly acidic as those of zinc chloride, in most cases their effect on the strength of the wood can be disregarded.

Table 3.2   Recommended Maximum Allowable Unit Stress Combinations for Structural Glued Laminated Hardwood Species[a]

Species	Stress Values, in pounds per square inch									
	F_b and F_t		F_c		$F_{c\perp}$		F_v		E[b]	
	Dry	Wet	Dry	Wet	Dry	Wet	Dry	Wet	Dry	Wet
Ash, black	1600	1300	1300	900	370	240	170	140	1.2	1.1
Ash, commercial white	2300	1800	2200	1600	610	410	230	210	1.6	1.5
Beech, American	2400	2000	2400	1800	610	410	230	210	1.8	1.6
Birch, sweet and yellow	2400	2000	2400	1800	610	410	230	210	1.8	1.6
Cottonwood, Eastern	1200	1000	1200	900	180	120	110	100	1.1	1.0
Elm, American and slippery	1800	1400	1600	1100	310	210	190	170	1.3	1.2
Elm, rock	2400	2000	2400	1800	610	410	230	210	1.4	1.3
Hickory, true and pecan	3100	2500	3000	2200	730	490	260	230	2.0	1.8
Maple, black and sugar	2400	2000	2400	1800	610	410	230	210	1.8	1.6
Oak, commercial red and white	2300	1800	2000	1500	610	410	230	210	1.6	1.5
Sweetgum	1800	1400	1600	1100	370	240	190	170	1.3	1.2
Tupelo, black and water	1800	1400	1600	1100	370	240	190	170	1.3	1.2
Yellow poplar	1400	1200	1400	1000	270	180	150	130	1.2	1.1

[a] Recommended in "Standard Specifications for the Design and Fabrication of Hardwood Glued Laminated Lumber for Structural, Marine and Vehicular Use" (Ref. 5).
[b] Multiply values by 1,000,000.

Table 3.3   Recommended Maximum Allowable Unit Stresses for Vertically Laminated Timber[a,b]

Species	Stress Values, in pounds per square inch							
	F_b		$F_{c\perp}$		F_v[c]		E[d]	
	Dry	Wet	Dry	Wet	Dry	Wet	Dry	Wet
Douglas fir (coast region)	2600	2000	450	305	165	145	1.80	1.60
Douglas fir and larch	2300	1800	385	260	195	175	1.82	1.51
West coast hemlock	1750	1600	365	240	115	100	1.54	1.40
Southern pine	2700	2200	450	305	200	175	1.80	1.60
California redwood	2200	1800	325	215	125	110	1.30	1.20

[a] Recommended in regional lumber association standard structural laminating specifications for three or more vertical laminations of the same grades of lumber. The allowable stresses for vertically laminated members made of combinations of grades of lumber will be the weighted averages of the grades.
[b] For allowable unit stress values for vertically laminated hardwood timber, see "Standard Specifications for the Design and Fabrication of Hardwood Glued Laminated Lumber for Structural, Marine and Vehicular Use" (Ref. 5).
[c] See special restrictions required in regional laminating specifications (References 4, 6–8) for these values to apply.
[d] Multiply values by 1,000,000.

In wood treated with highly acidic salts such as zinc chloride, moisture is the controlling factor in the corrosion of fastenings. Therefore, wood treated with highly acidic salts is not recommended for use under highly humid conditions.

The effect on strength of treatments other than preservative treatments should be investigated.

Size effect

Unit strength values, computed by the usual engineering methods from test data on wood bending members, decrease

Table 3.4   Design Factors for Duration of Loading

Duration of Load	Factor
Permanent	0.90
Normal	1.00
2 months (as for snow)	1.15
7 days	1.25
Wind or earthquake	1.33
Impact	2.00

as the size of the member increases. Current industry practice takes this decrease into consideration by modifying the allowable unit bending stress F_b for glued laminated beams over 12 in. in depth by a depth effect factor C_d:

$$C_d = 0.81 \left(\frac{d^2 + 143}{d^2 + 88} \right),$$

in which d = depth of member, in inches.

Recent research, reported in U.S. Forest Service Research Paper FPL 56,[16] indicates that this decrease in strength is an effect of size and stress distribution within the member rather than merely a depth effect.

Curvature factor

Stress is induced when laminations are bent to curved forms. Although much of this stress is quickly relieved, some remains and tends to reduce the strength of a curved member; therefore, the allowable stress in bending must be adjusted by multiplication by a curvature factor C_c:

$$C_c = 1 - 2000(t/R)^2.$$

The ratio t/R may not exceed 1/100 for hardwoods and Southern pine, nor 1/125 for softwoods other than Southern pine. The curvature factor is not applied to stresses in the straight portion of a member regardless of curvature in other portions.

The minimum radius to which lumber can be bent for use in curved laminated members varies with species. Table 3.5 gives the recommended minimum bending radii for various lamination thicknesses of several laminating species.

Form factor

When members with cross sections other than rectangular are used as beams, the allowable unit stress in bending, F_b, should be modified by a form factor C_f, as given in Table 3.6.

Table 3.6   Form Factors for Beams

Cross Section	Form Factor, C_f
Circular	1.18
Square section loaded diagonally	1.414
Box and I sections	$0.81\left[1 + \left(\frac{d^2 + 143}{d^2 + 88} - 1\right)C_g\right]$

in which $C_g = p^2(6 - 8p + 3p^2)(1 - q) + q$
p = ratio of depth of compression flange to full depth of beam
q = ratio of thickness of web or webs to full width of beam

Lateral support

Economy in glued laminated timber beam design usually favors a deep and narrow section. Such a section increases the likelihood of lateral buckling of the compression flange; therefore, in determining the stability of a deep-narrow glued laminated beam, the length of the compression flange is critical and should be contained in the governing parameters.

When the depth of a beam does not exceed its breadth, no lateral support is required. When the depth exceeds the breadth, lateral support is required, and the slenderness factor of the beam, C_s, is calculated from the formula

$$C_s = \sqrt{l_e d/b^2},$$

in which l_e = effective length of beam in inches (see Table 3.7).

When the slenderness factor C_s does not exceed 10, the full allowable unit stress in bending, F_b, adjusted in accordance with the provisions mentioned earlier in this chapter in the section "Modification of Stresses," is used and is defined as F_b'.

Table 3.5   Recommended Minimum Bending Radii for Various Lamination Thicknesses

Lamination Thickness inches	Douglas Fir, Larch, and West Coast Hemlock — Tangent Ends[a]	Douglas Fir, Larch, and West Coast Hemlock — Constant Curvature	Southern Pine	White Oak
$\frac{1}{4}$	2 ft 7 in. (31 in.)	2 ft 7 in. (31 in.)	2 ft 1 in. (25 in.)	1 ft 6 in. (18 in.)
$\frac{3}{8}$	4 ft 0 in. (48 in.)	4 ft 7 in. (55 in.)	3 ft 2 in. (38 in.)	2 ft 6 in. (30 in.)
$\frac{1}{2}$	6 ft 0 in. (72 in.)	7 ft 2 in. (86 in.)	4 ft 2 in. (50 in.)	3 ft 7 in. (43 in.)
$\frac{5}{8}$	7 ft 8 in. (92 in.)	9 ft 10 in. (118 in.)	5 ft 3 in. (63 in.)	4 ft 10 in. (58 in.)
$\frac{3}{4}$	9 ft 4 in. (112 in.)	12 ft 6 in. (150 in.)	6 ft 3 in. (75 in.)	6 ft 1 in. (73 in.)
1	15 ft 0 in. (180 in.)	20 ft 4 in. (244 in.)	8 ft 4 in. (100 in.)	8 ft 9 in. (105 in.)
$1\frac{1}{4}$	20 ft 8 in. (248 in.)	28 ft 0 in. (336 in.)	10 ft 5 in. (125 in.)	11 ft 8 in. (140 in.)
$1\frac{1}{2}$	27 ft 6 in. (330 in.)	35 ft 6 in. (426 in.)	12 ft 6 in. (150 in.)	14 ft 10 in. (178 in.)
$1\frac{5}{8}$	32 ft 0 in. (384 in.)	40 ft 0 in. (480 in.)	13 ft 7 in. (163 in.)	16 ft 0 in. (192 in.)
$1\frac{3}{4}$	36 ft 0 in. (432 in.)	45 ft 0 in. (540 in.)	14 ft 7 in. (175 in.)	18 ft 1 in. (217 in.)
2	45 ft 0 in. (540 in.)	56 ft 0 in. (672 in.)	16 ft 8 in. (200 in.)	21 ft 4 in. (256 in.)

[a] A member is considered to have tangent ends when the straight end distance is equal to or greater than 6 in. plus the depth of the member.

Table 3.7   Effective Length of Glued Laminated Beams

Type of Beam Span and Nature of Load	Value of Effective Length, l_e[a]
Single span beam, load concentrated at center	1.61l
Single span beam, uniformly distributed load	1.92l
Single span beam, equal end moments	1.84l
Cantilever beam, load concentrated at unsupported end	1.69l
Cantilever beam, uniformly distributed load	1.06l
Single span or cantilever beam, any load (conservative value)	1.92l

[a] Where l = unsupported length as defined in the text.

When the slenderness factor C_s is greater than 10 but does not exceed C_k, the allowable unit stress in bending, F_b', is calculated from the formula

$$F_b' = F_b[1 - \tfrac{1}{3}(C_s/C_k)^4],$$

in which $C_k = \sqrt{3E/5F_b}$.

When the slenderness factor C_s is greater than C_k but less than 50, the allowable unit stress in bending is calculated from the formula

$$F_b' = 0.40E/C_s^2.$$

In no case may C_s be greater than 50.

When the compression edge of a beam is so supported throughout its length as to prevent its lateral displacement, and the ends at points of bearing have lateral support to prevent rotation, the unsupported length l may be taken as zero.

When lateral support is provided to prevent rotation at the points of end bearing, but no other support to prevent rotation or lateral displacement is provided throughout the length of a beam, the unsupported length l is the distance between such points of bearing or the length of a cantilever.

When the beams are provided with lateral support to prevent both rotational and lateral displacement at intermediate points as well as at the ends, the unsupported length l may be taken as the distance between such points of intermediate lateral support. If lateral displacement is not prevented at these points of intermediate support, the unsupported length l must be defined as the full distance between points of bearing with adequate provision against rotation or as the full length of the cantilever.

Some of the means available for preventing rotation of a beam at its points of bearing are anchoring the bottom of the beam to pilaster and the top of beam to parapet; grounding the roof diaphragm to the wall; and providing a girt at the top of the wall and rod bracing for beams on wood columns with open sidewalls.

To provide continuous support of a compression flange, composite action between deck elements to create diaphragm action is essential for full lateral support. A plywood deck with edge nailing is one of the best examples of such composite action between deck elements. When plank decking is used, nailing patterns are most important so that nail couples will be created; having one nail per deck plank and no nails between planks fails to provide a system with adequate lateral support. If a wood deck is to supply such support, each piece must be securely nailed directly to the beam and to adjacent pieces to give a rigid diaphragm. If other kinds of decks are to provide such support, they must supply equivalent rigidity as a diaphragm.

If adjacent deck planks are nailed to each other so that little or no differential movement can occur between planks, then each plank will shift the same, namely zero, since planks over end supports cannot shift if torsional rotation is prevented at these end supports.

When joists have depth-breadth ratios of 8 or 9 (such joists have been in use for many years), the compression flange is continuously supported by a deck and the end bridging prevents torsional rotation. Intermediate bridging, in such cases, is provided to distribute concentrated loads to adjacent joists.

4.4 Standard sizes

The most efficient and economical production of glued laminated structural members results when standard lumber sizes are used for the laminates. Industry-recommended practice uses nominal 2-in. thick lumber of standard nominal width to produce straight members and curved members where the radius of curvature is within the bending radius limits for that thickness of the species (see Table 3.5). Nominal 1-in. thick boards are normally used when the bending radius is too sharp to permit use of nominal 2-in. thick laminations. These are standard practices subject to deviation to conform with specific job requirements and plant procedures. The use of nominal 1-in. and 2-in. thick laminations is generally the most economical and is, therefore, recommended for all normal uses. Exceptions should be made only when the shape of the structure requires nonstandard laminates.

Proper gluing procedures require surfaces planed uniformly smooth to exact thickness with a maximum allowable variation of plus or minus 0.008 in. Recommended standard practice is to surface nominal 2-in. laminations to a net $1\frac{5}{8}$- or $1\frac{1}{2}$-in. thickness, and nominal 1-in. laminations to a net $\frac{3}{4}$-in. thickness. Finished depths of members are thus increments of these net thicknesses. Laminations of special thicknesses may be used because of bending radius or the mixing of thicknesses for special purposes, thus resulting in net finished depths which may be nonstandard.

It is necessary to surface the wide faces of members to remove the glue squeeze-out and provide a uniformly smooth surface. Therefore, the net finished width of the glued laminated member is less than the net finished width of industry standard boards and dimension stock. Industry standard finished widths for glued laminated structural members are given in Table 3.8.

Table 3.8 Standard Widths of Structural Glued Laminated Timber Members

Nominal width, in.	3	4	6	8	10	12	14	16
Net finished width, in.	$2\frac{1}{4}$	$3\frac{1}{4}$	5 or $5\frac{1}{4}$	7	9	11	$12\frac{1}{2}$	$14\frac{1}{2}$

4.5 Loads and their application

Structural timber framing should be designed, as stipulated by the governing building code, to sustain dead load, live load, snow load, wind load, impact load, and earthquake load, and any other loads and forces which may reasonably affect the structure during its service life. In the absence of a governing code, the loads, forces, and combination of loads should be in accordance with accepted engineering criteria for the area under consideration.

Basic minimum roof load combinations

In designing, the most severe realistic distribution, concentration, and combination of roof loads and forces should be taken into consideration. Table 3.9 gives the basic minimum roof load design combinations which should be checked in all structures with these types of loads.

Each type of load should be determined individually and the effect of all possible combinations investigated. All possibilities must be investigated for unsymmetrical buildings.

Table 3.9   Basic Recommended Minimum Load Combinations for Design

Roof Load Combinations		Duration of Load	Stress Modification Factor for Duration
Wind→			
Windward Side	Leeward Side		
DL	DL	Permanent	0.90
DL + LL	DL + LL	7 days	1.25
DL[a]	DL + LL[a]	7 days	1.25
DL + SL	DL + SL	2 months	1.15
DL + ½SL[b]	DL + SL[b]	2 months	1.15
DL + WL	DL + WL	Wind or earthquake	1.33

[a] Full unbalanced loading.
[b] Half unbalanced loading.

NOTES

1. See "Duration of Load" provisions discussed earlier in this chapter for determining the governing duration factor when loads of different durations are applied simultaneously.

2. Special configurations, locations, and occupancies of structures may require investigation of (a) full unbalanced SL or (b) combinations of WL with LL and DL, or WL with partial SL and DL.

3. Where the magnitude of DL and other permanently applied loads is great, the effect of seismic loading should be investigated.

Special consideration should be given to structures of great span and/or height, and to trusses bearing on very long columns.

Structural members or systems should be designed to resist the stresses caused by partially or fully unbalanced live or snow loads and wind loads, including uplift, in combination with dead loads on the member or system, if such loading results in reversal of stresses or stresses greater in any portion than the stresses produced on the entire roof by the load combinations. The occupancy or use of the structure, its configuration, heating considerations, local climatic conditions, or other considerations may also cause partial or full unbalanced loading on the structure.

The following symbols are employed in Table 3.9:

DL = dead load, including weight of structure;
LL = live load;
SL = snow load, modified for roof slope; and
WL = wind load, modified for shape of roof, openings, and so on.

Ponding

When there is the possibility of water ponding, which may cause excessive loads and additional and progressive deflection, each component of the roof system, including decking, purlins, beams, girders, or other principal structural supports, should be designed accordingly. Continuous or cantilevered components should be designed for balanced or unbalanced load, whichever produces the most critical condition. In addition,

adequate drainage capacity, provision for heavy downpours, and proper construction details should be provided.

Roof beams should have a positive slope or camber equivalent to $\frac{1}{4}$ in. per ft of horizontal distance between the level of the drain and the high point of the roof, in addition to the recommended minimum camber given in Table 3.11 (see p. 84), to avoid the ponding of water.

When flat roofs have insufficient slope for drainage (less than $\frac{1}{4}$ in. per ft), when ponding of water is intentional, or when ponding occurs because of parapet walls, high gravel stops, plugged drains, coincidence of snow loads and rain, beams of insufficient stiffness, melting of snow and ice dams, and so on, deflection is initiated and continues at a progressive rate because the work energy of the ponding water is initially greater than the resistance of the beam. This deflection continues until equilibrium is reached between the added water load and the resistance of the beam, or it continues until failure occurs.

Beam deflection should not exceed the recommended limitations given in Table 3.10 (see p. 84). When flat roofs have insufficient slope for drainage (less than $\frac{1}{4}$ in. per ft), the stiffness of supporting members should be such that a 5-psf load will cause no more than $\frac{1}{2}$-in. deflection.

The effect of ponding of water on a flat roof system supported by glued laminated beams without camber is to magnify deflections and stresses under dead load, or dead load plus uniformly distributed applied load, or concentrated applied load (whatever loading is on the roof during ponding). Uniformly distributed applied loads may consist of snow loads or depths of water created by gravel stops, parapet walls, or ice dams. This magnification can be expressed by the formula

$$C_p = \frac{1}{1 - W' l^3/\pi^4 EI},$$

in which

C_p = factor for multiplying stresses and deflections under existing loads to determine stresses and deflections under existing loads plus ponding, and

W' = total load of 1-in. depth of water on the roof area supported by the beam or deck, in pounds.

The derivation of this magnification factor is based upon elementary theory of the mechanics of materials and is applicable to all engineering materials. It has been verified closely by experiments conducted at the U.S. Forest Products Laboratory. The analysis assumed elastic behavior and did not account for stresses or deflections caused by creep. The factor applies to each element of a roof system, including decking, purlins, beams, and girders. Allowable design stresses, as modified for duration of loading, depth-effect factor, and other applicable factors, and deflection limits may not be exceeded after application of the magnification factor to existing stresses and deflections.

This magnification factor may also be used for flat-roof systems with camber; however, if the effect of camber is disregarded, this method will yield conservative answers since the amount of ponding will always be less because of camber.

4.6 Column design

The usual formulas for determining the load-carrying capacities of wood columns may also be used for glued laminated timber columns. Wood columns are classified as simple solid or spaced.

Simple solid rectangular columns

Simple solid laminated columns consist of pieces properly glued together to form a single member. The l/d ratio for simple columns (see Fig. 3.2) may not exceed 50. The allow-

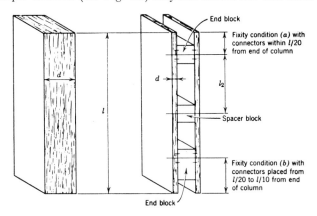

Simple Solid Column
d = dimension of least side of column, in.
l = overall unsupported length of column, in.

Spaced Column
d = dimension of least side of individual member, in.
l = overall unsupported length from center to center of lateral supports of continuous column, or from end to end of simple columns, in.
l_2 = distance from center of connector in end block to center of spacer block, in.

Figure 3.2   Wood Columns

able unit stresses for rectangular simple columns may be determined from the formula

$$F'_c = \frac{0.30E}{(l/d)^2}.$$

The formula is based on pin end conditions, but it is also applied to square end conditions.

The allowable stress values as determined from the formula may not exceed the values for compression parallel to grain, F_c, for the species, adjusted for service conditions and duration of loading.

The allowable unit stress values as determined from the formula are subject to the duration of loading adjustments given earlier in this chapter.

Spaced columns

Spaced columns consist of two or more individual members with their longitudinal axes parallel, separated at their ends and midpoints by blocking, and joined at the ends by connectors capable of developing the required shear resistance. For the individual members of a spaced column, l/d may not exceed 80, nor may l_2/d exceed 40 (see Fig. 3.2). To obtain spaced-column action, end blocks are required when the l/d ratio for the individual members exceeds

$$\sqrt{0.30E/F_c}.$$

When l/d ratios for the individual members of spaced columns do not exceed this value, the individual members are designed as simple solid columns. A multiplying factor is introduced in the spaced-column design formulas which depends on fixity conditions of the column (see Fig. 3.2).

The allowable unit stresses for individual members of a spaced column may be determined from the formulas

$$F'_c = \frac{0.75E}{(l/d)^2} \quad \text{for fixity condition (a)},$$

$$F'_c = \frac{0.90E}{(l/d)^2} \quad \text{for fixity condition (b)},$$

(see Fig. 3.2 for fixity conditions).

The individual members in a spaced column are considered to act together to carry the total column load. Each member is designed separately on the basis of its l/d ratio. Because of the end fixity developed in spaced columns, a greater l/d ratio than that allowed for simple solid columns is permitted. This fixity is effective only in the thickness direction; the l/d in the width direction is subject to the provisions for simple solid columns.

Connectors are not required for a single spacer block located in the middle tenth of the column length l. Connectors are required for multiple spacer blocks, and the distance between two adjacent blocks may not exceed one-half the distance between centers of connectors in the end blocks. When spaced columns are used as truss compression members, panel points that are stayed laterally are considered as the ends of the spaced column. The portion of the web members between the individual pieces of which the spaced column is comprised may be considered as the end blocks. In the case of multiple connectors in a contact face, the center of gravity of the connector group is used in measuring the distance from connectors to the end of the column (see Fig. 3.2).

The total load capacity determined by using the spaced-column formulas should be checked against the sum of the load capacities of the individual members taken as simple solid columns without regard to fixity; their greater d and the l between lateral supports which provide restraint in a direction parallel to the greater d should be used.

The values for F'_c determined by either of the formulas may not exceed the values for compression parallel to grain, F_c, for the species, adjusted for service conditions and duration of loading.

The values for F'_c computed by using the formulas are subject to duration of loading adjustments as given earlier in the chapter.

Spacer and end-block thicknesses may not be less than those of the individual members of the spaced column, nor may thickness, width, and length of spacer and end block be less than required for connectors of a size and number capable of carrying the computed load.

Round columns

The allowable stress for a round column may not exceed that for a square column of the same cross-sectional area or that determined by the formula

$$F'_c = \frac{3.619E}{(l/r)^2}.$$

The values for F_c' determined by the formula may not exceed the values for compression parallel to grain, F_c, for the species, adjusted for service conditions and duration of loading.

The allowable unit stress values as determined from the formula are subject to the duration of loading adjustments given herein.

Tapered columns

In determining d for tapered columns, the diameter of a round column or the least dimension of a rectangular column, tapered at one or both ends, is taken as the sum of the minimum diameter or least dimension and one-third the difference between the minimum and maximum diameters or lesser and greater dimensions.

4.7 Beam design

The usual formulas for designing solid sawn beams may also be used for the design of glued laminated timber beams, subject to the additional modification specified herein.

Simple span beams

The design procedure for simple span beams of constant rectangular cross section under uniform loads is as follows.

1. Determine size of beam required in bending using the formulas

$$M = wl^2/8 \text{ and } S = M/F_b.$$

The beam size determined in this manner is the required size if it meets shear and deflection requirements.

2. Determine size of beam required in shear using the formulas

$$R_V = wl_e/2 \text{ and } A = 3R_V/2F_v$$

Effective span length l_e is determined from the formula

$$l_e = l - 2d.$$

If shear governs, the size determined in this manner is the required size if it meets deflection requirements.

3. Determine size of beam required for deflection using the formula

$$I = 5wl^4/384E\Delta_L$$

in which $\Delta_L =$ deflection limit, in inches (see Table 3.10).

Member size as determined by deflection is usually limited by either applied load only or by applied load plus dead load, whichever governs, in accordance with Table 3.10. Applied load is live load, snow load, wind load, and so on.

For special uses, such as beams supporting vibrating machinery or carrying moving loads, more severe limitations may be required.

4. Determine actual beam deflection and required camber. The actual deflection Δ_A of a beam of a known size may be determined from the formula

$$\Delta_A = 5wl^4/384EI.$$

Camber for glued laminated beams is usually specified as some multiple of the dead-load deflection for appearance

Table 3.10 Deflection Limitations (Recommended by AITC)

Use Classification	Applied Load Only	Dead Load + Applied Load
Roof beams		
Industrial	$l/180$	$l/120$
Commercial and institutional		
Without plaster ceiling	$l/240$	$l/180$
With plaster ceiling	$l/360$	$l/240$
Floor beams		
Ordinary usage[a]	$l/360$	$l/240$
Highway bridge stringers	$l/200$ to $l/300$	
Railway bridge stringers	$l/300$ to $l/400$	

[a] Ordinary usage classification is intended for construction in which walking comfort, minimized plaster cracking, and the elimination of objectionable springiness are of prime importance.

purposes and/or to provide drainage. The camber for the actual dead load may be computed by multiplying the actual dead-load deflection, as determined by the formula for Δ_A, by the appropriate camber constant from Table 3.11.

Table 3.11 Recommended Minimum Camber for Glued Laminated Timber Beams

Roof beams[a]	$1\frac{1}{2}$ times dead load deflection
Floor beams[b]	$1\frac{1}{2}$ times dead load deflection
Bridge beams[c]	
Long Span	2 times dead load deflection
Short Span	2 times dead load $+ \frac{1}{2}$ of applied load deflection

[a] Roof beams. The minimum camber of $1\frac{1}{2}$ times dead load deflection will produce a nearly level member under dead load alone after plastic deformation has occurred. Additional camber is usually provided to improve appearance and/or provide necessary roof drainage. Roof beams should have a positive slope or camber equivalent to $\frac{1}{4}$ in. per ft of horizontal distance between the level of the drain and the high point of the roof, in addition to the minimum camber, to avoid the ponding of water.
[b] Floor beams. The minimum camber of $1\frac{1}{2}$ times dead-load deflection will produce a nearly level member under dead load alone after plastic deformation has occurred. On long spans, a level ceiling may not be desirable because of the optical illusion that the ceiling sags. For warehouse or similar floors where live load may remain for long periods, additional camber should be provided to give a level floor under the permanently applied load.
[c] Bridge beams. Bridge members are normally cambered for dead load only on multiple spans to obtain acceptable riding qualities.

Camber is built into a structural member by introducing a curvature, either circular or parabolic, opposite to the anticipated deflection movement. Camber recommendations vary with design criteria for various conditions of use and, in addition, are dependent on whether the member is of simple, continuous or cantilever span; whether roof drainage is to be provided by the camber; and other factors. Reverse camber may be required in continuous and cantilever spans to permit adequate drainage. (See "Deflection and Camber of Beams," Appendix B of AITC 102, included in "Timber Construction Standards," AITC 100.[17])

Tapered beams

Glued laminated beams are often tapered or curved to meet architectural requirements, to provide pitched roofs, or to provide a minimum depth of beam at its bearing. The most commonly used tapered and curved beams are simple span.

It is recommended that any sawn taper cuts be made on the compression face of tapered beams. It is also recommended that pitched or curved tension faces of beams not be sawn across the grain, but, instead, that the beam be so manufactured that the laminations are parallel to the tension face.

Camber is provided in beams which are taper cut on the compression face in a manner similar to the taper cut; that is, the minimum camber for dead-load deflection (center-line camber) is built into the members so that the laminations are parallel to the tension face. Also, it is customary to saw camber into the compression face of double tapered beams. This camber should equal one-fourth of the center-line camber if the center-line camber is over 2 in. If this camber is not provided, the compression face may give an optical illusion of sagging.

In the design of tapered beams, consideration must be given to the combined effects of bending, compression, tension, and shear parallel to grain, and to compression or tension perpendicular to grain at the tapered edge. Such analysis requires satisfaction of the following interaction formula at all points in the member:

$$\frac{f_x^2}{F_x^2} + \frac{f_y^2}{F_y^2} + \frac{f_{xy}^2}{F_{xy}^2} \leq 1,$$

where F_x, F_y, and F_{xy} are maximum allowable strength values and f_x, f_y, and f_{xy} are actual stress values encountered at the same point of the beam, and

$f_x =$ compression or tension stress parallel to grain,

$f_y =$ compression or tension stress perpendicular to grain (compression if the taper cut is on the compression face, and tension if on the tension face),

$f_{xy} =$ shear stress,

$F_x =$ allowable bending stress (F_b),

$F_y =$ allowable stress in compression perpendicular to grain ($F_{c\perp}$), or tension perpendicular to grain ($F_{t\perp} = F_v/3$), and

$F_{xy} =$ allowable horizontal shear stress (F_v).

The design of pitched and curved beams must provide for horizontal deflection at the supports. If it does not, that is, if the supports are designed to resist horizontal movement, the member must be designed as a two-hinged arch. It is usually preferable to provide for the relatively small amount of horizontal movement which occurs at the supports by detailing slotted or roller connections.

Cantilever beams

Cantilever beam systems permit longer spans or larger loads for a given-size member than do simple span systems, provided member size is not controlled by compression perpendicular to grain at the supports or by horizontal shear. In general, a uniform section should be used throughout the length of a cantilever system. Increasing the depth of section at supports

to meet shear requirements causes areas of stress concentration at the point of change and should be avoided. For economy, the negative bending movement at the supports of a cantilever beam should be equal in magnitude to the positive moment.

Consideration must be given to deflection and camber in cantilevered multiple spans. When possible, roofs should be sufficiently sloped to eliminate water pockets, or provision should be made to ensure that accumulation of water does not produce greater deflection and live loads than anticipated. Unbalanced loading conditions should be investigated for maximum bending moment, deflection, and stability. The lateral support provisions, covered under the heading "Modification of Stresses" elsewhere in this chapter, should be applied in cantilevered beam design.

4.8 Arch design

Three-hinged arches

Three-hinged arches, as the name implies, are hinged at each support and at the crown or peak. They may take shapes such as radial, gothic, A-frame, tudor, three-centered, or parabolic (see Fig. 3.3).

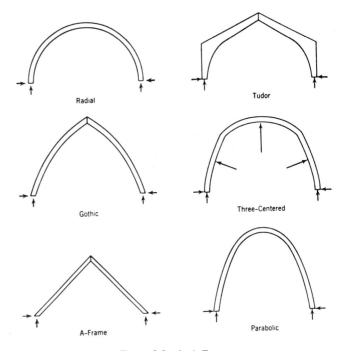

Radial

Tudor

Gothic

Three-Centered

A-Frame

Parabolic

Figure 3.3 Arch Types

The following design procedure may be used for simple three-hinged arches under usual loading conditions. This procedure is essentially a trial-and-error method, and revision of original values may be necessary. The procedure is given for a three-hinged tudor arch. The design of other three-hinged arch types is similar.

1. Determine the application of dead, live, or snow loads and wind loads.
2. Lay out to a convenient scale the arch outline indicating external architectural dimensions. Radius of curvature is also an architectural dimension, and it should not be less than the

minima recommended for various lamination thicknesses as given in Table 3.5.

3. Calculate and tabulate the approximate reactions for various combinations of loading conditions. Figure 3.4 indicates some of the more common loading conditions.

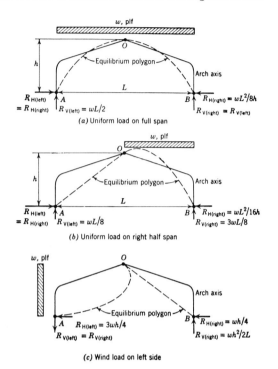

Figure 3.4   Three-hinged Arch Reactions and Equilibrium Polygons

4. Determine the minimum section size at the base.

5. As a first trial in the trial-and-error process, determine approximate depths d_c, of arch at crown and tangent points d_t. These are usually assumed to be

$$d_c = \tfrac{4}{3}b \text{ and } d_t = 1.25d_b$$

in which

  $b =$ arch width (assumed, based on section at base determined in step 4); and

  $d_b =$ arch depth at base (based on section at base determined in step 4 and assumption for b).

The section at the crown is often proportioned for architectural appearance or to meet the depth of purlins that frame into the arch at this point. In no case should the crown depth be less than the arch width. The final upper and lower tangent point depths should be approximately equal (within 10 per cent).

6. Lay out the arch to scale, using the approximate dimensions determined in steps 4 and 5 and the radius of curvature. Locate the arch axis at the midpoint of the arch depth.

Proceed with the design and analysis of a three-hinged arch in accordance with accepted structural theory. This is facilitated by using equilibrium polygons, which are moment diagrams for a specified loading drawn in such a position and to such a scale that they pass through the hinges. Equilibrium

polygons for each loading condition are illustrated in Figure 3.4.

7. Check arch section at several points along the arch axis using the formula for combined bending and axial compression. Points to be checked include the upper and lower tangent points and at least two points on the roof arm. More points on the roof arm should be checked if the two points investigated have stresses approaching the allowable values. Points on the vertical leg should be checked if it is long in relation to the length of the roof arm.

8. Check lateral stability considerations. When one edge of the arch is braced by decking fastened directly to the arch or braced at frequent intervals, as by girts or roof purlins, the ratio of tangent point depth to breadth of the arch, based on actual dimensions, should not exceed 4. When such lateral bracing is lacking, the ratio should not exceed 3, and the arch should be checked for column action in the lateral (width) direction.

The arch should be checked for column action in the lateral direction using the combined stress formula.

9. Check radial stress by the formula that has been given.

10. Determine deflection due to bending.

Two-hinged arches

Two-hinged arches are hinged at each base and are statically indeterminate. They may have a profile of any shape with any combination of straight, constant, or variable section depth. The horizontal thrust at the base must be resisted by some adequate means, such as tie rods, abutments, or foundations. After the reactions, moments, shears, and axial forces have been determined, the part of the design for determining the required arch section is similar to that for the three-hinged arch and follows accepted structural theory.

The following design procedure may be used for arches that are symmetrical about the center line. It neglects the effect of tie-rod elongation or differential settlement and spread of abutments.

1. Lay out one-half of the arch axis to a convenient scale and divide it into any number of equal divisions (the more divisions used, the more precise will be the results). Determine and tabulate the x and y distances for each point of division (see Fig. 3.5). Tabulate the values for y^2 also.

2. Determine vertical reactions.

3. Considering the entire arch as a simple span beam, compute and tabulate the bending moment M_S at each division point for dead loads, balanced and unbalanced vertical loads, and horizontal loads.

4. Multiply the M_S values by the corresponding y values and tabulate.

5. Compute horizontal reactions for the various loading conditions.

6. Determine and tabulate bending moment on the arch at each division point for each loading condition.

7. Determine axial thrust at each division point.

8. On the basis of the assumption of full lateral support, determine the section size required at each division point.

On the basis of these requirements, determine the required arch depths and widths at each division point. When one edge of the arch is braced by decking fastened directly to the arch or braced at frequent intervals, as by girts or purlins, the depth

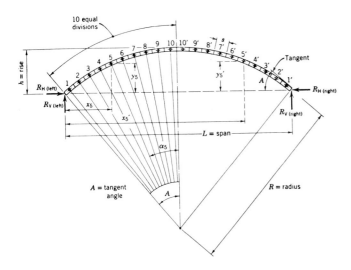

Figure 3.5   Two-hinged Arch Notations

to breadth ratio of the arch, based on actual dimensions, should not exceed 4. When such lateral bracing is lacking, the ratio should not exceed 3, and the arch should be checked for column action in the lateral (width) direction (see step 11).

9. Check the section for combined bending and actual compression using the combined stress formula.

10. Check the section for shear stress at the base and at the point of maximum shear.

11. Lateral stability considerations. As stated in step 8, when adequate lateral bracing is not provided (in which case the depth-breadth ratio should not exceed 3), the arch should be checked for column action in the lateral (width) direction, using the combined stress formula.

12. Check radial stress using the formula that was given earlier in this chapter.

13. Determine deflection due to bending.

Dome structures

Timber dome structures are commonly spherical in shape, with a low rise-to-span ratio. They may consist of a series of three-hinged or two-hinged arches having a common connection at the top, or crown, of the dome and with the bases arranged in a circle, or may be true domes made up of interconnecting members forming various patterns.

The design procedures for three-hinged and two-hinged arches can be followed for the design of domes consisting of arches joined at the crown. It should be pointed out, however, that the plane projection of the tributary area of load for each half of the arch is wedge-shaped, the apex being at the crown. Lateral support may be provided by purlins or joists spanning between adjacent arches, their span decreasing from supports to crown.

The radial rib-type dome is commonly comprised of glued laminated timber ribs, which are continuous but not necessarily one piece from base to crown; rings, which are made up of straight members spanning between the ribs; a tension ring, which resists outward thrust at the base; and diagonals, which are usually steel rods with turnbuckles for adjustment. The following principles apply to the design of radial rib domes:

(1) the ribs are in maximum stress when the entire dome is loaded; (2) a ring member is in maximum tension when all of the dome above the ring is fully loaded, and in maximum compression when all of the dome below the ring, as well as the ring itself, is fully loaded; and (3) the diagonals are not stressed when the dome is symmetrically loaded. The diagonals in a panel are in maximum stress when the dome on one side of a vertical plane passing through the center of that panel and through the crown is fully loaded and the other side unloaded.

4.9   Fastenings

The design of mechanical fastenings for joints in structural glued laminated timber members is similar to the design of fastenings for solid wood joints. Design of such fastenings is covered in References 10–14.

4.10   Design considerations

Fire safety

Neither building materials alone, nor building features alone, nor detection and fire extinguishing equipment alone can provide the maximum safety from fire in buildings. A proper combination of these will provide the necessary degree of protection for the occupants and for the property.

Wood, when exposed to fire, forms a self-insulating surface layer of char and thus provides its own fire protection. Although the surface chars, the undamaged wood below the char retains its strength and will support loads equivalent to the capacity of the uncharred section. Very often, heavy timber members will retain their structural integrity through long periods of fire exposure and still remain serviceable after the surface has been cleaned and refinished. This fire endurance and excellent performance of heavy timber is attributable to the size of the wood members and to the slow rate at which the charring penetrates.

In 1961, a comparative fire test of an unprotected structural glued laminated timber beam and an exposed structural steel beam was conducted at the Southwest Research Institute. The following is quoted from a report on the test, entitled, "Comparative Fire Test of Timber and Steel Beams".[18]

The wood beam continued to support its full design load, throughout the test, with a maximum deflection of only $2\frac{1}{4}$ inches at 30 minutes. The uniform deflection rate of the wood beam demonstrates the dependability of heavy timber framing under fire conditions.

At the conclusion of the test, the wood beam was sawed through at a representative section, revealing a depth of char penetration of approximately $\frac{3}{4}$ inch on each side and $\frac{5}{8}$ inch on the bottom ...

While penetration of char at the glue line was slightly greater, the deflection record demonstrates that the integrity of the casein adhesive bond was maintained during fire exposure.

Thus, after 30 minutes of fire exposure, during which temperatures in excess of 1500°F were recorded, 75 per cent of the original wood section remained undamaged and the beam continued to support its full design load.

It is also significant to note, in the case of all wet-use adhesives used by the structural timber laminating industry and required by CS253-63, that the glue line is not consumed by fire any faster than the wood.

How well a structure performs in protecting life and property in a fire, rather than the composition of the materials used, is therefore, the important criterion by which to judge the ultimate safety of the structure.

Durability

Wood has performed in service with satisfaction for centuries. When recognized principles of design and construction are observed, timber structures can be built with the assurance of durability and low maintenance costs.

Decay of wood is caused by fungi, which are low forms of plant life that develop and grow from spores just as higher plants do from seed. These microscopic spores are likely to be present wherever wood is used. The plantlike growth breaks down the wood substance, converting it to food required by the fungus for development. However, like all forms of plant life, these wood-destroying fungi must have air, suitable moisture, and favorable temperature as well as the food if they are to develop and grow. If deprived of any one of these four essentials, the spores cannot develop and the wood remains permanently sound, retaining its full strength. Wood permanently and totally submerged in water cannot decay because the necessary air is excluded. Wood will not decay when its moisture content is continuously less than 20 per cent. Temperatures above 100°F and below 40°F will essentially stop the growth of decay fungi. Progress will resume, however, each time favorable climatic conditions are restored. Moisture and temperature, which vary greatly with local conditions, are the principal factors affecting the rate of decay.

Figure 3.7   Glued Laminated Girders, Beams, and Columns Used in a School "All-purpose" Room

Among the design and construction principles that will assure long service and avoid decay hazards are

1. Positive site and building drainage;
2. Adequate separation of wood from known moisture sources; and
3. Ventilation and condensation control in enclosed spaces.

When recognized principles of design and construction will not provide adequate protection, wood can be treated with chemicals to resist attack by decay as well as by insects, marine borers, and fire.

The effectiveness of treatment is dependent upon the chemicals used and their retention and penetration. Retention and penetration must be adequate to give the required service life. It is for this reason that pressure treatment is the only method recommended for severe exposure. All pressure preservative treatments should meet the requirements of the AWPA Standards (Ref. 15).

Nonpressure methods of preservative treatment may provide adequate protection for moderate exposures. Nonpressure treatments should meet the requirements of the Vacuum Wood Preservers Institute (VWPI) Standards for the vacuum process, National Woodwork Manufacturers Association (NWMA) Standard for the dip process, and other established and accepted standards for hot and cold bath, cold soaking, and brush or spray methods.

Fire-retardant chemicals are those chemicals that, when impregnated in the wood with recommended retentions, lower the rate of surface flame spread and make the wood self-extinguishing if the external source of heat is removed. After proper surface preparation, the surface is paintable. These chemicals are recognized and accepted under several specifications, including Federal and Military. They are intended and recommended only for interior locations or locations protected against leaching. These treatments are sometimes used to meet a specific flame-spread rating for interior finish.

Water-repellent preservatives are those solutions containing water-repellent materials and a minimum of 5 per cent by weight of pentachlorophenol in a mineral spirits type of carrier as outlined in Federal Specification TT-W-572 or the

Figure 3.6   Prototype Laminated Wood Tower Being Tested for Use in High-voltage Power Transmission Lines. The crossarm is 87 ft above the base

Figure 3.8   Three-hinged Glued Laminated Church Arches

Figure 3.11   Treated Glued Laminated Timber Stringers Were Used over Three 60 ft Spans of This Railroad bridge

Figure 3.9   Modern Commercial Structure Illustrating the Combination of Glued Laminated Construction with Other Materials

Figure 3.12   U.S. Navy Minesweeper Has Treated Glued Laminated Timber Frame

Figure 3.10 Laminated Timber Members Being Erected for a Manufacturing Plant

Figure 3.13 Laminated Timber "Shade Trees" are Compact and Hurricane Proof

NWMA Standard. They offer a degree of dimensional stability to wood by retarding the rate of absorption of free water.

Economy

The economic success of the construction of a project may be greatly influenced by design. The designer must recognize that the entire building, not merely one component such as a beam or a truss, should be properly designed and engineered to obtain maximum economy. Standardized structural parts do have their place when used in designs where all components are designed with an eye to structural adequacy, efficiency, and economy. Often, it is less expensive to call for a fabricator's standard glued laminated arch, beam, or truss pattern than for a custom design. However, each structure should be analyzed so that its own requirements for utility and economy will be satisfied and so that it will not be forced arbitrarily to conform with a stereotyped structural framework design.

5 APPLICATIONS

Structural glued laminated timber is used in a great variety of applications, a few of which have been illustrated in this chapter. Among its structural applications are schools, churches, commercial buildings, industrial buildings, residences, and farm buildings and highway and railway bridges. Structural glued laminated timber is used by the electric utilities in light standards and as crossarms and poles for electrical transmission towers. Glued laminated timber is also used in the construction of ships, dredge spuds, and for various nonstructural applications such as diving boards and stadium seats. Figures 3.6 to 3.13 are typical illustrations of structures employing a variety of laminated wood elements.

REFERENCES

1. "Structural Glued Laminated Timber", U.S. Commercial Standard CS253-63, National Bureau of Standards, U.S. Department of Commerce, Washington, D.C., 1963.
2. A. D. FREAS and M. L. SELBO, "Fabrication and Design of Glued Laminated Wood Structural Members", U.S. Dept. Agr. Tech. Bull. 1069, U.S. Government Printing Office, Washington, D.C., 1954. (Out of print.)
3. "Standard Specifications for Structural Glued Laminated Douglas Fir (Coast Region) Timber", West Coast Lumbermen's Association, Portland, Ore., 1963.
4. "Standards for Structural Glued Laminated Members Assembled with WWPA Grades of Douglas Fir and Larch Lumber", Western Wood Products Association, Portland, Ore., 1966.
5. "Standard Specifications for the Design and Fabrication of Hardwood Glued Laminated Lumber for Structural, Marine and Vehicular Use," Southern Hardwood Lumber Manufacturers Association, Memphis, Tenn.; Appalachian Hardwood Manufacturers, Inc., Cincinnati, Ohio; and Northern Hardwood and Pine Manufacturers Association, Green Bay, Wis., 1959.
6. "Standard Specifications for Structural Glued Laminated West Coast Hemlock Lumber — Fabrication and Design", West Coast Lumbermen's Association, Portland, Ore., 1960.
7. "Standard Specifications for Structural Glued Laminated Southern Pine Timber," Southern Pine Inspection Bureau, New Orleans, La., 1965.
8. "Standard Specifications for Structural Glued Laminated California Redwood Timber," California Redwood Association, San Francisco, Calif., 1965.
9. "Inspection Manual," AITC 200, American Institute of Timber Construction, Washington, D.C., 1963.
10. American Institute of Timber Construction, *Timber Construction Manual*, John Wiley & Sons., Inc., New York, 1966.
11. "National Design Specification for Stress-Grade Lumber and Its Fastenings", National Forest Products Association, Washington, D.C., 1962.
12. W. F. SCHOFIELD and W. H. O'BRIEN, *Modern Timber Engineering*, Southern Pine Association, New Orleans, La., 1963.
13. Timber Engineering Company, *Timber Design and Construction Handbook*, McGraw-Hill Book Company, Inc., New York, 1956.
14. *Douglas Fir Use Book*, West Coast Lumbermen's Association, Portland, Ore., 1962.
15. "AWPA Standards," American Wood Preservers Association, Washington, D.C. (Revised annually.)
16. B. BOHANNAN, "Effect of Size on Bending Strength of Wood Members", U.S. Dept. Agr. Forest Service, Research Paper FPL 56, U.S. Forest Products Laboratory, Madison, Wis., 1966.
17. "Timber Construction Standards", AITC 100, American Institute of Timber Construction, Washington, D.C., 1965. (This publication also forms Part II of the "Timber Construction Manual" — see Reference 10.)
18. "Comparative Fire Test of Timber and Steel Beams", Technical Report No. 3, National Forest Products Association, Washington, D.C., 1961.

Chapter 4

Plywood

David R. Countryman
J. M. Carney
J. L. Welsh, Jr.

1 INTRODUCTION

1.1 Description

Plywood is a panel consisting of an odd number of plies of wood veneer with the grain of alternate plies at right angles (Fig. 4.1). These plies are bonded together under hydraulic

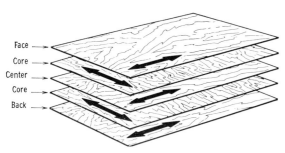

Figure 4.1   Panel Layup

pressure with either water-resistant or waterproof adhesives. The outside plies are called faces, and any interior plies with their grain running in the same direction as the face plies are called centers. The plies that are at right angles to the face plies are called cores. The types of plywood normally used in construction have plies ranging from $\frac{1}{12}$ in. to $\frac{1}{4}$ in. thick and have a total thickness of $\frac{1}{4}$ in. to $1\frac{1}{8}$ in. Thinner or thicker panels may be made for special uses. In all cases, an odd number of plies is used to give a balanced and dimensionally stable panel.

Nearly all plywood used structurally is made from trees of the softwood species such as Douglas fir, western hemlock, southern pine, and the true firs. Hardwood plywood is usually used for furniture and decorative panelling.

The first softwood plywood produced in this country was manufactured on the extreme West Coast, but plants rapidly spread inland through the far western states. In 1964, the first structural plywood was manufactured in the southern pine region along the gulf and southern Atlantic area. Manufacture of softwood plywood is now expanding into virtually all parts of the United States.

DAVID R. COUNTRYMAN is Assistant Director, Technical Services Division, American Plywood Association, Tacoma, Washington.

J. M. CARNEY is Head, Engineering Service, American Plywood Association, Tacoma, Washington.

J. L. WELSH, JR., is a Research Engineer, American Plywood Association, Tacoma, Washington.

1.2 Uses

The biggest percentage of plywood produced (52 per cent) is used in residential construction, the main uses being roof and wall sheathing, subfloor, and underlayment, with a small amount used for soffits. The second largest use of plywood (18 per cent) is in the industrial field, for such uses as crates, pallets, and boxcar lining. Next in line of importance (13 per cent) are uses for wall and roof sheathing of factories and warehouses, offices and institutions, with a large percentage used for concrete forming. Additional uses include home workshop projects and uses in the agricultural field, such as pallet bins, bulk storage structures, and animal shelters. Some applications of plywood in housing and heavy construction are shown in Figures 4.2–4.7.

Figure 4.2   Surety Federal Savings Bank, Detroit, Michigan. Roof composed of conical panels, 13 laminations of $\frac{5}{16}$-in. plywood outer portion, ribbed construction inner portion. Erected in 3 days

1.3 Manufacture

The transformation of raw timber into plywood starts when the logs are cut into desired lengths known as peelers, usually about 8 ft long. After the bark is removed, the peeler is placed in a giant lathe and rotated against a long knife that peels the wood into continuous thin sheets known as veneers. These veneers are then clipped to desired widths and are run through dryers that reduce the moisture content to about two or three per cent. After grading, the veneer goes through glue spreaders and is laid up into panel form. The panels are then pressed, trimmed to exact size and, if a smooth panel is desired, they are put through sanding machines. The panels are then graded and placed in a warehouse ready for shipment.

(a)

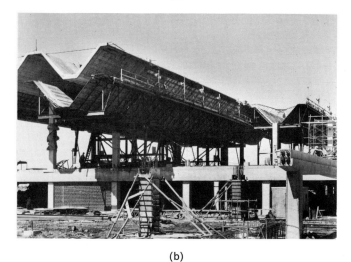

(b)

Figure 4.3   (a) Two plywood forms were used to make 17 plates of 300 × 480 ft roof for Minneapolis–St. Paul Terminal Building. Forms jacked to position. Insulation board laid in form created sound-deadening textured ceiling. (b) Form being lowered, stripping by its own weight. Two forms 150 ft long used for 300 ft width

Figure 4.4   Panels of tongue and grooved 2.4.1 flooring being installed in home. Panels 1⅛ in. thick are subfloor and underlayment combined. Two men install 1000 sq. ft in less than 4 hours

Figure 4.5   Roof consisting of vaults 56 ft. long and 11 ft 8 in. wide have top and bottom skins of 2 sheets of ¼-in. plywood glue stapled to lumber framework. Batt insulation inside

Figure 4.6   Self-supporting folded plates. Forty-foot and sixty-foot long plates 2⅜-in. thick span 34 ft and 48 ft. Plates are ⅝-in. plywood covers on 2 × 4, 2 × 6, 2 × 8 chords. Long V-shaped plates assembled on site

Figure 4.7   Supermarket at Phillipsburg, New Jersey

Softwood plywood is made in two basic types, interior and exterior, where the type refers to the exposure capabilities of the panel. Within each type there are several grades, which are established by the quality of the veneer in the panel. Interior-type plywood is a moisture-resistant panel, suitable for applications where it may be temporarily exposed to the elements but will, in its final use, be protected from moisture. Exterior-type plywood is bonded with completely waterproof adhesive and represents the ultimate in moisture resistance. It is suitable for permanent exterior exposures, such as boats, fences, siding, and signs.

The grades in each type of plywood are established by the quality of the veneer on the face and back of the panel. The grades in descending order of quality are designated as N, A, B, C, and D. Grade A-C, for example, has a grade A veneer on the face and a grade C veneer on the back. Grade C veneer is the lowest permitted in exterior-type plywood, and at least one face veneer must be C grade or better for interior-type plywood.

There are also many speciality panels manufactured. High-density and medium-density overlaid plywood have a resin-treated fiber surfacing on one or both sides. Medium-density overlaid panels provide an ideal paint base, while high-density overlaid panels may be permanently exposed to the elements with no finishing required. Marine-grade plywood is a premium panel that is specifically suited for bending into the shape of boat hulls. Specialty panels with decorative face treatments in the form of striations, grooving, embossing, brushing, and so on are exterior type, and may be used as combined sheathing and siding.

1.4 Product Standards

Nearly all softwood plywood is manufactured in accordance with a United States product standard. The purpose of a product standard is to establish quality criteria—standard methods of testing, grading, certification, and labeling a commodity — and to provide a uniform basis for fair competition.

Product standards originate with the proponent industry. The sponsors may be manufacturers, distributors, or users of a specific product. The U.S. Office of Commodity Standards assists the sponsor group in arriving at a tentative standard which is then referred to other elements of the same industry. The standard is then promulgated.

In order for a panel of plywood to carry the grade-trademark attesting to its conformance with a product standard, a sampling of the manufacturer's production is taken by an independent testing agency. Personnel employed by the testing agency check the glues, equipment, veneers, lay-up, and finished panels. Frequent random samples of the production are scientifically tested in laboratories as a continual check of glue-line durability and quality.

2 MISCELLANEOUS PROPERTIES

2.1 Sizes

The common plywood panel size is 4 ft wide by 8 ft long with the grain of the face and back plies oriented in the 8 ft direction. Panels 4 ft wide by 10 ft and 12 ft long are becoming more common and a few mills can manufacture 5 ft by 12 ft long. Any width or length may be obtained by either scarf-jointing or finger-jointing (Fig. 4.8).

Scarf Joint   Finger Joint

Figure 4.8   Sketch of Typical Scarf Joint and Finger Joint

2.2 Insulation

Plywood has the same coefficient of thermal conductivity as the species of wood from which it is made. For Douglas fir, the value commonly used is 0.78 Btu per hour per square foot per degree Fahrenheit temperature difference. Plywood, in addition to being a good insulation material, also provides an air-tight membrane that minimizes heat loss.

2.3 Acoustical

Tests have shown that plywood panels $\frac{1}{4}$ in. thick can effectively absorb sound in the lower frequency range (below 500 cycles per second). In a church, auditorium, or music hall, where reverberation is desired, thicker plywood panels designed for a minimum of vibration can be effective. Assemblies of common floor-ceiling material, including plywood floors, have shown good sound transmission and impact noise reducing properties.

2.4 Moisture content

The moisture content of plywood varies with the panel and also with the kind of press used in its manufacture. Exterior-type plywood comes from a hot press with a moisture content from 5 per cent to 7 per cent. Hot-press interior plywood accounts for approximately 90 per cent of total interior production, and normally leaves the mill at 7 per cent to 10 per cent. Cold-press interior plywood has an average moisture content around 13 per cent.

2.5 Dimensional stability

From a condition of oven-dry to fiber-saturation, plywood shrinks or swells approximately 0.2 to 0.3 of 1 per cent in the plane of the panel. The maximum shrinkage or swelling to be expected under ordinary atmospheric exposures is about 0.1 of 1 per cent. Shrinkage in thickness is approximately the same as that of wood.

On the basis of a three-ply Douglas fir plywood panel, the approximation for thermal expansion per degree Centigrade is 4.2×10^{-6} for the direction parallel to the face grain of the panel; 7.2×10^{-6} in the direction perpendicular to the grain of the face ply; and 27.9×10^{-6} in the direction perpendicular to the plane of the panel, or thickness.

3 DESIGN PROPERTIES

The plywood engineering properties and design methods presented in this chapter stem from extensive and continuing test programs conducted by the American Plywood Association, by other wood associations, by private industry, and by the

Table 4.1. Plywood Section Properties for Selected Constructions (All Properties Adjusted to Account for Reduced Effectiveness of Plies with Grain Perpendicular to Applied Stress; 12-in. widths)

1	2	3	4	5	6	7	8	9	10	11	12
		Effective Thickness for Shear		Properties for Stress Applied Parallel with Face Grain				Properties for Stress Applied Perpendicular with Face Grain			
Thickness (in.)	Approx. Weight (psf)	All Grades Using Exterior Glue	All Grades Using Interior Glue	Area for Tension and Compression (in.²)	Moment of Inertia I (in.⁴)	Effective Section Modulus KS (in.³)	Rolling Shear Constant I/Q (in.)	Areas for Tension and Compression (in.²)	Moment of Inertia I (in.⁴)	Effective Section Modulus KS (in.³)	Rolling Shear Constant I/Q (in.)

Section A. Face plies of different species group from inner plies. (Includes all Product Standard grades except those noted in Section B)

Rough panels

5/16 R	1.0	0.286	0.270	2.400	0.026	0.147	0.216	0.600	0.001	0.024	—
3/8 R	1.1	0.323	0.323	2.400	0.047	0.215	0.286	0.750	0.002	0.038	—
1/2 Rᵃ	1.5	0.471	0.410	3.000	0.099	0.336	0.409	1.200	0.016	0.115	0.215
5/8 R	1.8	0.527	0.498	3.493	0.171	0.466	0.539	1.500	0.031	0.180	0.269
3/4 R	2.2	0.585	0.585	3.500	0.261	0.591	0.659	2.250	0.070	0.316	0.347
13/16 R	2.4	0.615	0.629	3.502	0.313	0.655	0.720	2.250	0.096	0.388	0.401
7/8 R	2.6	0.730	0.673	4.650	0.418	0.813	0.662	2.250	0.110	0.397	0.502
1 R	3.0	0.788	0.760	3.876	0.531	0.903	0.800	3.320	0.204	0.625	0.585
1 1/8 R	3.3	0.848	0.848	4.620	0.724	1.093	0.869	2.625	0.262	0.688	0.722

Sanded panelsᵇ

1/4 S	0.8	0.241	0.210	1.680	0.013	0.091	0.179	0.600	0.001	0.019	—
3/8 S	1.1	0.305	0.305	1.680	0.040	0.181	0.309	1.050	0.004	0.053	—
1/2 S	1.5	0.450	0.392	2.400	0.080	0.271	0.436	1.200	0.016	0.115	0.215
5/8 S	1.8	0.508	0.480	2.407	0.133	0.360	0.557	1.457	0.040	0.214	0.315
3/4 S	2.2	0.567	0.567	2.778	0.201	0.456	0.687	2.200	0.088	0.366	0.393
7/8 S	2.6	0.711	0.655	2.837	0.301	0.585	0.704	2.893	0.145	0.496	0.531
1 S	3.0	0.769	0.742	3.600	0.431	0.733	0.763	3.323	0.234	0.682	0.632
1 1/8 Sᶜ	3.3	0.825	0.825	3.829	0.566	0.855	0.849	3.307	0.334	0.843	0.748

Section B. All plies from same species group. (Includes following grades: Structural I and II, Marine Exterior, all grades using Group 4 stresses.)

Rough panels

5/16 R	1.0	0.318	0.300	2.400	0.026	0.147	0.216	1.200	0.002	0.032	—
3/8 R	1.1	0.375	0.375	2.400	0.048	0.215	0.286	1.500	0.003	0.049	—
1/2 Rᵈ	1.5	0.574	0.500	3.600	0.100	0.339	0.410	2.400	0.029	0.183	0.215
5/8 R	1.8	0.662	0.625	4.586	0.175	0.477	0.546	3.000	0.056	0.286	0.269
3/4 R	2.2	0.750	0.750	4.600	0.266	0.603	0.663	4.500	0.132	0.509	0.348
13/16 R	2.4	0.794	0.813	4.605	0.319	0.668	0.722	4.500	0.182	0.628	0.403
7/8 R	2.6	0.949	0.875	6.900	0.474	0.922	0.594	4.500	0.207	0.643	0.507
1 R	3.0	1.037	1.000	5.354	0.574	0.976	0.728	6.639	0.391	1.020	0.589
1 1/8 R	3.3	1.125	1.125	6.840	0.815	1.231	0.776	5.250	0.502	1.122	0.729

Sanded panelsᵇ

1/4 S	0.8	0.276	0.240	1.680	0.013	0.091	0.179	1.200	0.001	0.027	—
3/8 S	1.1	0.375	0.375	1.680	0.040	0.182	0.308	2.100	0.007	0.079	—
1/2 S	1.5	0.574	0.500	3.120	0.081	0.277	0.438	2.400	0.029	0.183	0.215
5/8 S	1.8	0.662	0.625	3.135	0.135	0.367	0.557	2.914	0.076	0.345	0.316
3/4 S	2.2	0.750	0.750	3.876	0.207	0.470	0.691	4.400	0.168	0.597	0.394
7/8 S	2.6	0.949	0.875	3.994	0.329	0.639	0.637	5.786	0.279	0.811	0.535
1 S	3.0	1.037	1.000	5.520	0.498	0.846	0.673	6.646	0.454	1.119	0.636
1 1/8 S	3.3	1.125	1.125	5.978	0.664	1.003	0.746	6.660	0.650	1.385	0.753

ᵃ For 1/2 in. 3-ply use the following:

1/2 R	1.5	0.393	0.424	3.000	0.109	0.372	0.387	1.000	0.006	0.067	—

ᵇ Includes Touch-Sanded
ᶜ For 2-4-1 use the following:

1 1/8 2-4-1	3.3	0.832	0.832	5.360	0.655	1.000	0.746	4.894	0.322	0.084	0.642

ᵈ For 1/2 in. 3-ply use the following:

1/2 R	1.5	0.463	0.500	3.000	0.110	0.373	0.387	2.000	0.008	0.088	—

United States Forest Products Laboratory, and are backed by years of satisfactory experience. These technical data are presented as the basis for competent engineering designs. For any design to result in a satisfactory structure, of course, adequate materials and fabrication are also required. The properties explained later are to be used in conventional, mechanically fastened applications, and in the design of glued plywood-lumber structural assemblies.

3.1 Species

As mentioned previously in the introduction, there are many species of softwood used in the making of plywood. These species of wood have been divided into four groups according to stiffness. All species in any given group are assigned equal allowable stresses.

3.2 Section properties

The plywood section properties presented in Table 4.1 are for use with all species and grades of plywood, as noted in the table. These section properties are to be used in conjunction with the allowable unit stresses for the species group used in the faces (where known) as given in Table 4.2. The section properties presented in Table 4.1 include allowance for the contribution of all inner plies, adjusted for species and direction of grain. For panels designated by an identification index, for which species group is not known, use the section properties with stresses from footnote a in Table 4.2.

Due to the possibilities of various panel lay-ups, properties parallel to the face grain of the plywood are based on a panel construction giving minimum values in that direction. Likewise, properties perpendicular to the face grain are based on a different panel construction, giving the minimum values in that direction. Both values, therefore, are conservative.

Actual plywood thickness for calculation is equal to the nominal thickness, except in calculating shear. Since the type of glue and species of wood of the inner plies used in the manufacture of plywood directly affects the shear properties, the effective thickness to use in the calculation of shear-through-the-thickness is given in column 3 of Table 4.1 for all grades of plywood using exterior glue and in column 4 for all grades using interior glue.

The areas for direct stress values are effective areas that have been adjusted where necessary to reflect species of inner plies. These areas are intended for use with the allowable stresses for the face plies.

The section modulus is to be used in conjunction with the allowable stresses for the face plies. The I and S have been calculated to account for the reduced effectiveness of perpendicular plies, and are also corrected for permissible inclusion of lower strength species.

Rolling shear is a strength property that is not often found in other materials. It may best be described as a rolling of the wood fibers over one another when these wood fibers are at right angles to the principal shearing force. A rolling shear constant I/Q is used in conjunction with the allowable stress in rolling shear and is listed in Table 4.2.

3.3 Allowable stresses

The allowable unit stresses given in Table 4.2 are for normal (ten-year duration) loading conditions. When the duration of the full maximum load does not exceed the period indicated below, an increase in the allowable unit stresses may be taken as follows:

15 per cent for two months' duration, as for snow
25 per cent for seven days' duration
$33\frac{1}{3}$ per cent for wind or earthquake
100 per cent for impact.

These increases are not cumulative. They apply also to mechanical fasteners, but do not apply to the modulus of elasticity

Where a member is fully stressed to the maximum allowable level for many years, either continuously or cumulatively under maximum design load, a working stress 90 per cent of those shown in Table 4.2 should be used. This reduction again applies to mechanical fastenings but not to the modulus of elasticity.

Unit stresses shown in Table 4.2 apply to plywood used under conditions that are continuously dry, as in most covered structures. Where equilibrium moisture content in service will exceed 15 per cent (an average of 80 per cent relative humidity at normal temperature) decrease the dry location values as shown in Table 4.2.

The modulus of elasticity shown in Table 4.2 is for the species group of the face ply of the panels where known and includes an allowance for average shear deflection. Identification-Index panels use elastic moduli given in footnote a, Table 4.2. For cases where shear deflection may be large, as in heavily loaded, short spans, bending deflection and shear deflection should be computed separately, in which case an increase of 10 per cent to the published modulus of elasticity should be taken for bending deflection. This increase restores the allowance made when shear deflection is not computed separately. Shear deflection may be computed separately using the following formula:

$$\Delta_s = wCh^2l^2/1270E_eI$$

where

$\Delta_s =$ shear deflection (in.),
$w =$ uniform load (psf),
$h =$ panel thickness (in.),
$l =$ clear span between supports (in.),
$E_e =$ modulus of elasticity from Table 4.2 (psi),
$I =$ moment of inertia (in.^{4}/ft of width),
$C =$ constant, equal to 120 for panels applied with face grain perpendicular to supports, and 60 for panels with face grain parallel to supports.

This formula applies where all plies are from the same species group.

The allowable unit stresses in flexure for all grades apply to material with the load applied perpendicular to the plane of the panel because section properties have been adjusted to account for permitted species. With proper splicing, the same allowable stresses may be used for loads applied parallel to the plane of the panel, as in box beam webs, which are discussed later in this chapter.

Table 4.2. Allowed Stresses for Plywood

Types of Stress	Species Group	Exterior A-A, A-C, C-C,[a] Structural I A-A Structural I A-C Structural I C-C and Marine (Use Group 1 stresses)	Exterior A-B, B-B, B-C, C-C (Plugged) Concrete Form I[a] Concrete Form II[a] Structural I C-D (Use Group 1 stresses) Structural II C-D (Use Group 3 stresses) Standard Sheathing (Exterior Glue)[a] All Interior Grades with Exterior Glue	All other grades of Interior, including Standard Sheathing[a]
Extreme fiber in bending, Tension; Face grain parallel or perpendicular to span (at 45° to face grain use ⅛)	1	2000	1650	1650
	2, 3	1400	1200	1200
	4	1200	1000	1000
Compression parallel or perpendicular to face grain (at 45° to face grain use ⅓)	1	1650	1550	1550
	2, 3	1200	1100	1100
	4	1000	950	950
Bearing (on face)	1	340	340	340
	2, 3	220	220	220
	4	160	160	160
Shear in plane perpendicular to plies.[b] Parallel or perpendicular to face grain. (at 45° increase 100%)	1	250	250	230
	2, 3	185	185	170
	4	175	175	160
Shear, rolling, in plane of plies, parallel or perpendicular to face grain[c] (at 45° increase ⅓)	All	53	53	48
Modulus of Elasticity in bending. Face grain parallel or perpendicular to span	1		1,800,000	
	2		1,500,000	
	3		1,200,000	
	4		900,000	

[a] Exterior C-C and Standard Sheathing: The combination of Identification-Index designation and panel thickness determine the minimum species group and, therefore, the stress permitted, as follows:

       $\frac{5}{16}$ − 20/0, $\frac{3}{8}$ − 24/0, $\frac{1}{2}$ − 32/16, $\frac{5}{8}$ − 42/20, $\frac{3}{4}$ − 48/24 — Use Group 2 working stresses.

       All other combinations of C-C and Standard — Use Group 4 working stresses.

For Concrete form Class 1 — Use Group 1 stresses. For Concrete form Class 2 — Use Group 3 stresses. For 2-4-1 — Use Group 1 stresses.

[b] Shear-through-the-thickness stresses are based on the most common structural applications where the plywood is attached to framing around its boundary. Where the plywood is attached to framing at only two sides — such as in the heel joint of a truss — reduce the allowable shear-through-the-thickness values by 11% where framing is parallel to face grain and 25% where it is perpendicular.

[c] For Marine and Structural I use 75 psi. For Structural II use 56 psi.

Wet or Damp Location

Where moisture content is 16% or more, multiply the dry location values by the following factors.

For all grades of exterior and interior plywood with exterior glue:

     Extreme fiber in bending: 75%   modulus of elasticity: 89%

                 tension: 69%   shear: 84%

            compression: 61%   bearing: 67%

For all other grades of interior:

     Extreme fiber in bending: 69%   modulus of elasticity: 80%

                 tension: 69%   shear: 84%

            compression: 61%   bearing: 67%

The areas for tension and compression given in Table 4.2, when taken from the appropriate portion of this table, require no additional correction for grade or species group of the inner plies. The same stress is valid when applied to the area perpendicular to the face grain.

The allowable stresses for tension and compression at 45° to the face grain may be applied to the full thickness of the panel if all plies are of the same species group. If, however, the inner plies are not of the same species group, the total thickness must be adjusted in proportion to the actual modulus of elasticity and actual thickness of the inner plies.

As mentioned previously, when plywood is subjected to shear stresses, the wood fibers in the ply at right angles to the principal shearing force tend to roll and a so-called rolling shear is induced. The allowable stresses for rolling shear given in Table 4.2 are used for shear in the plane of the plies, applied to the contact area under stress. For some applications involving stress concentrations, such as stressed skin panels and box beams, a reduction in the allowable stress is required. Also note that the allowable shear stresses vary with the kind of glue used, rather than the type of panel.

For shear in the plane perpendicular to the plies, adjusted thickness values are given in Table 4.1, "Thickness for Shear." Here again the stresses depend on the kind of glue rather than the type of plywood.

4 FASTENERS

Fasteners and fastener systems on the market today are many and varied. For simplicity, they can be broadly categorized into two groups, mechanical and adhesive. Mechanical fasteners include nails, staples, bolts, screws, and metal plates. Each of these has its own characteristics that make it the appropriate fastener to use under certain conditions. Although adhesives are not quite as markedly different as the various mechanical fasteners, each particular glue does possess properties that make it suitable for certain types of applications.

As mentioned again later in this chapter, nails are sometimes used in combination with glue. In such an application, however, the nails are used only to provide adequate pressure on the two members until the glue has set. Since the structural glues commonly used for component fabrication possess very little or no elasticity, and nails require some differential movement of the members before they can take up any load, the glue line must fail completely before the nails go to work. For the adhesives used in component fabrication, full transfer of load between members is assumed.

4.1 Nails and staples

Nails and staples are probably the most common mechanical fastenings for temporary and permanent construction. In general, nails usually give higher allowable loads than staples. Nails give stronger joints when driven into the side grain of wood rather than into the end grain of wood. It would also hold true that nails have superior holding capacity when driven perpendicular to the plane of a plywood panel rather than into the end or edge of a panel. If at all possible, when attaching plywood to lumber members or plywood to plywood, nails should be stressed laterally rather than in direct withdrawal (see Tables 4.3 and 4.4). Direct withdrawal resistance of plywood, however, is more than adequate for such uses as fastening shingles to roof sheathing.

Nails may be driven very near the edge of a plywood panel due to plywood's superior splitting resistance.

There are several variations in nail configuration. In addition to common, smooth-shank nails, a variety of deformed shank nails are available such as ring-shank, spirally grooved, and barbed, as well as various coated nails such as cement-coated, zinc-coated, and chemical-etched. In addition there are hardened and nonhardened nails. In general, the hardened nail is superior to the nonhardened nail when lateral resistance is involved, and the deformed-shank nails are superior to the

Table 4.3. Allowable Withdrawal Loads for Common Nails, Normal Duration. Inserted Perpendicular to Grain in Wood

Specific Gravity G of Wood[a]	Allowable load in withdrawal, in pounds per inch of penetration of nail into the number holding the point Size of Nail (d = pennyweight)									
	6d	8d	10d	12d	16d	20d	30d	40d	50d	60d
0.32	9	10	12	12	13	15	16	18	20	21
0.34	10	12	14	14	15	18	19	21	23	24
0.35	11	13	15	15	16	19	21	23	25	26
0.36	12	14	16	16	17	21	22	24	26	28
0.37	13	15	17	17	19	22	24	26	28	30
0.38	14	16	18	18	20	24	25	28	30	32
0.40	16	18	21	21	23	27	29	31	34	37
0.41	17	20	22	22	24	29	31	33	36	39
0.42	18	21	23	23	25	30	33	35	38	41
0.43	19	22	25	25	27	32	35	38	41	44
0.44	20	23	26	26	29	34	37	40	43	46
0.45	21	25	28	28	30	36	39	42	46	49
0.46	22	26	29	29	32	38	41	45	48	52
0.47	24	27	31	31	34	40	43	47	51	55
0.48	25	29	33	33	36	42	46	50	54	58
0.51	29	34	38	38	42	49	53	58	63	68
0.53	32	37	42	42	46	54	59	64	69	74
0.59	42	48	55	55	60	71	76	83	90	97
0.62	47	55	62	62	68	80	86	94	102	110
0.64	51	59	67	67	73	87	93	102	110	119
0.66	55	64	72	72	79	94	101	110	119	128
0.67	57	66	75	75	82	97	105	114	124	133
0.68	60	69	78	78	85	101	109	119	128	138
0.71	65	76	86	86	94	111	120	130	142	152
0.80	89	103	117	117	128	152	163	178	193	208

[a] For specific gravity and wood grouping of various wood species, refer to the "National Design Specification for Stress-Grade Lumber and Its Fastenings" published by the National Forest Products Association.

Table 4.4. Allowable Lateral Loads for Common Nails, Normal Duration

Wood Group[a]	Allowable lateral loads (shear) in pounds for nails penetrating 10 diameters in Group I species; 11 diameters in Group II species; 13 diameters in Group III species; and 14 diameters in Group IV species into the member holding the point Size of Nail (d = pennyweight)									
	6d	8d	10d	12d	16d	20d	30d	40d	50d	60d
Group I	78	97	116	116	132	171	191	218	249	276
Group II	63	78	94	94	107	139	154	176	202	223
Group III	51	64	77	77	88	113	126	144	165	182
Group IV	41	51	62	62	70	91	101	116	132	146

[a] For specific gravity and wood groupings of various wood species, refer to the "National Design Specification for Stress-Grade Lumber and Its Fastenings" published by the National Forest Products Association.

common nails when straight withdrawal is involved, particularly where moisture content fluctuates. There are so many varieties of surface-coated nails that no general conclusions can be stated.

4.2 Screws

Screws are probably the next most common method of mechanially fastening plywood. Screws would be the logical fastener where direct withdrawal is the force to resist, since straight lateral resistance can be provided much faster and more easily with nails. Screws should be driven all the way with a screwdriver. They should never be started or driven with a hammer, because this practice tears the wood fibers and injures the screw threads, seriously reducing the load-carrying capacity of the screw.

4.3 Adhesives

As mentioned at the beginning of this discussion, there is also a great variety of glues on the market. In component fabrication, these glues may be loosely classified into three general categories: casein, synthetic resin, and epoxy. Each has its own particular properties that make it suitable for fabrication under certain conditions. Casein glue has a long and distinguished record for strength and durability in wood joints under service conditions where the wood remains dry. Casein glue is an excellent " gap filler," making it the glue to use where a precision fit of mating surfaces is difficult.

The synthetic resin glues that are the most common are the " room temperature-setting " resorcinol resins. The resorcinol resin glues produce strong joints on wood within the range of 2 per cent to 25 per cent moisture content and can stand long storage at room temperature. Their two possible drawbacks are that the mating surfaces to be glued must be held to close tolerances and curing temperature must be at least 70°F.

The third category of adhesive that is used in component construction is the epoxies. Epoxy resins are usually two-part systems which must be mixed immediately prior to application because of their limited pot life. These adhesives are capable of binding a wide variety of surfaces, including metals, plastics, rubbers, glass, ceramics, and wood, with minimum shrinkage and without emitting volatiles into the air. They are good gap fillers and require no more than contact pressure during curing. Some formulations, however, are toxic.

5 CONVENTIONAL CONSTRUCTION

Because of the rising cost of labor and materials, conventional construction from the smallest house to the largest building is highly engineered to minimize these costs. Plywood's light weight, large panel size, and predictable engineering properties have done a great deal to reduce costs in conventional construction. Through continued testing and research, the Plywood Association has arrived at minimum recommended thicknesses for given spacings of supports for floors, walls, and roofs. The recommended spans for floors and roofs appear on each sheathing panel in the form of an identification-index. The span designation incorporates allowed inner ply species

mixes, enabling a person to readily determine compliance with a specification without resorting to a working knowledge of species groups, panel thickness, or tables. The first number of the identification index refers to the floor span and the second number to the roof span. The recommended nailing schedule for floors is given in Table 4.5. Recommended nailing for a

Table 4.5. Recommended Nailing of Floors (d = pennyweight)

Application	Plywood Thickness (in.)	Nail Size and Type	Nail Spacing (in.)	
			Panel Edges	Intermediate
Subfloor	$\frac{1}{2}$	6d common or 6d deformed shank	6	10
	$\frac{5}{8}$ and $\frac{3}{4}$	8d common or 6d deformed shank	6	10
	$1\frac{1}{8}$	10d common or 8d deformed shank	6	6
Underlayment	$\frac{3}{8}$	3d common ring shank	6	8 each way
		16 ga. staples	3	6 each way

combination subfloor underlayment is the same as subflooring except that the deformed shank nail should be used and the nail head set $\frac{1}{16}$ in.

The recommended method of applying wall sheathing is outlined in Table 4.6.

Although the recommended roof support spacing is given on the panel, snow loads in some areas may dictate the use of a thicker panel or a shorter span. Table 4.7 gives the allowable loads for various spans for each indexed panel. These are minimum recommendations and FHA and local building code requirements may differ in some cases.

6 CONCRETE FORM

6.1 Description

Plywood is used extensively for concrete-form work. Its smooth surface and large panel size provide a satiny surface on the concrete that requires little or no finishing when the forms are removed. A wide variety of architectural design effects including curved surfaces is easily obtainable with plywood forms.

Concrete-form plywood is available in two grades. Class I is manufactured with all Group 1 faces whereas Class II has faces of other species groups. Both of these panels are fully exterior panels and can be expected to give equal service.

For the ultimate in service, high-density overlaid plywood is used for concrete forming. Up to 100 reuses have been recorded when care was taken in handling. Another advantage of high-density overlay is that the overlay tends to exclude moisture and dry stresses may be used.

Table 4.6. Wall Sheathing and Siding

Application	Recommended Thickness (in.)	Maximum Spacing of Supports, Center to Center (in.)	Nail Size and Type	Nail Spacing — Panel Edges (in.)	Nail Spacing — Intermediate (in.)
Sheathing	$\frac{5}{16}$	16[a]	6d common	6	12
	$\frac{3}{8}$–$\frac{1}{2}$	24	6d common	6	12
	$\frac{5}{8}$	24	8d common	6	12
Panel siding	$\frac{3}{8}$	16[b]	6d siding or casing[c]	6	12
	$\frac{1}{2}$ or thicker	24	6d siding or casing[c]	6	12
Lap siding or bevel siding	$\frac{3}{8}$[d]		6d siding or casing[c]	One nail per stud along bottom edge and 4 in. at vertical joints	8 in. vertical spacing on intermediate studs

[a] Apply with face grain perpendicular to supports for maximum strength and stiffness.
[b] Applies to panel siding with no sheathing; with sheathing $\frac{3}{8}$ in. panel siding can be used over supports spaced 24 in. o.c.
[c] Use noncorrosive (hot-dipped galvanized or aluminum) nails. Do not set.
[d] When separate sheathing is applied, $\frac{5}{16}$ in. overlaid plywood lap siding may be used over supports 16 in. o.c.

Table 4.7. Roof Sheathing Recommendations[a,b,c] (Plywood Continuous Over 2 or More Spans; Grain of Face Plies across Supports)

Panel Ident. Index	Plywood Thickness (in.)	Max. Span (in.)[d]	Unsupported Edge—Max. Length (in.)[e]	12	16	20	24	30	32	36	42	48	60	72
12/0	$\frac{5}{16}$	12	12	**100** (130)										
16/0	$\frac{5}{16}$, $\frac{3}{8}$	16	16	**130** (170)	**55** (75)									
20/0	$\frac{5}{16}$, $\frac{3}{8}$	20	20		**85** (110)	**45** (55)								
24/0	$\frac{3}{8}$, $\frac{1}{2}$	24	24		**150** (160)	**75** (100)	**45** (60)							
30/12	$\frac{5}{8}$	30	26			**145** (165)	**85** (110)	**40** (55)						
32/16	$\frac{1}{2}$, $\frac{5}{8}$	32	28				**90** (105)	**45** (60)	**40** (50)					
36/16	$\frac{3}{4}$	36	30				**125** (145)	**65** (85)	**55** (70)	**35** (50)				
42/20	$\frac{5}{8}$, $\frac{3}{4}$, $\frac{7}{8}$	42	32					**80** (105)	**65** (90)	**45** (60)	**35** (40)			
48/24	$\frac{3}{4}$, $\frac{7}{8}$	48	36						**105** (115)	**75** (90)	**55** (55)	**40** (40)		
2.4.1	$1\frac{1}{8}$	72	48							**175** (175)	**105** (105)	**80** (80)	**50** (50)	**30** (35)
$1\frac{1}{8}$ in. G 1 & 2	$1\frac{1}{8}$	72	48							**145** (145)	**85** (85)	**65** (65)	**40** (40)	**30** (30)
$1\frac{1}{4}$ in. G 3 & 4	$1\frac{1}{4}$	72	48							**160** (165)	**95** (95)	**75** (75)	**45** (45)	**25** (35)

Header for load columns: Allowable Roof Loads (psf)[f,g] — Spacing of Supports (in.) Center to Center.

[a] Applies to Standard, Structural I and II and C-C grades only.
[b] For applications where the roofing is guaranteed by a performance bond, recommendations may differ somewhat from these values. Contact American Plywood Association for bonded roof recommendations.
[c] Use 6d common smooth, ring-shank or spiral thread nails for $\frac{1}{2}$-in. thick or less, and 8d common smooth, ring-shank or spiral thread for plywood 1 in. thick or less (if ring-shank or spiral thread nails same diameter as common). Use 8d ring-shank or spiral thread or 10d common smooth-shank nails for 2-4-1, $1\frac{1}{8}$ in. and $1\frac{1}{4}$ in. panels. Space nails 6 in. at panel edges and 12 in. at intermediate supports except that where spans are 48 in. or more, nails shall be 6 in. o.c. at all supports.
[d] The spans shall not be exceeded for any load conditions.
[e] Provide adequate blocking, tongue and grooved edges or other suitable edge support such as PlyClips when spans exceed indicated value. Use two PlyClips for 48 in. or greater spans and one for lesser spans.
[f] Uniform load deflection limitation: 1/180th of the span under live load plus load, 1/240th under live load only. Allowable live load shown in boldface type and allowable total load shown within parenthesis.
[g] Allowable loads were established by laboratory test and calculations assuming evenly distributed loads. Figures shown are not applicable for concentrated loads.

7 DIAPHRAGMS

Plywood wall, roof, and floor diaphragms are used most commonly in wood-frame buildings. Plywood roofs and floors, as well as interior partitions, however, are frequently used as diaphragms in buildings with masonry or concrete walls. Plywood sheathing serves admirably as horizontal diaphragms with steel joists. Diaphragms may even be used to carry gravity loads, as in the case of folded-plate roof construction, discussed later.

7.1 Description

A diaphragm is a thin structural element, either flat or curved, and usually rectangular, capable of resisting shear parallel to its edges. Hence, if suitably designed and constructed, walls, partitions, floors, ceilings, and roofs may act as diaphragms while performing their other functions. Most diaphragms are either vertical or horizontal. Vertical diaphragms include walls and partitions, and are also called "shear walls"; horizontal diaphragms include floors and roofs. Curved or gabled roofs are so classed even though not strictly horizontal.

A diaphragm resists loads much like a plate girder (Fig. 4.9). These loads may be determined from the assigned wind or seismic forces that occur in any particular area. The sheathing acts as the fully shear-resistant web of the girder, while the plates, bond beams, or other boundary members carry all flexure acting as girder flanges in direct tension or compression. Intermediate framing members, such as joists and studs, stiffen the diaphragm against buckling and splice the plywood panel edges in shear. To limit deflections, a maximum ratio of span length to diaphragm width of 4 : 1 is recommended for unblocked horizontal diaphragms. Diaphragms having the plywood panel edges blocked, however, need have their dimensions limited only by permissible deflections of the attached walls. A maximum height-width ratio of $3\frac{1}{2}$: 1 is recommended for blocked vertical diaphragms.

Sizes of most members — sheathing, joists, and studs — are usually determined by vertical loads, although lumber framing thinner than 2 in. nominal is not recommended for nailed diaphragms. Boundary members may be of wood, steel, or reinforced concrete, designed and spliced to carry the direct chord stresses. As indicated, the sheathing must be attached to the boundary so as to transmit the shear and, in all-wood construction, shear can be effectively transmitted to the chord members by nailing. In steel and concrete buildings this shear may be transferred to the steel or concrete chords by bolts or anchors. Vertical diaphragms, especially those with large height-width ratios, may require carefully designed hold-down anchors attached to the outer studs to resist overturning. Careful foundation anchorage is particularly important with light wood-frame buildings subject to high winds.

Diaphragm capacity is normally rated as the shear strength in pounds per foot across its width, and is usually limited by the nailing schedule along the plywood edges. Although certain minimum plywood thicknesses are required to develop these strengths as shown in Table 4.8, plywood sheathing used in conventional thicknesses and spacings will normally have ample shear strength. Staggering of the joints parallel to the larger load increases the strength of the diaphragm.

Blocking may be omitted under panel edges when the shears are sufficiently low and other conditions do not require it. In this case diaphragm strength is usually limited by buckling of the free panel edges.

7.2 Design

To design a diaphragm, first calculate the loads and shears that will exist in the diaphragm. Normally, a diaphragm is designed for a load applied to the longest dimension of the building. A check, however, should be made for a load applied to the narrow dimension. To determine the loads on a roof diaphragm for a wind force at the eave line on a horizontal projection of the long side, first multiply the wind pressure in psf times one-half the wall height plus the height that the roof projects above the eave line. In the case of a flat roof, of course, there would be no roof projection above the eave line. This load will be in pounds per foot of building length.

As in any simple beam, the shear is highest at the ends of the beam. In the case of a diaphragm, the highest shear is at each end wall. This maximum shear is the load per foot of building length times one-half the length of the building. To determine the unit shear in the roof sheathing, divide the pounds of shear by the width of the building. This load will be in pounds per foot.

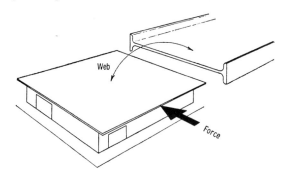

Figure 4.9   Diaphragm Acts As a Plate Girder

The chord stress caused by the bending moment in the roof diaphragm may be found by the following formula:

$$F = wL^2/8b,$$

where

$F =$ chord stress, pounds
$w =$ pounds per foot,
$L =$ diaphragm length (ft),
$b =$ diaphragm width (ft)

The maximum lateral deflection of a roof diaphragm may be calculated from the following formula:

$$d = \frac{5vL^3}{8EAb} + \frac{vL}{4Gt} + 0.94Le_n,$$

where

$d =$ deflection (in).,
$v =$ shear (plf),
$L =$ diaphragm length (ft),
$b =$ diaphragm width (ft),

A = area of chord cross-section (in.²),

E = elastic modulus of chords (psi),

G = shearing modulus of plywood ($\frac{1}{20}E$ for panels with exterior glue. Reduce 9 per cent for panels with interior glue.),

t = plywood thickness (in.),

e_n = nail deformation from Figure 4.10 at calculated load per nail, based on shear per ft divided by number of nails per ft.

This formula gives the deflection of the roof diaphragm at mid-length of the building. The first part of this formula is the bending deflection; the second, shear deflection; and the third, nail deformation.

To determine the load carried by the end walls, divide the previously calculated shear load by the length of the end wall. This length is not necessarily the width of the building since doorways may be in the ends of a building. It is the combined length of only the wall portions of the end of a building. The tensile force at each bottom corner of each wall caused by the overturning moment is then the pounds per foot of shear times the height of the building times the length of the end wall divided by the length of the end wall. This uplift force must then be resisted by some form of connection to the foundation.

8 FOLDED PLATES AND SPACE PLANES

8.1 Description

Although the standard diaphragm is rather an ordinary structure, the development of engineering data on roof diaphragms has led to more interesting roof shapes. The high shear strength and rigidity of plywood can be utilized efficiently in thin shell structures such as rectangular folded plates, radial folded plates, and space planes.

The basic concept of the folded plate involves the utilization as beams of the planes or "plates" of a pitched roof. Two plates with a common ridge chord act to carry vertical loads to the supports at their ends. Multiple-bay folded plates interact at the adjoining valley chord as well, so that such roofs can be produced with long spans parallel to the chords and limitless distances perpendicular to the chords, all without clutter at the eave line and with a significant reduction in the number of columns.

Space planes, like the rectangular folded plate, are also made possible by utilizing the high diaphragm strength of plywood. Unlike the rectangular folded plate, however, the chord members in any one plane intersect or tend to intersect, forming triangular plates. These plates, when erected, support each other providing the necessary strength. Seldom are individual plates structurally adequate in themselves. It is the forming of valleys and ridges at different angles when the plates are joined that gives space plane structures interesting shapes as evidenced by the accompanying photograph (Fig. 4.11).

8.2 Design

The folded-plate design approach following is general and can be applied to any folded plate whose chords are parallel or essentially parallel, and whose plates act in a manner similar to beams. Such roofs need not even be of the simple sawtooth

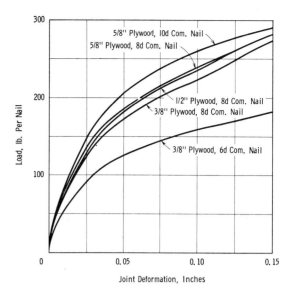

Figure 4.10 Lateral Bearing Strength of Douglas Fir Plywood Joints

form, although other profiles require special care in their analysis. This chapter will cover only the single-bay folded plate. The equations given should be considered as examples of the results obtainable using the general approach, and their restrictions should be carefully noted so that they will not be used in cases where they do not apply.

Figure 4.12 shows a single-bay folded plate. Dimensions and terms are illustrated on this sketch and in guide below.

L = length of building (ft)
   (or length of folded-plate span, where the building is more than one span long)

$2B$ = width of folded plate unit, valley to valley (ft)

H = total rise in roof (ft)

R = rise of roof, in. (inches of) rise per 12 in. (inches of) run

α = inclination of roof; cot $\alpha = B/H$

w = uniform load on horizontal projection of roof (psf)

T = horizontal thrust (lb)

P = compression force in posts (lb)

Figure 4.11 A clear span area of 32 × 110 ft was formed by the use of space planes in the Independent Congregational Church in St. Louis. Architects were Manske and Dieckmann

Table 4.8. Recommended Shears for Wind or Seismic Loading (lb per ft). (Plywood and Framing Already Designed for Perpendicular Loads.)

(a) Recommended Shear in Pounds Per Foot for Horizontal Plywood Diaphragms

Plywood Grade	Common Nail Size	Minimum Nominal Penetration in Framing (in.)	Minimum Nominal Plywood Thickness (in.)	Minimum Nominal Width of Framing Member (in.)	Blocked Diaphragms				Unblocked Diaphragms	
					Nail spacing at diaphragm boundaries (all cases) and continuous panel edges parallel to load (cases 3 and 4)[a]				Nails spaced 6 in. max. at supported edges[a]	
					6	4	2½	2	Load perpendicular to unblocked edges and continuous panel joints (case 1)	All other configurations (cases 2, 3, and 4)
					Nail spacing at other plywood panel edges					
					6	6	4	3		
Structural I	6d	1¼	$\frac{5}{16}$	2	185	250	375	420	165	125
				3	210	280	420	475	185	140
	8d	1½	$\frac{3}{8}$	2	270	360	530	600	240	180
				3	300	400	600	675	265	200
	10d	1⅝	½	2	320	425	640[b]	730[b]	285	215
				3	360	480	720	820	320	240
DFPA C-C Exterior, DFPA Structural II, DFPA Standard Sheathing and other DFPA grades	6d	1¼	$\frac{5}{16}$	2	170	225	335	380	150	110
				3	190	250	380	430	170	125
			$\frac{3}{8}$	2	185	250	375	420	165	125
				3	210	280	420	475	185	140
	8d	1½	$\frac{3}{8}$	2	240	320	480	545	215	160
				3	270	360	540	610	240	180
			½	2	270	360	530	600	240	180
				3	300	400	600	675	265	200
	10d	1⅝	½	2	290	385	575[b]	655[b]	255	190
				3	325	430	650	735	290	215
			$\frac{5}{8}$	2	320	425	640[b]	730[b]	285	215
				3	360	480	720	820	320	240

NOTE: Design for diaphragm stresses depends on direction of continuous panel joints with reference to load, not on direction of long dimension of plywood sheet.

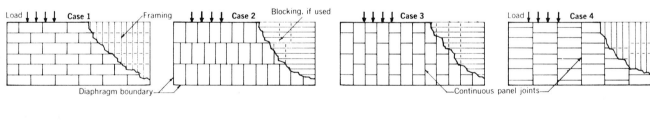

$n =$ allowable bearing load per nail
$F =$ chord stress (psi)
$v =$ plywood shear stress (psi)
$S = wB/2 \sin \alpha =$ vertical load component in plane of diaphragm.

A simple folded plate consists of two inclined planes, *AEFD* and *EBCF*. Members spanning from eave to ridge transfer vertical loads acting on the planes to points of action along the lines *AD*, *EF*, and *BC*. Beams or walls provide vertical support at lines *AD* and *BC* since the edge plates are incapable of resisting forces perpendicular to their own plane. Along line *EF* each plane supports the other. This action produces a thrust (*S*) in the tilted planes of the roof, which the diaphragm action carries to the ends of the building and hence to the four corners. Vertical supports and some means of resisting the

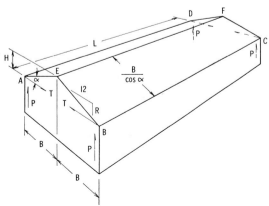

Figure 4.12   Single-bay Folded Plate

(b) Recommended Shear in Pounds Per Foot for Plywood Shear Walls [c]

Plywood Grade	Minimum Nominal Plywood Thickness (in.)	Minimum Nail Penetration in Framing (in.)	Plywood Applied Direct to Framing					Plywood Applied Over ½ in. Gypsum Sheathing				
			Nail Size (Common or Galvanized Box)	Nail Spacing at Plywood Panel Edges (in.)				Nail Size (Common or Galvanized Box)	Nail Spacing at Plywood Panel Edges			
				6	4	2½	2		6	4	2½	2
DFPA Structural I	$\frac{5}{16}$ or $\frac{1}{4}$	$1\frac{1}{4}$	6d	200	300	450	510	8d	200	300	450	510
	$\frac{3}{8}$	$1\frac{1}{2}$	8d	280	430	640	730	10d	280	430	640	730
	$\frac{1}{2}$	$1\frac{5}{8}$	10d	340	510	770	870	—	—	—	—	—
DFPA C-C Exterior,	$\frac{5}{16}$ or $\frac{1}{4}$[d]	$1\frac{1}{4}$	6d	180	270	400	450	8d	180	270	400	450
DFPA Structural II, DFPA Standard Sheathing,	$\frac{3}{8}$	$1\frac{1}{2}$	8d	260	380	570	640	10d	260	380	570	640
DFPA Panel Siding and other DFPA grades	$\frac{1}{2}$	$1\frac{5}{8}$	10d	310	460	690	770	—	—	—	—	—
			Nail Size (Galvanized Casing)					Nail Size (Galvanized Casing)				
DFPA Plywood Panel Siding	$\frac{5}{16}$[d]	$1\frac{1}{4}$	6d	140	210	320	360	8d	140	210	320	360
	$\frac{3}{8}$	$1\frac{1}{2}$	8d	160	240	360	410	10d	160	240	360	410

[a] Space nails 12 in. on center along intermediate framing members.

[b] Reduce tabulated allowable shears 10 per cent when boundary members consist of a single 2 in. lumber piece.

[c] All panel edges backed with 2-in. nominal or wider framing. Plywood installed either horizontally or vertically. Space nails at 12 in. on center along intermediate framing members.

[d] $\frac{3}{8}$ in. minimum recommended when applied direct to framing as exterior siding.

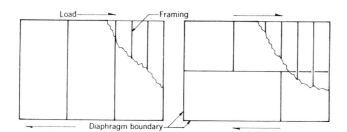

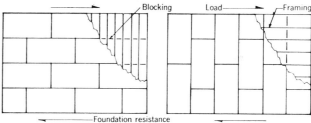

accumulated horizontal thrusts are necessary at points *A*, *B*, *C*, and *D*.

Each roof plate resists deflection parallel to its plane by beam action. Ridge and eave chords, similar to beam flanges, carry the compression and tension to the ends of the planes and the plywood "web" carries only shear.

The analysis of a multiple-bay folded plate differs slightly, in that at the interior valleys each plane again supports the other, so that no walls or vertical beams are required if the valley chords are adequately connected. Since these interior plates perform extra duty, they must carry extra stress.

Roofs are customarily designed for vertical loads, to allow for snow, ice, and so on, and horizontal loads to allow for wind. Since the planes of a folded plate are inclined, it is convenient to resolve these vertical and horizontal loads into components parallel to the planes of the plate. The applied uniform load *w* can be broken up as shown in Figure 4.13 for computation of loads in the plates.

In the figure, the portion of the uniform load which must be carried by each beam or wall is *wB*/2. That which is carried to the columns by diaphragm action of the plywood is *wB*.

This has been resolved into two equal components *S*, one acting in the plane of each plate. Since the forces carried by the beam or wall are independent of the folded-plate action, they are not considered further.

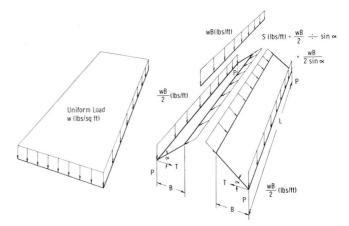

Figure 4.13 Forces Acting on Symmetrically Loaded Roof Planes

After the applied forces have been broken down into their components parallel to the plates, each plate can be analyzed as a diaphragm. Load, shear, and moment diagrams provide a direct and graphic method of visualizing the stresses in these diaphragms. If these diagrams are properly constructed, maximum shear and moment become obvious and the required plywood thickness, chord sizes, nailing, and so forth are readily computed. These diagrams are given in Figure 4.14 for a single-bay simple-span folded plate.

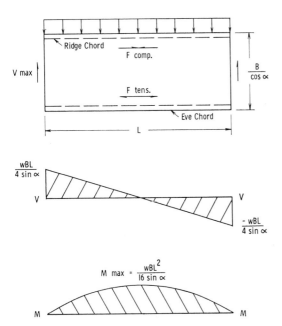

Figure 4.14   Shear and Moment Diagrams

In a multiple-bay roof, the two outer plates are designed as single-bay folded plates; and beams or walls are required, since there are no additional plates to provide support. Interior plates, because they support adjacent plates, carry twice as much vertical load as outer plates.

In the design of a plywood folded plate, the first item usually computed is the plywood thickness. First determine the thickness required to carry folded-plate shear, and then determine whether this thickness is adequate for the sheathing function which the plywood must also perform. The shear per foot of diaphragm width,

$$v_{\max}\,(\text{plf}) = \frac{wBL}{4 \sin \alpha} \div \frac{B}{\cos \alpha} = \frac{wBL}{4H}.$$

From the above shear equation it is evident that the shear in the roof diaphragm diminishes as the slope (H/B) of the roof increases and the length of the roof (L) decreases. Increasing the distance between valleys ($2B$), while maintaining the same slope, does not increase this shear.

If the rafter spacing has not been set, it may be determined by the plywood thickness.

In some cases, as with stressed skin panels, or plates where a smooth plywood underskin is desired, plywood skins both on top of and below the rafters may be functional. If both upper and lower skins are to resist stress, they must, of course, both be adequately connected to transfer this stress.

Wind loads seldom control the design, partly because an increase of one-third is allowable for stresses in wood structures subject to wind. This allowable increase is applicable to the total stresses caused by wind loads plus dead loads. A separate design, taking into account shears due to wind loads combined with those from dead loads should be compared with those from live load plus dead load shears to ensure that wind loads will not require a larger section.

Since the plywood diaphragms of the folded plate roof carry all shear, it can be seen that the rafter at the support must deliver the accumulated shear force from the diaphragm to the support. Thus this end rafter, or in the case of a cantilevered folded plate the rafter at the support, must carry a compressive stress as well as the bending stress carried by the other rafters. In the case of an end rafter not supported along its length by a shear wall this compression force varies from zero at the ridge to a maximum at the post equal to the total shear in the plywood diaphragm at this point. It is almost always necessary at least to double the end rafter and sometimes to supply even greater section in order to resist these combined forces and to reduce the compressive stress caused at the valley chords by the reaction from this rafter. It need not necessarily be framed into the side of the valley chord, but the most common details call for such a connection, thus bringing this heavy load into the chord, which must resist it in compression perpendicular to the grain.

As mentioned previously, to determine the chord sizes, assume that the chords take all moment stresses. For most folded plates it is sufficiently accurate to assume the total area of the chords to be concentrated at a point on the margin of the plate. For critical cases in narrower plates, a more accurate assumption of chord areas concentrated at their centroids should be used. Except in severe cases, chord centroids can be estimated.

For a simply supported plate of one fold, the eave chord force is

$$F_e = wBL^2/16H.$$

The ridge force is

$$F_r = wBL^2/8H.$$

Chords must be continuous or adequately spliced between supports. Do not neglect to take into account the reduction in chord area due to bolts or plates used for splicing.

Folded plates act as beams in their own inclined planes to carry imposed loads to the end supports. They do not have intermediate horizontal members to take out horizontal forces. These forces, therefore, accumulate at the end supports and wherever a folded plate is supported vertically there must be some means of resisting horizontal forces. It is customary to resist these horizontal thrusts either with tie rods across the ends of the building, or by taking the thrusts into the end walls below the gable. The horizontal thrust is equal in magnitude to the horizontal component of the force in the inclined plate at its support. For a single-bay folded plate of simple span this is

$$T = wB^2L/4H.$$

Plywood sheathing-to-rafter connections must resist the shear v at all points. Folded plates have an additional force on the ridge joint. This force is usually resisted by the plywood-to-ridge nailed connection, although the ridge-to-rafter joint could be designed to carry it. When the former is the case, the maximum force on the plywood-to-ridge joint is the result of the horizontal shear force and the ridge force. For uniform load along the span L, the shear stress would decrease to zero at mid-span, but to allow for partial loading use a conservative value. Thus, for the plate chosen, the number of nails per foot, N, varies from

$$N_1 = \frac{\sqrt{(wBL/4H)^2 + (wB/2 \sin \alpha)^2}}{n}$$

at the end of the diaphragm, to a value at the center of

$$N_2 = wB/2n \sin \alpha,$$

where n = allowable load per nail. The nailing at the gable ends, along the bottom chord, and at interior panel edges must resist only the horizontal shear. At the gable ends the required nailing is

$$N_3 = wBL/4Hn.$$

Nailing at other panel edges, N_4, may be decreased linearly from N_3 to a value at mid-span equal to two nails per foot or to $wBL/16Hn$, whichever is the greater. Panel edges should never have less than two nails per foot. Intermediate nailing, N_5, where plywood is continuous across supports, can be one nail per foot.

Strictly speaking, the deflection of a folded plate, even of the simple sawtooth configuration, is highly complicated since the plates must depart from a plane surface in order for the structure to deflect. For deflection computations, however, it is sufficiently accurate to assume that each plate deflects in its own plane. The deflection, $d \sin \alpha$, can be computed from the standard diaphragm equation discussed in the previous section. In a folded plate of one fold, deflection is downward at the ridge and outward at the walls (see Fig. 4.15). Vertical deflection of the ridge equals d, and the horizontal deflection of the eaves, h, to a very close approximation, equals $d \tan \alpha$. The value of d is not doubled for the deflection of the other side of the roof.

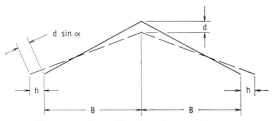

Figure 4.15   Deflection of Folded Plates

In most folded plates deflection is relatively small compared to the span, but may be noticeable from an appearance standpoint. If, on first trial, the deflection comes out unusually large, the easiest way to produce a folded plate with a smaller deflection for the same span is to increase the depth of the plates. This can be done either by increasing the valley-to-valley distance and maintaining the original slope, or by maintaining the valley-to-valley distance and increasing the slope. Either expedient will yield a stiffer structure.

9 GUSSET PLATES

The same two-way strength of plywood that makes it an ideal material for such structures as diaphragms, folded plates, and space planes also makes it an ideal material for use as gusset plates. Plywood gusset plate uses range from truss gusset plates to rigid-frame joints. As discussed in the conventional construction section, on-site labor costs in residential construction have increased so rapidly that most builders have gone to the use of one or more building components, and the most popular component in today's residential construction business is the trussed rafter.

Trusses with glued plywood gussets are strong and stiff. For some time, however, no accurate design method was available for design of trusses with glued gussets or any other type of rigid connectors. Instead, trusses were designed by empirical methods (often grossly inaccurate) and subsequently load-tested to check the validity of the design.

In 1961, Dr. S. K. Suddarth of Purdue University developed a design method that accurately predicts the performance of lumber trusses with rigid connections, including glued plywood gussets. This design procedure was later endorsed by the Joint Industry Advisory Committee on roof truss design sponsored by the National Association of Home Builders. Using this procedure, the American Plywood Association programmed for a computer a series of truss designs. Several trusses were then tested to verify the calculated stresses in the lumber members and plywood gussets.

A structural system that is becoming increasingly popular and that also uses plywood gusset plates is the plywood-and-lumber rigid frame. In principle, plywood rigid frames are arches formed by joining four straight pieces of lumber with nailed plywood gussets. Once assembled, they become rigid load-carrying units.

There are two shapes of rigid frames which have proved quite popular. One, the slant-legged version, looks much like a parabolic arch because its legs slant upward and inward from the foundation. The other, the vertical-leg design, produces a building similar in appearance to conventional framing. Of the two designs, the slant-legged version is the more popular because it makes the most efficient use of the member sizes and their spacing. If a building having the same loads were to be constructed as a similar vertical-leg design, heavier materials would frequently be required.

Plywood rigid frames are one of the most economical clear span structural systems available. They are practical on spans up to approximately 50 ft. Since rigid frames are essentially arches, less material is required to span a given distance than with conventional framing. Less labor is needed and usually the erection does not depend on highly skilled or specialized labor.

Both the plywood-gusseted truss and rigid-frame structure are indeterminate and must be designed by setting up trial dimensions and sizes of material, solving the forces, moments, and shears in the members and then analyzing the joints. If overstress is discovered, stiffeners may be added, plywood-grain orientation may be changed, or the gusset-plate thickness may be increased if nailed. Any further change requires complete rechecking of the entire structure, beginning with member stresses.

10 PLYWOOD BOX BEAMS

10.1 Description

A plywood-lumber box beam reacts to an imposed load in a manner quite similar to a steel I beam. In a box beam, lumber is used for the flanges which carry most of the bending (Fig. 4.16). The webs are of plywood and carry the shear. A glued or nailed joint between them transfers the stresses from one to the other. Vertical stiffeners set between flanges distribute concentrated loads and resist web buckling. Deflection resulting from shear is usually significant and must be added to the bending deflection. Flanges must be continuous the full length of the beam; this may be accomplished either by splicing the lumber or using scarfed joints. In addition, webs may require splicing, depending upon the imposed shear at that point.

Loads, spans, and allowable stresses, as well as desired appearance, determine the beam proportions. The depth and cross section may be varied along the length of the beam to fit design requirements, providing the resisting moments and shear capacity at all sections are adequate.

Camber may be provided opposite to the direction of anticipated deflection for purposes of appearance or utility. It has no effect on strength or actual stiffness. Where roof and floor beams are cambered, a recommended amount is 1.5 times the deflection due to dead load only. This provides a nearly level beam under conditions of no live load after set has occurred.

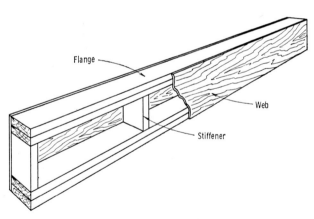

Figure 4.16   Cutaway View of Box Beam

10.2 Design

Bending

In plywood beams, one flange is stressed in compression and the other in tension. Since wood is stronger in tension than in compression, the allowable stress in compression parallel with the grain limits the design of symmetrical sections. Thus, the tension flange can theoretically be reduced in size or grade. Symmetrical cross sections are generally used for several practical reasons, including the possibility that either flange may be the compression flange when erected. All of the following equations assume a symmetrical section.

In a symmetrical section, allowable resisting moment may be calculated by the formula

$$M = c_{\parallel}\, I_n/0.5h,$$

where

$M =$ resisting moment (in.-lb),
$c_{\parallel} =$ working stress in compression parallel to the grain of the flange lumber (psi),
$I_n =$ net moment of inertia of continuous parallel-grain material in the section (in.^{4}),
$h =$ depth of beam (in.).

Butt joints in plywood webs are usually spliced to transmit shear only, with a splice plate only as deep as the clear distance between flanges. If such butt joints in webs are staggered 24 in. or more, one web may be used in computing the moment of inertia for bending stress. For joints closer than 24 in. the contribution of the webs should be neglected entirely in computing I_n. When webs are spliced full depth so as to carry direct "flange" stresses, both webs may be included in computing the moment of inertia, at the allowable stress given in Table 4.9.

Table 4.9.   Strength of Splice Plates

Plywood Thickness (in.)	Length of Splice Plate (in.)	Maximum Stress, psi				
		Struct. I A-C Struct. I C-D	Group 1 A-A, A-B, A-C, A-D, B-B, B-C, B-D	Struct. II	Standard grades using Group 2 Stresses	All grades using Group 4 Stresses
$\frac{1}{4}$ $\frac{5}{16}$ $\frac{3}{8}$ Sand. $\frac{3}{8}$ Rough	6 8 10 12	1500	1200	1100	1000	900
$\frac{1}{2}$ $\frac{5}{8}$ and $\frac{3}{4}$	14 16	1500 1200	1000 800	1100 850	950 750	900 700

When the cross section is not symmetrical about its center, the resisting moment may be calculated as above, except that the distance from the neutral axis to each flange is used in place of the value of $0.5h$, and the moment of inertia is calculated with due regard for the location of the neutral axis. The location of the neutral axis in this case is computed on the basis of the total cross section, without reduction for butt joints.

Horizontal shear

The allowable horizontal shear on a section can be calculated by the following formula:

$$V = \frac{v I_t \sum t}{Q},$$

where

$V =$ allowable total shear on the section (lb),
$v =$ allowable plywood shear stress (psi) through the panel thickness as given in the section of this chapter on properties,
$I_t =$ total moment of inertia about the neutral axis of all parallel-grain material, regardless of any butt joints (in.^{4}),
$\sum t =$ total thickness of all webs at the section (in.),
$Q =$ statical moment about the neutral axis of all parallel-grain material, regardless of any butt joints, lying above (or below) the neutral axis (in.^{3}).

The thickness of the webs may be varied along the beam length in proportion to the shear requirements. In doing so, however, consideration must be given to both shear through the panel thickness at the neutral axis, and rolling shear between flange and web. Where webs are discontinued, plywood or lumber shims may be glued to the flanges to maintain beam width as required for appearance or for gluing pressure.

Flange-web joints

Joints between flanges and webs at any section must be designed to resist the shear acting on that section. Maximum shear normally occurs at each end of the beam. The allowable flange-web shear on a glued, symmetrical two-web section may be calculated by the following formula:

$$V = 2sdI_t/Q_{fl}$$

where

V = allowable total shear on the section (lb),
s = allowable plywood rolling shear stress (psi) as given in the section on properties of this chapter, but reduced 50 per cent for shear concentration.
d = flange depth (in.),
I_t = total moment of inertia about the neutral axis of all parallel-grain material, regardless of any butt joints (in.^{4}),
Q_{fl} = statical moment about the neutral axis of all parallel-grain material, regardless of any butt joints, in the upper (or lower) flanges (in.^{3}).

Stiffeners

Lumber bearing stiffeners are required over reactions and where other concentrated loads occur, to distribute such loads into the beam. They should fit accurately against the flanges, and the webs should be securely attached to them. Bearing stiffeners at the ends of beams should have their dimension parallel to the beam span equal to or greater than that given by the following two considerations.

For compressive strength of the flange lumber, the thickness of stiffener must be at least equal to x in the following equation:

$$x = P/c_\perp b,$$

where

x = thickness of stiffener (parallel to beam span) (in.),
P = concentrated load or reaction (lb),
$c_\perp$ = allowable stress in compression perpendicular to grain for the *flange* lumber (psi),
b = flange width (in.).

To resist the rolling shear between the plywood and the stiffener, the thickness of the stiffeners should be at least equal to x in the following equation:

$$x = P/2hs,$$

where

x = thickness of stiffener (parallel to beam span) (in.),
h = depth of beam (in.),
s = allowable plywood rolling shear stress (psi) as given in the section on properties of this chapter, but reduced 50 per cent for shear concentration.

Intermediate stiffeners are required to stabilize the flanges, to space the flanges accurately during fabrication, to reinforce the webs in shear and prevent their buckling, and to serve as backing for gluing of web splice plates where prespliced or scarfed webs are not used. Such stiffeners are usually of 2-in. dimension lumber, and must be equal in width to the lumber flange between webs, allowing for splice plates, if any. Intermediate stiffener spacing should not exceed 48 in. on centers in order to develop all or nearly all the shear strength of a beam of normal proportions.

Deflection

The deflection of plywood beams is the sum of the deflections due to bending and shear. The bending deflection may be calculated by conventional engineering formulas, with due regard to loading condition and fixity of supports. Deflections due to several simultaneously applied loads may be calculated separately and added. In calculating the bending deflection, the elastic modulus of the flange lumber may be increased by 10 per cent over the values tabulated in the National Design Specification. The moment of inertia used for computing the bending deflection is the moment of inertia of all parallel-grain material in the section, regardless of butt joints.

Note that the moment of inertia used for computing shear deflection is that of *all* material in the section, including cross plies in the webs, regardless of any butt joints. The shear deflection may be calculated using the following formula:

$$\Delta_s = PlKh^2C/GI_g,$$

where

Δ_s = shear deflection (in.),
P = total load on beam (lb) ($=wl$ for uniform load),
l = span length (in.),
K = a factor determined by the beam cross section and shown in Figure 4.17,
h = depth of beam (in.),
C = a coefficient depending on the manner of loading, also shown in Figure 4.17,
G = shearing modulus of the webs (psi) ($\frac{1}{20} E$ for panels with exterior glue. Reduce 9 per cent for panels with interior glue.)
I_g = gross moment of inertia of all material in the section (in.^{4}).

11 STRESSED SKIN PANELS

11.1 Description

In a flat plywood stressed skin panel the plywood skins take most of the moment stresses while the lumber stringers take shear stresses (Fig. 4.18). Since stressed skin panels are usually relatively shallow, shear deformation between skins and webs would produce excessive vertical deflection. A rigid connection is therefore required between the plywood and the lumber. Such panels should therefore be assembled with glue, since mechanical fasteners require slip before they will carry a load.

Although it is possible to use laminated or scarf-jointed members for the stringers of stressed skin panels, such panels are usually restricted to single-lamination stringers, and their maximum length, therefore, is determined by the maximum

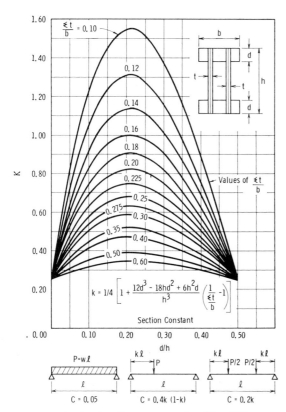

$$k = 1/4 \left[1 + \frac{12d^3 - 18hd^2 + 6h^2d}{h^3} \left(\frac{1}{\frac{\xi t}{b}} - 1 \right) \right]$$

Section Constant

Figure 4.17 Section Constant and Load Coefficients

are not normally considered in other forms of structural design. The first step in designing a panel is to assume a trial cross section. Stressed skin panels are designed by a "cut and try" method. A section must first be assumed and then checked for its ability to do the job intended. The whole 4-ft-wide panel is usually designed as a unit, in order to take into account edge conditions.

Basic spacing

In some cases the whole of the 4-ft width of the skins cannot be considered effective, since there is a tendency for the thin plywood to dish toward the neutral axis of the panel when stringers are widely spaced. A basic spacing, usually referred to as the *b* distance, is used to take this tendency into account. The *b* distance represents the amount of the skin which may be considered to act in conjunction with the stringer. Table 4.10 lists these *b* distances.

Table 4.10. Basic Spacing

Plywood Thickness (in.)	Basic Spacing *b* (in.)	
	Face grain parallel to longitudinal members (in.)	Face grain perpendicular to longitudinal members (in.)
$\frac{1}{4}$ Sanded	10.35	11.61
$\frac{5}{16}$ Rough	11.87	16.80
$\frac{3}{8}$ Rough (3-ply)	14.25	20.13
$\frac{3}{8}$ Sanded (3-ply)	16.43	16.43
$\frac{3}{8}$ Sanded (5-ply)	18.10	20.25
$\frac{1}{2}$ Rough and Sanded	23.25	28.5
$\frac{5}{8}$ Rough and Sanded	29.1	35.6
$\frac{3}{4}$ Sanded (5-ply and 7-ply)	38.2	38.2
$\frac{7}{8}$ Rough and Sanded	41.6	48.1
1 Rough	45.5	58.9
1 Sanded	54.5	47.9
$1\frac{1}{8}$ 2-4-1	53.4	57.3

Deflection

It will sometimes be necessary to compute one moment of inertia for deflection calculations and another for stress determination, since the two considerations are based on different aspects of panel behavior. Allowable stresses are determined by applying suitable reduction factors to the stresses obtained at ultimate; allowable deflections are arbitrarily set and applied to the behavior of the panel in its working range. Because of the previously mentioned "dishing" of the skins at high loads, therefore, a smaller net section may be effective at ultimate loads than is available in the working range. For thin skins and wide stringer spacings, the effective moment of inertia for bending may be less than that for deflection. As mentioned previously, the full 4-ft width of plywood is used for computing the neutral axis and moment of inertia for deflection calculations.

The usual system for figuring the neutral axis and moment of inertia involves taking moments about the plane of the bottom of the panel, using the actual net dressed cross sectional area of the lumber members after resizing.

To calculate deflection, the different moduli of elasticity of the skins and the stringers must be reconciled by use of a

length of lumber available. Headers at the ends of the panel, and blocking within the panel, serve to align the stringers, back up splice plates, support skin edges, and help to distribute concentrated loads. They may be omitted in some cases, but should always be used when stressed skin panels are applied with their stringers horizontal on a sloping roof. Without headers and blocking, panels so applied would tend to assume a parallelogram cross section.

Usually a panel with both a top and a bottom skin is the most efficient solution to a flooring or roofing problem. Owing to the high strength of plywood, calculations will often indicate that a thin bottom skin is structurally sufficient. There is some possibility, however, that $\frac{1}{4}$-in. bottom skins used with face grain parallel to stringers on 16-in. centers may assume a slight bow. Such a bow, although of no importance structurally, may be undesirable from an appearance standpoint. For stringers so spaced, $\frac{5}{16}$-in. plywood is a recommended minimum if appearance is a factor.

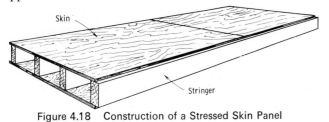

Figure 4.18 Construction of a Stressed Skin Panel

11.2 Design

Although a plywood stressed skin panel is not difficult to design, there are several factors that must be considered that

"transformed" section in computing both the neutral axis and the moment of inertia. For instance, if the modulus of elasticity of the stringers were half that of the skins, the correct neutral axis and moment of inertia could be computed, if the final modulus of elasticity were assumed to be equal to that of the skins, and if for computation the stringers were assumed to be only half as wide as they actually are. When computing the neutral axis, the cross sectional area of the plywood is taken as that of the parallel plies only and not of the crossbands.

After the neutral axis is located, the next step is to calculate the moment of inertia. Again, the transformed section is considered, and the I calculated is a true stiffness factor in bending alone.

The bending deflection may now be computed by conventional methods. Since the skins of stressed-skin panels are not subject to shear deflection, a 10 per cent increase to the modulus of elasticity of plywood may be taken. A value for shear deflection, computed separately, is then added to the first figure representing the moment deflection. The simple span deflection for uniform loading is given by the following equation:

$$\Delta_s = 1.8Pl/AG$$

where

Δ_s = shear deflection (in.),
P = total load on panel (lb),
l = span length (ft),
A = actual total cross-sectional area of all stringers (in.^{2}),
G = modulus of rigidity of stringers psi. G may be taken as 0.06 of the true bending modulus of stringers.

To combine equations for bending and shear deflections for a uniformly loaded simple span panel, the allowable load based on deflection is given by the following:

$$w = \frac{1}{Cl(7.5l^2/EI + 0.6/AG)},$$

where

w = total live plus dead load allowable (psf),
C = factor for allowable deflection, usually 360 for floors, 240 for roofs,
l = span length (ft),
EI = factor figured previously for a 4-ft-wide panel (lb-in.^{2}),
A = actual total cross-section area of all stringers (in.^{2}),
G = modulus of rigidity of stringers psi. G may be taken as 0.06 of the true bending modulus of stringers.

In addition to computing the deflection of the whole panel acting as a unit, the designer must also check the deflection of the top skin between stringers. Sections are usually selected such that only deflection must be checked for this top skin, but for unusual applications moment and shear should also be investigated. Section properties for span perpendicular to face grain should be used. For two-sided panels, this skin would be a fixed-end "beam" for which the equation is

$$\Delta = \frac{wl^4}{384 \, EI \, 12}$$

where

Δ = deflection (in.),
w = load (psf),

l = clear span between stringers (in.),
E = modulus of elasticity for top skin (psi),
I = moment of inertia of 1-ft width of top skin (in.^{4}).

Bending moment

As mentioned previously, for bending considerations it is sometimes necessary to figure on the reduced section provided by using the b distance to arrive at effective skin widths. If the clear distance between stringers is less than b, the effective width of skin is equal to the full panel width, and the neutral axis and moment of inertia are as figured previously for deflection. If the clear distance is greater than b, the effective width of skins equals the sum of the width of the stringers plus a portion of the skin extending a distance equal to $0.5b$ each side of each stringer (except, of course, for outside stringers where only part of that distance is available). In this case, a new neutral axis and moment of inertia for bending should be computed using the proper section. Only those plies whose grain is parallel to the span are considered.

As the spacing between framing members is increased, the allowable load is reduced in two ways. The first has been accounted for in considering the b distance. The second involves reductions in allowable stresses. The allowable stresses, both in tension and compression, for the grade of plywood used, should be reduced in accordance with the graph of Figure 4.19. This reduction is to provide against buckling of

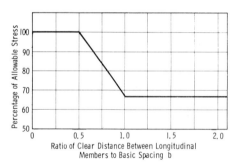

Figure 4.19 Stress Reduction Factor for Framing Member Spacing

the skins and applies to working stresses in both tension and compression parallel to the grain, but not to rolling shear stresses.

For stressed skin panels of unbalanced section, as are most panels with spaced stringers, it is necessary to figure the allowable load in bending as determined both by the top skin in compression and by the bottom skin in tension. The lower of these values then governs, unless there is a splice in an area of high moment. If there is, an additional check on the splice is required. For uniform loads the general equation $M = fI/c$ reduces to the following:

$$w = 8fI/cl^248,$$

where

w = allowable load (psf),
f = allowable stress (either tension or compression) (psi),
I = moment of inertia per 4-ft width (in.^{4}),
c = distance (in.) from neutral axis to extreme fiber (either tension or compression),
l = span (ft).

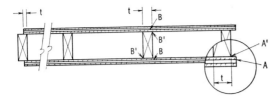

Figure 4.20   Location of Critical Rolling Shear Stress

11.3 Rolling Shear

The standard equation for shear, of course, is $v = VQ/Ib$. Since the Q of the top skin of a stressed skin panel is almost always larger than that of the bottom skin, it is generally sufficient to compute rolling shear for the top skin only. When the plywood skin has its face grain parallel to the longitudinal framing members, as is usually the case, the critical shear plane lies within the plywood, between the inner parallel ply and the adjacent perpendicular ply (at A or at B) in Figure 4.20. If the face grain of the skin is perpendicular to the framing members, the critical rolling shear plane lies between the inner perpendicular ply and the framing member (at A' or B').

It is convenient to compute a value for the sum of the allowable shear times the applicable shear width over all joints. Due to stress concentrations, the allowable stress in rolling shear for exterior joists is only half that for interior joists.

The allowable uniform load on a simple-span stressed skin panel as determined by rolling shear can now be found by using the following equation:

$$w = 2I(\textstyle\sum st)/Q4l,$$

where

    w = allowable load (psf),
    I = total moment of inertia of the 4-ft-wide panel (in.⁴),
    $\sum st$ = sum of the glueline widths over each stringer, each multiplied by its applicable allowable rolling shear stress (lb/in.),
    Q = the first moment of the parallel plies outside of the critical rolling shear plane, full 4-ft panel width (in.³),
    l = length of the panel (ft).

The final allowable load on the panel, of course, is the lowest of the figures obtained in any of the previous calculations.

12 SANDWICH PANELS

12.1 General

A structural sandwich panel is a component similar to the spaced-stringer stressed-skin panel. The sandwich panel is an assembly consisting of a lightweight core laminated between two thin faces of plywood. These panels may be used as roofs, floors, and walls, both load-bearing and non-load-bearing. Their primary advantages are light weight and high insulation qualities.

A variety of core materials may be used to form sandwich panels. Among these are polystyrene foams, polyurethane foams, and paper honeycombs. Of these core materials, the expanded foams provide the best insulation.

12.2 Design

Since in many cases the core thickness is determined by requirements for insulation rather than strength, the suggested design method is to select a given construction and to check it for all possible modes of failure under the design loads.

In general, the structural design of a sandwich panel may be compared to that of an I beam. Facings carry the compressive and tensile stresses and the core resists shear. This core should be thick enough to space facings so that they provide bending stiffness. It also supports the facings against buckling. Bending stiffness of the core is assumed to be insignificant.

The neutral axis may be found by using the following formula:

$$\bar{y} = \frac{A_1(h - \tfrac{1}{2}t_1) + A_2(\tfrac{1}{2}t_2)}{A_1 + A_2},$$

where

    A_1 = area of parallel grain material in top skin (in.²/ft of width),
    A_2 = area of parallel grain material in bottom skin (in.²/ft of width),
    t_1 = thickness of top skin (in.),
    t_2 = thickness of bottom skin (in.),
    h = panel depth (in.),
    $\bar{y}$ = distance to neutral axis from bottom of panel (in.).

The moment of inertia and the section modulus may be determined by using the following formulas:

$$I = \frac{A_1 A_2 (h + c)^2}{4(A_1 + A_2)}$$

where

$$c = \text{core depth (in.)},$$
$$S_{\text{top skin}} = I/(h - \bar{y}),$$
$$S_{\text{bottom skin}} = I/\bar{y},$$
$$S = \text{section modulus}.$$

Having found the moment of inertia and section modulus, the designer may find the strength and stiffness of a sandwich panel by using conventional engineering formulas.

It should be noted that because of the low shear modulus found in most core materials, deflection caused by shear may be significant and is sometimes greater than bending deflection. It can not be ignored as it commonly is with many materials having greater shear moduli.

13 CURVED PANELS

13.1 General

Curved panels may be designed as either of two structural types, depending upon whether horizontal thrust is developed. The curved flexural panel performs in flexure in the same way as a conventional flat panel, acting as a simple beam without developing horizontal thrust. No tie rods are required and to avoid damage to the supporting structure, bearing details must be designed with provision for horizontal deflection. Generally, thickness is greater than for arch panels of the same span, and construction is usually with spaced ribs to achieve necessary thickness. (See Fig. 4.21).

Arched panels are stressed both in compression and in

Figure 4.21   Attractive Curved Panel Roof

flexure. They exert horizontal thrust at the supports and therefore require tie rods or abutments. These panels are relatively thin and are generally of constant cross section and continuous from support to support, without fixity at the supports. Unlike the flexural panel, stresses are considerably affected by the radius of curvature of the arch. Also unlike the structural panel, the arched panel is "two-hinged" and is statically indeterminate.

Either type of panel may be made in any desired shape within the limits of fabrication of the materials used. Of course, no compound curvature is possible with plywood. For convenience in design and fabrication, circular curves are generally specified.

Panels of medium curvature (between perhaps 3 : 1 and 8 : 1 span-rise ratios) are most practical. With panels of low rise, outward thrusts become very high and small outward movements at the supports can cause undesirable deflections of the panel. Panels with very high rise must be especially designed for resistance to lateral forces such as winds, since they become less stable as they approach semicircular shape.

Plywood curved panels consist of full-length plywood skins top and bottom, with a structural core capable of resisting the shearing forces. For short spans, a solid panel is often used, with one or more layers of plywood, butt-jointed or full length, usually glued over the full area of the skins. In some cases, glue need be applied only in strips, making the attachment similar to that of a spaced-rib panel.

For any span, the core may consist of single-piece or laminated plywood or lumber ribs spaced as required, either preglued or glued during panel assembly. An advantage of this type of construction is that it can readily include blanket insulation. Where light weight is vital, the core material may consist of resin-impregnated-paper honeycomb or one of the foamed plastics.

13.2 Design

The first step in designing a curved panel of either flexural or arch type is to determine the curvature and the thickness of the top and bottom skin. The following table gives the radii for various thicknesses of plywood that have been found through experience to be appropriate minimums when bent dry. Of course, shorter radii can be developed by wetting or steaming.

Panel Thickness	Across Grain	Parallel to Grain
$\frac{1}{4}$ in.	2 ft	5 ft
$\frac{5}{16}$ in.	2 ft	6 ft
$\frac{3}{8}$ in.	3 ft	8 ft
$\frac{1}{2}$ in.	6 ft	12 ft
$\frac{5}{8}$ in.	8 ft	16 ft

Next, using accepted engineering methods, compute the neutral axis, moment of inertia, and section modulus. The stressed skin panel method is used for spaced ribs and the sandwich panel method is used when a foam or honeycomb core is used. Figure two values of section modulus, one for top, and one for bottom, if the neutral axis is not at the centerline of the panel depth.

The design of a spaced-rib panel (Fig. 4.22) is similar to that of a flat stressed skin panel. For a curved sandwich panel, however, a point not covered in the design of a flat panel is the skin buckling stress. The maximum combined stress, when computed, must be less than $c_\parallel$, and less than one-third of the value of c_{cr} as found below.

$$c_{cr} = 0.5 \sqrt[3]{EE_c G_c}$$

where

E = modulus of elasticity of plywood (psi)
E_c = modulus of elasticity of core (psi)
G_c = modulus of rigidity of core (psi)
c_{cr} = critical skin buckling stress (psi).

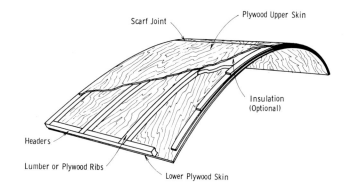

Figure 4.22   Cutaway View of Spaced Rib Curved Panel

This critical stress represents the value at which the skin may be expected to suffer local buckling, crushing the core or causing the core to fail in tension normal to the skins.

At this point, one more item may be determined which is applicable to both the flexural panel and the arched panel. When a curved member is subjected to bending moment, a radial stress is induced. When the moment is in the direction tending to decrease curvature (increase the radius) the stress is in tension. When this moment is in the direction tending to increase curvature (decrease the radius) the stress is in compression. This radial stress is given by the following equation and must be less than one-half the allowable stress for rolling shear in the plane of the plies for radial tension and less than the allowable stress for compression perpendicular to the grain for radial compression. For sandwich panel core material it should be less than one-third of the allowable shear value of the core.

$$f_R = \frac{3}{2}\frac{M}{Rbh},$$

where

f_R = radial stress (psi) due to moment M,
M = bending moment (in.-lb) on a width of panel equal to the rib spacing for spaced-rib panels, or a 1-ft width for sandwich panels,
R = radius of curvature at centerline of member (in.),
b = width of rib (in.) for spaced-rib panels or 12 in. for sandwich panels,
h = over-all depth of panel (in.).

Note: For the curved portion of members the allowable unit stress in bending and in tension and compression parallel to the grain must be multiplied by the following curvature factor:

$$1 - 2000(t/R)^2$$

in which

t = distance between extreme fibers of plies parallel with the stress for one lamination (in.),
R = radius of curvature of a lamination (in.),
t/R = shall not exceed $1/125$.

13.3   Curved flexural panels

The design of spaced-rib panels is the same as that of flat stressed skin panels as outlined previously except that the horizontal deflection at the supports must be calculated. The deflection at the free support of a panel that is free to deflect horizontally at one end may be determined from the following equation:

$$\Delta_H = 4y\Delta_v/l,$$

where

Δ_H = horizontal deflection at support (in.),
y = rise of curved panel at mid-span (in.),
Δ_v = vertical deflection of curved panel at mid-span due to vertical load (in.),
l = horizontal span length (in.).

This equation, while approximate, is sufficiently accurate to determine the amount of space to allow at supports for the horizontal deflection to take place, so as to avoid developing thrust that is not provided for in the design.

13.4   Arch panels

Two-hinged arch panels may be designed using the following procedure. The solution for three-hinged arches is the same as that outlined for two-hinged arches, except that the horizontal thrust is calculated by statics instead of by use of the graph, Figure 4.17.

First, calculate the vertical reactions and horizontal thrusts of the arch for the various conditions of loading. The graph of Figure 4.23 will simplify this step. Notice that since the thrust-coefficient graph is set up for loads at the tenth points of the span, it is necessary to break up the uniform load into equivalent concentrated loads acting at these points. Ignore the load

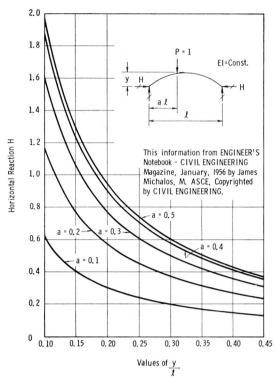

Figure 4.23   Thrust Coefficients for Two-hinged Circular Arches

on 0.05 of the span adjacent to each support since it has negligible influence on the thrust and divide the remainder into equal sections with concentrated loads at their midpoints. The graph then gives thrust coefficients, which, when multiplied by a load, yield the thrust produced by that load.

Calculate by statics and plot the moments at intervals along the arch for the various conditions of loading. A convenient interval for use with the thrust graph is obtained by dividing the span into ten equal parts. By inspection of the moment curves, the most severe loading condition and the magnitude of the maximum moment may be determined. The worst case for maximum moment will usually occur either with dead load plus concentrated erection load, or with dead load plus half-span live load. In either case it will be near the quarter-point of the span.

By resolving acting forces into components tangential to the slope of the arch, calculate the direct stress at the point of maximum moment. Guided by the moment and direct stress, assume a trial cross section for the arch and calculate its weight and section properties.

The combined stresses due to bending and compression may be calculated by the following formula:

$$f = \frac{P}{A} \pm \frac{M}{S},$$

where

f = stress in extreme fiber (psi), tension or compression,
P = direct force (lb per ft width of arch),
A = area of arch cross section (in.² per ft width of arch),
M = bending moment (in.-lb),
S = section modulus (in.³ per ft width of arch).

For spaced-rib and plywood-core panels, the shear stress must not exceed rolling shear values for plywood. Normally maximum shear occurs at the ends of the arch for both dead load and full-span uniform live load, and maximum shear for half-span occurs at the center of the span, with shears at the ends almost as high for most span-to-rise ratios. Maximum shear due to concentrated erection load will probably occur at an end, when the load is placed near that end. Maximum shear may be calculated by the following formula:

$$v = VQ/Ib,$$

where

$v =$ shear stress (psi), either rolling shear in plywood or horizontal shear in rib,

$V =$ shear acting normal to the slope of the arch (lb per ft of arch width, or per rib section),

$Q =$ first moment about neutral axis of area of parallel-grain material from panel face inward to plane at which shear stress is to be calculated (in.^{3} per ft width of arch panel, or per rib section),

$I =$ moment of inertia of arch panel (in.^{4} per ft width of arch panel, or per rib section),

$b =$ width of rib (in.) for spaced-rib panel; width of glued strip, for strip-glued solid-core panel; or 12 in. for solid-plywood-core panel glued over its total area.

For honeycomb or foamed-plastic core, shear stress in the core should not exceed one-third the ultimate shear strength. It may be computed by the following formula, which allows credit for that portion of the shear resisted by wood edge members:

$$q = 2V_2/(h + c)b,$$

where

$q =$ shear stress in core (psi),

$V_2 = V - V_1 =$ total shear acting normal to the slope of the arch minus allowable shear carried by wood framing members (lb per panel width),

$h =$ total depth of panel (in.),

$c =$ thickness of core (in.),

$b =$ width of core per panel (in.).

After calculating the radial stresses as done previously, we may now calculate the maximum deflection of the arch for various conditions of loading. The following equation, which ignores the minor portions of deflection due to shear and direct stress, may be used:

$$\Delta_a = \int \frac{Mm\,ds}{EI},$$

where

$\Delta_a =$ deflection at point a in same direction as unit load (in.),

$M =$ moment due to the load or loads which cause the deflection (in.-lb per ft of panel width),

$m =$ moment due to unit load at point a (unit load-in.),

$s =$ length along arch (in.),

$E =$ modulus of elasticity of plywood faces in arch (psi),

$I =$ moment of inertia of arch panel (in.^{4} per ft of panel width).

As with the stressed skin panel, if any of the allowable stresses or the allowable deflection has been exceeded, a new trial cross section must be taken and the previous calculation repeated.

REFERENCES

1. U.S. Product Standard PS1-66.
2. "Plywood Design Specification," American Plywood Association, Tacoma, Wash.
3. "Plywood Construction Systems," American Plywood Association, Tacoma, Wash.
4. "Plywood Folded Plate Design Method," Lab. Bull. 58-B, American Plywood Association, Tacoma, Wash.
5. "Design Considerations for Details of Plywood Folded Plate Roofs," Lab. Bull. 60-B, American Plywood Association, Tacoma, Wash.
6. "Hurricane-Resistant Plywood Construction," American Plywood Association, Tacoma, Wash.
7. "National Design Specification for Stress-Grade Lumber and Its Fastenings," National Forest Products Association, Washington, D.C.
8. "Nail-Glued Residential Trussed Rafters," Lab. Rept. 102, American Plywood Association, Tacoma, Wash.
9. "Design and Fabrication of Plywood Beams," American Plywood Association, Tacoma, Wash.
10. "Design and Fabrication of Plywood Stressed Skin Panels," American Plywood Association, Tacoma, Wash.
11. "Load-Bearing Plywood Sandwich Panels," Lab. Rept. 93, American Plywood Association, Tacoma, Wash.
12. "Design and Fabrication of Plywood Curved Panels," American Plywood Association, Tacoma, Wash.

Chapter 5

High-Pressure Laminates

M. G. Young

INTRODUCTION

Between the years 1908 and 1921 plastic laminates developed into an industry that is currently of major proportions. These laminates wrestled through the usual initial period of research, patents, and legal suits during which they established a reputation as a quality electrical insulation which evolved into today's high-pressure laminate industry. Initially high-pressure industrial laminates took the place of "mica" as a quality insulation in motors; today they are being utilized in communications equipment and computers and are designed to possess the uniformity of product required for the technology of the space age.

By 1921 patents no longer restricted the manufacture of laminates as an important insulation or as a material for mechanical components. The raw materials became available to everyone interested, and it became apparent that if markets were to be opened further, industry standards had to be developed and supported.

Thus in 1921 several companies producing laminates set up a section to develop standards within the Associated Manufacturers of Electrical Supplies, the forerunner of the present-day National Electrical Manufacturers Association (NEMA), which was formed in 1926.

This NEMA Section was, of course, devoted to industrial applications of plastic laminates. It was not until the late twenties that the potential of plastic laminates as a decorative material became apparent. By this time more knowledge had been gained about resin systems and reinforcing materials and new resins had been formulated. By inserting a resin-impregnated, printed sheet near the surface of the multilayer buildup, a new and practical material was born for decorative surfaces. This was in addition to the "decorative" black phenolic laminates already in vogue for the fronts of radios.

In 1947 a separate NEMA Section was set up devoted to the new "decorative laminates." Through its trade association, the NEMA Decorative Laminate Section and the NEMA Industrial Laminate Section develop product standards, are represented on international standards committees, work in close association with numerous government agencies and other associations such as the Underwriters Laboratories, American Society for Testing and Materials, Electronic Industries Association, and The Society of the Plastics Industry. The NEMA Standards of these Sections (LD1-1964 Decorative and LI1-1965 Industrial) are published and kept up to date by the Association, not only to facilitate transactions between the

customer and seller, thus creating economies which are reflected in purchases, but also to safeguard the fabricator and ultimate consumer against inferior material.

The high-pressure laminated plastics industry has progressed so rapidly that the present versatility of plastic laminates allows the engineer, the architect, and the designer to build into his product almost any combination of products required. Components can be literally "custom built" at an economical cost with high-pressure laminated plastics. Space-age science has benefited tremendously from plastic laminates which have made possible the miniaturization of many industrial products such as computers, communication equipment, and a whole host of instruments.

1 INDUSTRIAL HIGH-PRESSURE LAMINATES[1]

A high-pressure laminate for industrial use is made from a layup of sheets of paper, cloth, asbestos, synthetic fiber, or glass in a sandwich construction, each bonded with a suitable resin such as epoxy, tetrafluoroethylene, melamine, polyester, silicone, or phenolic, all balanced to achieve the desired properties. The impregnated sheets are stacked so as to obtain the desired thickness of the final laminate and pressed between metal pressure plates at both a temperature and pressure necessary to ensure completion of the polymerization process. The laminate may be formed as sheets, rods, tubes, or specially molded shapes to meet the application needs. The completed laminate contains a high percentage of resin (nearly 50 per cent). Care is exercised in the fabrication process to ensure that the material is free of voids and dry areas which adversely affect the electrical and mechanical properties of the finished material. A copper or other conductive layer may be bonded to the laminated structure on either one or both faces to form a clad laminate for use in printed circuit applications. The number of layers of the laminate in the composite may vary from few (for applications in multilayer circuitry) to many, depending on the requirements necessary to provide the proper physical and electrical properties for the application. The study of these laminated plastics embraces the fundamentals of chemistry, physics, and engineering.

Some of the decisions that must be made in the choice of an industrial laminate for a specific application are

1. Environmental requirements of temperature, humidity, radiation effects, thermal expansion, pressure.
2. Tolerances to be maintained.
3. Costs involved versus design complexity.
4. Weight.
5. Unclad versus clad (single or double sided).

M. G. YOUNG is Professor and Chairman, Department of Electrical Engineering, and Director, University of Delaware–NEMA Facility, University of Delaware, Newark, Delaware.

6. Choice of material; for example: paper based, cloth based, glass based, all with a suitable binder such as phenolic, epoxy, melamine, silicone, polyester or tetrafluoroethylene.
7. Thickness.
8. Packaging factors such as mounting problems and hardware.
9. Stresses that the laminate may have to withstand in many industrial and space applications.
10. Ease of fabrication, punchability, shearing, drilling, and so on.

1.1 Effect of elevated and subnormal temperatures on physical and electrical properties[2]

From the standpoint of temperature the environmental conditions that the laminate may have to withstand can vary from cryogenic temperatures, at which the laminate must maintain minimum specific properties such as flexural strength, electric strength, and close dimensional tolerances, to temperatures so high that the laminate possesses ablative properties and still maintains adequate electrical and physical characteristics. Although no single laminate possesses these qualities over such an extreme temperature range, modifications in its structure, manufacturing process, and binding resin can provide the desired property values over a limited temperature range within these extremes.

The laminate must also be designed to have low water absorption properties where high humidities are to be encountered in its application. Under these conditions low water absorption assists the laminate in maintaining its desired physical, electrical, and dimensional characteristics.

Laminated plastics have a place as high-temperature materials for service in the range of 500°F, and they suffer no loss in strength at cryogenic temperatures. They are tough and withstand rough handling and impact loads, and they are stable and do not creep within a widely useful range of stress and temperature. Laminated plastics are not subject to plastic deformation or introduction of internal stresses accompanying plastic flow.

Exposure to elevated temperature tends to cause a permanent continuing change in properties of plastic laminates, with increased effect at higher temperatures. This is the result of continued curing of the resin. Initial exposure to heat normally produces a reduction in strength, followed by an increased strength as curing reaches completion. Of course, if the temperature is high enough, continued exposure results in degradation of the resin or base material. Steady-state values of mechanical strength are obtained only after a prolonged heating at a constant temperature. During this period transient-state values may be of more design significance than the ultimate steady-state values. Properties depend not only on the temperature of operation but also on the exposure time, cycling rate or direction of heat cycle. Therefore, property values must be clearly identified with the test procedures when using test results as a guide in selection of material and safe design stress.

1.2 Effect of neutron radiation on physical and electrical properties[3]

When applications of the laminate expose it to radiation, the quantity of radiation must be ascertained and the laminate must maintain adequate property values when exposed to this radiation dosage. Most of the laminates listed in this chapter do not have their physical or electrical properties changed significantly by multiple neutron radiation dosages of the order of 10^{14} *nvt* (neutrons $\times$ velocity $\times$ time). With the usual plastic laminates, radiation of this order has no measurable effect on the flexural strength except on some low strength nylon phenolic and asbestos phenolic materials. Even with these materials, the average change is less than 15 per cent. Some differences exist between specimens of glass silicone and glass polyester, but the loss in flexural strength is almost insignificant. Insulation resistance, using the ASTM D-257 bullseye pattern, shows no significant effect when tests are made at 50 per cent relative humidity. Some materials show minor increases in insulation resistance, while others show minor decreases. At a humidity level of 90 per cent, however, the neutron irradiation effects show a significant improvement in the insulation resistance on glass melamine laminates. In so far as other electrical characteristics are concerned, the dielectric breakdown resistance parallel to laminations, tested in oil, shows only a minor change and similar minor changes are noted for the dielectric constant and dissipation factor of most laminates. Changes that occur in the dielectric constant and dissipation factor of the usual grades of plastic laminates are all on the increasing side, with nylon phenolic showing the greatest increase.

Paper-base grades[4]

Grade X

Sheets. Primarily intended for mechanical applications where electrical properties are of secondary importance. Should be used with discretion when high-humidity conditions are encountered. Not equal to fabric-based grades in impact strength.

Rolled Tubes. Good punching and fair machining qualities. Low power factor and high dielectric strength under relatively dry conditions.

Rods and Molded Tubes. This grade is not recommended in these forms.

Grade XP

Sheets. Primarily intended for hot punching. More flexible and not as strong as Grade X. Intermediate between Grades X and XX in moisture-resistance and electrical properties. With good punching practice, sheets having a thickness up to and including $\frac{1}{16}$ in. may be punched cold. When heated to between 120°C and 140°C, sheets having a thickness up to and including $\frac{1}{8}$ in. can be punched.

Tubes and Rods. This grade is not recommended in these forms.

Grade XPC

Sheets. Primarily intended for cold punching and shearing. More flexible and higher cold flow but lower in flexural strength than Grade XP. With good punching practice, sheets up to and including $\frac{1}{8}$ in. in thickness can be punched at a room temperature of approximately 23°C. In general, at the same temperature and with a sharp power squaring shear sheets up to $\frac{3}{32}$ in. in thickness can be sheared in $1\frac{1}{2}$-in. wide strips in both the lengthwise and cross wise direction without developing surface cracks.

Tubes and Rods. This grade is not recommended in these forms.

Grade XX

Sheets. Suitable for usual electrical applications. Good machinability.

Rolled Tubes. Good machining, punching and threading qualities. Not as strong mechanically as Grade X rolled tubes but better in moisture resistance. Better for low dielectric losses, particularly on exposure to high humidity.

Molded Tubes. Better in moisture resistance than Grade XX rolled tubes. Good machining and good electrical properties, except in thin walls where the dielectric strength may be low at the molded seams.

Rods. Characteristics are similar to those for sheets except as limited by inherent differences in construction and shape.

Grade XXP

Sheets. Better than Grade XX in electrical and moisture-resisting properties and more suitable for hot punching. Intermediate between Grades XP and XX in punching and cold-flow characteristics.

Tubes and Rods. This grade is not recommended in these forms.

Grade XXX

Sheets. Suitable for radio-frequency work and for high-humidity applications. Has minimum cold-flow characteristics.

Molded Tubes and Rods. Characteristics are similar to those for sheets except as limited by inherent differences in construction and shape.

Rolled Tubes. Best electrical properties for rolled paper-base tubes under humid conditions.

Grade XXXP

Sheets. Better in electrical properties than Grade XXX and more suitable for hot punching. Intermediate between Grades XXP and XX in punching characteristics. This grade is recommended for applications requiring high insulation resistance and low dielectric losses under severe humidity conditions.

Tubes and Rods. This grade is not recommended in these forms.

Grade XXXPC

Sheets. Similar in electrical properties to Grade XXXP and suitable for punching at lower temperatures than Grade XXXP. With good punching practice, sheets up to and including $\frac{1}{16}$-in. thickness may be punched at a temperature not less than 23°C (73.4°F) and, in thicknesses over $\frac{1}{16}$ in. up to and including $\frac{1}{8}$ in., when warmed to temperatures up to 60°C (140°F). This grade is recommended for applications requiring high insulation resistance and low dielectric losses under severe humidity conditions.

Tubes and Rods. This grade is not recommended in these forms.

Fabric-base grades

Grade C

Sheets. Made from cotton fabric weighing over 4 ounces per square yard and having a count of not more than 72 threads per inch in the filler direction nor more than a total of 140 threads per inch in both the warp and filler directions as determined from an inspection of the laminated sheet. A strong, tough material suitable for gears and other applications requiring high impact strength. The heavier the fabric base used, the higher will be the impact strength but the rougher the machined edge; consequently, there may be several subgrades in this class adapted for various sizes of gears and types of mechanical service. This grade does not have controlled electrical properties and its use for electrical applications is not recommended.

Rolled Tubes. Made from a cotton fabric with the same weight and thread-count limits as sheets of this grade.

Molded Tubes. No standards developed for this grade in this form.

Rods. Made from a cotton fabric with the same weight and thread-count limits as sheets of this grade. In general, characteristics are the same as those for sheets except as limited by inherent differences in construction and shape.

Grade CE

Sheets. Made from a cotton fabric with the same weight and thread-count limits as Grade C. Suitable for electrical applications requiring greater toughness than provided by Grade XX or for mechanical applications requiring greater resistance to moisture than provided by Grade C. This grade is not recommended for primary insulation* for electrical applications involving commercial power frequencies at voltages in excess of 600 volts.

Rolled Tubes. No standards contemplated for this grade in this form.

Molded Tubes. Made from a cotton fabric with the same weight and thread-count limits as Grade C. Suitable for use where a tough, dense, fabric-base material having good mechanical properties and good moisture resistance is required. This grade is not recommended for primary insulation* for electrical applications involving commercial power frequencies at voltages in excess of 600 volts. Electric strength may be low at molded seams, especially in thin walls.

Rods. Characteristics are similar to those for molded tubes except as limited by inherent differences in construction and shape.

Grade L

Sheets. Made from fine-weave cotton fabric weighing 4 ounces or less per square yard. The minimum thread count in any ply is 72 threads per inch in the filler direction and a minimum total of 140 threads per inch in both the warp and filler directions as determined from an inspection of the laminated sheet. For purposes of identification, the surface sheets have a minimum thread count of 75 threads per inch in either the warp or filler direction, and a minimum total of 152 threads per inch in both the warp and filler directions. This grade is suitable for small gears and other fine machining applications, particularly in thicknesses under $\frac{1}{2}$ in. Not quite so tough as Grade C. This grade does not have controlled electrical properties and its use for electrical applications is not recommended.

Rolled Tubes. No standards developed for this grade in this form.

Molded Tubes. Made from fine-weave cotton fabric weighing 4 ounces or less per square yard. The minimum thread count in any ply is 72 threads per inch in the filler direction and a total of 140 threads per inch in both the warp and filler directions as determined from an inspection of the molded tube. Has high density and good moisture resistance. Primarily suitable for mechanical applications where finer machined appearance is required than is secured with Grade CE molded tubes or where tougher material than Grade LE molded tubes is required. This grade does not have controlled electrical properties and its use for electrical applications is not recommended.

Rods. Made from a cotton fabric with the same weight and thread-count limits as molded tubes of this grade. In general,

---

* "Primary insulation" means insulation which is in direct contact with terminals, conductors or other current-carrying members. Laminated insulation used for its mechanical or thermal properties, such as armature slot wedges, spacers, structural members, and switchboard panels where terminals have separate insulation, is not considered "primary insulation."

characteristics are similar to those for molded tubes except as limited by inherent differences in construction and shape.

Grade LE

Sheets. Made from a cotton fabric having the same weight and thread-count limits as Grade L sheets. Suitable for electrical applications requiring greater toughness than provided by Grade XX. Better in machining properties and appearance than Grade CE, and available in thinner sizes. Good in moisture resistance. This grade is not recommended for primary insulation (see footnote, p. 116) for electrical applications involving commercial power frequencies at voltages in excess of 600 volts.

Rolled Tubes. Made from a cotton fabric having the same weight and thread-count limits as Grade L molded tubes. Suitable for use where the seams of a molded tube may be objectionable and where the application requires good machining qualities, together with fair electrical and good mechanical properties. This grade is not recommended for primary insulation (see footnote, p. 116) for electrical applications involving commercial power frequencies at voltages in excess of 600 volts.

Molded Tubes. Made from a cotton fabric having the same weight and thread-count limits as Grade L molded tubes. Has excellent machining and moisture-resisting characteristics. For use in restricted electrical applications where a tougher material than Grade XX tube is required at some sacrifice of electrical properties. Electric strength may be low at molded seams, especially in thin walls. Better electrically than Grade CE molded, but not quite as tough. This grade is not recommended for primary insulation (see footnote, p. 116) for electrical applications involving commercial power frequencies at voltages in excess of 600 volts.

Rods. Made from a cotton fabric with the same weight and thread-count limits as for molded tubes of this grade. In general, characteristics are similar to those for molded tubes except as limited by inherent differences in construction and shape.

Asbestos-base grades

Grade A-Asbestos Paper Base

Sheets. More resistant to flame and slightly more resistant to heat than cellulosic laminated grades because of high inorganic content. Not recommended for primary insulation (see footnote, p. 116) for electrical applications involving commercial power frequencies at voltages in excess of 250 volts. Small dimensional changes when exposed to moisture.

Rolled and Molded Tubes. Characteristics are similar to those for sheets except as limited by inherent differences in construction and shape.

Rods. This grade is not recommended in this form.

Grade AA-Asbestos Fabric Base

Sheets. More resistant to heat and stronger and tougher than Grade A. Not recommended for primary insulation (see footnote, p. 116) for electrical applications at any voltage. Small dimensional changes when exposed to moisture.

Rolled and Molded Tubes. Characteristics are similar to those for sheets except as limited by inherent differences in construction and shape.

Rods. This grade is not recommended in this form.

Glass-base grades

Grade G-2 — Staple-Fiber-Type Glass Cloth.
　　　　Electrical and Heat-Resistant Grade

Sheets. Good electrical properties under high humidity conditions. Mechanically it is the weakest of the glass-base grades. Lower in dielectric losses than other glass-base grades except silicone. Good dimensional stability.

Tubes and Rods. This grade is not recommended in these forms.

Grade G-3 — Continuous-Filament-Type Glass Cloth.
　　　　General-Purpose Grade

Sheets. High impact and flexural strength. Bonding strength is the poorest of the glass-base grades. Good electrical properties under dry conditions. Electric strength perpendicular to laminations is good. Good dimensional stability.

Rolled Tubes. Characteristics are similar to those for sheets except as limited by inherent differences in construction and shape.

Molded Tubes. This grade is not recommended in this form.

Rods. Characteristics are similar to those for sheets except as limited by inherent differences in construction and shape. Mold seams are weak points mechanically and electrically.

Grade G-5 — Continuous-Filament-Type Glass Cloth
　　　　with Melamine Resin Binder

Sheets. Highest mechanical strength and hardest laminated grade. Good flame resistance; second only to silicone laminates in heat and arc resistance. Excellent electrical properties under dry conditions. Low insulation resistance under high humidities. Good dimensional stability.

Rolled Tubes. Characteristics are similar to those for sheets except as limited by inherent differences in construction and shape. Especially high internal bursting strength.

Molded Tubes. This grade is not recommended in this form.

Rods. Characteristics are similar to those for sheets except as limited by inherent differences in construction and shape. Mold seams are weak points mechanically and electrically.

Grade G-7 — Continuous-Filament-Type Glass Cloth
　　　　with Silicone Resin Binder

Sheets. Extremely good dielectric loss and insulation resistance properties under dry conditions and good electrical properties under humid conditions, although the percentage of change from dry to humid conditions is high. Excellent heat and arc resistance. Second only to Grade G-5 in flame resistance. Good impact and flexural strength. Electric strength perpendicular to laminations is the best of the silicone grades. Meets IEEE requirements for Class 180 insulation.

Rolled Tubes. Characteristics are similar to those for sheets except as limited by inherent differences in construction and shape.

Molded Tubes and Rods. No standards developed for this grade in these forms.

Grade G-9 — Continuous-Filament-Type Glass Cloth
　　　　with Heat-Resistant Melamine Resin Binder

Sheets. High mechanical strength and one of the hardest laminated grades. Good flame resistance. Second only to silicone laminates in heat and arc resistance. Excellent electric strength properties under wet conditions. Good dimensional stability.

Rolled Tubes. Characteristics are similar to those for sheets except as limited by inherent differences in construction and shape. Especially high internal bursting strength.

Molded Tubes. This grade is not recommended in this form.

Rods. Characteristics are similar to those for sheets except as limited by inherent differences in construction and shape. Mold seams are weak points mechanically and electrically.

Grade G-10 — Continuous-Filament-Type Glass Cloth
　　　　with Epoxy Resin Binder

Sheets. Extremely high mechanical strength (flexural, impact and bonding) at room temperature. Good dielectric loss and electric strength properties under both dry and humid conditions. Insulation resistance under high humidity is better than Grade G-7.

Rolled Tubes. Characteristics are similar to those for sheets except as limited by inherent differences in construction and shape.

Molded Tubes. This grade is not recommended in this form.

Rods. Characteristics are similar to those for sheets except as limited by inherent differences in construction and shape. Mold seams are weak points mechanically and electrically.

Grade G-11 — Continuous-Filament-Type Glass Cloth with Heat-Resistant Epoxy Resin Binder.

Sheets. Properties similar to those of Grade G-10 at room temperature and, in addition, the material retains at least 50 per cent of its room-temperature standard flexural strength when measured at 150°C after one hour at 150°C. Insulation resistance is similar to Grade G-10.

Rolled Tubes. Characteristics are similar to those for sheets except as limited by inherent differences in construction and shape.

Molded Tubes. This grade is not recommended in this form.

Rods. Characteristics are similar to those for sheets except as limited by inherent differences in construction and shape. Mold seams are weak points mechanically and electrically.

Nylon-base grade

Grade N-1

Sheets. Nylon cloth base with phenolic resin binder. Excellent electrical properties under high humidity conditions. Good impact strength, but subject to flow or creep especially at temperatures higher than normal.

Rods and Tubes. No standards developed for this grade in these forms.

Polyester grade

Grades GPO-1 and GPO-2

Grade GPO-1 and GPO-2 sheets are made from a mat of random-laid glass fibres which is saturated with polyester resin combined with suitable fillers, cured by applying heat and pressure and meets the performance requirements specified in NEMA LI 1-1965, Part 11.

Flame-resistant grades

Grade FR-2

Sheets. Paper-base laminates with a phenolic resin binder so modified as to be self-extinguishing after the source of ignition is removed. Similar in all other properties to Grade XXXPC.

Tubes and Rods. This grade is not recommended in these forms.

Grade FR-3

Sheets. Paper-base laminate with epoxy resin binder having higher flexural strength than Grade XXXPC and so formulated as to be self-extinguishing after the source of ignition is removed. Has low dielectric loss properties with good stability of electrical properties under conditions of high humidity. With good punching practice, sheets up to and including $\frac{1}{16}$ in. in thickness may be punched at temperatures not less than 27°C (80°F) and, in thicknesses over $\frac{1}{16}$ in. up to and including $\frac{1}{8}$ in., when warmed to a temperature not exceeding 65.5°C (150°F).

Tubes and Rods. No standards developed for this grade in these forms.

Grade FR-4

Sheets. Continuous-filament glass cloth with an epoxy resin binder similar to Grade G-10 but self-extinguishing after the source of ignition is removed. Similar in all other properties to Grade G-10.

Rolled Tubes. Characteristics are similar to those for sheets except as limited by inherent differences in construction and shape.

Molded Tubes. This grade is not recommended in this form.

Rods. Characteristics are similar to those for sheets except as limited by inherent differences in construction and shape. Mold seams are weak points mechanically and electrically.

Grade FR-5

Sheets. Continuous-filament glass cloth with an epoxy resin binder similar to Grade G-11 but self-extinguishing after the source of ignition is removed. Similar in all other properties to Grade G-11.

Rolled Tubes. Characteristics are similar to those for sheets except as limited by inherent differences in construction and shape.

Molded Tubes. This grade is not recommended in this form.

Rods. Characteristics are similar to those for sheets except as limited by inherent differences in construction and shape. Mold seams are weak points mechanically and electrically.

Post-forming grade CF

Grade CF

Grade CF sheets used for post-forming are made from a cotton fabric weighing more than 4 ounces per square yard and having a thread count of not more than 72 threads per inch in the filler direction nor more than a total of 140 threads per inch in both the warp and filler directions as determined from an inspection of the laminated sheet. The fabric shall be impregnated with a thermosetting phenolic type of synthetic resin and cured under heat and pressure.

Test Methods for Physical and Electrical Properties of Laminated Thermosetting Sheet, Tubing, and Rod: NEMA LI 1-7.04 (Sheets); LI 1-7.05 (Tubes); LI 1-7.06 (Rods)[4]

To assure quality manufacture and continuous development of industrial laminates, these materials must be subjected to many tests and meet rigid standards. Some of the typical tests and test methods required for evaluation of these laminates are indicated in Table 5.1. In each case the latest published revisions or changes in ASTM methods, and the NEMA modification where it applies, should be consulted.

NEMA Standards Publication No. LI 1-1965 lists the physical and electrical properties of a number of NEMA Grades in Parts 4, 5, 6 for unclad sheets, rods and tubes and Part 10 covers the copper-clad materials. The values in these sections designate maximum or minimum standards. These values represent the average for the number of specimens specified for the particular test under the given conditions and serve as the basis for determining whether or not the requirements of these standards have been met. Since these tables are published in their latest revised form in the NEMA publication No. LI 1-1965 they are not repeated here.

1.3 Applications of industrial laminated thermosetting products[4,6]

Applications

Considered on the basis of strength-weight ratio, laminated phenolic is one of the strongest materials known. With a density for cellulose-base grades of approximately 1.35 g/cc, only half

Table 5.1. ASTM or NEMA Test Methods[4,5]

Test	Sheets	Tubes	Rods	Copper-Clad Laminates (Sheets)
Flexural strength	D790	—	D349	D790 Modified by Par. C LI-1-7.07
Impact strength	D256 Method A	—	—	—
Bonding strength	D229	—	—	—
Water absorption	D229	D348	D570	D229 and LI 1-10.15
Dielectric breakdown parallel to laminations-step-by-step test	D229 Modified by LI 1-7.07	D229 Modified by Par. D of LI 1-7.07	—	D229 Modified by LI 1-7.07
Dissipation factor and dielectric constant — 1 megacycle per second	D150 Modified by LI 1-7.07	D348	—	D150 and LI 1-10.17
Flame resistance	D229	—	—	D229
Flame resistance (Switchgear applications)	LI 1-7.08	—	—	—
Density	—	D348	D349	—
Compressive strength in the axial direction	—	D348	D349	—
Electric strength perpendicular to laminations	—	D348	—	—
Solder float	—	—	—	LI 1-10.11
Peel strength After solder float	—	—	—	LI 1-10.12
After elevated temperature	—	—	—	LI 1-10.13
Volume resistivity and surface resistance	—	—	—	LI 1-10.14 and D257
Oven blister test	—	—	—	LI 1-10.21

that of aluminum, the mechanical grades find large application in the aircraft and other structural fields.

Because of their high strength, resilience, good wearing and quiet running qualities, gears cut from either laminated phenolic plate or molded blanks are used in thousands of industrial applications ranging from the tiny gears in electric clocks to 8-in. to 10-in. face gears in rolling mills.

The high strength, excellent resistance to moisture and heat, and good electrical properties of laminated phenolic, combined with the fact that it is readily machined, account for its use in large volumes in all branches of the electrical industry.

The resistance of laminated phenolic to corrosion makes it suitable for many applications in the various chemical industries, particularly where organic solvents, organic acids in any concentration or dilute inorganic acids are encountered. In general, laminated phenolic is not suitable for use in alkaline media, although certain grades are more resistant to alkalies than others and are used for special applications in dilute alkaline solutions.

Paper-base epoxy laminates are used for those applications that require greater mechanical strength than provided by Grade XXXPC. They combine the stable electrical properties of epoxies under humid conditions with self-extinguishing characteristics after the source of ignition is removed.

Glass-base phenolic grades are used for motor insulation and in other applications where high strength and good electrical properties are required even at fairly high temperatures. Glass-base melamine grades are used primarily for their high mechanical strength and resistance to arc and flame and are particularly suitable for power equipment in marine applications.

Glass-base silicone grades are resistant to high temperatures up to 200°C (392°F) and have especially low dielectric losses. These grades extend the upper temperature range of laminated material to a new high.

Glass-base epoxy laminates are used primarily in electronic applications at room temperature where their stable electrical properties are particularly desirable. They are often used in printed-circuit applications. Caution should be used in applications of Grades G-10 and FR-4 involving high mechanical stresses at elevated temperatures. However, Grades G-11 and FR-5 have excellent hot strength. These laminates have excellent dimensional stability over a wide temperature range.

Glass-base epoxy laminates having a high degree of flame resistance can be produced. Such materials are used in electronic data-processing equipment, missiles, space vehicles, and other electronic equipment where resistance to burning is required.

Nylon-base laminates are suitable for application in the electronic and high-frequency fields and provide superior insulation resistance under high humidities. Their high flow or creep, particularly under hot conditions, requires special handling and design considerations.

From simple toys to precision instruments used in our billion dollar space exploration programs, plastic laminates have opened the door to new efficiency and economy.

Laminated ball bearing retainers are machined to perfection for use in the inertial guidance systems of space-craft. Here, where human lives must depend on proven reliability, plastic laminate technology has produced retainers that weigh only half as much as aluminum, yet are more serviceable than steel and can operate in extreme temperature ranges.

Ball bearing retainer rings of a similar type are used in air-driven dental drills that operate at 400,000 rpm.

The extreme toughness of plastic laminates has also led to less delicate uses, such as a "cap" to withstand the brutal onslaught of thousands of blows of a pile driver. The pile driver cap is made up of alternate layers of canvas-reinforced laminate and aluminum. The aluminum dispels the heat while the phenolic resin impregnated fabric transmits the shock waves.

The very clothes we wear are an indirect result of plastic laminate development. The high-speed shuttles used on textile looms offer almost unbelievable impact resistance and ensure wearability at high frictional temperatures. The longer the shuttle is used, the smoother its nonsnagging surface becomes.

Laminates contribute to our safety in other ways, too. A molded-glass polyester guard is now used, for instance, to shield a portable chain saw, a job once considered impossible. The guard is light, strong, durable and weather resistant — and meets its challenge perfectly.

Low cost, long wearing, extremely high impact strength, dimensional stability, light weight — these are just a few of the keys to the popularity of thermosetting laminates in an endless variety of applications.

Manufacturers of laminates are equipped today to build a future that has no horizon. This is an industry where customer challenges are welcomed.

Most problems can be solved because of the boundless flexibility, the built-in changeability, that is the very nature of a laminate.

Only in plastic laminates can these electrical, mechanical and chemical properties — in any combination — be found; flame retardancy; unusual strength-to-weight ratio, even where it weighs half as much as aluminum; high impact strength; low coefficient of friction; good abrasion resistance; outstanding corrosion resistance and low water absorption; ease of fabrication (machining, drilling, punching); ease of forming; good ablative ability, when desired; low dielectric loss; high dielectric strength; and good insulation resistance.

Industrial laminates are evolutionary — they are the modern materials that can unlock whole new areas for industrial use, and open the door to economical, high-quality modifications or innovations for many products in any industry.

Some applications may require the laminate to be exposed to extreme pressure changes from several atmospheres to vacuum conditions as low as 10^{-11} mm mercury in a satellite environment. If the laminate is metal clad and serves as a current-carrying member, the current-carrying capacity under vacuum conditions such as those encountered in space will be lower for a given line width than is acceptable under normal atmospheres. This is due to the heat dissipation problem, and special heat sinks and passive cooling methods must be provided for successful application.

2 LAMINATED THERMOSETTING DECORATIVE SHEETS[7]

Laminated thermosetting decorative sheets consist essentially of layers of a fibrous sheet material, such as paper, impregnated with a thermosetting condensation resin, and consolidated under heat and pressure. The top layers have a decorative color or a printed design. The resultant product has an attractive exposed surface which is durable and resistant to damage from abrasion and mild alkalies, acids, and solvents.

Sheets are available in a wide variety of colors, decorative designs, and surface finishes. They are used for good appearance and functional performance under hard service in such applications as counter and table tops, bathroom and kitchen work surfaces, furniture and cabinets, wall paneling and partitions, and doors.

There are four types of laminated thermosetting sheets currently listed in the NEMA Standard LD1-1964, namely (1) general-purpose type, (2) vertical-surface type, (3) post-forming type, and (4) hardboard-core type. Each type is designed for a specific application, as follows.

General-purpose type

General-purpose-type laminate is designed for both horizontal and vertical applications and is used where good appearance, durability, resistance to stains and resistance to heat from ordinary sources are required. The backs of the sheets may be sanded to permit bonding with adhesives to some suitable base material for mechanical support, such as plywood.

Vertical-surface type

Vertical-surface-type laminate is designed specifically for vertical applications where good appearance, durability, resistance to stains, and resistance to heat from ordinary sources are required. Because of its inherent properties, it should be used only in accordance with the laminate manufacturer's recommendations.

Post-forming type

Post-forming-type laminate is similar to general-purpose-type laminate but it can be formed under controlled temperature and pressure in accordance with the laminate manufacturer's recommendations.

Hardboard-core type

Hardboard-core-type laminate has a surface similar to that of the general-purpose-type laminate molded to a hardboard core which is generally self-supporting.

Although NEMA officially lists only the four types of laminated thermosetting sheets, fire-retardant decorative laminates are generally manufactured on order. These laminated products are specifically designed to meet building-code requirements in application where low flame spread ratings are called for. Since flame-spread, fuel-contributed, and smoke-developed requirements may vary widely throughout a building, depending on application, general-purpose laminates, as well as fire-retardant laminates, are usually satisfactory to meet all restrictions. Fire retardant laminates should be tested by officially recognized test methods. NEMA standards for fire-retardant laminates are currently under consideration.

Backer sheets are phenolic laminates for balancing construction on the under side of core material to resist moisture and increase stability. They are in essence a laminate from which the decorative face has been omitted. The most popular thickness is 0.030 in. Like other laminates, one side of a backer sheet is sanded to afford an easy gluing surface. The other side is normally the natural brown color of kraft paper.

Edge-banding type laminate has essentially the same thickness and properties as vertical-surface type laminate and may be bent to a 3-in. radius at room temperature and to a $\frac{3}{4}$-in. radius with heat between 325°F and 360°F. It is used for self-edging countertops, tables, desks, window stools, mirror frames, and so on.

2.1 NEMA decorative test methods[7]

High-pressure thermosetting, decorative laminated plastic sheets are tested for a number of properties, the specific tests being dependent upon the application of the particular type of laminate. Some typical properties that are evaluated are as follows.

Abrasion resistance (NEMA LD 1-2.01)

This test evaluates the ability of the surface of the laminated decorative material to maintain its original design or color under abrasive wear such as may be encountered in service.

Heat resistance (NEMA LD1-2.02, LD1-2.03)

NEMA offers two tests for rating the resistance to heat of a decorative surface. These are the boiling water test and the hot wax test which evaluate the resistance of the surface of the laminated thermosetting material to high temperature (180°C) over a local service area such as may be encountered in kitchen service. For the boiling water test, boiling water is poured onto the surface and then the flat-bottom aluminum vessel, filled with water (still boiling vigorously), is placed in the puddle. No detectable surface change is allowed upon removal of the water and vessel. On the other hand, a high-gloss finished panel may show a slight loss of gloss when a similar flat-bottom vessel containing hot wax at 356°F is placed upon it. But there must be no burning, blistering, cracking, or other defect noted.

Cigarette burn resistance (NEMA LD1-2.04)

This test evaluates the resistance of the surface of laminated thermosetting decorative material to spot heating, such as is comparable to a fast-burning cigarette. The test device provides a concentrated, intense heat of measurable and controllable temperature, which is always the same distance from the decorative face under test. The test panel is clamped to a $\frac{3}{4}$-in. Douglas Fir Plywood backing since the heat dissipation is a function of the type of substrate behind the laminate.

Stain resistance (NEMA LD1-2.05)

High-pressure laminates are highly resistant to staining. These laminates are tested with groups of substances (reagents) which are found in the normal household for their staining (or lack of) power. The reagents suggested in the NEMA Standards are divided into three categories, namely (1) materials that have no effect upon high-pressure laminated sheets; (2) materials that leave a mark requiring soap or mild abrasive to clean away; and (3) reagents that harm the laminates when exposure is of sufficient time duration.

Color fastness (NEMA LD1-2.06)

This test evaluates the relative resistance to change in color of the surface of laminated thermosetting decorative sheets when exposed to a source of light having a frequency range approximately that of sunlight. The test, however, is not intended to show the resistance of these materials to continuous exposure under out-of-doors weathering conditions.

Colors on laminates possess fastness to light equal to or exceeding normal requirements for interior decorative materials. When tested in accordance with the NEMA Standard there shall be no more than a slight color change allowed and no surface deterioration. The destructive rays of sunlight are simulated in a device known as the Atlas Fadeometer, or equivalent. However, such rays are concentrated to make the test of shorter duration and every attempt is made to keep the rays uniform. So, testing can be made on a sunny day as well as on a cloudy day, a condition which would vary greatly were actual sunshine required.

For control purposes, part of the sample is covered so as to represent the "as received" condition. Once the required time duration has expired, the cover is stripped away. If the exposure has deteriorated the unprotected surface, a sharp line of demarcation will be evident.

Moisture absorption (NEMA LD1-2.07)

High-pressure thermosetting laminated plastic sheets have a high resistance to water absorption through their melamine surface. This property is evaluated by the NEMA test which calls for samples being boiled for two hours in water. The specimens are allowed a specified gain in weight or thickness, the value depending on the type, but there shall be no crazing, chalking, or delamination of the specimens as a result of this test. As examples, NEMA Standards allow the specimens to pick up no more than 10 per cent water in either weight or thickness and not less than 2 per cent in weight for general-purpose type decorative laminates and no more than 12 per cent in either weight or thickness for post-forming type decorative laminates. Moreover, the samples tested must not show surface deterioration, blistering, or any delamination.

Advantage is taken of this property when high pressure decorative laminates are used as surfacing around lavatories and showers. Common building materials, such as particle board, hardboard, and plywood, are thus adequately protected from surface wetting by the laminate. However, sealing in the back of a laminate assembly is important in this particular application to prevent water absorption through the core material.

Dimensional stability (NEMA LD1-2.08)

Design of fabricated assemblies should take into account the change of dimension with relative humidity. Atmospheric moisture seeks to come to a point of equilibrium with the plastic, resulting in a change of physical dimensions of the plastic sheet. Although this change is slight, proper installation will take this factor into consideration. Parallel examples in other materials can be seen when allowances are made for metal expanding due to temperature or seams being placed in concrete sidewalks to compensate for movement due to heat and/or water.

Since both loss of moisture or gain of moisture content will affect the dimensions of laminated plastic, the NEMA test prescribes measurements under both humid and dry conditions. Each condition is recorded separately and as the algebraic difference. For example, if a panel shrinks 0.1 per cent due to moisture lost and gains 0.2 per cent due to moisture gain, it can be seen that this panel will work through a distance of 0.3 per cent of its measured dimension. There is a difference in dimensional stability across a panel and along the length of the panel. Consequently, dimensional stability is reported CD and MD (cross direction and machine direction) at both wet and dry conditions and the gross dimensional change. Thus, NEMA Standards establish the maximum allowable limits for $\frac{1}{16}$-in. general-purpose laminates, as an example, as 0.5 per cent MD gross change and 0.9 per cent CD gross change. As a double check, the test samples are not allowed to show physical deterioration as a result of the exposure to severe climates maintained during testing. It should be emphasized that laboratory conditions, however, never occur under field service.

Flexural strength, modulus of elasticity, and deflection at rupture (NEMA LD1-2.09)

Since decorative laminated plastic must be machined and handled, it is necessary that it have certain minimum physical properties. These are defined as flexural strength, deflection at rupture, and modulus of elasticity in flexure. Test methods adopted by NEMA for these properties are those specified in the latest revision of the ASTM "Method of Test for Flexural Properties of Plastics," D790. Again, using general-purpose-type decorative laminates as an example, NEMA requires minimum values as follows: flexural strength, 18,000 psi when the face of the decorative sample is in compression (while bending) or 12,000 psi when the face is in tension; modulus of elasticity, 800,000 psi; and deflection to rupture, 0.02 in. in tension or 0.03 in. in compression.

Flexural strength tests are made by supporting the ends of the plastic sample firmly and applying a bending force to the center. Whether the sample is bent around the decorative face (in compression) or away from it (in tension) is important since the melamine face is more brittle than the phenolic underside. Modulus figures are computed mathematically from values received for flexural strength and deflection at rupture with consideration being made for length, width, and thickness. Deflection is measured simply as the distance the center of the test specimen moves (from the place wherein the ends are supported) before breaking occurs.

Inspection of appearance (NEMA LD1-2.10)

The standards for the visible quality of high-pressure decorative laminates are such that when the panels are inspected under controlled light conditions at a prescribed height, none of the defects listed as type A, B, or C shall be evident. The defects which are not allowed are:

Type A defect — Smudges, smears, fingerprints or streaks;
Type B defect — Any foreign particle that has an area of 0.60 square millimeter or larger and which is visible at a distance of 7 ft from the spot;
Type C defect — Any group of three or more foreign particles, each having a minimum area of 0.30 square millimeter, which occur within a 12-in. diameter circle and which are visible at a distance of 5 ft. If only two foreign particles occur within a 12-in. diameter circle, each particle shall be judged separately in accordance with the criteria for Type B defects.

Formability (bend test) (NEMA LD1-2.11)

This test covers the bending characteristics of thermosetting materials when heated to a temperature recommended by the manufacturer.

Test specimens are cut to size and sanded smooth on the edges to remove any minute cracks that may have developed in the cutting operation. The specimens, after being conditioned for 24 hours at 23°C and 50 per cent relative humidity, are placed on the heating jig so that the heat is applied to the decorative face. After the specimens reach the suggested temperature, as determined by marking their backs with a temperature-sensitive composition, they are removed from the heating trough and within 5 seconds placed in the bending jig which is fitted with male and female blocks of the appropriate radius. The bending jig is closed within 1 second after the specimen is placed in it and the formed specimen allowed to cool in the jig for 1 minute.

In reporting the results of the forming test the following information is supplied: Radius of male forming block; inside or outside bend; direction of specimen (machine or cross-machine direction); thickness of specimen; temperature used for forming; heating time in seconds; and results of forming. If failure occurs, the type of defect (edge cracks, cracks all the way across, blisters, and so on) are recorded. For materials that are of the post-forming type, such as post-forming grades and Class F vertical-surface laminates, the material must post-form satisfactorily without blistering, cracking, or delaminating when tested with the appropriate radius on the forming block.

Water swell (NEMA LD1-2.12)

This test is employed to determine the percentage increase in weight and thickness of hardboard-core-type decorative laminates. The test specimens are in the form of bars with smooth edges free from cracks, with a small hole in the center of the face of each specimen. After preconditioning in an oven at 50°C for 24 hours, the specimens are cooled in a desiccator, weighed, and their thickness measured with the aid of a micrometer covering the hole. The specimens are then immersed in a water bath at 50°C for 18 hours after which they are weighed again and the thickness measured over the hole.

Results of the test prescribe the percentage increase in thickness and weight and any observations as to a change in the appearance of the specimens. Requirements for hardboard-core-type decorative laminates are such that the average of three test specimens shall show a gain of not more than 25 per cent in either weight or thickness and the laminate shall show no crazing, chalking, or delamination.

Surface finish (NEMA LD1-2.13)

This method of test covers the comparative determination of 60° specular gloss of laminated thermosetting materials for the purpose of classifying the surface finishes. Finishes are found to fall within certain categories, easily distinguished from one another. When tested in accordance with this specification, laminated thermosetting decorative sheets having satin (a low gloss), furniture (an intermediate gloss), mirror (a high gloss) are to be rated in accordance with Table 5.2, using the average values determined for each direction.

Table 5.2. Gloss (see Note 1)

	Machine Direction	Cross-Machine Direction
Satin finish	15–34	5–19
Furniture finish	36–60	21–35
Gloss-finish	80–100	80–100
Other finishes	Values not established (see Note 2)	

Note 1: Gloss should not be confused with reflectance or apparent reflectance. Reflectance is the ratio of the total quantity of light which is reflected from a surface to the total quantity of incident light on the surface, regardless of direction. Apparent reflectance, of which gloss is a special kind, refers to a specified condition of view or reflection. Gloss, as determined by this test method, refers to the light that is specularly reflected, that is, light measured when the angle of incidence is equal to the angle of reflection. Gloss is determined by the smoothness of the surface; it describes the "mirror effect." Reflectance is determined by color and shading of the opaque laminate and is independent of surface finish.
Note 2: Other finishes are available, but because of the nature of the surface texture, no standard test method has yet been developed which adequately characterizes these surfaces.

Tensile strength perpendicular to surface (internal bond strength) (NEMA LD1-2.14)

This test method utilizes specimens 2-in. square with each specimen glued flatwise between two metal blocks 2-in. square and approximately 1-in. thick. Each specimen bonded between the metal blocks is placed in a tensile testing machine in such a manner that the direction of loading is perpendicular to the faces of the specimen, with the center of the load passing through the center of the specimen. The specimen is stressed by separation of the self-aligning heads of the test machine until failure occurs. The average bond strength of hardboard-core-type decorative laminates, to which the test specifically applies, shall be not less than 300 psi. The test specimen shall show 95 per cent or more fiber failure to be accepted as valid.

Impact strength (NEMA LD1-2.15)

This method of test covers the resistance of the laminated thermosetting decorative material to a falling ball which represents the effect of spot impact in either a horizontal or vertical direction.

Decorative laminated plastic is commonly used as a covering for horizontal surfaces. Thus, it is subject to impact loads from falling objects. Naturally, the extent to which the plastic can endure the impact stress is also a function of the material used as a core or substrate beneath the plastic laminate. In the test to establish a working standard for impact resistance, the NEMA method specifies that the laminate must be held flat against the surface of Lauan-faced Douglas Fir five-ply plywood.

In testing, a steel ball weighing 8 ounces is dropped onto the decorative face of the laminate clamped in its test jig from varying heights until permanent damage of the plastic occurs. The test jig is shifted each time the ball is dropped so that it does not hit the same spot on the decorative face of the plastic.

According to the Standards, the laminate must not fracture when the steel ball is dropped from a height of 36 in. for general-purpose-type decorative laminates and from a height of 12 in. for the vertical-surface-type decorative laminates. Since this represents a blow of considerable force, it may be seen that high-pressure laminates have appreciable resistance to impact.

2.2 Applications of laminated thermosetting decorative sheets[7]

Applications of laminated thermosetting decorative sheets vary widely. A few of the outstanding applications are the following.

Countertops are perhaps the most widely known application of fabricated laminates. Application of laminates to core material and installation may have some variation but should follow certain basic rules such as the following: the proper substrate must be chosen for support of the laminate top; the substrate must be properly supported; the decorative laminate must be firmly secured to the substrate using quality adhesives and techniques; cut-outs must be rounded, never under-cut; nails or screws should never be driven through the plastic itself; and inside corners should be rounded and have proper support.

Vertical wall paneling offers functional beauty with the lowest maintenance requirements. Like countertops, it should be

properly installed. Design considerations should be such that wide spans can be avoided when possible, seams should allow for normal dimensional movement, holes (if used) should be drilled, and rounded corners on cut-outs should be employed.

Doors surfaced with high-pressure decorative laminate are a perfect example of how this material offers both functional and decorative values. Because they resist hand prints and kick marks, laminated doors are ideal for public and institutional buildings. Construction of doors must follow the basic rule that they be symmetrically constructed from an imaginary center line. Doors are available with solid core, with lead-filled core for X-ray rooms, with laminated hardboard in lieu of cross banding, fire rated, and so on. It is recommended, however, that general-purpose laminate (of nominal $\frac{1}{16}$-in. thickness or thicker) be used. Also, in design of frames, allowances must be made so that stresses exerted upon hinges and latch are not carried by the plastic.

Contract furniture such as desks, cabinets and the like is another excellent use for high-pressure decorative laminates. Such furniture may be of either period or contemporary design. Two thoughts should be given consideration when working with general-purpose laminates to fabricate contract furniture: (1) Thermosetting resins should be used whenever feasible. For institutional furniture such as school desks or tables this is almost a must. (2) construction for furniture as named above should be balanced, using backer sheets.

Another type of furniture may be constructed using only open framework to which hard-board-type decorative high-pressure plastic is adhered. This type of furniture finds its outlet in occasional tables, end tables, coffee tables, and the like. It is relatively inexpensive, lightweight and, of course, the tops are resistant to most household liquids which may be spilled on them.

Baseboards and window stools are an especially appropriate application for high-pressure laminates, because they provide an aesthetic product. The laminate is adhered to the baseboard in the great majority of cases with contact cement. Window stools, on the other hand, can usually be prefabricated, in which case a thermosetting-type adhesive is preferable.

Although general statements of recommended practices for fabricating and applying laminated thermosetting decorative sheets for each of the more common applications may be made, it is suggested that the designer secure detailed information on the specific and necessary techniques for handling the material from either the commercial fabricator or the material suppliers, in order to ensure the best end use of the laminate. The NEMA Standard Publication LD1-1964 lists, in Part 7, recommendations as a guide for fabricating and applying laminated thermosetting decorative sheets. This part of the NEMA Standards considers such factors as assembly details, adhesives, core material, waterproofing, fabrication tooling using standard woodworking or metalworking machinery with the proper precautions, and field working of bonded assemblies.

REFERENCES

1. F. M. CLARK, *Insulating Materials for Design and Engineering Practice*, John Wiley & Sons, Inc., 1962, Part II, Resin Laminated Structures, pp. 787–855.
2. MILTON G. YOUNG, "How Temperature Affects Plastic Laminates," *Prod. Eng.* **35**, 67–72 (July 20, 1964).

3. "These Plastics Defy Radiation," *Prod. Eng.* **37**, 106 (March 28, 1966).

4. "Industrial Laminated Thermosetting Products," NEMA Standards Publication No. LI 1-1965 (with latest revisions), National Electrical Manufacturers Association, New York.

5. ASTM Standards Publications, American Society for Testing and Materials, Philadelphia, Pa.

6. *Industrial Laminates — Key to Product Innovations*, A Film Presentation of the Industrial Laminate Section, National Electrical Manufacturers Association.

7. "Laminated Thermosetting Decorative Sheets," NEMA Standards Publication No. LD 1-1964 (with latest revisions).

Chapter 6

Mechanical Behavior of Fiber-Reinforced Plastics

L. J. Broutman

1 INTRODUCTION

This discussion of fiber-reinforced plastics has as its goal a more complete understanding of the mechanical behavior of this class of composite materials. Specifically, the materials

L. J. BROUTMAN is Associate Professor of Mechanics and Materials, Illinois Institute of Technology, Chicago, Illinois.

to be considered will include both thermosetting (such as epoxies, polyesters, and phenolics) and thermoplastic (such as nylon, polystyrenes, acrylics, and polyethylene) polymers which may serve as the matrix or binder for continuous fibrous reinforcements such as glass, metallic wire, or ceramics. Table 6.1 is included to demonstrate the variety of fibrous

Table 6.1. Properties of Reinforcements

Filament Type	Density (gm/cc)	Tensile Strength (10^3 psi)	Specific Strength (10^6 in.)	Young's Modulus (10^6 psi)	Specific Modulus (10^7 in.)	Typical Cross-Section Dimensions (microns)
Continuous Fibrous Reinforcements						
Glass						
E glass	2.55	500	5.4	10.5	11.4	10
S glass	2.50	650	7.2	12.6	14.0	10
SiO$_2$	2.19	850	10.8	10.5	13.3	35
Polycrystalline						
Al$_2$O$_3$	3.15	300	2.6	25	21.9	—
	3.8– 3.9	100	0.7	60	43	75–200
ZrO$_2$	4.84	300	1.7	50	28.6	75–200
Carbon/Graphite						
(Thornel 25)	1.50	350	6.5	30	55.6	5
BN	1.90	200	2.9	13	18.8	7
Multiphase						
(tungsten substrate)						
B	2.36	400	4.7	55	64.7	115
B$_4$C	2.36	330	3.9	70	82.4	115
SiC	4.09	300	2.0	70	47.3	70
TiB$_2$	4.48	15	0.09	74	45.6	115
Metal						
Be	1.83	185	2.8	44	77.5	127
W	19.4	580	0.8	59	8.5	13
Mo	10.2	320	0.9	52	14.1	25
Steel (low carbon)	7.74	600	2.1	29	10.3	125
Stainless steel	7.74	275	1.0	29	10.3	4–25
René 41	8.26	290	1.0	24	8.1	25
Discontinuous Fibrous Reinforcements (Whiskers)						
Ceramic						
Al$_2$O$_3$	3.96	3000	21.2	62	43.4	3–10
BeO	2.85	1900	18.4	50	48.5	10–30
B$_4$C	2.52	2000	21.9	70	76.9	—
SiC	3.18	3000	26.1	70	60.8	1–3
Si$_3$N$_4$	3.18	2000	17.4	55	47.8	3–10
Graphite	1.66	2845	47.4	102	170	—
Metal						
Cr	7.20	1290	5.0	35	13.4	—
Cu	8.92	427	1.3	18	5.6	—
Fe	7.83	1900	6.7	29	10.2	—
Ni	8.98	560	1.7	31	9.6	—

reinforcements available for reinforcing polymers. Many of these are commercially available; others are in the laboratory stage. The table is divided into two major sections: discontinuous and continuous fibers. By discontinuous fibrous reinforcements we refer to fibers whose lengths are usually no greater than 5000 times their diameters, which is typical of metal and ceramic whiskers and chopped-glass reinforcement. The table is also subdivided by class or type of materials into inorganic glasses; polycrystalline ceramics; multiphase, since the major portion of the fiber is deposited by chemical vapor deposition onto a substrate·such as tungsten; and metals. The ability of these fibers to reinforce a polymer is not only dependent upon the fiber properties but upon the interface formed between the fiber and matrix and upon the thermal and mechanical compatibility of the matrix with the fiber. This chapter will primarily treat the behavior of glass-fiber composites. The behavior of composites made with metal wire or ceramic fibers will follow the same general principles discussed in this chapter.

This chapter does not attempt to treat the general theory of laminated materials or orthotropic analysis as this subject is treated elsewhere in this book. The investigations in micromechanics* of fibrous composites have been numerous in recent years and have been undertaken in order to better understand the observed mechanical properties of glass-reinforced plastic laminates. The purpose here is not to discuss micromechanics studies but to discuss the observed behavior of fiber-reinforced plastics, although results of some micromechanics investigations may be discussed when they add to the understanding of a particular phenomenon.

2 MECHANISM OF REINFORCEMENT IN FIBER-REINFORCED PLASTICS

2.1 Continuous fibers

Before describing the types and properties of fiber-reinforced plastics in later sections it is desirable to review briefly some of the fundamental considerations regarding the mechanism of reinforcement for both continuous and discontinuous fibrous composites. The fibers are usually found dispersed throughout the composite and for purposes of this discussion we will assume they are uniform, continuous, unidirectional, and firmly gripped by the matrix so that no slippage can occur at the interface between reinforcement and matrix. As indicated in Figure 6.1 the total load or composite load P_c is shared between the fiber load P_f and the matrix load P_m. Therefore,

$$P_c = P_m + P_f \qquad (1)$$

or in terms of stresses

$$\sigma_c A_c = \sigma_m A_m + \sigma_f A_f, \qquad (2)$$

$$\sigma_c = \sigma_m V_m + \sigma_f V_f, \qquad (3)$$

where A represents area and V represents a volume fraction.

* Studies of internal stresses, discontinuities, fracture and other phenomena on a microscale which takes into consideration individual fibers, fiber spacing, arrangement, etc.

Since no slippage is allowed at the interface the strain experienced by the composite is equal to the fiber strain and also the matrix strain:

$$\varepsilon_c = \varepsilon_m = \varepsilon_f. \qquad (4)$$

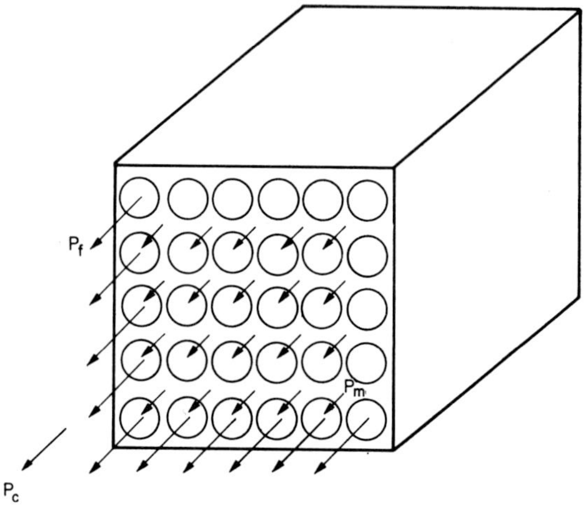

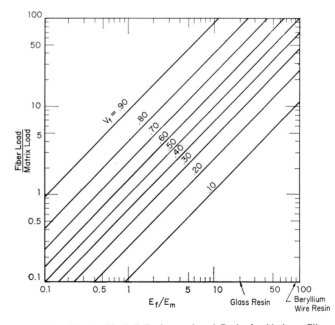

Figure 6.1   Load Distribution in Fibrous Composites

Then equation 3 can be rewritten as

$$\sigma_c = E_m \, \varepsilon_c \, V_m + E_f \, \varepsilon_c \, V_f \qquad (5)$$

or

$$\sigma_c = E_m \, \varepsilon_c \, V_m + E_f \, \varepsilon_c (1 - V_m)$$

since $V_f + V_m = 1$. The ratio of the load carried by the reinforcement to the load carried by the matrix is

$$\frac{E_f \, \varepsilon_c (1 - V_m)}{E_m \, \varepsilon_c V_m} = \frac{E_f}{E_m} \left(\frac{1 - V_m}{V_m} \right). \qquad (6)$$

This ratio of reinforcement load to matrix load is plotted in Figure 6.2 as a function of the ratio between fiber and matrix

Figure 6.2   Elastic Moduli Ratio vs. Load Ratio for Various Fiber-volume Ratios

modulus and as a function of fiber volume per cent. A ratio can also be formed between the reinforcement load and the total load carried by the composite and this relationship is plotted in Figure 6.3.[1] It is evident that to attain high stresses in the fibrous reinforcement and therefore to use high-strength

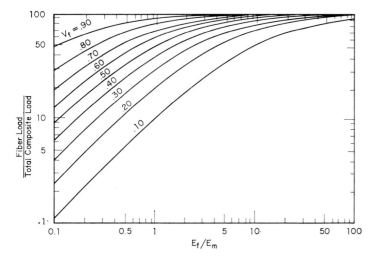

Figure 6.3 Elastic Moduli Ratio vs. Percentage Load Assumed by the Fibers for Various Fiber-volume Ratios

reinforcements most efficiently it is necessary for the fiber modulus to be much greater than the matrix modulus. The volume fraction of fibers in the composite must also be maximized if the proportion of the composite load carried is to be maximized. Although the maximum volume per cent of cylindrical fibers which can be packed into a composite is almost 91 per cent, above 80 per cent by volume the composite properties usually begin to decrease because of the inability of the matrix to wet and infiltrate the bundles of fibers. This results in poorly bonded fibers and voids in the composite.

The excellent strengths and strength-to-weight ratios achieved by glass-fiber-reinforced plastics are a result of the high strength possessed by the glass fibers and the ability of the composite to utilize this strength since the ratio E_f/E_m is approximately 20. Even at 10 per cent by volume of glass fibers the glass assumes 70 per cent of the total load.

The deformation of a glass-reinforced plastic composite containing uniaxially aligned continuous filaments when stressed parallel to the fibers may proceed in two stages:

1. Both fibers and matrix deform elastically;
2. Fibers continue to deform elastically but the matrix now deforms plastically or in a nonlinear elastic manner (the matrix deformation is also time and temperature dependent).

In stage 1, the composite modulus E_c can be predicted accurately by the "rule of mixtures":

$$E_c = E_f V_f + E_m V_m. \tag{7}$$

Experimental results have agreed well with this equation.

Stage 2 may occupy the largest portion of the composite stress-strain curve and in this stage the matrix stress-strain

curve is no longer linear, so that the composite modulus must be predicted at each strain level by

$$E_c = E_f V_f + \left(\frac{d\sigma_m}{d\varepsilon_m}\right)_{\varepsilon_f} V_m, \tag{8}$$

where $(d\sigma_m/d\varepsilon_m)_{\varepsilon_f}$ is the slope of the stress-strain curve of the matrix at the strain ε_f of the fibers. Since $E_f \gg E_m$ for glass-reinforced plastics the nonlinearity in the matrix stress-strain curve is hardly noticeable and the composite modulus is not sensitive to small changes in the matrix modulus.

The ultimate strength of a fibrous composite containing more than a certain volume fraction ($V_{\min}$) of continuous fibers is ideally reached at a total strain equal to the strain of the fibers at their ultimate tensile strength assuming that the fibers are uniform and all have the same tensile strength. The ultimate strength of the composite is then given by

$$\sigma_{cu} = \sigma_{fu}^* V_f + (\sigma_m)_{\varepsilon_{fu}}(1 - V_f); \; V_f \geqslant V_{\min}, \tag{9}$$

where σ_{fu}^* is the ultimate tensile strength of the fibers after placement in the composite and $(\sigma_m)_{\varepsilon_{fu}}$ is the matrix stress when the fibers are strained to their ultimate tensile strain.

For small values of V_f the behavior of the composite may not follow that predicted by equation 9. This is because there are an insufficient number of fibers to effectively restrain the elongation of the matrix so that the fibers are rapidly stressed to their fracture point. If we assume that the fibers all break at the same stress, the composite fails unless the remaining matrix $(1 - V_f)$ can support the full composite stress. Therefore, failure of all the fibers results in immediate failure of the composite only if

$$\sigma_{cu} = \sigma_{fu}^* V_f + (\sigma_m)_{\varepsilon_{fu}}(1 - V_f) \geqslant \sigma_{mu}(1 - V_f) \tag{10}$$

This defines a minimum volume fraction $V_{\min}$, which must be exceeded if the strength of the composite is to be given by equation 9. From equation 10

$$V_{\min} = \frac{\sigma_{mu} - (\sigma_m)_{\varepsilon_{fu}}}{\sigma_{fu}^* + \sigma_{mu} - (\sigma_m)_{\varepsilon_{fu}}}. \tag{11}$$

Although this volume fraction can be large for certain combinations of metal fibers reinforcing metal matrices,[2] it is typically quite small for high-strength fibers such as glass and polymer matrices. The plastic flow and work-hardening capability of a metal matrix can make the numerator of equation 11 a large term while the inability of most polymers to flow and work harden results in a small number for this difference between the matrix ultimate stress and the matrix stress at composite failure. For example, if we assume the strength of a glass-fiber reinforcement to be 400,000 psi ($E_f = 10^7$ psi), the composite strain at failure would be 4 per cent so that $\sigma_{mu} - (\sigma_m)_{\varepsilon_{fu}}$ for typical epoxy resins may range from 1000 to 4000 psi. $V_{\min}$ would therefore range from 0.25 per cent to 1.00 per cent for continuous-fiber reinforcement.

2.2 Discontinuous fibers

The behavior of composites reinforced with fibers of finite length l cannot be described by relations such as equations 3 and 9 unless the length l is much greater than a critical length l_c, to be defined below. The difference is that the value σ_f or

V_f in equations 3 or 9 requires adjustment. A portion of the end of each finite-length fiber is stressed at less than the maximum fiber stress of a continuous fiber as is shown in Figure 6.4. If we denote the ineffective end portion (critical

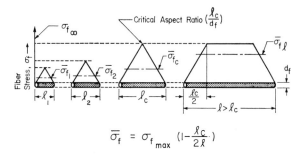

$$\bar{\sigma}_f = \sigma_{f\,\text{max}}\left(1 - \frac{l_c}{2l}\right)$$

Figure 6.4 Average Fiber Tensile Stress, $\bar{\sigma}$, for Various Fiber Lengths

fiber transfer length) of each fiber over which the fiber stress is decreased from $\sigma_{f\text{max}}$ to zero by $\frac{1}{2}l_c$, a quantity β can be defined as the ratio of the area under the stress distribution curve over the length $\frac{1}{2}l_c$ to the area of the rectangle represented by the product $\sigma_{f\text{max}} \times \frac{1}{2}l_c$. The ineffective fiber ends, $\frac{1}{2}l_c$, can either be viewed as supporting a reduced average stress, $\beta\sigma_{f\text{max}}$, or equivalently, the fiber ends $\frac{1}{2}l_c$ correspond to an effective length $\frac{1}{2}\beta l_c$ subjected to stress $\sigma_{f\text{max}}$. The average stress in the discontinuous fiber (Fig. 6.4) is then less than the breaking stress σ_{fu}^* when a fiber is extended to failure. The average stress is

$$\frac{1}{l}\int_0^l \sigma_f\, dl.$$

Integrating this expression yields the fiber average stress:

$$\bar{\sigma}_f = \sigma_{f\text{max}}[1 - (1 - \beta)l_c/l]. \tag{12}$$

For an ideal plastic matrix* which has a yield stress τ_{my} the fiber stress increases linearly from the end and β equals $\frac{1}{2}$, so equation 12 can be rewritten as

$$\bar{\sigma}_f = \sigma_{f\text{max}}(1 - l_c/2l). \tag{13}$$

The critical fiber length required to reach the maximum fiber stress when the matrix behaves as an ideal plastic material can be represented as follows :[2]

$$\frac{l_c}{d_f} = \frac{\sigma_{f\text{max}}}{2\tau_{my}}. \tag{14}$$

If interfacial failure occurs first τ_{my} in equation 14 must be replaced by the interfacial shear strength.

A similar equation has been developed which assumes elastic behavior of the fiber and matrix. Sutton and Chorne[3] have summarized the original work of Dow[4] and the critical equations can be represented as follows:

$$\frac{l_x}{d_f} = \frac{\cosh^{-1}[1/(1 - \Phi)]}{\Gamma}, \tag{15}$$

*An ideal plastic material is one for which the elastic portion of the stress-strain curve can be neglected and the material yields at a constant stress.

where l_x = distance along the fiber from one end, $\Phi = \sigma_f/\sigma_{f\text{max}}$, and

$$\Gamma = \left[\frac{24\left(\dfrac{G_f}{E_f}\right)\left[1 + \left(\dfrac{V_f}{V_m}\right)\left(\dfrac{E_f}{E_m}\right)\right]}{1 - 3\dfrac{G_f}{G_m} + 2\dfrac{G_f}{G_m}\left(\dfrac{V_f^{-3/2} - 1}{V_f^{-1} - 1}\right)}\right]^{1/2}.$$

The critical fiber length for an elastic matrix, $(l_c)_{\text{el}}$, can be arbitrarily defined as the length of a short fiber necessary to pick up 97 per cent of the stress of a similar fiber infinitely long. Hence, $(l_c)_{\text{el}}$, corresponds to $\Phi = \sigma_f/\sigma_{f\text{max}} = 0.97$. Since $(l_c)_{\text{el}} = 2l_x$ by symmetry, equation 15 can be expressed as follows:

$$\frac{(l_c)_{\text{el}}}{d_f} = \frac{8.40}{\Gamma}. \tag{16}$$

Using Dow's solution, Sutton[3] also illustrates that the average fiber stress can be written as follows for an elastic matrix:

$$\bar{\sigma}_f = \sigma_f\left[\frac{1 - \sin^{-1}\tanh\left(\Gamma l/d_f\right)}{\Gamma(l/d_f)}\right]. \tag{17}$$

A comparison of the critical aspect ratios for both boron and glass fibers in an epoxy resin matrix in which both elastic and plastic behavior of the matrix are assumed is shown in Table 6.2. The critical aspect ratio, where ideal plastic

Table 6.2. Critical Aspect Ratios for Fibers Embedded in Ideally Elastic and Plastic Matrices[a]

| | Critical Aspect Ratio, l_c/d_f | | | |
| | Glass in Epoxy | | Boron in Epoxy | |
V_f	Elastic Matrix Behavior	Plastic Matrix Behavior	Elastic Matrix Behavior	Plastic Matrix Behavior
0.10	14.2	20	16	15
0.25	6.7	20	6.2	15
0.50	2.3	20	2.3	15

[a] Properties assumed to calculate aspect ratios are the following:

Epoxy Resin: $E_m = 450 \times 10^3$ psi
$G_m = 169 \times 10^3$ psi
$\tau_{my} = 10{,}000$ psi

Boron Fiber: $E_f = 60 \times 10^6$ psi
$G_f = 24.6 \times 10^6$ psi
$\sigma_{fu}^* = 300{,}000$ psi

Glass Fiber: $E_f = 10 \times 10^6$ psi
$G_f = 4.1 \times 10^6$ psi
$\sigma_{fu}^* = 400{,}000$ psi

behavior is assumed, is independent of the volume fraction of fibers. The critical aspect ratios for both glass and boron fibers are nearly identical for the elastic matrix and are only dependent on the fiber strength in the plastic analysis. Another interesting comparison is the average stress in a short fiber as a function of fiber length for elastic and ideally plastic matrices. From Figure 6.4 and equation 11 it can be seen that the average stress increases as the fiber length increases (plastic matrix), and when $l = 10l_c$ the average stress is 0.95 of the fiber maximum stress. By inspection of Table 6.2 it can be observed that the average fiber stress for an elastic matrix is in most cases greater than the average stress for a plastic matrix since the critical transfer length ($\frac{1}{2}l_c$) is shorter.

The ultimate strength of a composite containing uniaxially aligned discontinuous fibers can be represented by equation 9 with $\bar{\sigma}_f$ from equation 12 substituted for σ_{fu}^*:

$$\sigma_{cu} = \sigma_{fu}^* V_f[1 - (1 - \beta)l_c/l] + (\sigma_m)_{\varepsilon fu}(1 - V_f); \quad V_f > V_{\min}.$$
(18)

It is useful to let $\alpha = l/l_c$ and to rewrite equation 18 as

$$\sigma_{cu} = \sigma_{fu}^* V_f[1 - (1 - \beta)/\alpha] + (\sigma_m)_{\varepsilon fu}(1 - V_f); \quad V_f > V_{\min}.$$
(19)

Comparison of equations 9 and 19 shows that discontinuous fibers always strengthen a composite to a lesser degree than continuous ones. If we assume $\beta = \frac{1}{2}$ (plastic matrix), the ratio of the strength of a discontinuous fiber composite to a continuous fiber composite is,

$$\frac{(\sigma_{cu})_{\text{disc}}}{(\sigma_{cu})_{\text{cont}}} = 1 - \frac{1}{2\alpha\left[1 + \frac{(\sigma_m)_{\varepsilon fu}}{\sigma_{fu}^*}\left(\frac{1}{V_f} - 1\right)\right]}.$$
(20)

A simple way of illustrating this relationship is to set $V_f = 1$ and the value of $(\sigma_{cu})_{\text{disc}}/(\sigma_{cu})_{\text{cont}}$ given in equation 20 is then the ratio of the extrapolated values of σ_{cu} at $V_f = 1$.[2] This ratio is plotted as a function of α in Figure 6.5. It is a limiting

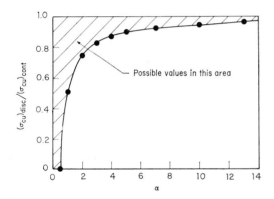

Figure 6.5   The Ratio of the Strengths of Composites Containing Discontinuous and Continuous Fibers as a Function of $\alpha(l/l_c)$ (Ref. 2)

ratio, since, for any given α, $(\sigma_{cu})_{\text{disc}}/(\sigma_{cu})_{\text{cont}}$ with $V_f < 1$ is greater than $(\sigma_{cu})_{\text{disc}}/(\sigma_{cu})_{\text{cont}}$ with $V_f = 1$. When $\alpha = 10$ the strength ratio is 0.95 and it is not until $\alpha < 5$ that the strength is significantly reduced.

As was the case for continuous-fiber composites (equation 11) a minimum volume fraction of discontinuous fibers must be exceeded for strengthening to occur. This can be determined in an analogous manner:

$$\sigma_{cu} = \sigma_{fu}^* V_f[1 - (1 - \beta)/\alpha] + (\sigma_m)_{\varepsilon fu}(1 - V_f) > \sigma_{mu}(1 - V_f).$$
(21)

This results in a minimum volume fraction for discontinuous fibers, $(V_{\min})_{\text{disc}}$:

$$(V_{\min})_{\text{disc}} = \frac{\sigma_{mu} - (\sigma_m)_{\varepsilon fu}}{\sigma_{fu}^*[1 - (1 - \beta)/\alpha] + \sigma_{mu} - (\sigma_m)_{\varepsilon fu}}.$$
(22)

The minimum volume of discontinuous fibers required for reinforcement is greater than for continuous fibers and as α increases this difference becomes less.

2.3 Efficiency of reinforced plastics

Before discussing particular types of reinforced plastics and their mechanical behavior it is well to summarize the efficiencies of typical reinforced plastic structures. The efficiency as used here denotes the percentage of the fiber strength which is utilized in the composite. This is determined by dividing the calculated fiber stress at failure by an assumed fiber strength. Table 6.3 summarizes the results of such calculations. Only those fibers parallel to the uniaxial tensile stress direction are accounted for in the calculation of fiber stress. It should be realized that this efficiency rating is an oversimplified calculation and one of the major errors is in the choice of the fiber tensile strength. The strength that should be used is the *in situ* fiber strength or strength of the fiber after being incorporated into the composite. Unfortunately this strength is not usually known, so the efficiencies in Table 6.3 were calculated by assuming a 400,000-psi strength level for the E glass reinforcement. The low efficiencies exhibited by the fabric laminates may be a result of using too high a fiber strength in the calculation, since fiber damage most likely occurs during the fabric weaving. This combination of increased filament handling and the fact that the fibers are crimped and not straight makes this composite less efficient than the nonwoven cross-plied laminate. The importance of the glass-surface treatment in increasing the fiber efficiency can readily be observed by comparing the Volan* treated and HTS* treated glass fabric.

The discontinuous fiber composite that consists of randomly dispersed fibers (it is assumed that $\frac{1}{3}$ of the fibers are aligned in the load direction) has an efficiency almost equal to the fabric laminates. This is an indication that the fiber lengths are great enough to act virtually as infinite length fibers.

3 DISCONTINUOUS-FIBER-REINFORCED PLASTICS

3.1 Whisker-reinforced plastics

Discontinuous-fiber-reinforced plastics, by our definition, refer to plastics whose reinforcing fibers have length-to-diameter (l/d) ratios varying between 100 and 5000. The whisker reinforcement that will be discussed has l/d ratios between approximately 150 and 2500. Discontinuous-glass-fiber-reinforced plastics which consist of premix thermosetting resins, chopped-strand mat-reinforced thermoset resins, and fiber-reinforced thermoplastic resins have fibers with l/d ratios between 150 and 5000. It has been demonstrated earlier that the strength and modulus of short-fiber-reinforced composites can approach the values for continuous-fiber composites providing that the short filaments can be aligned unidirectionally and that their length is much greater than the critical length (l_c) required for shear-stress transfer.

*These are trade names referring to coupling agents applied to the glass surface. HTS refers to high tensile strength.

Table 6.3. Fiber Efficiency of Reinforced Plastics

Material (epoxy resin matrix)	Fiber Orientation	Fiber Length	Fiber Volume Fraction (V_f)	Composite Strength (psi)	Calculated[a] Fiber Stress, psi	Fiber[b] Efficiency
Nonwoven Unidirectional Laminate		Continuous	0.47	160,000	327,000	82%
Nonwoven Cross-ply Laminate		Continuous	0.47	75,000	288,000	72%
E Glass Fabric, Style 181, Volan Finish		Continuous	0.60	55,800	180,000	45%
E Glass Fabric, Style 181, HTS Finish		Continuous	0.59	72,200	232,000	58%
Chopped Fibers		$\frac{1}{2}$ in.	0.48	27,000	169,000	42%

[a] Fiber stress calculated from the volume percent of fibers in the load direction.
[b] Fiber Efficiency = calculated fiber stress at failure/assumed fiber strength (400,000 psi).

The properties of whiskers have been summarized in Table 6.1. Parratt[5] has investigated the feasibility of reinforcing an epoxy resin with silicon nitride whiskers. These whiskers were approximately 2 μ in diameter* and 300 μ in length so that their l/d ratio was approximately 150. A mat of whiskers was impregnated by a solution of epoxy resin in acetone and the impregnated mat was hot pressed at pressures up to 4000 psi depending upon the desired whisker content. The mat structure produces a two-dimensionally reinforced composite whose properties are nearly isotropic in the plane of the mat since the whiskers are probably randomly oriented. The elastic modulus of this composite in any direction in the plane of the mat is only $\frac{1}{3}$ of the modulus attainable if the whiskers were all aligned in one direction.[6]

The maximum tensile strength reported for this composite was 40,000 psi which was achieved with 30 volume per cent whiskers. At higher whisker loadings the composite strength was reduced because of damage to the whiskers during compaction at the higher required pressures. A modulus of 5×10^6 psi was reached at a whisker loading of approximately 37 volume per cent. A modulus of 15×10^6 psi would have been realized for unidirectionally oriented filaments.

Sutton *et al.* have investigated alumina-whisker-reinforced epoxies.[7] The length of the whiskers varied between 0.3 and 0.6 in., while the effective diameter ($\sqrt{area}$) probably varied between 5 to 9 μ. This results in l/d ratios from 1000 to 2500. This composite was prepared so that the whiskers were aligned parallel by packing them into a capillary tube and infiltrating them with a low-viscosity epoxy resin. The tensile fracture strength of this composite (14 volume per cent

* μ = micron = 10^{-4} centimeters.

whiskers) was 113,000 psi which gives an average fiber stress of 750,000 psi. The whisker strength was given as 1,000,000 psi which gives a utilization of 75 per cent of the whisker's strength, indicating the potential effectiveness of whisker-reinforced plastics.

The whisker composites described above were made in the laboratory, and many problems still remain to be solved before useful composites can be fabricated in larger quantities. Proper resin matrices must be selected not only by their mechanical properties but by the viscosity of the uncured resin, wetting, characteristics, adhesion to the whisker surface, and cure characteristics. Resins can be formulated with the proper mechanical properties to optimize composite behavior, but the processing of the composites will be the most difficult problem. There are not yet adequate techniques for achieving high whisker contents without damaging the whisker strength and for obtaining parallel alignment of the whiskers. Milewski[8] suggests various techniques for production of unidirectionally reinforced plastics with whiskers, such as (1) extrusion of melt incorporating whiskers; (2) centrifugal casting; (3) impregnation of highly oriented paper; and (4) electrostatic or magnetic alignment of the whiskers.

The dimensions of the whiskers (l/d) must also be sufficient to allow for stress transfer into the filament so that the average stress in the discontinuous filament approaches the stress that can be transferred to an infinite length filament. Referring to previously developed equations 13 and 14 for elastic fibers in elastic-plastic matrices, we can calculate the length-to-diameter ratios needed to load the whisker to its ultimate strength and to provide an average stress $\bar{\sigma}_f$ equal to 0.995 $\sigma_{f\,max}$. This is accomplished when $l/l_c = 100$ from equation 13. If it is assumed that the yield strength of the epoxy is

8000 psi then l_c/d for the alumina whiskers ($\sigma_f = 1,000,000$ psi) should be 62.5 in order to load the whisker to its maximum stress at least at one point. Since the actual l/d ratio varies from 1000 to 2500, the composite behaves almost as if it were reinforced by continuous filaments. The l/d ratio would have to be 6250 before the discontinuous filament has an average stress equal to 0.995 of the filament strength, and at the l/d ratio of 2500 the average stress in the whisker is 0.9875 of the filament strength. The silicon nitride whiskers ($\sigma_f = 500,000$ psi) require an l/d ratio of 31 and had an actual ratio of 150. The average stress in the filament is therefore 0.9 the stress of an infinite-length filament.

3.2 Glass-fiber-reinforced thermoplastics

The incorporation of glass fibers into thermoplastic resins is not typical of the techniques used for reinforcing thermosetting resins. The thermoplastic resins undergo no chemical cure during molding so that the fabrication technique is basically one of heating the material until it is soft and flows, and then cooling it when it has been molded to the final shape. The reinforced thermoplastics are usually made by hot-impregnating continuous glass roving or strand and then chopping. The material received by molders consists of small cylindrical pellets in which the reinforcement is contained. This material may be processed by techniques normally used for thermoplastics such as injection molding or compression molding. The fiber diameter usually varies from 0.00035 to 0.0005 in. and up to 40 per cent by weight of $\frac{3}{8}$ to $\frac{1}{2}$-in. length strands are added as reinforcement. Filament lengths reduced to 0.030 in. are now available which allow materials to be molded with up to 70 per cent by volume of glass included in the plastics.[9]

Some of the thermoplastics which are reinforced with glass fibers include nylon, polyethylene, polypropylene, polycarbonate, polystyrene, styrene-acrylonitrile, and the acetals.[10] The mechanical properties for three of the thermoplastics are summarized in Table 6.4. The technology is still in its infancy and the properties given for these materials by various manufacturers as found in the literature[9-12] vary because of differences in glass surface treatments, fiber dispersion, fiber

strength, and fabrication procedures. Tensile strengths are increased providing that the surface of the glass is treated so that good bonding between the resin matrix and filament is achieved. The greatest improvement in strength has been observed for the nylons whose strength goes up three times with the incorporation of 40 per cent glass fibers. The tensile modulus is also dependent upon filament surface treatment and increases as great as 800 per cent have been observed for nylon-reinforced resins. The elongation is considerably reduced as the filament volume is increased since the resin becomes restrained by the filaments. The change in impact strength of the reinforced thermoplastics is not well understood. The impact strength of brittle thermoplastics such as polystyrene is greatly improved by the addition of glass fibers. The ductile, tough thermoplastics such as polycarbonate suffer a decrease in impact strength upon the addition of fibers because of the great decrease in elongation without a corresponding improvement in strength.[13] The fatigue characteristics of reinforced polycarbonates have been investigated, and the endurance limit at 2×10^6 cycles increases from 1000 psi for unreinforced plastic to over 7000 psi for plastic filled with 40 per cent glass fibers.[10]

Murphy[12] has indicated that the strength of the E glass fibers which become incorporated into the thermoplastic may be less than 150,000 psi compared to E glass virgin strength of nearly 500,000 psi. Maintaining a higher percentage of this virgin strength, which is accomplished for continuous-fiber-reinforced plastics, will cause a significant improvement in the efficiency of this material. The length of the glass fiber required to allow the fiber stress to reach the ultimate fiber strength of 150,000 psi can be calculated from equation 14. If we assume a yield stress that can vary from 3000 psi to 8000 psi for unreinforced thermoplastics, the critical length for a 0.0005-inch-diameter filament varies between 0.013 and 0.005 in., respectively. The minimum length of fiber incorporated into the plastic is 0.030 in. which is well above both these values so that theoretically effective reinforcement can result. (However, during processing of the material the original fibers are usually broken and the fiber length is degraded.) In order for the discontinuous-fiber-reinforced plastic to have properties approaching the continuous-fiber plastic the fiber length

Table 6.4. Properties of Unreinforced and Reinforced Thermoplastics (Refs. 9–11)

Property	6/6 Nylon				Polycarbonate			Polystyrene	
	Unreinforced	20%[a]	40%[a]	70%[b]	Unreinforced	20%[a]	40%[a]	Unreinforced	30%
Specific Gravity	1.14	1.31	14.1	—	1.20	1.31	1.44	1.05	1.28
Tensile Strength (psi)	10,000	22,000	29,000	30,000	9,500	15,500	19,000	6,500	14,000
Elongation %	60	5	5	4	100	7	2	2	1.1
Tensile Modulus (psi)	4×10^5	1.18×10^6	1.60×10^6	3.10×10^6	3.2×10^5	8.7×10^5	1.5×10^6	4×10^5	1.2×10^6
Notched Izod Impact (ft-lb/in.)	1.0	2.0	2.5	1.5	4	2	1.2	0.3	2.5
Heat Distortion Temperature (264 psi)°F	150	>465	>465	>465	280	300	305	183	220

[a] $\frac{1}{4}$-in.-length sized fibers.
[b] 0.030-in.-length fibers.

should be at least 10 times greater than the critical length ($\sigma_{\text{ave}} = 0.95\sigma_f$) which would result in fiber lengths of 0.130 and 0.05 in. for resins with yield strengths of 3000 and 8000 psi, respectively. Therefore, a plastic that had a yield stress of 3000 psi should be reinforced with $\frac{1}{8}$-inch filaments if 150,000 psi is the fiber strength.

Lomax and O'Rourke[10] have measured the variation in properties of a 20 per cent reinforced acetal resin with fiber length. The effect of the fiber length on tensile strength and modulus is indicated in Figure 6.6. It appears that at least a

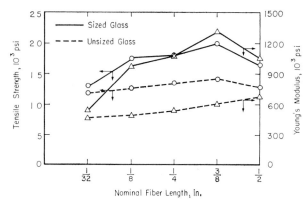

Figure 6.6   Effect of Fiber Length on Tensile Strength and Modulus for Fiber-reinforced Thermoplastics (Ref. 10)

$\frac{1}{4}$-in. length filament is required for adequate reinforcement, which is greater than the length estimated by theory. However, the dispersion and orientation of the fibers are affected by their length as the processability of the material is dependent upon fiber length. This makes it very difficult to apply any quantitative theory to this material in its current stage of development. Figure 6.6 indicates that the strength is decreased in going from $\frac{3}{8}$- to $\frac{1}{2}$-in. fibers which may well be due to improper mixing and inability to disperse the longer fibers. The unsized fibers or untreated fibers do not bond well with the matrix and offer no effective reinforcement.

Murphy and Hauck[13] have described the effect of glass-fiber content on the various mechanical properties, and Figure 6.7 shows this effect on the tensile strength of several thermoplastics. The specific effect is dependent upon the matrix material and is also a function of fiber length. For some of the plastics like styrene acrylonitrile the increase of strength levels off with increasing fiber content after 30 per cent is reached. This again must be a result of poor fiber orientation or dispersion due to the difficulty in processing this highly loaded material. Also the fiber lengths and strengths may be degraded during processing.

4   CONTINUOUS-FIBER-REINFORCED PLASTICS

Continuous-fiber-reinforced plastics typically include those filaments wound from a continuous length of nonwoven reinforcement (usually in roving form) or fabricated from a glass fabric or mat. Although the reinforcement fibers in these cases usually span the dimensions of the structure that they reinforce, noncontinuous fibers can also simulate continuous-length fibers providing they have a sufficiently great length-to-diameter ratio ($l/d > 5000$). The discussion in this section will

center around woven-glass-fabric laminates and nonwoven-glass laminates made of glass roving or a glass tape. Fabrication methods will not be considered in this chapter as they are treated in more detail elsewhere in the book. In addition several other books treat fabrication and manufacturing technology in great depth.[14-17]

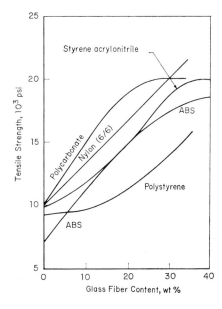

Figure 6.7   Tensile Strength for Various Glass-fiber-reinforced Thermoplastics as a Function of Glass-fiber Content (Ref. 13)

4.1   Stress-strain relations

4.1.1   Nonwoven composites

Nonwoven unidirectional reinforced plastics are the least complex of the reinforced-plastic materials and therefore we will begin with a discussion of their behavior. These materials can be produced directly by filament winding from roving or tapes in such geometries as cylindrical tubes or rings. Unidirectionally reinforced preimpregnated resin tape or ribbon, consisting of a single layer of filament ends or strands, is made in sheet sizes up to 48×60 inches. This material typically contains 65 to 80 per cent by weight of glass. Some typical mechanical properties for laminates fabricated of this material using both E and S glass* are shown in Table 6.5.[18-20]

Davis and Zurkowski[18] have presented a thorough report on the stress-strain characteristics of nonwoven-glass-fiber-reinforced plastics. The stress-strain curves for unidirectionally reinforced material (56 per cent glass by volume) in both tension and compression are shown in Figure 6.8(a). When stressed parallel to the filaments the curves appear linear to failure for both tensile and compressive stresses. It may be expected for tensile loading that nonlinearity in the curve at high stresses will result from individual filament breaks, but very precise determination of the stress-strain curve may be required to detect this nonlinearity. The tensile

---

* The tensile strength and Young's modulus of E glass (density = 2.54 g/cc) are 500,000 psi and 10.5×10^6 psi and the values for S glass (density = 2.48 g/cc) are 700,000 psi and 12.4×10^6 psi.

Table 6.5. Some Properties of Nonwoven Unidirectional Reinforced Plastics (Refs 18–20)

Material (epoxy resin matrix)	Glass Content (% by volume)	Specific Gravity	Percent of Glass Parallel To Load	Tensile Strength (psi)	Tensile Modulus (psi)	Compression Strength (psi)
E glass	73.3	2.17	100	238,000	8.1×10^6	—
E glass	56.5	1.97	100	149,000	6.2×10^6	87,000
E glass	56.5	1.97	54.5[a]	75,000	3.5×10^6	71,000
E glass	56.5	1.97	0[a]	5,000	1.47×10^6	20,000
S glass	71.6	2.12	100	275,000	9.6×10^6	200,000
S glass	71.6	2.12	0[a]	8,000	—	20,000

[a] Cross-ply composite so that remainder of glass is perpendicular to load.

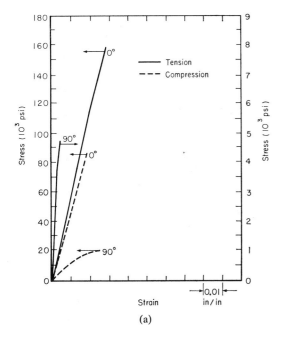

(a)

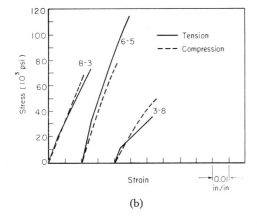

(b)

Figure 6.8 Stress-strain Curves for Nonwoven-glass-filament-reinforced Epoxy Plastics (Scotchply 1002, 35 per cent resin content by weight) (Ref. 18). (a) Tension and compression curves for unidirectional composites stressed at 0° and 90° to the fiber direction. (b) Tension and compression curves for 11-ply cross-ply composites. Three types of composites are shown with various number of plies in the load direction. For example, 8-3 refers to 8 plies in the load direction and 3 normal to the load

. and compression curves for composites stressed at 90° to the filament direction are also shown in Figure 6.8(a) and while the compression curve is not linear the tensile curve is nearly linear. In this direction the resin is the continuous phase and the stress-strain curve for the composite reflects the behavior of the resin. Tensile failure occurs at very low stresses, approximately 5000 psi, for these composites stressed at 90° to the filament direction. There is good reason to believe that failure results from the stress concentrations or strain concentrations at the filament-resin interfaces. Complex triaxial stresses exist in the resin phase making analysis of this strength very difficult.

A more complex composite results by alternating the orientation of adjacent plies by 90° to form a so-called cross-ply laminate. The properties of this type of composite can be easily altered by varying the number of plies oriented in each direction. These composites are still relatively simple since the axes of the plies are only rotated by 90°. A more nearly isotropic composite can be fabricated by using layers with alternating 45° plies and, as can easily be imagined, an almost infinite number of variations exist which allow a designer considerable freedom. Stress-strain curves for cross-ply composites with a varying number of plies parallel to the load are shown in Figure 6.8(b). The most interesting feature of these curves is the change in slope or knee in the curves at low stresses which did not clearly exist for the unidirectional composites. The knee in the stress-strain curves only occurs for tensile loadings and has not been observed for compressive stresses. The existence and the location of this change in slope of the stress-strain curve can be explained rationally. The knee in the stress-strain curve for nonwoven-fiber-reinforced composites represents the fracturing of the plies oriented 90° to the load direction. In Figure 6.8(b), for 11-ply composites with 6 plies and 3 plies oriented in the load direction, the knee in the stress-strain curve occurs at approximately 0.004 to 0.005 in./(in. of strain). The knee in the curve for the laminate with 8 plies oriented in the load direction is more difficult to observe because of the smaller change in slope which occurs when only the three 90° plies fracture. In Figure 6.8(a) it can be observed that the failure strain for the unidirectional composite stressed at 90° to the filament direction is indeed 0.004 in./in. or 0.4 per cent. The knees in the stress-strain curves for cross-ply composites are a result of failure of the 90° plies and this is governed by a strain criterion.

4.1.2 Woven-fabric composites

The use of fabrics in composites allows one to achieve a more balanced structure in a single plane by weaving together yarns in an orthogonal pattern. The degree of anisotropy of a fabric can be controlled by varying the number of yarns in the warp direction (lengthwise yarns on weaving looms) and the fill direction (crosswise yarns on weaving looms). In addition, the yarns themselves can be varied by altering the number of strands used to form the yarn, the number of filament ends per strand and the filament diameter in the strand. All of these parameters give the designer great freedom in preparing a composite with specific properties in any direction. One of the important factors which distinguishes the behavior of a composite reinforced with a woven fabric from one reinforced with a nonwoven reinforcement is the crimp of the fabric yarns which is a measure of the displacement of the yarn in passing over and under yarns in the orthogonal direction. For example, McGarry and Desai[21] have proposed that the straightening out of the crimped fibers may cause high stresses in the matrix which could result in failure of the composite. Another consideration is that the strength of the glass fibers that are woven into fabric structures is probably degraded by additional mechanical handling, so that the ultimate tensile or flexural strengths of fabric laminates does not approach those for nonwoven laminates.

The mechanical properties of typical glass-fabric composites are summarized in Table 6.6. The changes in properties with the use of different glasses is obvious as well as with different types of fabrics. The properties of a greater variety of fabric composites can be found in Reference 22. The type of resin used also affects the properties, as shown in Table 6.6 for epoxy and polyester resins.

The stress-strain characteristics of fabric-reinforced laminates are more complex than those already considered for non-woven reinforced laminates. The addition of twisted yarns and crimped fibers introduces complexities not present even in cross-ply composites of nonwoven reinforcement. Tensile stress-strain curves for three types of fabrics and two types of matrix materials are shown in Figure 6.9. Curves for loading at 0°, 45°, and 90° to the warp direction are shown for each case. It is often observed that the stress-strain curves for fabric composites possess two or more knees or changes in slope during loading. These knees are indicated by circles on the curves in Figure 6.9.

In the case of nonwoven reinforced composites the knee in the stress-strain curve observed for cross-ply composites was attributed to fracturing of the layers whose filament axes were positioned 90° to the load direction. Although the mechanism must also occur for fabric-reinforced composites, another effect has also been postulated for these composites. McGarry and Desai[21] proposed that upon tensile loading the crimped fibers would tend to straighten, thereby applying a tensile stress to the resin matrix between the filaments in a direction normal to the filament axes. When this stress exceeds the local strength of the resin matrix, microcracking would occur. Experimental verification for the existence of both of these effects during loading of composites was obtained by measuring the water absorption of samples that had been stressed to various load levels.[21,23] This will be referred to again later when we consider the effect of moisture on the strength properties of a laminate.

Fabric laminates behave as shown in Figure 6.10 when

Table 6.6. Mechanical Properties of Fabric Laminates

| Laminate Description | | | | Tensile Properties | | | | Compression Properties | | | | Interlaminar Shear | |
Glass Type	Glass Surface Treatment	Fabric Style	Resin	Resin Content Percent by Weight	Strength (10³ psi) Warp Direction	Strength (10³ psi) Fill Direction	Modulus (10⁶ psi) Warp Direction	Modulus (10⁶ psi) Fill Direction	Strength (10³ psi) Warp	Strength (10³ psi) Fill	Modulus (10⁶ psi) Warp	Modulus (10⁶ psi) Fill	Warp	Fill
S	HTS	181[a]	Epoxy	28.5–33.6	97.7	95.3	3.15	3.09	67.4	64.0	4.60	4.53	3040	3120
S	HTS	143[a]	Epoxy	30	139.5	31.8	5.52	0.74	79.8	34.2	5.85	2.36	4950	2250
E	Volan[b]	181	Epoxy	28	55.8	52.6	3.16	2.89	59.2	52.5	4.22	3.94	3330	3110
D	HTS	181	Epoxy	33.9	35.0	33.3	2.69	2.57	52.7	49.2	2.85	2.79	3710	3720
E	Volan[b]	181	Polyester	35	48.0	45.3	2.80	—	38.0	36.3	—	—	—	—

[a] Fabric Construction: Style 181: 57 × 54 ends and picks/inch; 0.0085 inches thick; 8 harness satin weave; warp and fill yarn, 225½.
Style 143: 49 × 30 ends and picks/inch; 0.009 inches thick; crowfoot satin weave; warp yarn, 225½; fill yarn, 450½.

[b] Trade name for methacryatochromic chloride.

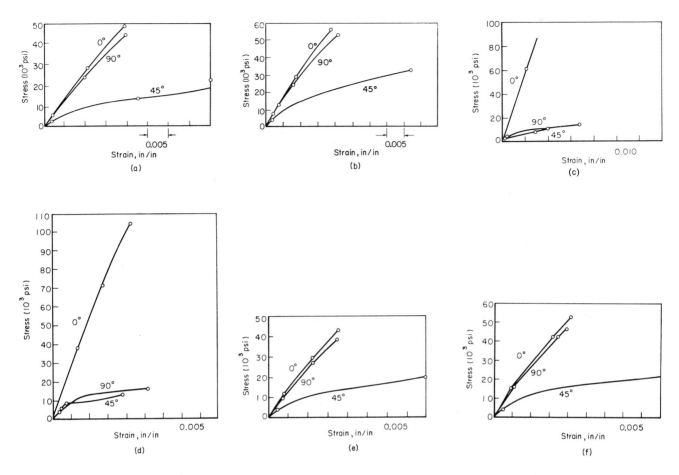

Figure 6.9 Typical Stress-strain Curves for Glass-fabric Laminates (Ref. 22). (a) Typical tensile stress-strain curves for parallel laminate made with 181 glass fabric and polyester resin. (b) Typical tensile stress-strain curves for parallel laminate made with 181 glass fabric and an epoxy resin. (c) Typical tensile stress-strain curves for parallel laminate made with 143 glass fabric and polyester resin. (d) Typical tensile stress-strain curves for parallel laminate made with 143 glass-fabric and epoxy resin. (e) Typical tensile stress-strain curves for parallel laminate made with 112 glass-fabric and polyester resin. (f) Typical tensile stress-strain curves for parallel laminate made with 112 glass-fabric and epoxy resin

subjected to edgewise shear stresses. Edgewise shear stresses refer to the application of forces to the edge of a panel, resulting in shear deformation in a plane parallel to the individual laminations as shown in the sketch in the corner of Figure 6.10. Edgewise shear should not be confused with interlaminar shear which occurs when shear forces are applied parallel to the surface of the panel, resulting in shear deformations in the plane perpendicular to the individual laminations. It can be seen from Figure 6.10 that the type of fabric does not greatly influence the shear characteristics since the fibers are not being loaded directly. The stress-strain curves are therefore more characteristic of the resin matrix behavior and the composite shear properties can be expected to be greatly influenced by the properties of the matrix.

As already indicated by the tensile stress-strain curves for nonwoven and woven laminates, these materials are not only anisotropic but their degree of anisotropy can be easily altered. For example, consider the variation in the tensile modulus and strength with the direction in the laminate. This

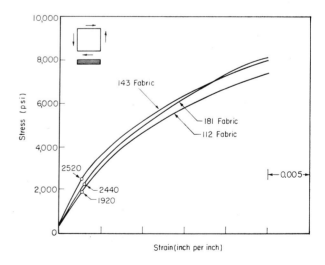

Figure 6.10 Stress-strain Curves in Edgewise Shear for Polyester-resin Glass-fabric Laminates (Ref. 22)

relation is shown in Figure 6.11 for a parallel laminate of 143 glass fabric for which the warp direction is the same in each ply and a cross-laminate in which the warp direction alternates from ply to ply. Although the properties at 45° remain similar for parallel and cross-laminated composites, the cross-laminated composite is balanced since the 0° and 90° properties are similar. The prediction of moduli and strength properties for single-ply and cross-laminated composites as well as angle-ply composites has recently been treated by Tsai.[24,25] In addition, Chapter 1 of this book treats this subject in greater detail.

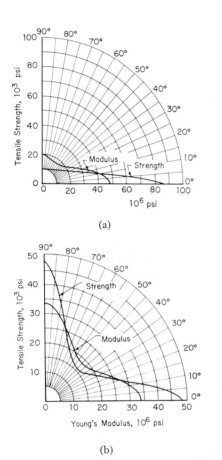

(a)

(b)

Figure 6.11   Directional Properties of Glass-fabric Laminates (Ref. 22). (a) Directional properties in tension of parallel laminated 143 glass-fabric laminate made with epoxy resin. (b) Directional properties in tension of cross-laminated 143 glass-fabric laminate made with epoxy resin

Not much discussion is usually given to the subject of the difference between the compressive and tensile stress-strain characteristics of fiber-reinforced plastics. Although there is certainly a difference between the tensile and compression strength of this material, contradictory evidence is available in the literature concerning the difference between the compressive and tensile elastic moduli. Tensile and compression stress-strain curves have already been shown in Figure 6.8. The compressive modulus is typically reported to be greater than the tensile modulus as indicated for fabric laminates in Table 6.6. However, Abbott and Broutman[26] have utilized a stress-wave-propagation technique which allowed them to determine

both the compressive and tensile modulus on one specimen at one time. This was accomplished by determining the velocity of the stress wave propagating along a rectangular bar before reflection (compression) and after reflection (tension) from the free end of the bar. The experiments were performed on non-woven-filament-reinforced bars with different ratios of filaments along the bar axis. The results are shown in Table 6.7.

Table 6.7.   Comparison of Tensile and Compressive Moduli of Elasticity for Glass-Filament-Reinforced Epoxies Determined by Stress-Wave Propagation[26]

Filament Orientation	Percentage of Filaments along Bar Axis	Approximate Stress Level (psi)	Tensile Modulus of Elasticity (10^6) psi	Compressive Modulus of Elasticity (10^6 psi)
2:1	67	11,600	6.27	6.27
1:2	33	5,000	4.61	4.67
1:0	100	20,000	8.68	8.43
0:1	0	5,700	3.92	3.82

There was essentially no difference found between the compressive and tensile modulus for these materials at the approximate stress levels shown in Table 6.7. It is possible for differences to occur if the stress is great enough to exceed one of the knees in the stress-strain curves.

4.2 Interlaminar properties

The interlaminar planes that exist in fiber-reinforced plastics between laminations of fabrics or layers of filament-wound tape or roving are weak planes in the material. Failure of fiber-reinforced plates or beams subjected to flexural loads often occurs by interlaminar shear failure due to the high interlaminar shear stresses created at the beam's neutral axis. In fact, beams with short span-to-depth ratios are used as a test method to measure interlaminar shear strength since the shear stresses at the neutral axis exceed the interlaminar shear strength before the outer fiber stresses exceed the compression or tensile strength of the material. Microscopic views of interlaminar planes are shown in Figure 6.12 for the case of nonwoven reinforcements. These sections were taken from a glass-fiber-reinforced epoxy plate which was filament wound from preimpregnated tape. The interlaminar planes between orthogonally oriented fiber layers are shown in Figure 6.12(a) and the interlaminar plane formed between parallel layers of tape is shown in Figure 6.12(b). These interlaminar planes are essentially resin-rich boundaries between laminations and are often subject to the accumulation of voids that have a very detrimental effect on the shear properties as will be shown shortly.

4.2.1   Tension and compression

The factors influencing interlaminar tension and compression properties are primarily the resin matrix properties, the matrix-fiber interface strength, and voids present in the interlaminar planes. The tension and compression characteristics have been measured by Kimball[27] for fabric- and nonwoven-reinforced composites and typical stress-strain curves for laminates stressed normal to the interlaminar planes are shown in Figure 6.13. A summary of the properties is presented in Table 6.8. The tensile strength is much less than the strength

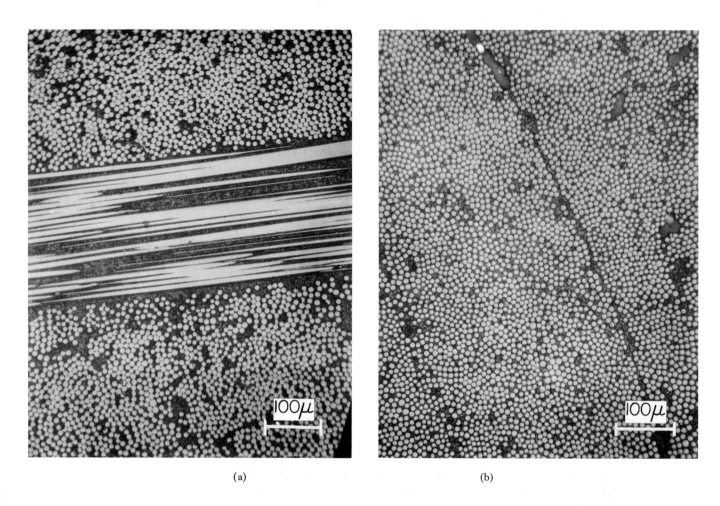

(a)

(b)

Figure 6.12   Interlaminar Planes in Nonwoven-glass-reinforced Plastics (Magnification 150×). (a) Orthogonally reinforced laminate. (b) Unidirectionally reinforced laminate

Table 6.8.   Tensile and Compressive Properties Normal to the Interlaminar Plane[27]

Material			Tension		Compression	
Resin	Reinforcement	Resin Content (Wt%)	Modulus of Elasticity (psi)	Maximum Stress (psi)	Modulus of Elasticity (psi)	Maximum Stress (psi)
Epoxy	181 fabric (Volan finish)[a]	31.8	1.35×10^5	2230	1.42×10^5	50,000
Epoxy	Unidirectional glass filaments	35.2	1.42×10^5	3590	1.33×10^5	21,600
Epoxy	Crossplied (90°) glass filaments	31.1	1.91×10^5	3380	1.84×10^5	93,800
Phenolic	181 fabric (A1100 finish)[b]	27.5	1.13×10^5	710[c]	2.07×10^5	80,400

[a] Trade name for methacryatochromic chloride.
[b] Trade name for γ-aminopropyltriethoxysilane.

[c] These values may be too low because of machining of tensile specimen. For unnecked specimens the modulus was 2×10^5 psi and the strength was 2790 psi.

of the resin alone because of the stress concentrations produced at the glass-resin interface, the lack of perfect adhesion, and the presence of voids. In compression tests, the glass fibers actually reinforced the resin because the interlaminar compression strength is greater than the resin compression strength.

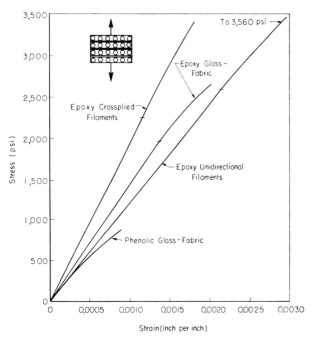

Figure 6.13   Typical Tension Stress-strain Curves for Reinforced Plastic Laminates Stressed Normal to the Interlaminar Planes (Ref. 27)

Failure of these compression specimens usually occurs by shear failures passing through the layers of reinforcements at an angle of about 45° to the direction of load. However, the failure of the unidirectional filament composite occurs by spreading of the filaments and the compressive strength appears no greater than the strength of the resin alone.

4.2.2   Shear

The interlaminar shear strength, similar to the tensile and compression strength, is influenced mostly by the resin matrix strength, the matrix-fiber interface strength, and voids that may be present in the interlaminar areas. Many test methods are available to measure the interlaminar shear strength and these as well as methods for other properties have been summarized by Fried.[28] Interlaminar shear strengths measured by investigators at Forest Products Laboratory[27,29] are presented in Table 6.9. The shear strength is shown to depend on the resin used for the matrix and on the angle at which the laminate is stressed with respect to the fiber direction (0° refers to the axis parallel to the principal fiber direction). The polyester resin laminates yield low interlaminar shear strengths because of the typically low interface strength between this resin and the glass-fiber surface. The orientation of the filaments affects the shear strength and an all unidirectional laminate possesses a higher shear strength than one cross-plied with alternate 90° layers. The effect may be attributed to the better nesting of the fiber for the unidirectional laminate and therefore the greater difficulty

Table 6.9.   Interlaminar Shear Strengths for Fiber-Reinforced Plastics[27,29]

Material		Resin Content (Wt%)	Shear Strength (psi)[a]		
Resin	Reinforcement		0°	45°	90°
Epoxy	181 fabric (Volan Finish)[b]	31.8	6260		
Phenolic	181 fabric (A1100 Finish)[c]	27.5	6160		
Polyester	181 fabric (Volan Finish)[b]	32.2	4780	6230	
Polyester	143 fabric (Volan Finish)[b]	28.2	4320	5110	
Epoxy	Unidirectional glass filaments	31.6	6720	5460	5190
Epoxy	Crossplied (90°) glass filaments	31.9	6050	6470	
Polyester	Mechanically bonded mat	62.8	4870	4510	

[a] A block shear test was the test method used for all of the data reported in this table.
[b] Trade name for methacryatochromic chloride.
[c] Trade name for γ-aminopropyltriethoxysilane.

for an interlaminar crack to be generated. This same effect can explain the higher interlaminar shear strengths for fabrics stressed at 45° to the warp direction. In addition, a crack would have to pass around a greater number of fibers in this direction.

Since the interlaminar shear strength is often a mode of failure for plates or beams subjected to flexural loading and also appears to be a common mode of failure in axial compression,[30] investigators have been interested in means of improving the interlaminar shear strength.[30-34] Prosen et al.[31] have shown that the interlaminar shear strength is a function of glass content and begins to decrease when glass contents greater than 74 per cent by weight are reached for filament-wound rings or cylinders. Voids in a composite represent a degree of unreproducibility which it would be desirable to eliminate and can limit the permanence of these materials by providing paths for environmental penetration into and through the composite. Paul and Thompson[34] demonstrated by filament winding in a vacuum environment to eliminate all voids that they could increase the capability of a given resin to provide higher interlaminar shear strengths. In fact, if voids are discounted there appears to be a linear correspondence between resin tensile strength and interlaminar shear strength. The influence of void content on interlaminar shear strength is shown in Figure 6.14. Here the results of two investigations are indicated. The curve representing the higher shear strengths was obtained by measuring short beam shear strengths of Naval Ordnance Laboratory (NOL) rings. The rings were wound from preimpregnated epoxy roving and the variation in void content was achieved by varying the roving tension during winding.[33] Hand's[35] results, represented by the lower curve in Figure 6.14, were obtained on short-beam shear samples removed from an orthogonally reinforced flat plate. Although the absolute values of the shear strength differ because of variations in filament orientation and fabrication conditions, the curves are nearly parallel indicating an identical influence of voids. Hand, through a regression analysis of his data, has developed the following equation:

$$S_s = 9600 - 785V, \qquad (23)$$

where S_s is interlaminar shear strength and V is the void content.

An alternative means of increasing the shear strength by addition of fillers to the resin matrix has been investigated at the Naval Ordnance Laboratory.[32] NOL rings were used to provide test specimens and the shear strengths were determined using the short beam shear test. It was concluded that the interlaminar shear strength was improved by the random incorporation of small fibrous additives (up to 4 per cent of the resin weight) to the resin matrix prior to impregnation of the reinforcement. High modulus sapphire (Al_2O_3) whiskers and 400-μ length ground glass fibers appear the most promising of the fillers evaluated. Interlaminar shear strengths have been increased from 10,300 psi to 11,700 psi by the addition of 4 per cent sapphire whiskers (1 to 30 μ in diameter) to the epoxy resin matrix.

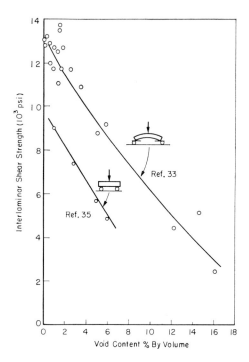

Figure 6.14 The Effect of Void Content on Interlaminar Shear Strength of Nonwoven-glass-reinforced Epoxy Resin (Refs. 33, 35)

4.3 Strength properties of glass-fiber-reinforced plastics

4.3.1 General considerations

The purpose of this section is not to further an understanding of the mechanics or micromechanics of failure of glass-fiber-reinforced plastics. This subject has been well reviewed recently by Corten.[36] This section will treat those material and environmental parameters that affect the strength properties of a fiber-reinforced plastic.

One of the most important material variables in this composite material is the resin matrix. The function of the resin matrix in a fibrous composite varies depending on how the composite is stressed. For compressive loading, the matrix prevents the fibers from buckling[37] and is, therefore, a very critical part of the composite since without it the reinforcement could carry no load. On the contrary, a bundle of continuous fibers could sustain high tensile loads in the direction of the filaments without a matrix. However, in the case of glass fibers, the resin matrix prevents environmental corrosion of the glass and loss of strength due to abrasion. The resin also provides a stress-transfer medium so that when an individual fiber breaks it does not lose its load-carrying capability; interfacial shear stress acts at the broken filament ends and the axial fiber stress is built up again after a critical length from the fiber end is exceeded. If this same fiber failed in an unsupported bundle, it would obviously lose its load-carrying capability and the stress in the remaining fibers would be increased.

The resin physical properties that most influence the behavior of a fibrous composite include the following:

1. Shrinkage during cure;
2. Modulus of elasticity;
3. Ultimate elongation (elastic and plastic strain);
4. Strength (tensile, compression, shear);
5. Fracture toughness.

In addition, factors relating to the processability of the resin such as viscosity-time, viscosity-temperature, and wettability relations may have a significant effect upon the ultimate composite properties and the reproducibility of these properties because of the influence of these resin properties on interface bonding and flaws such as voids in the composite. The composite strength and modulus (equations 5 and 8) should be directly proportional to the volume fraction of reinforcement and matrix, varying between the resin properties and reinforcement properties for 0 per cent and 100 per cent reinforcement material by volume, respectively. Experimental results over wide ranges of resin contents do not agree with this relation. Brown[38] has shown that the flexural strength and modulus for 181 style glass-cloth laminates increase with decreasing resin content until approximately 30 per cent by weight resin content is reached. Below 30 per cent resin both the strength and modulus of these laminates decreases. This is due to the inability of the resin matrix to properly distribute itself throughout the composite and results in voids and poorly bonded or resin-starved regions. The effect also exists for other types of reinforcements but the resin content at which the mechanical properties decline differs for each type of reinforcement. A nonwoven unidirectional glass-reinforced composite continues to show strength increases until 15–20 per cent resin content by weight is reached.

4.3.2 Effects of laminate thickness

The thickness of a laminate influences the tensile, compressive, and flexural strengths, and for certain reinforcement-resin systems there appears to be one thickness that produces an optimum strength.[39-41] Typical results for tensile and compressive strengths of cloth laminates with a variety of resin systems are illustrated in Figures 6.15 and 6.16.[39]

The tensile strength of thin laminates, 0.020 to 0.080 in., increases at a rapid rate as the thickness increases. This is opposed to the size effect in materials whose strength can be traced to the influence of flaws in the material. If a distribution of small flaws is present in a material, the severity of the most severe flaw tends to increase with the volume of the material; thus, the strength generally decreases as the volume increases. A reasonable hypothesis exsists to explain the behavior of

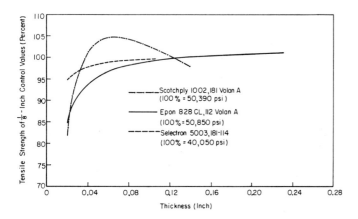

Figure 6.15 The Effect of Thickness on the Tensile Strengths of Epoxy (Scotchply and Epon 828) and Polyester (Selectron) Laminates Expressed as a Percentage of the $\frac{1}{8}$-in.-thick Control Values (Ref. 39)

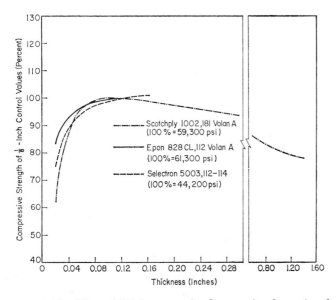

Figure 6.16 Effect of Thickness on the Compressive Strengths of Epoxy (Scotchply and Epon 828) and Polyester (Selectron) Laminates Expressed as a Percentage of the $\frac{1}{8}$-in.-thick Control Values (Ref. 39)

fiber-reinforced plastics. The crimped fibers in glass-cloth plies located at the surface experience less lateral constraint than do similar plies located at the interior of a laminate. This lack of constraint leads to fiber-resin separation and matrix cracking due to straightening of the crimped fibers under tensile loads at stress levels less than are required for center plies. In addition, surface plies may become physically damaged during laminate fabrication. Thus for a laminate that is made very thin, say two plies (181 glass fabric is approximately 0.008 in. per ply), each ply acts as a surface ply

and exhibits low strength. As the number of plies is increased, the two weak surface plies remain but interior plies experiencing constraint from both sides are introduced between the surface plies. As the number of interior plies increases, the fraction of the load carried by the weak surface plies decreases and the strength of the laminate levels off at a nearly constant value for laminate thicknesses that exceed approximately $\frac{1}{16}$ in. For one of the laminates in Figure 6.15, the tensile strength decreased as the thickness increased from $\frac{1}{16}$ to $\frac{1}{8}$ in. This corresponds to the expected behavior for homogeneous brittle materials but the difference in behavior between this laminate and the others shown in Figure 6.15 is not explained.[39]

The influence of laminate thickness on compressive strength is even greater than on the tensile strength. As shown in Figure 6.16, the compressive strength of a 0.020-in.-thick laminate is reduced by 50 per cent from $\frac{1}{8}$-in.-thick laminate. The crimped fibers in the laminate surface plies subjected to compression tend to buckle outward especially for a two-ply laminate without interior restraint. Ekvall[42] has demonstrated this localized buckling effect for 181 fabric laminates and one can observe this visibly as a pattern of white spots on the laminate surface. The above hypotheses suggest therefore that the surface layers of cloth laminates are inherently weaker than the interior plies.

4.3.3 Effect of stressing at an angle to the fibers

For isotropic materials failure envelopes can be constructed from theory or experiment to indicate the strength of the material when subjected to biaxial or triaxial stresses. In this plot the failure envelope represents the limiting strength of the material and the axes or coordinate system represents the stresses in each of the orthogonal directions. In the case of an anisotropic material, such as fiber-reinforced plastic, failure envelopes must first be constructed to represent the strength of the material as a function of fiber direction for simple single-ply (unidirectional fibers) and cross-ply or angle-ply composites. Stowell and Liu[43] analyzed this problem by assuming a maximum stress theory to govern failure. Three maximum stresses were chosen, the governing one dependent upon the angle of the fibers, θ, to the applied load direction. The three strengths are represented by

σ_L = composite strength parallel with fibers,
σ_T = composite strength normal to fibers,
τ = matrix or interface shear stress.

Failure therefore occurs by one of three mechanisms, each characterized by a separate function and each dominant in a particular range of θ. If composite failure is based on tensile failure parallel to the fibers then

$$\sigma = \sigma_L/\cos^2 \theta, \qquad (24)$$

where σ is applied stress. At intermediate values of θ failure can take place by shear of the matrix or interface on a plane parallel to the fibers. The fracture stress is then given by

$$\sigma = \tau/\sin \theta \cos \theta. \qquad (25)$$

At high values of θ the composite strength is governed by transverse failure so the necessary applied stress is given by

$$\sigma = \sigma_T/\sin^2 \theta. \qquad (26)$$

The above equations are represented in Figure 6.17 and it is seen that the strength rises to a maximum at some θ_{max} determined by equating equations 24 and 25:

$$\tan \theta_{max} = \tau / \sigma_L. \quad (27)$$

This angle is approximately 4° for a typical high-strength unidirectional glass-plastic laminate.

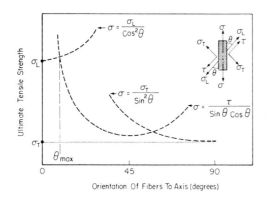

Figure 6.17   Variation of Ultimate Strength of a Uniaxially Stressed Fiber-reinforced Material with Fiber Orientation (Ref. 46)

Comparison of the above maximum stress theory with experiment as determined by Tsai[44] is shown in the right half of Figure 6.18 for a nonwoven unidirectional epoxy laminate having the following properties:

$\sigma_L = 150,000$ psi, $\sigma_T = 4000$ psi, $\tau = 8000$ psi

$\nu_{LT} = 0.3$ (Poisson's ratio),

$E_L = 8 \times 10^6$ psi, $E_T = 2.7 \times 10^6$ psi, $G_{LT} = 1.25 \times 10^6$ psi,

Results are also included for compression strength measurements (transverse compression strength approximately 20,000 psi) for angles greater than 30° and the theoretical curves are raised since σ_T for compression is greater than σ_T for tension.

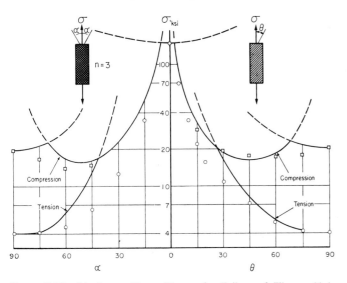

Figure 6.18   Maximum Stress Theory for Failure of Fibrous Unidirectional and Angle-ply Composites. Experimental points are shown for compression and tension strengths of nonwoven-glass epoxy composites (Ref. 44)

Also shown in Figure 6.18 are Tsai's[44] results for a three-ply angle-ply composite. The angle α defines the angle of the plies and the number of plies is denoted by n. The agreement between the experimental results and this maximum-stress theory is not sufficient between 15° and 45°, although Cooper[45] and Jackson and Cratchley[46] have had better agreement with this theory for fiber-reinforced metal composites.

Recently Tsai[25] has proposed a single function to cover all values of θ or α, assuming a Von Mises type of failure criterion. Earlier, Norris[47] had also suggested the use of a single function obtained by considering the energy of distortion. The resulting equation can be expressed as follows:

$$\frac{1}{\sigma^2} = \frac{\cos^4 \theta}{X^2} + \left(\frac{1}{S^2} - \frac{1}{XY} \right) \sin^2 \theta \, \cos^2 \theta + \frac{\sin^4 \theta}{XY}, \quad (28)$$

where X, Y, and S represent the strength parallel to the fibers, normal to the fibers, and the shear strength, respectively; and σ is the strength at any angle θ to the natural axes of the composite. Experimental results for plywood and 143 type glass-fabric laminates agreed very well with equation 28. Tsai also used a distortion energy or work criterion but essentially extended the work of Norris to distinguish between homogeneous composites and laminated composites. The theory could also then be used to predict the strength of cross-ply or angle-ply composites. This distortion-energy theory is shown in Figure 6.19 along with the same experimental points

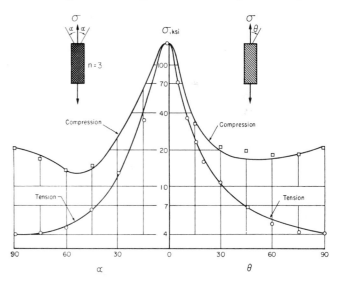

Figure 6.19   Distortion Work Theory for Failure of Fibrous Unidirectional and Angle-ply Composites (Ref. 44)

shown in Figure 6.18. It can be concluded that this theory is superior to the maximum-stress theory for the prediction of failure strengths at angles to the natural axes of the laminations in a composite.

4.3.4   Effect of moisture

The specific role of moisture on the composite strength is dependent upon several factors such as time of exposure, temperature of exposure, type of stress and stress magnitude during exposure, and material construction including interface treatments. The effect of the prestress magnitude coupled with humidity cycling has been considered by Brelant et al.[48] for filament-wound pressure vessels, while Youngs[49] and Krolikowski[23] considered its effect on the properties of fabric laminates. Brelant's results as indicated in Table 6.10 show

Table 6.10. Humidity Effects on Balanced Biaxial
Filament-Wound Pressure Vessels[48]

Prestress Per cent	Treatment	Average Burst Pressure	Percentage Difference
0	None	3280	0
40	None	3275	0.1
80	None	3150	4.0
0	MIL-E-5272[a]	2960	9.7
40	MIL-E-5272	2530	22.9
80	MIL-E-5272	2660	18.9

[a] Temperature cycle between 68 and 160°F at 95 per cent relative humidity.

that these filament-wound composites (burst by internal pressurization) when prestressed as low as 40 per cent of ultimate burst undergo subsequent rapid environmental degradation and fail at approximately 80 per cent of the original strength. Prestress by itself did not appear to damage the composite seriously and it was necessary that the prestress be followed by exposure to an adverse environment for damage to occur.

The explanation for the above behavior can best be explained by considering the results of McGarry and Desai.[21] They have measured the amount of water absorption for laminates subjected to various prestress levels from 0 to 90 per cent of the ultimate strength. Figure 6.20 represents the weight gain after sufficient water soaks as a function of the prestress level expressed as a percentage of the ultimate load. It can be seen in this figure that both polyester and epoxy fabric laminates undergo a sudden change in water absorption between 50 and 70 per cent of the ultimate stress for the epoxy resin and 30 to 60 per cent of the ultimate stress for the polyester resin. The increase in water absorption is attributed to the opening of cracks throughout the composite. These stress levels at which

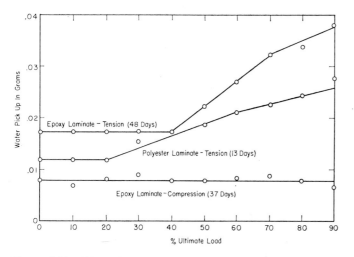

Figure 6.20 Water-absorption Rates for Prestressed Glass-fiber-reinforced Plastics (Ref. 21)

water absorption rates increase correspond to the previously discussed knees in the stress-strain curves. The internal cracking is a result of the tendency to straighten crimped fibers in fabrics and stress concentrations produced by fibers normal to the load direction. Compression stressing does not result in any change of water absorption (Figure 6.20) which indicates that microcracking does not occur. The occurrence of the knee in a compressive stress-strain curve for a fabric laminate is much less common and evident than for tensile loading and when it does occur it is probably associated with an effect such as fiber buckling rather than microcracking in the matrix.

Water absorption measurements performed by McGarry and Desai[21] on nonwoven laminates (Scotchply) also agree with the previous discussion on their stress-strain behavior. The stress-strain curve for unidirectional laminates does not usually possess a knee nor does the water absorption change for any prestress level. However, a cross-ply (1 : 1) laminate begins to gain water at a stress of approximately 40 per cent of the ultimate strength. The introduction of filaments normal to the load direction causes microcracking of the matrix because of the strain concentrations in the resin matrix between the filaments.

In order to represent in the laboratory the effects of long-time water immersions on the strength of reinforced plastics, an arbitrary test method has been developed. In early tests of polyester-glass fabric laminates, a one-month immersion in water was found to be represented approximately by two hours in boiling water. Hence, the two-hour water boil test came to be used as a standard means of comparing laminate properties. The two-hour boil test could be misleading in attempts to provide long-term performance data from short-term tests. Rawe,[50] for example, has measured the flexural strength of glass fabric-epoxy laminates for various resin contents and exposure times in boiling water. As shown in Figure 6.21 from the results of short-term boil tests (4–6 hours), the erroneous conclusion could be drawn that a 25 per cent resin content is significantly better than any other, whereas in the long term the resin content is unimportant. Rawe[50] has also shown that the strength retentions for typical epoxy-glass fabric laminates are approximately 60 to 70 per cent when pre-immersed in water for up to six months (in an unstressed condition).

One of the naturally occurring questions is whether the effects of water immersion are reversible and the strength of the laminate restored after complete drying. Krolikowski,[23] as well as other investigators,[14,50] have proven that this phenomenon is reversible so that after drying, the strength is restored. Krolikowski showed that a 13 per cent reduction in strength occurred for a polyester-glass fabric laminate prestressed above the knee in the stress-strain curve and then immersed in water for 100 days. However, when this laminate was dried before testing, the strength retention was 99 per cent.

One popular use for the two-hour water boil test has been in the evaluation of coupling agents that have been developed in large part to increase the strength retention of laminates subjected to immersion in water or exposure to high relative humidities. The particular subject of coupling agents used to improve the interface properties of glass-reinforced plastics and the reasons for the improvement have been very well discussed in the literature[51-60] and need not be repeated here. However, it is important to point out the influence of the interface treatment on the strength retention after immersion for 8 hours in boiling water (Table 6.11). This data represents only a small portion of that which has been accumulated, but it is sufficient to demonstrate the effectiveness of a coupling agent and the specific nature of its enhancement with a given polymer matrix. Significant improvements have been made in the permanence properties (wet strength) of glass-reinforced plastics with over 200 per cent enhancement occurring in some cases. The dry flexural strength also increased by as much as 43 per cent in one case (Table 6.11), which illustrates the important influence of the glass-polymer interface on the fail ure strength of the composite material under certain loading conditions, in this case flexural loading. Tensile strength and compressive strength can also be increased by nearly 100 per cent for polyester laminates with silane finishes.[54]

4.3.5 Effect of temperature and strain rate

The fact that the properties of fiber-reinforced plastics are affected by temperature and the rate of stressing is a result of the time-dependent mechanical properties of the resin matrix. The glass-fiber reinforcement compared to the resin is not affected by the rate of stressing. Furthermore, the mechanical properties of the glass fibers do not change significantly until

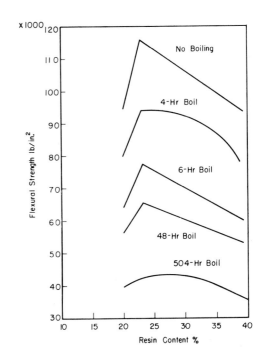

Figure 6.21   Effect of Resin Content on Flexural Strength of Epoxy Laminates (Ref. 50)

Table. 6.11.   Strength Retentions of Laminates with Silane Coupling Agents after Exposure to Moisture (Ref. 50)

Material	Flexural Strength (10^3 psi)		Percentage Improvement	
	Dry	Wet (8-hr Boil)	Dry	Wet (8-hr Boil)
Glass Cloth (181) *Reinforced Polyester Resins (Paraplex-P43)*				
Control	61	23	—	—
Y-4086[a]	71	58	16	152
O				
S $-CH_2CH_2Si(OCH_3)_3$				
Y-4087[a]	76	58	24	152
$CH_2-CHCH_2OCH_2CH_2CH_2Si(OCH_3)_3$				
O				
A-172	69	61	13	165
$CH_2=CHSi(OCH_2CH_2OCH_3)_3$				
A-174 CH_3 O	87	79	43	243
$CH_2=C$——C—$OCH_2CH_2CHSi(OCH_3)_3$				
Glass Cloth (181) *Reinforced Epoxy Resins*				
Control	78	29[b]	—	—
A-1100	92	67	18	130
$NH_2CH_2CH_2CH_2Si(OC_2H_5)_3$				
Y-4086[a]	81	51	4	76
Y-4087[a]	97	60	24	107
Y-2967[a]	87	55	12	90
$(HOCH_2CH_2)_2NCH_2(CH_2)_2Si(OC_2H_5)_3$				

[a] Commercial Identification Numbers.
[b] 72-hour Boil Used for Epoxy Resins.

temperatures over 600°F are reached so that changes in properties at lower temperatures can be attributed to changes in the resin matrix.

McAbee and Chmura[61,62] have measured the tensile properties of epoxy, polyester, and polyurethane laminates at high rates of stressing[61] and have also reported changes of interlaminar shear strength with rate of stressing.[62] The stress-strain characteristics for epoxy and polyester fabric laminates which are loaded to failure in times differing almost by a factor of 10^4 are shown in Figure 6.22. Not only do the laminate

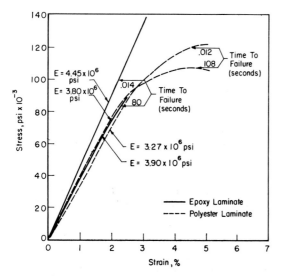

Figure 6.22 Effect of Rate on Tensile Properties of Epoxy and Polyester Resins Reinforced with 181 Glass Fabric ("S" Glass) (Ref. 61)

strengths increase at higher loading rates but the ultimate elongation for certain of the laminates increases. However, this effect is dependent upon the specific resin used in the laminate. The work to fracture measured as the area under the stress-strain curve would then also increase at higher loading rates. The interlaminar shear strength measured with a notched flat shear specimen increased by about 50 per cent when the failure time was reduced from approximately 100 seconds to 6×10^{-3} second. Again, this is a result of the change in the resin properties.

It has been found that viscoelastic materials such as the resins used for the matrices of reinforced plastics respond to low temperatures and high strain rates in an analogous fashion. In other words, the mechanical properties increase as either the strain rate is increased or the temperature decreased. Toth[63] has shown that the strength properties and moduli of reinforced plastics as well as the tensile elongation increase as the temperature is decreased to 20°K. This study at cryogenic temperatures was performed on S-glass-reinforced epoxy resins.

The strength properties of reinforced plastics at elevated temperatures have been studied by many investigators.[64–67] The strength decreases as the temperature increases at first because of the gradual loss in strength and stiffness of the matrix but as the temperature is increased further, the resin undergoes thermal degradation and the laminate properties degrade more quickly. Boller[64] has measured tensile, compressive and interlaminar shear strengths for several resin-

reinforcement combinations. The strengths are not only affected by the test temperature but the exposure time at this temperature, especially in the temperature range where thermal degradation of the resin is occurring. Some of the strength properties are shown in Figure 6.23 for various temperatures and exposure times. The retention of tensile strengths at the

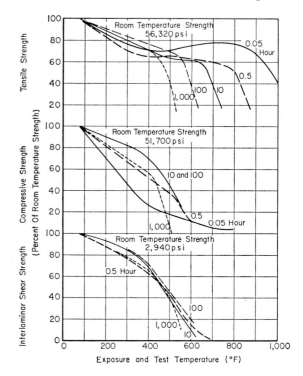

Figure 6.23 Strength Properties at Elevated Temperatures and Various Soak Periods for Epoxy (Epon 1031) Resin 181-Volan A Glass Fabric (Ref. 64)

various elevated temperatures is greater than for the other strength properties since the resin matrix does not have as great a role in determining the tensile strength as it does for either the compression or interlaminar shear strength. For example, the interlaminar shear strength is reduced by 40 per cent after $\frac{1}{2}$ hour at 400°F while the tensile strength can withstand a temperature of 700°F at this exposure time for the same strength retention. It appears that for this epoxy resin system at temperatures less than 400°F, the tensile strength is a minimum for $\frac{1}{2}$-hour soak or exposure and can be optimized when left exposed for 100 hours. At greater times the strength is again reduced so this appears to be a critical exposure time. The reason for this behavior is not well understood but must be from the competing effects of thermal degradation and perhaps the beneficial effects of residual stress relief or other physical property changes of the resin which would cause an increase in composite strength.

Recently, Strauss[66] and Mackay[67] have presented results of strength retentions at elevated temperatures for some of the newer high-temperature resin systems such as the polyimides and polybenzimidazoles. These resin systems expand the temperature capabilities of reinforced plastics. For example, the strength retention of polybenzimidazoles reinforced with 181 glass fabric (S glass) is 75 per cent for the interlaminar shear strength after exposure to 800°F for 30 minutes.[67]

4.4 Fatigue behavior of fiber-reinforced plastics

The permanence or long-term behavior of materials includes such considerations as their response to fatigue loadings and establishment of an endurance limit, or response to long-term constant loads and establishment of creep deflections or stress rupture life. The above measurements have all been made on fiber-reinforced plastics and the one type of behavior which appears more critical than the others and, indeed, may be a current limitation of the material is its response to cyclic loads or its fatigue life. Most of the fatigue studies on glass-fiber-reinforced plastics used axial tensile or compressive loading, although flexural fatigue results have been reported[68,69] and more recently biaxial compressive fatigue and interlaminar shear fatigue results have been reviewed.[70] Material parameters such as resin matrix type, resin content, reinforcement type, and reinforcement orientation have all been studied.[71] In addition, other influencing factors such as effects of notches, environments, mean stress levels and pre-cyclic stresses have been studied.

One effect that should be noted and that has significance in fatigue testing of fiber-reinforced plastics is the heat generation in the specimen during cyclic fatigue loading. The heat generation is a result of the high internal friction of the resin and its low thermal conductivity which results in a temperature increase in the specimen. Heywood[72] reports internal temperatures as high as 90°C before failure for cyclic loading rates of 2770 cycles per minute. Boller used cyclic rates of 900 cycles per minute in his fatigue studies[73–75] and reports that at a composite stress level of 14,500 psi a 10°F rise occurs during the test but that just prior to failure the rise is about 70°F. At higher stress levels the temperature rise would be even greater so that in a typical fatigue (S-N) curve for a glass-reinforced plastic the various stress levels investigated also represent different test temperatures. These changes in temperature influence the modulus, elongation, and strength of the resin as well as providing a healing mechanism for internally generated cracks.

Boller[71] has investigated the effect of matrix materials on the fatigue strengths of glass-reinforced plastic laminates. The S-N curves are shown in Figure 6.24, and all the resins were reinforced with style 181 E-glass fabric. These measurements were made before 1955 and since then improved glass coupling agents have been developed for some of the specific resins shown in the figure. These composites may have improved strength values compared to those shown in Figure 6.24 but the fact remains that epoxy resins are superior to all others with regard to fatigue endurance. Since the resin mechanical properties are not considerably different, the higher strength levels at 10^6 or 10^7 cycles must be due to the better adhesion and interface formed between the epoxy resin and glass reinforcement.

The effect of filament orientation for nonwoven reinforcements has also been investigated by Boller.[71,74] In Figure 6.25 the effect of orienting glass normal to the load direction is shown. Maximum single cyclic strengths occur when all the glass is parallel to the load, and as some of the glass plies are oriented normal to the load direction the strength decreases. However, for alternating stresses, at 10^7 cycles, the laminate with all the glass parallel to the load direction exhibits the

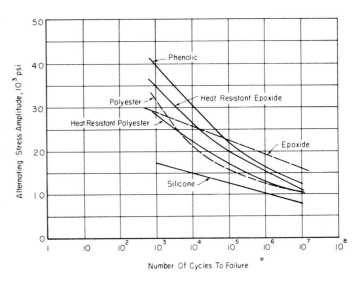

Figure 6.24 *S-N* Curves of Unnotched Specimens of Laminates Made of 181 Glass-fabric and Reinforced with Various Types of Resins. Tests were made parallel to warp, at 73°F and 50 per cent relative humidity, 900 cycles per minute and at zero mean stress (Ref. 71)

lowest fatigue strength. A similar effect is shown in Figure 6.26 for the case of alternate fiber plies oriented symmetrically at small angles to the load direction. The fatigue strengths are highest when the plies are oriented at 5° and 10° to the load axis and at 15° the fatigue strength becomes less than the 0°-oriented composite. The low-cycle portion of the 15° curve was determined by experiment while the dashed lines in Figures 6.25 and 6.26 shown for the other curves were

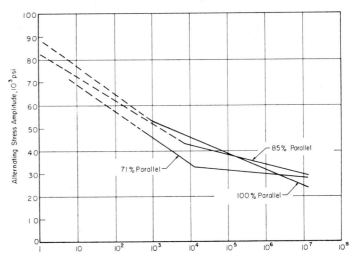

Figure 6.25 *S-N* Curves of Unnotched Specimens of Epoxy (Scotchply Type 1002 Resin) Reinforced with Unidirectional Glass Fibers Oriented with Various percentages parallel to the principal Axis. Tested at 73°F and 50 per cent relative humidity and zero mean stress (Ref. 71)

estimated. The resin contents for the composites shown in Figures 6.25 were approximately 34 per cent. A summary of these fatigue strengths is given in Table 6.12.

The effects of notches of various geometries have been investigated by Kimball.[76] Results for a circular $\frac{1}{8}$-in.-diameter

Table 6.12. Strength of Laminates with Various Percentages of Unwoven Glass Parallel to Load Axis and Strength of Laminates with Alternating Plies at Various Angles to Load Direction (Zero Mean Stress)

Resin Content (wt. %)	Percentage Glass Parallel to Load Direction	Fiber Orientation Angle	Tensile Strength (10^3 psi)	Compressive Strength (10^3 psi)	Fatigue Strength at 10^7 cycles (10^3 psi)	Fatigue Strength / Compressive Strength
34.6	100	0°	120.4	89.4	25.0	28%
34.5	85	0°, 90°	116.9	83.5	24.0	29%
33.1	71	0°, 90°	106.9	81.3	28.2	35%
32.4	—	± 5°	117.1	93.0	36.0	39%
35.1	—	±10°	96.5	85.4	26.4	31%
35.4	—	±15°	89.5	83.8	20.0	24%

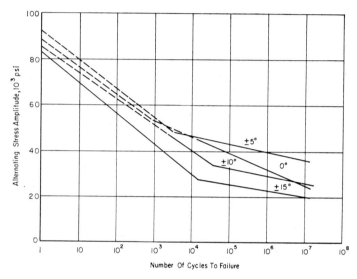

Figure 6.26 *S-N* Curves of Unnotched Specimens of Epoxy (Scotchply Type 1002 Resin) Reinforced with Unidirectional Glass Fibers That Are Oriented with Alternate Plies at 0°, ±5°, ±10°, and ±15° to the Principal Axis. Tested at 73°F and 50 per cent relative humidity and zero mean stress (Ref. 71)

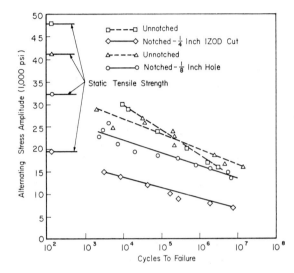

Figure 6.27 *S-N* Curves of Epoxy Laminates Reinforced with 181-Volan A Glass Fabric Showing Effect of Notches. Tests were made at 0° to warp, 73°F and 50 per cent relative humidity at zero mean stress and at 900 cycles/minute (Ref. 76)

hole and a $\frac{1}{4}$-in. Izod edge notch that has a 45° angle of cut and a 0.01-in.-radius tip are shown in Figure 6.27. The circular hole results in a uniform lowering of the stress along the entire fatigue curve. The sharp notch produces a more drastic reduction of the stress level, especially at a small number of cycles, but as the number of cycles increases the unnotched and notched failure stresses come closer together.

All of the *S-N* curves presented thus far appear to have negative slopes even at 10^7 cycles, which implies the absence of an endurance limit or stress limit below which cyclic failure will not occur. Although there is a knee in many of the curves (Figs. 6.25 and 6.26) resulting in a second curve with much smaller negative slope, a horizontal line still does not result. Boller[73] does report that glass-fabric laminates (181 glass fabric) stressed at 45° to the warp direction possess a true endurance limit which begins at approximately 40,000 cycles. However, these same laminates stressed in the warp direction possess no endurance limit through 10^7 cycles. For the laminates stressed at 45°, the resin matrix carries much of the load in shear between the glass fabric layers and the fatigue

curve may be more indicative of the resin behavior than for the laminates stressed in the fiber direction.

Fatigue *S-N* curves have also been established for non-woven S-glass laminates.[74] These curves closely parallel those established for E glass except that they are displaced vertically to higher strength levels. For example, unidirectional S glass laminates (38.1 per cent resin content by weight) have given fatigue strengths (10^7 cycles) of 39,000 psi which represents 38 per cent of the compressive strength of the material. Cross-ply composites (38.6 per cent resin content by weight) yield strengths of 27,000 psi at 10^7 cycles which represents 37 per cent of the compressive strength.

4.5 Creep and stress rupture behavior of fiber-reinforced plastics

The creep deformations experienced by glass-fiber-rein-forced plastics are quite dependent upon the type of loading (for example, tension, compression, shear, or flexure). While the resin matrix properties are time dependent, the properties of the glass at room temperature are essentially independent of time (although glass exhibits some creep it is negligible compared to the resin matrix creep) and glass does not exhibit readily measured creep strains at constant loads. However, the glass fibers may be geometrically arranged so that the reinforcement structure, such as a fabric, may be able to deform as the matrix deforms. Although not affected by creep, the strength of the glass is dependent upon time, and glass fibers are subject to static fatigue or a loss in strength with increasing time of loading.[77]

Creep strains for a polyester resin reinforced with a 181 fabric and a nonwoven mat are illustrated in Table 6.13.[22] The total tensile strains for a 181-fabric-reinforced polyester after 1000 hours are only approximately 10 per cent greater than the initial strain when the composite is loaded to over 50 per cent of its static strength in a direction parallel to the warp. This increase in strain with increasing time must be a result of the resin creep and the ability of the woven fabric to change its geometry by stretching of the twisted and crimped yarns forming the textile structure.

Flexural creep is greater than tensile creep because of the ability of the fabrics to deform or sag with increasing resin deformations. The same effect would also occur with compressive creep because the glass reinforcement offers essentially no resistance to deformation if standing by itself. Therefore, as the resin creeps the glass reinforcement can easily deform by collapse of the fibers. The results shown in Table 6.13 for the fabric loaded at 45° to the warp direction indicate large creep deformations for reasons similar to those discussed for flexural or compressive loadings.

Boller[78] investigated the long-term creep properties of several glass-reinforced plastics and applied a quantitative relationship previously used by Findley to describe the creep of plastics and laminates.[79] The equation can be expressed as follows:

$$\varepsilon_t = \varepsilon_0' \sinh \sigma/\sigma_\varepsilon + m't^n \sinh \sigma/\sigma_m,$$

where ε_t is total strain in inches per inch, σ is applied stress level in psi, t is time in hours and n, ε_0', σ_ε, m', and σ_m are experimentally determined constants. For an epoxy resin reinforced with a 181 glass fabric Boller found the constants

to have the following values:

$$\varepsilon_0' = 0.0057, \sigma_\varepsilon = 25,000, m' = 0.0005, \sigma_m = 50,000,$$

$$n = 0.160.$$

In addition to the apparent modulus of the glass-reinforced plastic changing with time, the strength decreases with increasing time of loading which is referred to as stress rupture or creep rupture. This is not surprising since both components of this composite, the glass fibers and the resin matrix, are subject to stress rupture. In addition, the interface between the resin and fiber reinforcement is subject to mechanical breakdown, especially in the presence of an environment such as moisture. Boller[78] has measured creep rupture characteristics of several types of epoxy and polyester laminates tested at 50 per cent humidity and in water. The tensile stress is plotted in Figure 6.28 as a function of the logarithm of time in hours for epoxy and polyester laminates. The data has shown that the curves could be represented by

$$\sigma_R = \sigma_0 - M \log_{10} t,$$

where σ_R is the tensile rupture stress, t is time in hours, and σ_0 and M are experimental constants. The values determined by Boller for these constants are presented in Table 6.14 for both dry and wet conditions of test.

Goldfein[80,81] has constructed master stress rupture curves for glass-reinforced plastics by applying the Larson and Miller parameter, $K = T(20 + \log_{10} t)$, where $T =$ absolute temperature (°K) and $t =$ time in hours, which previously gave good results for metals at high temperatures.[82] Master rupture curves were plotted as tensile or flexural stress on the ordinate and this parameter K as the abscissa. Application of this relation allowed the use of steady-load short-time tests at elevated temperatures to determine long-time data at room temperature. Good agreement has been obtained between calculated and observed tensile rupture strengths for times up to 20,000 hours.[81]

Results of tensile stress rupture experiments on laminates made from S glass 181 fabric with an epoxy resin have recently been presented by Mettes and Lockwood.[83] Stress rupture

Table 6.13. Tensile and Flexural Creep for Plastic Laminates Reinforced with Glass Fibers and Tested at 73°F and 50 Percent Relative Humidity (Ref. 22)

Material	Proportion of Applied Stress to Static Strength[a]	Ratio of Initial Strain to Strain at		
		10 hours	100 hours	1000 hours
Polyester Resin, 181 Glass Fabric tested Parallel to Warp	50 T	0.935	0.908	0.878
	60 T	0.938	0.914	0.883
	50 F	0.916	0.944	0.758
	60 F	0.910	0.852	0.790
Polyester Resin, 181 Fabric tested at 45° to Warp	65 T	0.459	0.340	—
	75 T	0.492	0.375	—
Polyester Resin, 1½ oz. Mat,	60 T	0.789	0.700	0.594

[a] $T =$ Tensile Stress
$F =$ Flexural Stress

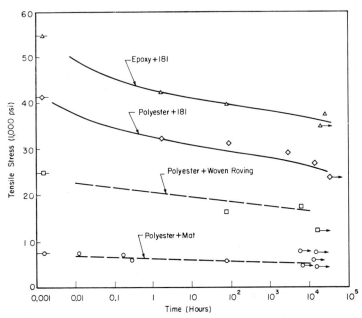

Figure 6.28 Tensile Stress Rupture of Glass-reinforced Plastic Laminates Tested at 73°F and 50 per cent Relative Humidity (Ref. 78)

Table 6.14. Summary of Constants Describing Tensile Stress-Rupture Curve for Various Plastic Laminates (Ref. 78)

Material		σ_0		M	
Resin	Glass	Dry	Wet	Dry	Wet
		(10^3 psi)	(10^3 psi)	(10^3 psi)	(10^3 psi)
Epoxy	181 Fabric	43.5	—	1.500	—
Polyester	181 Fabric	32.56	27.70	1.640	3.05
Polyester	Woven Roving	21.64	17.70	1.060	1.80
Polyester	Mat	6.40	—	0.275	—
Epoxy	Unidirectional non-woven fibers	90.70	75.20	3.250	4.92

curves were included for laminates tested both dry and wet and for laminates tested both at high and low temperatures. Cornish *et al.*[84] have investigated the biaxial compressive stress rupture properties of both E and S glass reinforcements by external pressurization of hollow filament-wound cylinders. The interest in biaxial compressive stress rupture experiments is a result of the potential application of this material as a hull or shell for deep ocean submergence structures.

5 CONCLUSIONS

An understanding of the mechanical behavior of fiber-reinforced plastics is progressing at a rapid rate spurred by the desired of investigators to combine micromechanics studies and macromechanics studies and to work more closely with the designers who must know how to take advantage of these materials. Materials scientists are also developing new fibrous materials such as boron and graphite and new high-strength and high-temperature resins to further improve the properties of fibrous composites. In fact, a comparison of the strength-

and stiffness-to-weight ratios for two advanced composites and conventional engineering materials in Table 6.15 indicates the gains already made even in this early stage of composite-materials development.

The future for fibrous-reinforced plastics looks very bright considering the potential properties that could be achieved by future continuous fibrous reinforcements, some of which are illustrated in Figure 6.29 where the specific strength (strength-to-weight ratio) is plotted against specific modulus. Only a few of these are in commercial use; others are limited to critical applications where cost is secondary, or are in the laboratory stage. The picture can be expected to change as production techniques and properties are more fully explored.

The properties of the fibers only have been plotted in Figure 6.29 as we are assuming that composites made of these fibers will eventually be able to utilize a constant proportion of the fiber strength and modulus. It seems certain that in the near future other new fibers could be added to this figure and the points now representing most of the fibers could probably be shifted to higher strengths.

Table 6.15. Comparison of Strength-to-Weight and Stiffness-to-Weight Ratios for Reinforced Plastics and Homogeneous Engineering Materials

Material	Percentage Volume of Matrix	Tensile Strength to Density Ratio (10^6 psi)	Mod. of Elasticity to Density Ratio (10^6 psi)
S Glass fibers unidirectionally oriented in epoxy resin	35	3.53	103
Boron fibers unidirectionally oriented in epoxy resin	35	2.70[a]	486
Beryllium	—	1.30	600
Titanium	—	1.50	100
Aluminum	—	0.90	0.95
Steel	—	2.00	100

[a] This is based on average fiber strength of 400,000 psi.

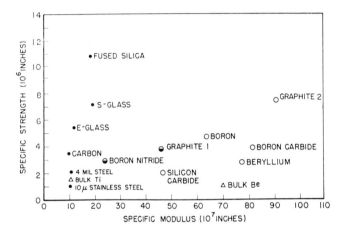

Figure 6.29   Specific Strength vs. Specific Modulus for Continuous Fibrous Reinforcement

REFERENCES

1. J. R. TINKLEPAUGH *et al.*, "Metal Fiber Reinforced Ceramics," WADC Technical Report 58–452, Part III, November 1960.

2. A. KELLY and G. J. DAVIES, "The Principles of the Fibre Reinforcement of Metals," *Met. Rev.* **10**, 37 (1965).

3. W. H. SUTTON and J. CHORNE, "Potential of Oxide-Fiber Reinforced Metals," *Fiber Composite Materials*, American Society for Metals, Metals Park, Ohio, 1965.

4. N. F. DOW, "Study of Stresses Near a Discontinuity in a Filament-Reinforced Composite Metal," General Electric Company Rept. TIS R63SD61, 1963.

5. N. J. PARRATT, "Reinforcing Effects of Silicon Nitride Whiskers in Silver and Resin Matrices," *Powder Met.* **7**, No. 14, 152–167 (1964).

6. H. L. COX, "The Elasticity and Strength of Paper and Other Fibrous Materials," *Brit. J. Appl. Phys.* **2**, No. 72, 72-79 (1952).

7. W. H. SUTTON, B. W. ROSEN, and D. G. FLOM, "Whisker-Reinforced Plastics for Space Applications," *Soc. Plastics Engrs.* **20**, No. 11, 1203–1209 (1964).

8. J. V. MILEWSKI, "How to Use Whiskers in Reinforced Plastics," *Reinforced Plastics '65*, Regional Technical Conference, Society of Plastics Engineers, Inc., July 14, 1965 Society of Plastics Engineers, Stamford, Conn., pp. 149–166.

9. T. P. MURPHY, "Reinforced and Filled Thermoplastics," *American Chemical Society Division of Organic Coatings and Plastics Chemistry, Papers Presented at the Atlantic City Meeting*, Vol. 25, No. 2, American Chemical Society, New York, Sept. 1965, pp. 76–90.

10. J. Y. LOMAX and J. T. O'ROURKE, "Engineering Plastics Updated: Greater Functional Design with Filled Thermoplastics," *21st Annual Technical Conference*, Society of Plastics Engineers, Inc., Stamford, Conn., 1965, X-5.

11. "Fiberglass Reinforced Thermoplastics," Technical Bulletin, Form No. 2393–64, Fiberfil, Inc., Evansville, Ind.

12. T. P. MURPHY, "Fiberglass Reinforced Thermoplastics," *Reinforced Plastics '65*, pp. 199–211.

13. T. P. MURPHY and H. E. HAUCK, "How to Select Reinforced Thermoplastics," *Mater. Design Eng.* **61**, No. 3, 99–103 (1965).

14. R. H. SONNEBORN, *Fiberglass Reinforced Plastics*, Reinhold Publishing Corporation, New York, 1954.

15. P. MORGAN, *Glass Reinforced Plastics*, Iliffe Books Ltd., London, 1961.

16. D. V. ROSATO and C. S. GROVE, *Filament Winding—Its Development, Manufacture, Applications, and Design*, John Wiley & Sons, Inc., New York, 1964.

17. S. S. OLEEKSY and J. GILBERT MOHR, *Handbook of Reinforced Plastics*, Reinhold Publishing Corporation, New York, 1964.

18. J. W. DAVIS and N. R. ZURKOWSKI, "Put the Strength and Stiffness Where You Need It," Technical Report, Reinforced Plastics Division, Minnesota Mining and Manufacturing Company.

19. J. W. DAVIS, "Pre-Impregnated Epoxy Nonwoven Filament Materials," *Reinforced Plastics '65*, p. 27.

20. J. W. DAVIS and G. R. MODIG, "Rupture and Static Properties of Unidirectional Glass Filament Reinforced Pre-Impregnated Construction," *Proc. 19th Annual Technical Conference, Vol. IX*, Society of Plastic Engineers, Stamford, Conn., 1963, Section IV-3.

21. F. J. McGARRY and M. B. DESAI, "Failure Mechanisms in Fiberglass Reinforced Plastics," *Proc. 14th SPI Reinforced Plastics Division*, Society of the Plastics Industry, New York, 1959, Section 16-E.

22. *Plastics for Flight Vehicles, Part I, Reinforced Plastics*, MIL-HDBK-17, U.S. Government Printing Office, Washington, D.C., Nov. 5, 1959.

23. W. KROLIKOWSKI, "Stress-Strain Characteristics of Glass Fiber Reinforced Polyester," *Soc. Plastics Engrs.* **20**, No. 9, 1031 (Sept. 1964).

24. S. W. TSAI, "Structural Behavior of Composite Materials," NASA CR-71, National Aeronautics and Space Administration, Washington, D.C., July 1964.

25. S. W. TSAI, "Strength Characteristics of Composite Materials," NASA-CR-224, National Aeronautics and Space Administration, Washington, D.C., Apr. 1965.

26. B. W. ABBOTT nad L. J. BROUTMAN, "Determination of the Modulus of Elasticity of Filament Reinforced Plastics Using Stress Wave Techniques," *Proc. 21st SPI Reinforced Plastics Division*, Society of the Plastics Industry, New York, 1966, Section 5-D.

27. K. E. KIMBALL, "Interlaminar Properties of Five Plastic Laminates," U.S. Dept. Agr. Forest Serv., Forest Prod. Lab. Rept. 1890, Forest Products Laboratory, Madison, Wis., Dec. 1962.

28. N. FRIED, "Survey of Methods of Test for Paralleled Filament Reinforced Plastics," *Symposium on Standards for Filament-Wound Reinforced Plastics*, ASTM Spec. Tech. Publ. 327, American Society for Testing and Materials, Philadelphia, Pa., 1963, pp. 13–40.

29. F. WERNER and B. G. HEEBINK, "Interlaminar Shear Strength of Glass Fiber Reinforced Plastic Laminates," U.S. Dept. Agr. Forest Serv., Forest Prod. Lab. Rept. 1848, Forest Products Laboratory, Madison, Wis., Sept. 1955; No. 1848A, Nov. 1956.

30. N. FRIED, "The Response of Orthogonal Filament-Wound Materials to Compressive Stress," *Proc. 20th SPI Reinforced Plastics Division*, Society of the Plastics Industry, New York, Feb. 1965, Section 1-C.

31. S. P. PROSEN, C. E. MUELLER, and F. R. BARNETT, "Interlaminar Shear Properties of Filament Wound Composite Materials for Deep Submergence," *Proc. 19th SPI Reinforced Plastics Division*, Society of the Plastics Industry, New York, Feb. 1964, Section 9-D.

32. C. E. MUELLER, S. P. PROSEN, and F. R. BARNETT, "Means for Increasing Interlaminar Shear Resistance in Filament Wound Structures," U.S. Naval Ordnance Lab. Rept. 66–85, June 8, 1966.

33. S. BRELANT, "The Relationship of Voids, Process and Material Parameters, and Performance of Filament-Wound Pressure Vessels," presented at SPE Retec, *Reinforced Plastics '65*.

34. J. T. PAUL, JR. and J. B. THOMPSON, "The Importance of Voids in the Filament-Wound Structure," *Proc. 20th SPI Reinforced Plastics Division*, Section 12-C.

35. W. HAND, "Quality Control of Filament-Wound Materials for Deep Submergence Vessels," *Proc. of 20th SPI Reinforced Plastics Division*, Section 1-E.

36. H. T. CORTEN, "Micromechanics and Fracture of Fiber Reinforced Plastics," *Composite Materials*, L. J. Broutman and R. H. Krock, Editors, Addison-Wesley Publishing Company, Inc. (in preparation).

37. B. W. ROSEN, "Mechanics of Composite Strengthening," *Fiber Composite Materials*, American Society for Metals, pp. 37–75.

38. G. BROWN, "Physical Properties of Prepreg Laminates," *Proc. 13th SPI Reinforced Plastics Division*, Society of the Plastics Industry, New York, Feb. 1958, Section 2-A.

39. K. E. KIMBALL, "Relationship Between Thickness and Mechanical Properties of Several Glass-Fabric-Base Plastic Laminates," U.S. Dept. Agr. Forest Serv., Forest Prod. Lab. Rept. 1885, Forest Products Laboratory, Madison, Wis., May 1962.

40. R. L. YOUNGS, "Effect of Thickness on the Mechanical Properties of Glass-Fabric-Base Plastic Laminate," U.S. Dept. Agr. Forest Serv., Forest Prod. Lab. Rept. 1873, Forest Products Laboratory, Madison, Wis., May 1960.

41. K. H. BOLLER, "Effect of Thickness on Strength of Glass-Fabric-Base Plastic Laminates," U.S. Dept. Agr. Forest Serv., Forest Prod. Lab. Rept. 1831, Forest Products Laboratory, Madison, Wis., May 1954.

42. J. C. EKVALL, "Geometric Factors Affecting the Strength of Monofilament Composites," Presented at AIAA/ASME Seventh Structures and Materials Conference, Cocoa Beach, Fla., Apr. 1966.

43. E. Z. STOWELL and T. S. LIU," On the Mechanical Behavior of Fiber Reinforced Crystalline Materials," *J. Mech. Phys. Solids* **9**, 242 (1961.)

44. S. W. TSAI, Private Communication, July 1966.

45. G. A. COOPER, "Orientation Effects in Fibre-Reinforced Composites," *J. Mech. Phys. Solids* **14**, 103 (1966).

46. P. W. JACKSON and D. CRATCHLEY, "The Effect of Fibre Orientation on the Tensile Strength of Fibre-Reinforced Composites, *J. Mech. Phys. Solids* **14**, 49 (1966).

47. C. B. NORRIS, "Strength of Orthotropic Materials Subjected to Combined Stresses," U.S. Dept. Agr. Forest Serv., Forest Prod. Lab. Rept. 1816, Forest Products Laboratory, Madison, Wis., May 1962.

48. S. BRELANT, I. PETKER, and K. W. SMITH, "Combined Effect of Pre-Stress and Humidity Cycling upon Filament-Wound Internal Pressure Vessels," *Soc. Plastics Engrs. Journal*, 1964, p. 1019.

49. R. L. YOUNGS, "Effects of Tensile Preloading and Water Immersion on Flexural Properties of a Polyester Laminate," U.S. Dept. Agr. Forest Prod. Lab. Rept. 1856, Forest Products Laboratory, Madison, Wis., June 1956.

50. A. W. RAWE, "Environmental Behavior of Glass-Fibre Reinforced Plastics," *Plastics Inst.* (London), *Trans. J.* Feb. 1962, (pp. 27–37).

51. S. STERMAN and J. G. MARSDEN, "The Newer Silane Coupling Agents," Regional Technical Conference of the Society of Plastics Engineers, Inc., Cleveland, Oct. 1, 1963, Society of Plastics Engineers, Stamford, Conn., p. 67.

52. H. A. CLARK and E. P. PLUEDDEMANN, "Bonding of Silane Coupling Agents in Glass-Reinforced Plastics," *Mod. Plastics* **40**, No. 10, 133 (June 1963).

53. N. M. TRIVISIONNO and L. H. LEE, "The Effect of Glass Finishing Agents on the Strength of Polyester-Fiberglass Laminates," *Proc. 12th SPI Reinforced Plastics Division*, Society of the Plastics Industry, New York, Feb. 1957, Section 16-B.

54. K. ITO, "Evaluation of Surface Treatment of Glass Fiber in Fabric-Reinforced Plastics," *J. Poly. Sci.* **45**, 155 (1960).

55. B. M. VANDERBILT, "Effectiveness of Coupling Agents in Glass-Reinforced Plastics," *Mod. Plastics* **37**, No. 1, 125 (Sept. 1959).

56. S. STERMAN and H. B. BRADLEY, "A New Interpretation of the Glass Coupling Agent Surface Through Use of Electron Microscopy," *Proc. 16th SPI Reinforced Plastics Division*, Society of the Plastics Industry, New York, Feb. 1961, Section 8-D.

57. E. P. PLUEDDEMANN *et al.*, "Evaluation of New Silane Coupling Agents for Glass Fiber Reinforced Plastics," *Proc. 17th SPI Reinforced Plastics Division*, Society of the Plastics Industry, New York, Feb. 1962, Section 14-A.

58. L. J. BROUTMAN, "Glass-Resin Joint Strengths and Their Effect on Failure Mechanisms in Reinforced Plastics," *Polymer Engineering and Science* **6**, No. 3, 263 (July 1966), presented at SPE Retec *Reinforced Plastics '65*, Seattle, Washington, July 14, 1965.

59. J. A. LAIRD and F. W. NELSON, "Glass Surface Chemistry Relating to the Glass-Finish-Resin Interface," *Proc. 19th SPI Reinforced Plastics Division*, Section 11-C.

60. N. M. TRIVISSONNO, L. H. LEE, and S. M. SKINNER, "Adhesion of Polyester Resin to Treated Glass Surfaces," *Ind. Eng. Chem.* **50**, 912 (1958).

61. E. McABEE and M. CHMURA, "Static and High Rate Tensile Behavior of Various Resin-Reinforcement Combinations," *High Speed Testing*, Vol. V, Interscience Publishers, New York, 1965, pp. 85-98.

62. M. CHMURA and E. McABEE, "Effect of Rate of Loading on Tensile Properties of Glass-Reinforced Polyester," Tech. Rept. FRL-TR-39, Picatinny Arsenal, Aug. 1961.

63. L. W. TOTH, "Properties Testing of Reinforced Plastic Laminates Through the 20°K Range," *Proc. 20th Annual SPI Reinforced Plastics Division*, Section 7-L.

64. K. H. BOLLER, "Effect of Elevated Temperatures on Strength Properties of Reinforced Plastic Laminates," Tech. Doc. Rept. AS-TDR-62-629.

65. H. L. JONES *et al.*, "Performance of Epoxy Filament-Wound Units at Elevated Temperatures and the Relationship to Resin Elevated Temperature Properties," *Proc. 19th Annual SPI Reinforced Plastics Division*, Section 14-A.

66. E. L. STRAUSS, "High Temperature Structural Properties of Polybenzimidazole, Polyimide and Phenolic Laminates," *Proc. 22nd Annual Technical Conference*, Society of Plastics Engineers, Stamford, Conn., March 1966.

67. H. A. MACKAY, "Evaluation of Polybenzimidazole Glass Fabric Laminates," *Mod. Plastics* **43**, No. 5, 149 (Jan. 1966).

68. B. B. PUSEY, "Flexural Fatigue Strengths of Reinforced Thermosetting Laminates," *Proc. 12th SPI Reinforced Plastics Division*, Feb., 1957, Section 5-C.

69. H. R. NARA, "Some Fatigue Characteristics of Glass Reinforced Plastics," *Proc. 12th SPI Reinforced Plastics Division*, Section 5-D.

70. J. F. FREUND and M. Silvergleit, "Fatigue Characteristics of Glass Filament Reinforced Plastic Material," *Proc. 21st Reinforced Plastics Division*, Section 17-B.

71. K. H. BOLLER, "Resumé of Fatigue Characteristics of Reinforced Plastic Laminates Subjected to Axial Loading," Tech. Doc. Rept. ASK-TDR-63-768, July, 1963.

72. R. B. HEYWOOD, "Present and Potential Fatigue and Creep Strengths of Reinforced Plastics," Tech. Note Chem. 1337, Royal Aircraft Establishment, Ministry of Supply, London, Oct. 1958.

73. K. H. BOLLER, "Fatigue Tests of Glass-Fabric-Base Laminates Subjected to Axial Loading," U.S. Dept. Agr. Forest Serv., Forest Prod. Lab. Rept 1823, Forest Products Laboratory, Madison, Wis., Aug. 1958.

74. K. H. BOLLER, "Repeated Normal Stresses. A Fatigue Evaluation of Filament Reinforced Plastic Laminates," SPE Regional Technical Conference, *Reinforced Plastics '65*, July 1965.

75. K. H. BOLLER, "Effect of Pre-Cyclic Stresses on Fatigue Life of RP Laminates," *Mod. Plastics* **42**, No. 8, 162 (Apr. 1965).

76. K. E. KIMBALL, "Supplement to Fatigue Tests of Glass-Fabric-Base Laminates Subjected to Axial Loading, Effect of Notches," U.S. Dept. Agr. Forest Serv., Forest Prod. Lab. Rept. 1823-C, Forest Products Laboratory, Madison, Wis., October 1958.

77. W. H. OTTO, "Properties of Glass Fibers at Elevated Temperatures," Filament Winding Conference, Society of Aerospace Material and Process Engineers, March 1961.

78. K. H. BOLLER, "Effect of Long Term Loading on Glass-Reinforced Plastic Laminates," *Proc. 14th Annual SPI Reinforced Plastics Division*, Section 6-C.

79. W. N. FINDLEY *et al.*, "The Effect of the Creep of two Laminated Plastics as Interpreted by the Hyperbolic-Sine Law and Activation Energy Theory," *Proc. Am. Soc. Testing Mater.* **48**, 1217 (1948).

80. S. GOLDFEIN, "Determination of Long Time Rupture and Impact Stresses in Glass Reinforced Plastics from Short Time Static Tests at Different Temperatures," *Proc. 12th Annual SPI Reinforced Plastics Division*, Section 1-C.

81. S. GOLDFEIN, "Long Term Rupture Strength in Glass Reinforced Plastics," *Proc. 13th Annual SPI Reinforced Plastics Division*, Section 5-A.

82. F. R. LARSON and J. MILLER, "A Time-Temperature Relationship for Rupture and Creep Stresses," *Trans. Am. Soc. Mech. Engrs.* **74**, 765–775 (July 1952).

83. D. G. METTES and P. A. LOCKWOOD, "The Mechanical Properties of Laminates with High Performance Glass Fiber Fabric," *Proc. 21st Annual SPI Reinforced Plastics Division*, Section 4-G.

84. R. H. CORNISH, H. R. NELSON, and J. W. DALLY, "Compressive Fatigue and Stress Rupture Performance of Fiber Reinforced Plastics," *Proc. 19th Annual SPI Reinforced Plastics Division*, Section 9-E.

Chapter 7

Composite-Glass Structures

George B. Watkins

Glass composites or laminates consist of two or more layers of glass bonded to one or more layers of plastic material to produce a composite structure.

The laminated construction is employed to improve the safety characteristics of glass as, for example, in automobile windshields; to produce resistance to penetration by missiles, as in bullet-resisting glass; or to achieve special optical properties, as in light filters.

Another type of laminate consists of plates or sheets of glass enclosing a layer of air for insulating purposes. In this construction, the glass plates are bonded to a separator or fused together at the margins, thus forming an air cell of predetermined thickness.

1 GLASS COMPONENTS

1.1 Composition, classification, and properties

Although it is recognized that the composition of glass can vary widely, depending on the particular application for which it is designed, the types of glass most commonly used for laminating purposes are referred to generically as "flat glass," and have a soda-lime-silica base composition. Therefore, only soda-lime-silica glasses will be considered in this discussion unless specific mention is made to the contrary.

Flat glass used for composite structures or laminates may be broadly classified as transparent or translucent.

1. TRANSPARENT GLASSES. Transparent glasses, used extensively for glazing purposes, include
   (a) *Sheet or Window Glass.* Sheet or window glass is characterized by its fire-finish surface, formed as the sheet is drawn from the molten pool of glass.
   (b) *Plate Glass.* Plate glass is characterized by a machined surface, produced by mechanically grinding and polishing the surfaces of the plate-glass blanks after forming. The advantages of plate glass over sheet or window glass that justify the grinding and polishing operations are not necessarily the improved quality of finish, but rather flatness and parallelism of the surfaces, which serve to minimize distortion of objects viewed through the glass.
   (c) *Float Glass.* Produced by continuously flowing glass in a molten or semisolid state on a bath of molten metal. The glass is cooled until rigid on the molten metal. The bottom surface of the glass may be treated with a polishing agent.

GEORGE B. WATKINS was Consultant, Libbey-Owens-Ford Glass Company, Toledo, Ohio. He died in 1966.

This chapter has been revised by JOSEPH D. RYAN, Consultant, Libbey-Owens-Ford Glass Company, Toledo, Ohio.

   (d) *Colored Glasses.* Colored, transparent glasses, produced by adding small percentages of certain materials to glass batches, find application for light filtering and decorative purposes.
2. TRANSLUCENT GLASSES. Translucent glasses, widely used for glazing purposes, include
   (a) *Figured Glasses.* The translucent characteristics of figured glass are due to light scattering, caused by the roughened or figured pattern placed on the glass surface during the process of manufacture.
   (b) *Opal Glass.* The translucent characteristics of opal glass are due to light scattering; however, the light scattering is caused by opaque particles dispersed in the glass matrix, rather than by surface effects as in the case of figured glasses.

1.2 Physical and chemical characteristics

Following are some of the more important physical and chemical properties of the flat glasses having a soda-lime-silica base composition:

Specific gravity (70°F)	2.46 to 2.53
Weights per sq. ft. per in. thickness	12.8 to 13.2 lb
Flexural strength (modulus of rupture)	6000 psi
Modulus of elasticity (Young's modulus)	10,500,000 psi
Hardness:	
(a) Mohs' scale (talc = 1; diamond = 10)	5.5 to 6.5
(b) Knoop (Tukon)	400 to 600
Softening point	1330°F to 1360°F
Thermal conductivity (k) (0°F to 120°F)	6 Btu in./sq ft hr °F
Specific heat (32°F to 212°F)	0.20
Coefficient of linear expansion	4.5 to 5.5 $\times$ 10^{-6} per °F

At normal temperatures all soda-lime-silica base glasses are resistant to all acids except hydrofluoric acid. They are, however, subject to attack by alkaline materials.

Transparent ordinary plate and sheet glasses have a total visible light transmittance of 88 to 92 per cent in thicknesses up to $\frac{1}{4}$ in.

1.3 Heat-strengthened glass

The strength characteristics of soda-lime-silica glass can be increased materially by subjecting the glass to a special heat-treating operation, which places the outside layers in a state of high compression.

Such a heat-treating operation serves to increase the strength of glass having a thickness of $\frac{1}{8}$ in. and greater approximately three to five times that of ordinary annealed glass of the same thickness.

The results of strength tests conducted on regular, on annealed, and on heat-strengthened plate glass, 12 in. × 12 in., 18 in. × 18 in., and 24 in. × 24 in. (free opening) and having a nominal thickness of $\frac{1}{4}$ in., are recorded in Table 7-1. In these tests the samples were glazed by clamping at the margin with a gasket; the loading was carried out at the rate of 5 psi per minute. The breaking pressures recorded are the average of at least six specimens.

Table 7.1. Strength Tests of Annealed versus Heat-Strengthened $\frac{1}{4}$ in. Plate Glass at a Test Temperature of 75°F

| Glass | Average Breaking Pressure, psi Glass Size (Free Opening) | | |
	12 in. × 12 in.	18 in. × 18 in.	24 in. × 24 in.
$\frac{1}{4}$-in. Annealed Plate	14	7	4.5
$\frac{1}{4}$-in. Heat-Strengthened Plate	60	30	15

1.4 Variation of strength of glass with temperature

The National Bureau of Standards[1,2] has conducted tests on the effect of time and temperature on the strength of annealed and heat-strengthened glass laths. With the exception of one group of tests, as noted below, the laths were subjected to a simulated service condition by lightly sandblasting the surface. The results are recorded in Table 7.2 and Figure 7.1.

From these results it may be seen that the strength of annealed glass is only slightly affected by temperature, whereas heat-strengthened glass is appreciably weakened if heated to

Table 7.2. Average Modulus of Rupture Values for Sandblasted Plate Glass Specimens (Breaks at Surface)

| Testing Temp. (°F) | Exposure Time (Hr) | Modulus of Rupture in psi | |
		Annealed	Heat-Strengthened
75	1	6900 psi	21,990 psi
75[a]	1	(11,320)	(29,600)
300	1	6620	24,010
400	1	6330	20,820
400	500	—	—
550	1	5870	20,840
550	500	—	—
700	1	6550	21,180
700	500	7260	13,310
870	1	6990	19,370
870	500	7980	11,580
963	1	5900	15,200

[a] These specimens were not sandblasted but had the original ground and polished surface.

temperatures over 800°F. One also notes from the two sets of data taken at 75°F that the light sandblasting decreased the original strength of the glass appreciably.

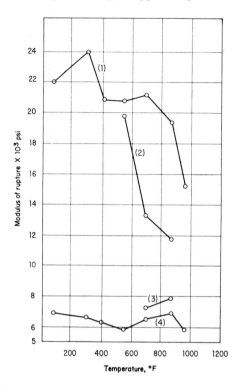

Figure 7.1 Modulus of Rupture of Sandblasted Plate Glass Specimens at Different Temperatures. (1) Tempered specimens 1 hr at temperature. (2) Tempered specimens 500 hr at temperature. (3) Annealed specimens 500 hr at temperature. (4) Annealed specimen 1 hr at temperature

1.5 Chemically tempered glass

Recently a chemical process has been developed for producing compressive forces in the surface layer of glass. Special glass formulations have been developed which form, under proper heating conditions, a low-expansion layer on the surface of the glass. Upon cooling the glass, this outer layer is placed in a high state of compression.

The more important thermal and mechanical properties of a typical chemically tempered glass are given[3] as follows:

Specific gravity (75°F)	2.44
Flexural strength (modulus of rupture) for abraded specimens	45,000 psi
Modulus of elasticity (75°F)	11,600,000 psi
Annealing point	1146°F
Strain point	1058°F
Thermal conductivity 75°F	8 Btu in./sq ft hr °F
Specific heat (75°F)	0.20
Coefficient of linear expansion (32°F–575°F)	3.9×10^{-6} per °F

1.6 Electrically conducting glass

For many applications laminated glass is manufactured with one or more of the glass layers covered with a transparent layer that conducts electricity. This layer is usually made up of tin oxide which has been applied to the glass at an elevated temperature. In aircraft applications this type of construction is used to prevent ice formation on the window or to prevent condensation of moisture on the inner ply. For these and other heating applications, power densities from 15 to 600 watts per square foot may be used. For power densities above 400 watts per square foot, tempered glass or low expansion glass is used as the substrate. The Illuminant C light transmittance of moderate thickness electrically conducting glass is usually 65 to 85 per cent. The commercial minimum resistivity of this type of film is about 50 ohms per square (resistance of a square area with electrodes on two parallel edges). Resistivities up to the megohm range are available. For control purposes, especially in aircraft use, a resistance-type temperature sensor is usually positioned inside the laminate. Electrode materials are frequently made by using finely divided silver in a ceramic frit or in an organic resin binder. At times electrically conducting glass is used to prevent an electrostatic charge from building up on the glass surface. Other uses are for radio-frequency shielding and as transparent electrodes for electroluminescent lamps.

A second type of electrically conducting glass utilizes a transparent metallic film as the conducting material. Gold is commonly used because of its high light transmittance at usable resistivity. The film, with adhesor and protective layers, is usually applied by either vacuum evaporation or sputtering, permitting its application to optical surfaces without risk of distorting the glass. Resistivity of gold film is usually 6 to 30 ohms per square, and the light transmittance varies somewhat with the resistivity. The uses of the gold-filmed glass are similar to those of the oxide filmed glass. In addition to the electrical uses mentioned, both types, but particularly the gold film, have appreciable infrared reflectivity.

1.7 Interlayer components

Interlayer materials may vary from very thin layers to sheets of considerable thickness. They are capable of bonding glass to glass or glass to dissimilar materials to form composite structures. By far the most important materials used as interlayers for laminated safety glass are the transparent organic plastics.

During the past four decades, various types of transparent plastics have been used for making glass-plastic laminates. The earlier interlayer materials included pyroxylin plastic, cellulose acetate plastic, and plasticized acrylic resin.

At the present time, however, these plastics have been almost entirely supplanted for use in automotive safety glass by an interlayer of plasticized vinyl resin. More specifically, the vinyl resin employed is a poly-vinyl butyral resin plasticized with the proper amount and kind of plasticizer.

The three commonly used plasticizers are 3-GH (triethylene glycol dihexoate), dibutoxyethyl adipate, and dibutyl sebacate, in the approximate amounts of 20 to 30 per cent. Different percentages of plasticizer may be employed depending upon the application to be made of the finished glass composite. For example, in glass laminates made with vinyl butyral plastic for use as aircraft windshields, the interlayer plasticizer content can be as low as 17 per cent, whereas in automotive windshields the amount generally employed is approximately 30 per cent.

For aircraft glazings that will be exposed to cabin pressurization loads and to the high temperatures resulting from aerodynamic heating in supersonic flight, new interlayer materials based on transparent silicone elastomers have been developed in recent years. Vinyl butyral interlayers are unsatisfactory for this application because of their thermoplastic character. The new transparent silicone materials make possible the manufacture of aircraft windshields having long-term service life at temperatures as high as 265°F.

2 TRANSPARENT GLASS-PLASTIC LAMINATES

2.1 History

The art of combining laminae of glass and other materials to form composite structures was recognized and practiced as early as 1885, when Arthur Thomas Fullicks[4] of England obtained different coloring effects in one composite pane of glass by carefully arranging pieces of differently colored glass in pattern form and cementing this pattern with suitable adhesives between two plates or sheets of clear glass.

As far as public records reveal, the idea of reinforcing glass for safety purposes by laminating with plastics was first conceived by an Englishman, John Crewe Wood,[5] who obtained British and United States patents in 1905 and 1906, respectively, describing a method for the manufacture of safety glass by cementing with Canada balsam a sheet of transparent celluloid plastic between two panes of glass. Wood's venture was without success, presumably because of the high cost of materials, poor quality of product, and small demand.

In 1910 a Frenchman named Benedictus[6] obtained French and British patents for the manufacture of laminated safety glass, employing the same general principle as Wood, except that he proposed gelatin and other materials, instead of Canada balsam, for adhering the glass-plastic layers. Benedictus named his product "Triplex," and the French firm that manufactured it was called La Société du Verre Triplex. The latter part of 1912, the Triplex Safety Glass Company, Ltd., purchased patent rights from the French company and commenced the manufacture of laminated safety glass at Willesden, England, in the middle of 1913.

It was not, however, until 1914 at the time of World War I, that any additional real progress was made. The demand for laminated glass for goggle lenses, gas masks, and windshields for motor vehicles and aircraft served to establish its manufacture as an industry during World War I.

Although the merits of laminated safety glass had been established, for some years following the war little progress was made to improve the quality and stability of the celluloid plastic interlayer (the only plastic commercially available for the purpose). Thus, the industry remained at a standstill for a decade.

The trend in the motorcar industry from open- to closed-model cars in 1925 presaged increased use of glass for automobile glazings. The glass and plastic manufacturers, realizing the worthwhile contribution in the way of increased safety that a well-made safety glass would afford the motoring public, made substantial investments in research and manufacturing facilities which in a few years resulted in a much improved safety-glass product from the standpoints of clarity and stability.

Milestones in the safety-glass industry resulting from organized research were

1. The development in 1931 of the autoclave to replace platen presses for uniting the glass-plastic layers.
2. The development in 1931 of roll equipment for applying adhesives continuously.
3. The development in 1932 of the more stable cellulose acetate plastic and adhesives for bonding the glass-plastic layers.
4. The development in 1932 of extruding equipment for producing plastic sheeting in a continuous manner.
5. The development in 1937 of a polyvinyl acetal resin plastic that can be bonded directly to glass by the application of heat and pressure, thereby making available commercially a superior safety glass from the standpoint of resistance to impact, and weathering, as compared to plastic laminations previously produced. This plastic is widely used at the present time for safety-glass manufacture.
6. The introduction in 1950 of shaded glare reducing laminated glass for automotive windshields and other glazings.

2.2 Classification of transparent glass-plastic laminates

Transparent glass laminates of commercial importance include

1. Laminated safety glass for glazing motor vehicles, railway passenger and Pullman cars, and all other applications where the breaking of glass constitutes a hazard.
2. Aircraft safety glass.
3. Bullet-resistant glass.
4. Laminated glass filters.

3 LAMINATED SAFETY GLASS

Laminated safety glass usually consists of two lights of sheet or plate glass having a nominal thickness of 0.095 to 0.115 in. bonded or united together with an interposed layer of vinyl plastic (polyvinyl butyral resin) plasticized with 28 to 30 per cent plasticizer. However, glass of lesser or greater thickness than the range mentioned can be and is used for laminated safety glass for special applications.

3.1 Manufacturing

Briefly, the process for the manufacture of laminated safety glass involves the assembly of a clean dry sheet of plastic between two lights of clean dry glass, subjecting the resulting glass-plastic assembly or sandwich to a preliminary heating and pressing action to remove noncondensible gases such as air and to provide sufficient adhesion between the glass-plastic layers to permit handling as a composite. The partially bonded structure is then placed in an autoclave where the glass-plastic layers are firmly united by the further application of heat and pressure.

3.2 Steps in manufacture
Plastic processing

The vinyl plastic widely used for safety-glass laminates is quite hygroscopic, absorbing as much as 4 per cent moisture when at equilibrium in an atmosphere of 100 per cent relative humidity. Because of the tacky characteristics of the plastic sheeting, its surfaces are coated by dusting with soda (sodium bicarbonate) to permit shipment in rolls and to facilitate handling, such as, drying cutting, and washing, in the laminator's plant.

Drying

Important to obtaining the required level of adhesion between the glass and plastic layers as well as to good stability of the finished product, is control of the moisture content of the vinyl plastic at the time of laminating. The practice of drying and handling the vinyl plastic to hold the moisture content to 0.5 per cent maximum at the time of assembly between glass sheets has proved satisfactory in controlling adhesion and stability.

The moisture content of the plastic sheeting can be reduced to the required level by carrying the sheet on roll mechanism or on a belt through suitable drying ovens at temperatures ranging from 140°F to 160°F. The time for drying varies from 6 to 15 min., depending on the moisture content of the plastic as received, and also on the humidity of the air used in the drying oven. The plastic as discharged from the oven should be stored in a conditioned atmosphere having a relative humidity of 25 per cent or lower until it is laminated.

Washing

Before assembly between glass plates, it is necessary to remove all traces of the soda which remained on the sheet during the drying and cutting operations. This is accomplished by passing the plastic through special washing machines where an application of hot water, followed by cold water, dissolves and flushes the soda from the plastic surfaces.

Preparation of glass

For many applications of laminated glass, particularly for the automotive industry, it is necessary to bend the glass before lamination. To carry this out, pairs of glass of the desired thickness are subjected to a carefully controlled thermal cycle —the contour being obtained by the use of a mold. The glass may be cut to the final size and shape either before or after this bending process.

The pattern-size glass is usually washed mechanically in specially designed machines which convey the glass through detergent solutions and rinse sprays. The glass coming from the washing machine is rinsed and dried and then conveyed to the assembly room.

Assembly

In the assembly room, the plastic interlayer sheet is placed between pairs of glass and the sandwich is conveyed to the pre-pressing machine, where it is subjected to heat and slight roll pressure to remove noncondensible gases and to effect a preliminary bonding.

Final pressing

Coming from the prepressing machine, where the continuous-line operation terminates, the partially bonded assemblies are stacked in suitable racks, which are transferred to large autoclaves. There the final bonding is consummated by the application of pressures ranging from 200 to 250 psi and temperatures ranging from 240°F to 300°F. The holding time at the maximum temperature and pressure for ordinary $\frac{1}{4}$-in. laminated safety glass varies from 6 to 30 min.

The composite or laminate coming from the autoclave is washed, edge-finished to specifications, inspected, and boxed for shipment.

3.3 Physical properties

Although many of the physical properties of $\frac{1}{4}$-in. laminated safety glass, such as the index of refraction, light transmission, and surface hardness, are, for all practical purposes, the same as the glass components used for making the composite structure, the flexural strength is lowered appreciably by the presence of the nonrigid plastic interlayer.

Pressure tests

The results of pressure tests conducted on $\frac{1}{4}$-in. regular plate and laminated safety plate glass in sizes 12 in. × 12 in., 18 in. × 18 in., and 24 in. × 24 in. recorded in Table 7.3 are the average of ten specimens and will serve to illustrate the reduction in strength due to the plastic interlayer.

Resistance to impact

The safety features of glass-plastic laminates depend largely on the ability of the adherent plastic layer to hold the composite together and thereby reduce the hazard of flying glass when the lamination is cracked or broken. It follows that the resistance to impact and, therefore, the safety feature of laminated safety glass depends on the plastic thickness used in the lamination, the glass components, and the temperature at the time of impact since the vinyl plastic used is a thermoplastic.

The results of impact tests conducted at 0°F, 75°F, and 120°F, with $\frac{1}{2}$- and 2-lb steel spheres on 12-in. × 12-in. specimens of $\frac{1}{4}$-in. laminated safety glass, made with annealed and heat-strengthened $\frac{7}{64}$-in. plate glass and vinyl plastic interlayers 0.015, 0.030, and 0.045-in. thick are recorded in Table 7.4.

The "critical distance" recorded in Table 7.4 resulted from testing at least 20 specimens and is that distance from which the steel sphere was allowed to fall from rest to produce failure of 50 per cent of the specimens tested. Specimens were classified as failures when a hole or continuous shear in the plastic

interlayer exceeded $1\frac{1}{2}$-in., even though the impacting sphere failed to penetrate the lamination completely.

Similar impact tests conducted on 12-in. × 12-in. specimens of regular $\frac{1}{4}$-in. plate glass (annealed) with a $\frac{1}{2}$-lb steel sphere show the plate glass to have a critical distance of about 2 ft over the temperature range of 0°F to 120°F.

Temperature stability

To meet heat-stability requirements as specified in the American Standards Association Safety Code Z-26.1, laminated safety-glass specimens, 12 in. × 12 in. in size, are required to withstand immersion in boiling water for a period

Table 7.4. Summary of Impact Tests on Laminated Safety Glass Made with Vinyl Plastic Containing 28 per cent Dibutyl Sebacate

Glass Components Used	Plastic Thickness (in.)	Weight of Steel Ball (lb)	Critical Distance (ft)		
			0°F	75°F	120°F
$\frac{7}{64}$-in. plate	0.015	$\frac{1}{2}$	18	Greater than $33\frac{1}{2}$	20
$\frac{7}{64}$-in. heat-strengthened plate	0.015	$\frac{1}{2}$	22	Greater than $33\frac{1}{2}$	27
$\frac{7}{64}$-in. plate	0.015	2	$2\frac{3}{4}$	$5\frac{1}{4}$	$4\frac{1}{4}$
$\frac{7}{64}$-in. heat-strengthened plate	0.015	2	$5\frac{1}{2}$	8	7
$\frac{7}{64}$-in. plate	0.030	2	$7\frac{1}{2}$	22	11
$\frac{7}{64}$-in. heat-strengthened plate	0.030	2	10	33	15
$\frac{7}{64}$-in. plate	0.045	2	12	Greater than 33	33
$\frac{7}{64}$-in. heat-strengthened	0.045	2	21	Greater than 33	33

of 2 hr without showing noticeable discoloration or bubble formation in the lamination.

Although vinyl plastic properly made and laminated will withstand the temperature of boiling water for short periods of time without showing apparent change, sustained temperatures much above 150°F will adversely affect the plastic

Table 7.3. Pressure Tests: $\frac{1}{4}$-in. Plate and $\frac{1}{4}$-in. Safety Plate Glass. Temperature of Test Specimens: 75°F; Pressure Loading: 5 psi per min.

Glass Specimen	Method of Glazing	Average Breaking Pressure psi Glass Size (Free Opening)		
		12 in. × 12 in.	18 in. × 18 in.	24 in. × 24 in.
$\frac{1}{4}$-in. Plate Regular	Clamped with gaskets at margin	14	7	4.5
$\frac{1}{4}$-in. Safety Plate ($\frac{7}{64}$-in. glass, 0.015-in. plastic, $\frac{7}{64}$-in. glass)	Clamped with gaskets at margin	11	6	3.5

interlayer, thereby decreasing its usefulness as a safety-glass product.

4 AIRCRAFT SAFETY GLASS

Although the general principle of laminating glass and plastic for automobile and aircraft glazings is the same, the functional requirements of safety glass for the modern aircraft differ widely from those used for glazing motor vehicles operating on land highways. Greater resistance to impact and pressure loads are required of aircraft safety glass. Also for streamlining purposes curved sections and different methods for glazing or mounting in the aircraft are necessary to afford flush joints that will be weather- and pressure-tight.

4.1 Construction

The strength requirements for aircraft glazings and, therefore, their construction vary widely with the type, design, and performance specifications of the aircraft. For example, trainer planes and smaller aircraft function very satisfactorily with laminated safety glass of the same configuration used for glazing automobiles; however, the modern jet aircraft operating under conditions of high speed, high elevations, or rapid change in elevation, require safety-glass glazings able to withstand severe stresses imposed by wind loading and cabin pressurization. To meet these severe requirements, aircraft safety glass is usually made with heat-strengthened glass united with vinyl plastic four to ten times the thickness used for automobile safety glass.

For many aircraft applications it is necessary that the glazing be heated to prevent frosting or icing of the windows. To accomplish this, one or more lights of glass may be coated with an electrically conducting layer as discussed earlier in this chapter. An alternate approach is to embed fine wires in the plastic interlayers. The heat generated by passing an electric current through these gives the desired defogging or deicing effect.

4.2 Edge construction

Important to the proper functioning of aircraft glazings is the method of mounting the laminated structure in the plane openings. Using plastic of substantial thickness and allowing it to extend marginally beyond the edge of the glass plates affords means for glazing the safety-glass laminate by bolting or clamping through the extended plastic layer. An important modification of this type of edge construction comprises reinforcing the extended plastic flange by the insertion of a flexible metal collar inwardly $\frac{1}{4}$ to $\frac{3}{8}$ in. from the edge of the glass. The metal collar is laminated in the approximate center of the plastic flange and serves to reduce the extensibility of the plastic flange under load due to plastic flow.

The mountings just described afford flexibility that minimizes the strain in the glass structure due to the twisting or warping of the plane and also serves to compensate for the dimensional tolerances necessary in the manufacture of glass and plane openings. Also, since streamlining of aircraft generally requires curved-glass sections, particularly for pressurized

cabins, the extended-plastic mountings serve to carry a portion of the pressure load in tension.

The two types of mountings, extended plastic and the metal-reinforced extended plastic, are shown in Figures 7.2 and 7.3.

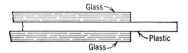

Figure 7.2   Extended Plastic Edge

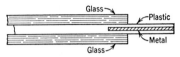

Figure 7.3   Extended Plastic Edge Reinforced with Metal

4.3 Physical characteristics of aircraft safety glass

Impact resistance

The resistance of glass-plastic laminates to impacting objects will, of course, depend on the number and thickness of glass-plastic layers used to make the composite. In times of war, aircraft glazings are the target for all types and kinds of impacts, including the steel dart used in armor-piercing ammunition, shrapnel, and gun blast. During peacetime, however, perhaps the greatest impact hazard confronting aircraft is that of collision in flight with migratory birds, such as wild ducks and geese. Therefore, to guard against such hazards, aircraft windshields must be constructed to resist the impacts of wild fowl weighing as much as 12 lb at speeds equal to or even greater than the cruising speed of the aircraft.

The Civil Aeronautics Administration[7] has, with the co-operation of the glass and plastic manufacturers, done considerable research on this problem, and their results indicate that double-glazed windshields comprising $\frac{1}{4}$-in. laminated safety glass as the outboard light and laminated heat-strengthened glass with $\frac{3}{8}$-in.-thick vinyl plastic as the inboard light will resist the impact of a 4-lb bird carcass at a velocity of approximately 400 mph.

Strength in tension of plastic and plastic-reinforced edges for mounting

The results of tensional loading of the extended-plastic and the metal-reinforced extended-plastic edges for mounting aircraft safety glass made with $\frac{1}{8}$-in.-thick vinyl plastic interlayer (containing 20 per cent dibutyl sebacate plasticizer) are given in Table 7.5. As the results in Table 7.5 show, the metal reinforcing materially reduces the extensibility of the edge mounting under load.

Pressure loading

The results of pressure-loading glass-plastic laminations of aircraft construction and having the extended-plastic and metal-reinforced extended-plastic edges for glazing are recorded in Tables 7.6, 7.7, and 7.8. The composite structures were made by laminating flat plates of heat-strengthened glass, with $\frac{1}{8}$-in.-thick vinyl plastic interlayer (containing 20 per cent of dibutyl sebacate plasticizer).

Table 7.5. Tensional Loading of Aircraft Safety Glass at 75°F.

Type of Construction	Kind and Thickness of Metal Reinforcing	Load in lb per Linear in., Edge Support	Elongation under Load (in.)	Remarks
Extended plastic ⅛ in. thick	Plastic only	9.5	1½	Plastic flowed
Extended plastic ⅛ in. thick, reinforced with continuous metal collar	Soft aluminum 0.025 in. thick	160	⅛	Aluminum frame sheared

Table 7.6. Pressure Tests. Temperature of Test Specimen: 75°F; Rate of Pressure Loading: 1 psi per 5 min.

Glass-Plastic Construction	Type of Edge Mounting	Glass Size (Free Opening) Average Breaking Pressure, psi		
		12 in. × 12 in.	18 in. × 18 in.	24 in. × 24 in.
⅛-in. glass ⅛-in. plastic ⅛-in. glass	Extended plastic Extended plastic, metal-reinforced	13 19	9 12	7 8
⅛-in. glass ⅛-in. plastic ³⁄₁₆-in. glass	Extended plastic Extended plastic, metal-reinforced	17 31	11 22	9 12

Table 7.7. Pressure Tests. Temperature of Test Specimens: 75°F; Rate of Pressure Loading: 5 psi per min.

Glass-Plastic Combination	Type of Edge Mounting	Glass Size (Free Opening) Average Breaking Pressure, psi		
		12 in. × 12 in.	18 in. × 18 in.	24 in. × 24 in.
⅛-in. glass ⅛-in. plastic ⅛-in. glass	Extended plastic Extended plastic, metal-reinforced	23 27	15 15	10 10
⅛-in. glass ⅛-in. plastic ³⁄₁₆-in. glass	Extended plastic Extended plastic, metal-reinforced	33 44	20 21	15 15

If the breaking pressures recorded in Tables 7.6 and 7.7 are compared, it becomes apparent that the type of edge construction used for glazing or mounting aircraft safety glass is an influencing factor in predicting the pressure load that glass-plastic laminates of this construction will safely carry.

The lower breaking pressures attained with the extended-plastic edge compared with those of similar specimens having the metal-reinforced plastic edge are attributable to greater plastic flow of the former. Also the lower breaking pressures recorded in Table 7.6 compared to those in Table 7.7, are likewise attributable to increased plastic flow, due to the slower rate of loading the specimens reported in Table 7.6.

With reference to Table 7.8, the substantial increase in the breaking pressure of glass-plastic laminates as the temperature is lowered from 120°F to −60°F is due to the greater resistance to plastic flow at low temperatures and to the increase in flexural strength with decreasing temperatures.

Figure 7.4 shows a test specimen of aircraft safety glass

pressure-tested to destruction. The construction of the glass-plastic laminate comprised two lights of ⅛-in. heat-strengthened glass and ⅛-in.-thick vinyl plastic with a metal-reinforced extended-plastic edge for mounting in the pressure-test chamber. As the picture shows, although the glass plates in the lamination are broken, the edge mounting provides firm anchorage of the plastic interlayer. Aircraft safety glass of this construction affords protection from the atmospheric elements

Figure 7.4 Aircraft Safety Glass Tested to Destruction

and the sudden release of internal pressure should the glass become cracked or broken.

Table 7.8. Temperature of Test Specimens: 120°F, 75°F, and −60°F; Rate of Pressure Loading: 5 psi per min. Glass Size: 18 in. × 18 in. (Free Opening)

Glass-Plastic Combination	Type of Edge Mounting	Temperature of Test Specimen and Average Breaking Pressure, psi		
		120°F	75°F	−60°F
⅛-in. glass ⅛-in. plastic ⅛-in. glass	Extended plastic, metal-reinforced	10	15	30
⅛-in. glass ⅛-in. plastic 3⁄16-in. glass	Extended plastic, metal-reinforced	17	21	35

Flexural strength

The results of flexural strength tests (ultimate strength in bending), conducted on symmetrical glass-plastic laminates made with heat-strengthened glass and vinyl plastic (containing 18 per cent dibutyl sebacate plasticizer) in different thicknesses and tested at different temperatures are recorded in Table 7.9. The test specimens were 3¾-in. wide and 11-in. long; distance between supports was 10 in. Each result recorded is the average of six determinations. The "Apparent Modulus of Rupture" values were calculated by using the total glass-plus-plastic thickness. As the results show, the ultimate strength in bending of laminated glass decreases with increasing plastic thickness and increasing temperature.

Table 7.9. Bending Tests

Glass Thickness (in.)	Plastic Thickness (in.)	Apparent Modulus of Rupture, psi			
		−60°F	0°F	80°F	125°F
7⁄64	0.075	24,900	21,600	10,600	6570
	0.120	24,300	18,500	8340	5190
⅛	0.075	27,100	21,600	11,700	7030
	0.120	26,000	20,100	9610	5710
3⁄16	0.075	28,600	24,300	11,030	8120
	0.120	22,700	17,940	8240	7120

Temperature stability

Aircraft safety glass made with vinyl plastic will withstand temperatures up to 150°F for long periods of time without bubble formation, discoloration, or loss in light transmission. Although higher temperatures for short intervals of time may be used for deicing or defrosting purposes, it should be remembered that the plastic interlayer is organic and to serve the purpose intended must remain transparent. Accordingly, elevated temperatures much above 150°F are to be avoided.

5 BULLET-RESISTANT GLASS

Bullet-resistant glass is a composite structure consisting of multiple layers of glass bonded together with alternate layers of vinyl plastic to produce a structure of the required thickness. The same general principle employed in the manufacture of laminated safety glass is applicable for bullet-resistant glass, except that, as the thickness of the glass laminates is increased, the heating and cooling periods of the laminating cycle are materially increased, so as to avoid temperature differentials in the laminated structure that will cause glass breakage. Also, the vinyl plastic interlayers usually contain about 20 per cent plasticizer, compared to 28 to 30 for regular safety glass. Further, the interlayers of plate glass making up the bullet-resistant structure are usually much heavier than thin plate glass used for the outside layers or for regular ¼-in. laminated safety glass.

Bullet-resistant glass is readily available in thicknesses of ½ to 3 in. and for special purposes has been furnished 6-in. thick. For calculating the weight of bullet-resistant glasses, it may be assumed that 13 psf per inch of thickness is average and representative. Bullet-resisting characteristics and light transmission of glasses of different thicknesses are given in Table 7.10.

6 LAMINATED GLASS FILTERS

By laminating combinations of clear or colored glasses with interposed clear or colored plastic layers, it is possible to obtain a variety of glass-plastic laminates which, in effect, serve as light filters to reduce or screen out certain portions of the spectrum, such as the ultraviolet and infrared radiation normally present in the solar spectrum.

6.1 Colored glass

The color and light-transmitting characteristics of soda-lime-silica glasses can be varied widely by incorporating small amounts of certain compounds in the glass batch.

Depending on the colorants used, these glasses are effective in selectively absorbing bands in the solar spectrum. This principle is employed in the manufacture of the commercially available heat-absorbing and glare reducing glasses, and also the blue, green, and red glasses used for light filters and decorative purposes.

Spectral-transmission curves which are typical for low iron plate, regular plate, heat-absorbing plate (blue-green), grey plate, and bronze plate are shown in Figures 7.5–7.9. Figure 7.8 also includes the transmission curve for two lights of grey plate laminated together.

6.2 Colored plastic

The incorporation of coloring agents in vinyl plastic or other plastics that can be bonded to glass presents another method for the preparation of laminated glass filters. The coloring agent may consist either of a dyestuff or a dispersed pigment in the plastic layer.

A wide variety of colors can be achieved by using various dyestuffs to color the interlayer. Of particular interest are shades of blue and green which are attractive for automotive as well as architectural applications. The spectral transmission curve for a "Standee Window" green plastic laminated with single strength sheet glass is shown in Figure 7.10.

Although several types of dispersed pigment colorations have

Table 7.10. Characteristics of Bullet-Resistant Glasses of Various Thicknesses. Specimen Size 12 in. × 12 in.

Nominal Thickness (in.)	Visible Light Transmission %	Bullet-Resistant Property				
		Degree of Obliquity	Type of Gun	Type of Ammunition	Firing Distance (ft)	Results of Firing Tests
$\frac{1}{2}$	82 to 84	0	0.38 Hammerless Harrington & Richardson	Standard	20	Impacted with 5 shots in an 8-in.-diameter circle; 3 of the bullets punctured the structure
$\frac{3}{4}$	81 to 83	0	0.38 Hammerless Harrington & Richardson	Standard	20	Impacted with 5 shots in an 8-in.-diameter circle; no bullets punctured the structure; little or no glass was spalled from back of structure
$\frac{7}{8}$	81 to 83	0	Colt D.A. 0.45	Standard	15	Of 5 shots in an 8-in.-diameter circle, 1 punctured the structure; sample bulged badly opposite points of impact
$1\frac{1}{8}$	81 to 82	0	Colt D.A. 0.45	Standard	15	Of 5 shots impacting within an 8-in.-diameter circle, none punctured the structure; little or no glass spalled from back of sample
$1\frac{1}{8}$	81 to 82	0	Smith & Wesson 0.357 Magnum	0.357 Magnum	20	Of 3 shots within an 8-in.-diameter circle, 1 bullet punctured the structure; considerable glass spalled from back of sample
$1\frac{1}{2}$	80 to 82	0	Smith & Wesson 0.357 Magnum	0.357 Magnum	20	5 shots impacting in 8-in.-diameter circle failed to puncture the structure; very little if any glass spalled from back opposite points of impact
$1\frac{1}{2}$	80 to 52	45	0.30-caliber Springfield Service Rifle	Armor-piercing	60	The bullet failed to puncture the structure; considerable glass was spalled from back opposite point of impact
$1\frac{1}{2}$	80 to 82	45	0.30-caliber Springfield Service Rifle	Armor-piercing	600	The bullet failed to puncture the structure; little if any glass spalled from point opposite that of impact, complying with the tests set up in British Specification DTD 402, regarding penetration of cellophane screen
2	79 to 80	0	0.348 Winchester	Standard 150-gr bullet	60	Bullet failed to puncture the structure; only a small amount of glass spalled from back of sample
3	73 to 75	0	0.30-caliber Springfield Service Rifle	Armor-piercing	60	Bullet failed to puncture the structure; only a very small amount of glass spalled from back of sample
3	73 to 75	45	0.50-caliber machine gun	Armor-piercing	125	Bullet failed to puncture the structure; small amounts of glass spalled from sample opposite point of impact
6	59 to 61	0	0.50-caliber machine gun	Armor-piercing	125	Bullet failed to puncture the structure, spalling small amounts of glass from sample opposite points of impact

been used, the carbon black dispersions have had the most commercial interest. These give the laminate a range of colors from amber to neutral gray, and they are also effective in reducing heat transmission. Typical spectral transmission curves for laminates made up with carbon black pigmented

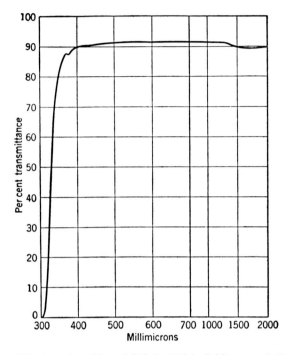

Figure 7.5   Low Iron Plate 0.256-in. Thick, 0.02 per cent Fe₂O₃

plastic between two sheets of regular plate glass are given in Figure 7.11. The different levels of transmission are achieved by varying the amount of pigment in the plastic. Intermediate levels of transmittance can be attained by using two or more layers of the various types of plastic.

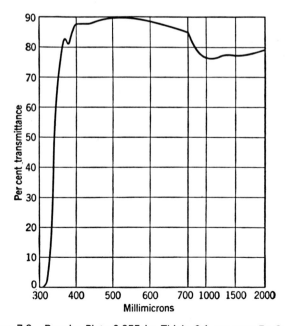

Figure 7.6   Regular Plate 0.255-in. Thick, 0.1 per cent Fe₂O₃

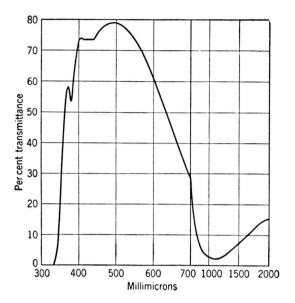

Figure 7.7   Heat Absorbing Plate 0.247-in. Thick

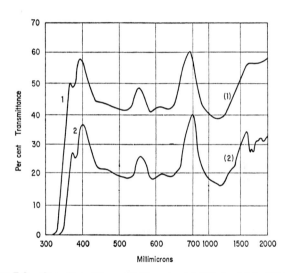

Figure 7.8   Grey Plate Glass. (1) One sheet 0.250-in. thick. (2) Two sheets 0.250-in. thick laminated with 0.015-in. plastic

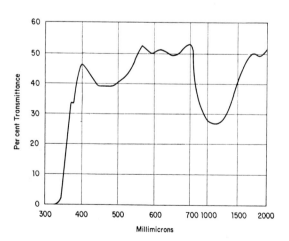

Figure 7.9   Bronze Plate 0.250-in. Thick

For certain specialized uses where an opaque laminated glass is desired, titanium dioxide pigments may be incorporated into the interlayer. The resulting laminates are white in color and are opaque.

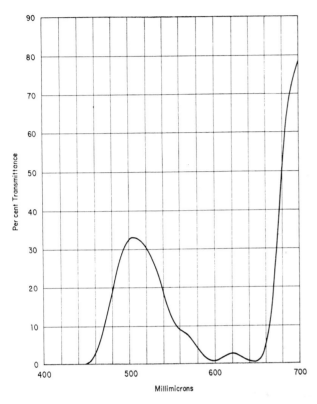

Figure 7.10   Lamination of Green Plastic with Single Strength Sheet Glass

6.3   Colored glass and plastics

Obviously, a wide variety of light filters can be manufactured by combining the light-filtering characteristics of colored glasses and colored plastics through the process of lamination.

Transmission curves for laminates made up with combinations of heat-absorbing glass and dyed-plastic interlayers are given in Figure 7.12.

Colored glasses may also be combined with pigmented interlayers. Spectral transmission curves for two such laminates are shown in Figure 7.13.

6.4   Translucent glass-plastic laminates

Laminates of figured glasses are not commercially available owing to the difficulties involved in bonding the irregular surfaces. However, opal glasses are readily laminated, employing the same procedure as used for the manufacture of laminated safety glass.

Laminates of opal glass and plastics find application in scientific instruments and for glazing certain openings where nondirectional lighting is desired and the safety feature of laminated safety glass is a requirement.

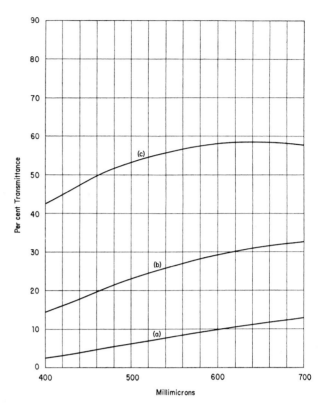

Figure 7.11   Laminations of Plate Glass with Various Carbon Pigmented Plastics. (a) 10 per cent visible light transmission. (b) 30 per cent visible light transmission. (c) 55 per cent visible light transmission

7   COMPOSITE GLASS-AIR-CELL STRUCTURES

Multiple layers of glass separated by air cells of predetermined thickness are referred to as insulating-glass windows. Although composite structures of this type are laminates comprising layers of glass and air, their construction varies widely from glass-plastic laminations. In the latter, the layers of unlike materials are bonded firmly together throughout the entire area, whereas the glass layers in the insulating window, separated a predetermined distance, are united or bonded at the margin only to form an air cell.

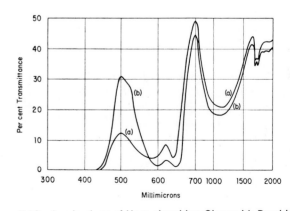

Figure 7.12   Laminations of Heat-absorbing Glass with Dyed Interlayer Plastics. (a) Dense green plastic with heat-absorbing plate glass. (b) Green plastic with heat-absorbing sheet glass

7.1 Construction

Multiple-glass units for window-insulating purposes are available in various combinations of glass and air spaces. The number of lights of glass and air spaces and also the glass and air-space thicknesses determine the insulation properties of the unit.

Regular plate and sheet glass are generally used in the production of multiple-glass units to give high light transmission. However, by using one or more lights of different types of glass in the composite, special requirements such as the control of transmitted solar energy, greater strength and safety, and

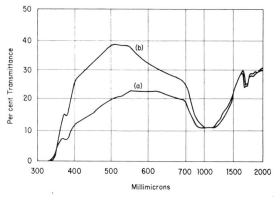

Figure 7.13 Laminations of Heat-absorbing Glass with Carbon Black Pigmented Interlayers. (a) 23 per cent visible light transmission. (b) 35 per cent visible light transmission

heat-producing units can be realized. Some examples are

1. Types of glass found effective in reducing the transmission of solar energy include the heat-absorbing glasses, filmed glass having reflecting properties, and laminated glass incorporating colored or shaded plastic interlayer.
2. Heat-strengthened and chemically strengthened glasses and laminated safety glass provide increased protection and safety.
3. Glass coated with electrically conducting films is effective for deicing and defogging of glass.
4. Colored glasses offer decorative possibilities.
5. Patterned glass provides both light transmittance and translucency.

Although the fabricators of multiple-glass units, in general, have the same objectives in obtaining a glass-air-cell composite structure in which the air space or spaces are tightly sealed at the glass margin to exclude the infiltration of extraneous materials such as moisture and dust, the methods for constructing the marginal seal vary widely with the different fabricators.

The three major systems of joining the lights of glass at the margin to form the air cell are the metal-to-glass or metallic seal, the mastic or organic seal in combination with a metallic spacer, and the fused glass seal. Figures 7.14, 7.15, and 7.16 show three commercially available insulating-glass units supplied under the trade names of "Bondermetic Thermopane," "Twindow," and "GlasSeal Thermopane."

Figure 7.14 "Bondermetic Thermopane" Insulating Unit

Figure 7.15 "Twindow" Insulating Unit

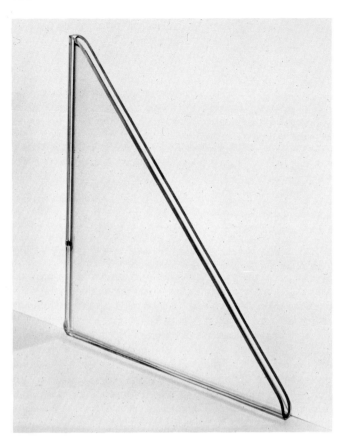

Figure 7.16 *"GlasSeal Thermopane"* Insulating Unit

7.2 Thermal insulation

The thermal insulation of multiple-glass glazings is due to the combined resistances of glass and the relatively still layers of air to the transfer of heat by conduction and convection. Since the resistance to heat flow through any path equals the reciprocal of the conductivity over the same path, the thermal transmittance or over-all coefficient of heat transmission U through a multiple-glass glazing is equal to $1/R$, where R is the over-all resistance; and is equal to the sum of the resistances of the glass, the air space, and the air films (outside and inside).

The over-all heat transmission coefficient varies with the temperature range at which the coefficient is determined. For winter conditions an outside temperature of 10°F, an inside

temperature of 70°F, and a wind velocity of 15 mph and a room temperature of 75°F to 80°F are assumed. Table 7.11 gives the U values using the above criteria and assuming natural convection inside the room.

Table 7.11. Coefficients of Transmission (U) of Windows
Btu per hr (sq ft) °F

Description	Winter	Summer
Flat Glass		
single glass	1.13	1.06
Insulating glass-double		
$\frac{3}{16}$-in. air space	0.69	0.64
$\frac{1}{4}$-in. air space	0.65	0.61
$\frac{1}{2}$-in. air space	0.58	0.56
Insulating glass-triple		
$\frac{1}{4}$-in. air spaces	0.47	0.45
$\frac{1}{2}$-in. air spaces	0.36	0.35
Storm windows		
1 in.–4 in. air space	0.56	—

These values may be greatly influenced by other factors involved with the installation. Air movement across the interior surface of a window will increase the U value. On the other hand the presence of shading devices close to the glass will decrease the value of U appreciably since they restrict the air movement across the glass surface. As a rough approximation it can be assumed that a tight-fitting device reduces the U value by 25 per cent.

REFERENCES

1. MATTHEW J. KERPER and THOMAS G. SCUDERI, "Mechanical Properties of Glass at Elevated Temperatures," *Bull. Am. Ceram. Soc.* **42**, 735–740 (1963).
2. "Properties of Glasses at Elevated Temperatures," Interim Report, WADC-TR-56-645, Pt. 7, Aeronautical Systems Division, Wright-Patterson AFB, Ohio, March 1962, 90 pp.
3. CHARLES B. KING, "Chemical Tempering of Glass Induces Residual Stresses for Diverse Applications," *Research/Development* **15**, No. 4, 20–23 (1964).
4. A. T. FULLICKS, British Patent 15,303 (Aug. 20, 1885).
5. J. C. WOOD, British Patent 9972 (1905).
6. EDUARD BENEDICTUS, French Patent 405,881; British Patent 1790 (1910).
7. G. L. PIGMAN, *Aeronaut. Eng.* **4**, 9 (Jan. 1945).

Chapter 8

Structural-sandwich Construction

R. T. Schwartz
D. V. Rosato

1 INTRODUCTION

There has been a growing interest in many applications of composite or sandwich constructions. These developments are based on the concept, not of using or favoring any one material, but rather of employing, as occasion requires, all available materials. It goes even further in that it sets up, for certain uses, desired requirements as to mechanical and physical properties that are not yet available in any material. These objectives provided a target, as it were, for further development.[1]

Another feature of sandwich construction is the opportunity, through efficient structural design, of stressing each material to the practical limit of its possibilities.

The American Society for Testing and Materials (ASTM) definition for a structural sandwich is a construction comprising a combination of alternating dissimilar simple or composite materials, assembled and intimately fixed in relation to each other so as to use the properties of each to specific structural advantages for the whole assembly.[2] Sandwiches are not only of the edible variety; the sandwich principle is of great importance in many stationary and moving structures. Structural sandwiches are a special form of laminated composite in which thin, strong, stiff, hard, but relatively heavy facings are combined with thick, relatively soft, light, and weaker cores to provide a light-weight composite much stronger and stiffer in most respects than the sum of the individual stiffness and strengths.

Thin strips of metal are easily bent, and a block of plastic foam is easily broken, but when the metal strips are bonded to opposite faces of the plastic foam the resulting sandwich strongly resists bending and requires a heavy load to break it.

The basic principle is much the same as that of an I beam, which is an efficient structural shape because as much as possible of the material is placed in flanges situated farthest from the center of bending or neutral axis. Only enough material is left in the connecting web to make the flanges act in concert and to resist shear and buckling. In a structural sandwich the facings take the place of the flanges, and the core takes the place of the web. The difference is that the core of a sandwich is a different material from the facings, and it is spread out instead of concentrated in a narrow web. The facings act together to form an efficient internal stress couple or resisting moment counteracting the external imposed bending moment. The core resists the shear stresses set up by the external loads, and it has the further important function of stabilizing the facings against wrinkling or buckling. It must therefore be strong and stiff enough to resist transverse tension and compression set up by the facings as they try to wrinkle.

In direct compression the same supporting action is important. A thin plate buckles quickly when an axial compressive stress is applied. If it is made a part of a sandwich, lateral restraint against buckling is provided by the adjacent core, and the plate can withstand considerably higher compressive stresses.

Evidently, the bond between facings and core must be strong enough to resist the shear and tensile stresses set up between them. The adhesive used to bond facings and core together is of critical importance.

The over-all subject of sandwich construction can be discussed broadly under several headings, as follows:

> Materials
> Design criteria
> Fabrication
> Test methods
> Data
> Applications

The great majority of panel constructions in the past have not been structural or load-supporting sandwiches. They have been heavy high-strength faces, often mechanically fastened together, strong and rigid enough to perform a desired job by themselves. Designers and engineers, however, have become increasingly aware of the advantages to be gained by use of the more efficient structural composite of facing, core, and adhesive. This is not to imply that the structural sandwich is entirely new.

The basic principle of spaced facings was discovered about 1820 by a Frenchman named Duleau and panels utilizing asbestos board skins with vegetable fiberboard cores were used as early as World War I. During World War II the trend to more efficient use of labor and materials, particularly in aircraft, resulted in increasing use of panels. However, the development or adaptation of new materials, the majority of which are plastics, has made an impact on the field of sandwich construction. An example of their potential is the use of foams[3,4] in building construction.[5]

The application of sandwich materials to aerospace structures is important and provides one of the major areas for new developments. Possibilities of choosing widely different core and facing materials, advantages and disadvantages of bonded

R. T. SCHWARTZ is Chief, Nonmetallic Materials Division, Air Force Materials Laboratory, Dayton, Ohio.

D. V. ROSATO is Technical Editor of *Plastics World*.

sandwich structures, and emphasis on weight optimization in aerospace design will be cited. Aluminum and stainless steel honeycombs and glass, cotton, paper, foam, balsa and reinforced-plastic honeycombs are compared in terms of specific strength, density, and heat transfer coefficient at room temperature. Shear and compressive properties for aluminum and for glass-fabric-reinforced plastic honeycomb are given. Advantages of sandwich-type construction include higher strength-to-weight ratios, smooth surfaces, better stability, absence of potential leaks, high load-carrying capacity, increased fatigue life, and high sonic fatigue endurance. Applications of sandwich construction include their use in aircraft wings, helicopter rotor blades, and such other helicopter parts as flooring and deck panels, fire walls, access doors, and fuselage primary structures.[6]

Structural building panels, in which plastic materials are the greatest part of the panel, are now being used in schools, hospitals, and industrial building construction. Architects are recognizing the fact that curtain wall panels that utilize large parts of plastics and bonding resins are particularly well suited for curtain wall applications.[5] A typical panel widely used is in widths to 4 ft and lengths up to 10 to 20 ft.

2 MATERIALS

2.1 Facings

Almost any sheet material — paper, wood, metal, plastic, hardboard, and asbestos board, reinforced plastic laminate — can be used as a facing material (see Table 8.1).

Table 8.1. Typical Properties of Facing Materials

Material	Elastic Modulus, 10^6 psi
Plastic	
Phenolic–Asbestos Laminates	5–6
Epoxy–Glass Cloth Laminates	3.0–4.6
Polyester–Glass Fiber Laminates	1.0–3.5
Metal	
Aluminum	10
Steel	
Carbon	30
Stainless	29
Titanium	15–18
Wood	
Plywood (3-ply)	1.1–1.7
Hardboard	0.4–1.3
Hardwood	1.3–2.2

It is important to give consideration to the effects if the same material, or materials of approximately equal coefficients of thermal expansion, are not used for the same faces, since the composite structure may warp with temperature changes.[7] Also, dissimilar materials in core and facings may lead to corrosion problems. If dissimilar metallic materials are used, special pretreatments are required.[8]

Metal facings are usually formed before bonding to eliminate core deformations while forming. However, some assemblies are formed after bonding one face to the core. Many panels are formed or partially formed at the time of bonding by locally compressing the core or by precutting the core to contour. Face gages of from 0.012 in. to 0.125 in. are normally used in aircraft; however, other thicknesses are possible. As an example, with a face gage as low as 0.008 in. the mechanical durability of the sandwich is jeopardized. Bare (nonclad) sheet in gages less than 0.020 in. thick is generally not used.

Many different types of facing materials can be used to combine structural strength and decorative appeal. However in most applications a decorative top panel is just used over the basic structural sandwich panel. Decorative surfaces can consist of inlaid tile, concrete, vinyl, acrylic, or just paint.

2.2 Cores

Different core materials are used. Aluminum foil honeycomb is probably the most widely used at this time in the primary structural applications. Foamed or cellular low-density plastics are principally used in many new building constructions with paper, honeycomb, or fiberboard to provide an efficient combination of strength and heat insulation. However the different available cores generally provide different inherent characteristics which make them useful in specific applications.

Honeycomb

Honeycomb has come of age since 1945 and is being used successfully in commercial and military applications. Large-scale use of honeycomb sandwiches in military applications provided the initial motivation for the commercial development of the material.

Honeycomb cores can be made of a wide variety of materials; the most frequently used are paper, cotton, glass cloth, and aluminum foil, sometimes combined with plastic foam. A plastic impregnant is used with the first three of these to make the material rigid enough for use as a sandwich core. A 20 per cent, by weight, phenolic resin is usually used with paper, and up to 50 per cent phenolic with cotton. Polyester, phenolic, or nylon-phenolic resins are principally used with glass-fiber cloth.

Various types of phenolics are used in paper core. A water-soluble resin has a high degree of effectiveness, but makes the impregnated paper somewhat too brittle for some applications. An alcohol-soluble resin does not penetrate the paper as completely as does the water-soluble type, but the treated paper is far more flexible.

Plastics are also used in honeycomb sandwiches to provide the bond between the honeycomb core and face sheets.

Cell sizes on commercial aluminum cores are $\frac{1}{8}$, $\frac{3}{16}$, $\frac{1}{4}$, and $\frac{3}{8}$ in. across the flats, with node (bond lines) of 0.072, 0.108, 0.144, and 0.218 in.; foil thickness is from 0.0007 to 0.006 in. Cell sizes above $\frac{1}{4}$ in. often present problems in intercell buckling, and therefore are not normally recommended.[8]

The density of honeycomb core refers to its weight per cubic foot (pcf) in the nominal expanded condition (cells of true hexagonal shape). The density of the core may be altered by varying the degree of expansion, although the mechanical performance is proportionately biased toward greater orthotropism. Density of the core ranges from 1.6 to 10.5 pcf. It is normally recommended that usage be limited to densities above 3.5 pcf in an area of stress concentration.

Aluminum core can be used at service temperatures up to 500°F for short exposure periods with heat resistant adhesives, such as epoxy-phenolic. Government specification MIL-C-7438 pertains to aluminum foil core materials and specifies the minimum acceptable requirements for shear strengths as well as other requirements. The core shear strength is a function of ribbon direction (the long continuous direction of the foil, going from node to node, refers to the ribbon direction, density, core thickness and face thickness. It must also be corrected for panel configurations other than those of standard test specimens.

Stainless-steel foil cores are available in 17-7PH, AISI 321, A 286 and AM 3500 alloys, with foil thicknesses from 0.001 to 0.006 in. Several cell configurations are available commercially. The configuration is bonded at the nodes with a thin (0.001 to 0.003 in.) layer of organic adhesive, and is available in cell sizes from $\frac{1}{8}$ to $\frac{3}{8}$ in. The hexagonal cell has greater strength in its longitudinal direction.

The inorganic core materials are usually used where thermal insulation or heat resistance is required. Glass used as lightweight foamed material or as glass fabrics is the best known. The glass fabrics are impregnated with a thermosetting resin and formed into cellular core material. The selection of the resin stabilizing material is critical and varies with the end application.

Cotton fabric impregnated with up to 45 per cent, by weight, phenolic resin is useful where repeated impact loading is a service condition and thermal insulation is desired. This material is furnished commercially in the honeycomb pattern. Where thermal resistance alone is required, glass fabric is a better choice.

The long-time use of paper, corrugated and glued to paper faces to form the well-known container board and boxes, was an early type of sandwich material, although it was not so designated. The more recent development is the paper honeycomb core. This honeycomb core is produced by impregnating kraft paper with phenolic resin. After curing, the paper is bonded together and expanded to form the core. Cell sizes across the flats vary but are commonly from approximately 0.3 to 0.7 in., and densities from 2.0 to 4.0 pcf.

The hexagonal honeycomb core is predominantly used since it provides maximum structural efficiency. In addition to the square cell, there are also corrugated and figure-eight types. Experimental types of practically any shape that can be imagined have been produced. Cores have also been developed experimentally to evaluate different materials of construction.[9]

Balsa wood

For over 30 years balsa wood has been employed as a structural core in different applications such as aircraft and boat sandwich structures. This use of balsa has been extended in the past years to include many applications in reinforced plastics where thick sections are required for thermal insulation, rigidity or both. Most of these reinforced-plastics applications are nonmilitary and enter into daily life in the form of tank trailers that transport bulk milk and fruit juices; ladders for use in telephone line service; and others.[10]

Balsa generally in its weight range possesses the highest strength-to-weight ratio of all core materials (see Table 8.2 and Figure 8.1). The differences are particularly appreciated

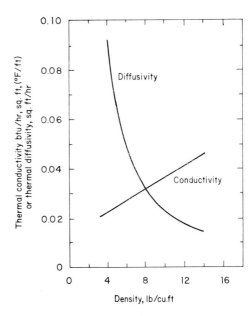

Figure 8.1   Thermal Properties of Balsa. Thermal properties of Balsa are dependent on density. Moisture content also affects them, and these conductivities are for balsas containing 12 per cent water. Diffusivities were estimated by assuming specific heat is equal to apparent specific gravity, a reasonable assumption for woods

after a part subjected to shock or vibration has been in service for some time and the effects of fatigue are felt.

As a core material balsa can be completely sealed. Any variation in moisture content of the core material after lamination depends primarily on skin integrity. It has been shown that kiln-dried balsa cores pick up moisture only if the facing is porous and continuously immersed.

In 25 years of use in reinforced plastics, the balsa industry has stated that there has not been a single reported case of fungus attack. The balsa core is sterilized at the time of kiln drying and resterilized during curing by the resin heat of reaction. Exposure to 170°F is reported by the balsa industry to be sufficient to kill fungal spores and tissue. This sterilization is an important reason for kiln drying the wood. Any subsequent attack must come from penetration through the facings. Studies to date indicate that even porous reinforced plastics do not present passages large enough for fungal spores.

Cellular plastics

Foam or cellular plastics in general have been on the market long enough to have established themselves as unparalleled thermal insulators. They owe their existence to their very low thermal conductivity. However, it was apparent that any significant future growth in markets for such foams would depend upon their usage for purposes other than or in addition to insulation.

There are a number of other important properties or advantages of foams which help determine utility. Chief among these has been their usefulness for structural purposes.

Compared with traditional materials of construction such as steel, wood, and stone, the foams have relatively low strength

Table 8.2. Properties of Balsa Cores.[11] (Data for Pieces Averaging 12 per cent Moisture Content)

Weight in pounds per cubic foot	6	11	15½
Specific gravity	0.0962	0.176	0.248
Compressive Strength (pounds per square inch)			
(A) *Parallel to grain (end grain)*			
— Stress at proportional limit	500	1,450	2,310
— Maximum crushing strength	750	1,910	2,950
— Modulus of elasticity	330,000	768,000	1,164,000
(B) *Perpendicular to grain (flat grain)*			
— Stress at proportional limit			
high strength value	84	144	198
low strength value	50	100	145
— Modulus of elasticity			
high strength value	16,000	37,000	55,000
low strength value	5,100	13,000	19,900
Bending Strength (pounds per square inch)			
Static bending			
— Stress at proportional limit	825	1,725	2,535
— Modulus of rupture	1,375	3,050	4,525
— Modulus of elasticity	280,000	625,000	925,000
Tensile Strength (pounds per square inch)			
(A) *Parallel to grain (end grain)*			
— Maximum	1,375	3,050	4,525
(B) *Perpendicular to grain (flat grain)*			
— Maximum — high strength value	112	170	223
— low strength value	72	118	156
Toughness (inch pound per specimen)			
— high strength value	125	310	475
— low strength value	120	267	400
Shear (pounds per square inch)			
— high strength value	180	360	522
— low strength value	158	298	425
Hardness (pounds)			
Load required to embed a .444-in. ball to one half its diameter			
(A) *Parallel to grain (end grain)*	102	250	386
(B) *Perpendicular to grain (flat grain)*			
— high strength value	50	120	186
— low strength value	47	103	151
Cleavage (pounds per inch of width)			
Load to cause splitting			
— high strength value	56	70	87
— low strength value	37	63	86
Thermal conductivity (Btu/hr/sq ft/in./°F)	0.25	0.35	0.45

properties. However, when used as cores in sandwich structures, foam composites exhibit useful properties.[3,12]

Certain properties characterize most of the plastic foams (Table 8.3 and Figure 8.2). They are light weight, ranging from 0.1 to 70 pcf. They have low thermal conductivity, from 0.12 Btu/sq ft/hr/°F and up. In general they have good water resistance, high strength-to-weight ratios, fire resistance, low moisture permeability, buoyancy, and corrosion resistance. Their resistance to chemical agents and other properties varies with the type of basic plastic, cell structure of foam, and method of foaming.

Some of these foams are thermoplastics (becoming soft when heated), others are thermosetting (infusible when heated), and others are hard compounds of natural or synthetic rubber. All have one feature in common; a cellular or noncelluar foam structure produced by the release of a gas while the material is in a formable state. In some of the compositions, the release of the gas within the material and the cure of the core can be made to take place in one operation, so that the foam structure can be developed as the core material expands, filling the space between the facings. This process is designated as "foamed-in-place" core material.

The majority of foams or cellular cores used are those made in block or extruded forms (profiles). The block of foamed

Table 8.3. Typical Properties of Plastic Foams[a]

Type	Available form	Density (lb/cu ft)	K factor[b] at 75°	Compressive strength (psi)	Tensile Strength (psi)	Maximum Continuous Service Temperature (°F)	Coefficient of Expansion (10^{-5}) (in./in./°F)
Acrylonitrile styrene	expandable beads, billets	1	—	10	—	160	3
Cellulose acetate	boards, rods	6–8	0.30	100–125	120–170	350	2
Epoxy	blocks, spray powder	2–20	0.12	14–1000	20–500	300	1
Methylmethacrylate styrene	boards	1.7	0.50	10	—	175	3
Phenolic	foam-in-place, beads	$\frac{1}{10}$–25	0.20	2–1000	10–250	300	1
Polyethylene	boards, rods, sheets	1.8–60	0.35	20–3500	20–1800	160	3
Polypropylene	slabs, liquid	3–20	—	—	—	250	—
Polystyrene	boards, planks, logs, beads, etc.	1–10	0.23	20–460	40–600	175	2–4
Polyvinyl chloride	sheets, shapes, liquids, paste	3–26	0.20	10–80	200–1200	170	5
Silicone	powder, paste, liquid	3–16	0.28	200–900	30–400	600	—
Urea formaldehyde	boards, shredded	$\frac{1}{3}$–1	0.18	200	120	150	—
Urethane	boards, shapes, foam-in-place	$1\frac{1}{2}$–70	0.11	50–600	400–8000	300	3

[a] Properties depend on manufacturing process, composition and density.
[b] Lowest thermal conductivity listed (Btu/sq ft/hr/°F/in.) other insulation K-factors are:

Powdered gypsum	—0.50	Rock wool	—0.32	Glass fiber	—0.23
Glass foam	—0.40	Cork	—0.28	Urethane foam (2 pcf)	—0.21
Sawdust	—0.36	Hair felt	—0.24	CO_2	—0.21
				fluorocarbon	—0.11

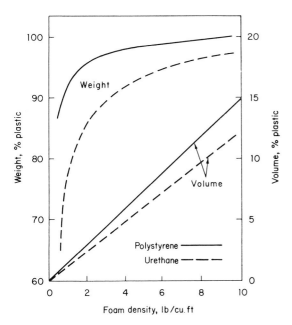

Figure 8.2   Weight and Volume of Foam versus Density

plastics can be sawed or machined to the core shape required. Most organic cores can be used in the lower temperature range of −100°F to 260°F, although some foams under development are being used for limited applications up to 600°F.

Formulators are able to control versatility of foams by varying the density, cell geometry and size, strength-to-weight ratios, thermal and electrical insulation, mechanical flexibility or rigidity, corrosion resistance, color processability, degree of adhesion to other materials, and sound or microwave absorption. Parts can perform multifunctional uses: insulation and load bearing, insulation and ease of application, or buoyancy and structural rigidity. For example, urethane foamed-in-place in the hull of hydrofoils makes the vehicle virtually unsinkable and reduces noise level and structural vibration.[12]

Compared with other plastics, polystyrene foam is the construction industry's principal industrial wall and cold storage insulation. Total foam penetration in this huge market is comparatively small. To help get a bigger share of this market, cost reductions have been brought about by production research that reduces polystyrene density without adversely affecting properties.

The use of rigid urethanes is expanding in sandwich structural-insulation applications in trucks and different transportation vehicles. This market is expanding since urethane foam provides the lowest K factor for insulation. Designs are possible to produce extremely thin sandwich panels.

Even though polystyrenes and urethanes are the present favorites for growth in building and construction, another new rigid foamable polymer, available at much lower cost, could be developed. There are an infinite number of resinous compositions from which foams can be made, as all polymers can be expanded by a gas.

Hundreds of foams are on the market with many different characteristics (Figure 8.3). Forms include slabs, preformed logs, sheets, rods, tubes, choppings, netting tapes, and particles. The foamed-in-place types are liquid, beads, molding powders, and plastisols.

Like all materials, they have limitations. No foam is fireproof, but many of them can be made nonflammable. Phenolics and silicones have excellent heat resistance but could crumble when subjected to vibrational stress.

Another type of plastic foam is identified as syntactic. It provides higher structural strength — at higher densities. Microballoons, or spheres, ranging in diameter from 30 to 100 microns of phenolic, urea, glass, or silica combined with epoxy, polyester, phenolic, or urea resin produce these unique composites. These 8 to 50 pcf foams are used as void fillers in boat structures, cores for aircraft sandwich structures, refrigerators, microwave absorbers, high-frequency communication antennas, and deep submergence sandwich-structure vessels.

Cellular metals

The process for foaming metals by the introduction of metallic hydrides has been studied for several years. The current status is that aluminum alloys, zinc, and lead can be reliably foamed or made into cellular structures within certain limits. They offer unique uses in the area of sandwich structures, including applications where shock-absorption and shock-protection are important, and where weight saving or increased stability of a specific metal is desired. Thermal conductivity is lowered by two orders of magnitude.[13]

The foamed metal alloy can be cast into intricate shapes with reasonable control of densities between 12 per cent and 50 per cent of the base material. In the process a skin, generally impervious, is formed where the foam makes contact with the mold, providing increased localized strength (see Figure 8.4). It is possible to vary this thickness to total imperviousness.

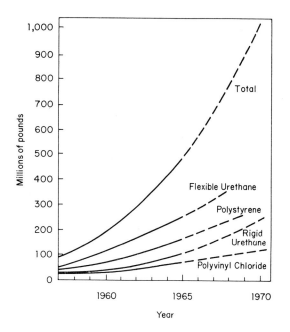

Figure 8.3   Plastic Foam Consumption in U.S., Growth and Forecast (*Plastics World* survey, Ref. 12)

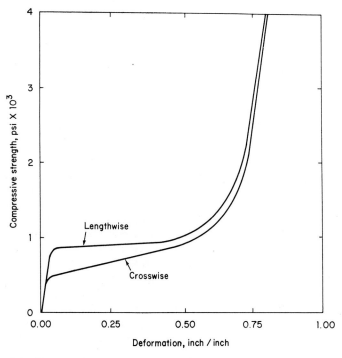

Figure 8.4   Foamed Aluminum Core. Density: 22 lb per cubic ft ±10 per cent

Cell sizes can be varied within controllable limits, as well as their interconnections.

The low-density material can be worked somewhat like wood, using similar tools. It can be machined, sawed, nailed, bonded and to a limited degree it can be formed. Considerable work has been done on two aluminum alloys: Almag (7 per cent magnesium) and aluminum-silicon (11.5 per cent silicon) in 1 to 2 in. panel forms. The following compressive strength values can be used for design purposes:

| Density, pcf | Initial crushing strength, psi | |
	Crosswise	Lengthwise
22	500	900
27	1000	1500
32	2000	2500
37	3500	4000

This material has good energy-absorbing properties without springback.

Research and development previously conducted proved the feasilbility of foaming metals by means of a miscible gas former. Progress has been made in the design and construction of prototype equipment for the continuous production of low-density foams from aluminum-magnesium alloys.[14] Important innovations have occurred such as improved quality of foam through the introduction of air or oxygen into the foamed mixture prior to solidification and the introduction of aluminum oxide coated steel, a material with high corrosion resistance toward molten aluminum, magnesium, and their alloys.

2.3 Bonding, adhesives

The nature of the facing and core materials determines the method and type of bonding used. All-metal constructions, made of the same material or combinations of steel, titanium, beryllium, aluminum, and so on can use welding, brazing, or adhesive techniques.[15–17] Brazing bonds of stainless steel produces sandwich structures with high thermal resistance capabilities. The metals also use adhesives, such as plastic-based, ceramic, and ceramic-braze.[18]

Plastic-based adhesives are principally used in bonding constructions made from plastics, wood, and other nonmetals (see Table 8.4). They are also used to bond nonmetals and metal composites. The section of the optimum adhesive depends upon the service conditions. Normally for maximum performance it is in a liquid or film form. The liquid can be used as a primer on metal surfaces, and for brushing or rolling the face (cell edges) of the core. The film is normally available as a partially cured film of adhesive unsupported or supported by a carrier of glass, nylon, or cotton fabric.

Present and future industries can use these adhesives as important "building blocks" for composites. They can be planned to provide a variety of special properties. Their multi-functional capabilities make them even more important. Advantages include efficiency in application, reliability, compactness, high ultimate fatigue resistance, and aesthetic appeal (see Figure 8.5). They can combine materials and parts which otherwise could not be combined by other techniques. When these adhesives appear to be high in material cost when

Table 8.4. Typical Adhesive Strengths[12]

Type	Tensile Lap-Shear Strength, psi
Alkyd	1000–1500
Acrylate	800–1200
Casein	10– 100
Cellulose-nitrate	50– 200
Cellulose-vinyl	300– 400
Epoxy	1000–5000
Epoxy–novalac	1600–4200
Epoxy–phenolic	2000–3600
Epoxy–polyamide	4000–5200
Epoxy–polysulfide	3000–4500
Epoxy–silicone	1000–2400
Melamine-formaldehyde	2500–3200
Phenolic	1000–5000
Phenolic-butadiene acrylonitrile	400–1600
Phenolic-neoprene	2500–3500
Phenolic-neoprene	3000–4800
Phenolic-vinyl	2000–5300
Polyacrylonitrile	900–1400
Polyamide	2000–4200
Polyethylene	900
Polyimide	1000–2600
Polyvinyl-acetate	50– 350
Polyvinyl-butyral	900
Resorcinol-phenol formaldehyde	1800–2000
Silicone	300–1000
Vinyl butyral-phenolic	1800–2000
Vinyl Copolymers	1700–2200
Urea-formaldehyde	2500–3000
Urethane	4000–9000

This test represents only one of numerous tests that can be used for evaluation, control, etc. The tensile shear test with $\frac{1}{2}$-in. lap (ASTM) is predominantly used for strength control. These data are typical room temperature range of values based on bond line of 1 to 8 mils in thickness and bonding to different adherends. Note even though certain adhesives listed above may appear similar they will differ based on other properties—moisture resistance, peel strength, solvent resistance, temperature of application, etc.

compared to competitive attachment methods, the completed assembly cost should be considered.

Adhesives are generally applied in a low-viscosity fluid form to wet the adherend surface. Transition from fluid to

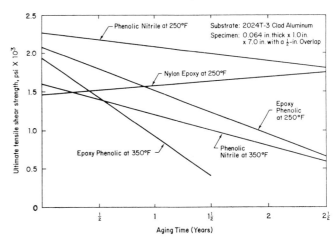

Figure 8.5 Effect of Long-time Aging on the Different Classes of Structural Adhesives (Ref. 60)

the useful solid adhesive form is accomplished by polymerization, solvent removal, or cooling of hot-melt. The polymerization systems, which have provided the most rapid technological progress, involve thermosetting resins, vulcanized elastomers, and certain thermoplastic resins such as polymethyl methacrylate and cyanoacrylate esters. Adhesives are cured by heat, special catalysts for room-temperature curing, or activation by light, oxygen, or even defrosting. Systems with no external heating, required for certain production runs, use formulations of resorcinal formaldehyde, unsaturated polyesters, methyl methacrylate, cyanoacrylate esters, epoxies, and urethanes. All polymerization reactions are exothermic so that bond lines may produce temperatures ranging from 90°F to 400°F.[19]

One of the most important factors from the designer's standpoint is effect of environment — temperature, humidity, tension, shear, peel, damping, and so on. Adhesives are available to meet these different conditions. Bonds can perform in cryogenic (−423°F, see Figure 8.6) to ultrahigh (over

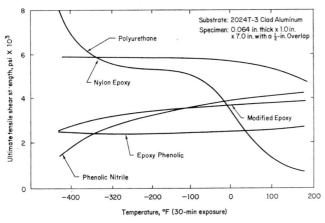

Figure 8.6   Effect of Cryogenic Temperatures on Different Classes of Structural Adhesives (Ref. 60)

1200°F) temperatures from vacuum to atomic blast pressures, and from extreme dryness to complete immersion in water.

To retain strength but increase operating temperature from 350°F, polyimide and polybenzimidazole, among others, are presently being evaluated.

Silicone adhesives have become useful because they remain rubberlike over a wide temperature range, and they are resistant to certain solvents. New one-part systems cure on contact with moisture in the air. However, they have followed other adhesive growth patterns by developing markets based on their inherent characteristics.[19]

Silicone compounds include dimethylsilicone elastomers, polysiloxane, vinylalkoxysilanes, and cyanoalkylsilicone. The vinylalkoxysilane homopolymers can be used as plasticizers and modifiers for synthetic resins. They provide toughening and insoluble characteristics. Silicone pressure-sensitive tapes have been prepared using siloxane resin.

Present dry bonds are possible by the individual or combined properties of newer resins such as polyamide, acrylate, polyvinyl acetate, epoxy, polysulfide, and urethane. These plastics have been copolymerized or alloyed with other resins to provide properties required by different industries.

Resins such as polyimide, polysulfone, polyphenylene oxide, polybenzimidazole, and others will provide more advanced adhesives for tomorrow. They will improve or simplify method of application, improve adhesion, and increase heat resistance.

2.4 Insulation boards

Insulation boards continue to be used in building constructions. The boards are well suited for use as cores in sandwich-type constructions because of their light weight, availability in relatively large areas, ease of laminating, low thermal conductivity, and low cost. Insulation boards may be used as cores in a number of ways. For example, several layers may be laminated together to the required thickness resulting in what will be called flat-grain cores. Other types of cores may be made by cutting strips of insulation board to a width equal to the thickness of core desired and then placing these side by side on their edges. The resultant cores are referred to as end-grain cores.

The conventional method for using insulation boards in sandwich constructions consists of gluing together several layers of insulation board to which are then laminated higher-density face materials. Such materials as sheet metal, lignocellulose hardboard and cement-asbestos board have been used as facings. Sheet-metal faces on conventionally laminated cores have often proved unsatisfactory because they can be readily peeled off the core. Cement-asbestos-faced boards, on the other hand, have been widely and successfully used.[20]

The cement-asbestos-faced slab provides insulation and the exterior and interior finish of the building. It requires no painting and contributes structural strength to the framing. This slab is used for exterior walls, partitions, and ceilings in permanent and temporary housing, tourist cabins, farm buildings, and small business structures. On industrial buildings it is used for the exterior as a curtain wall or as an inside lining material and as roof decking.

Sandwich-type constructions having ordinary insulation-board cores do not have high bending strength. This weakness is owing to the low shear strength parallel to the surface of the board which causes the board to fail in shear when it is bent before the maximum strength of the outer layers is reached. Weakness in shear also brings with it low stiffness.

As papermaking machinery is generally used for felting the fibers to building-board thicknesses, most of the fibers lie in planes parallel to the faces of the board. This fiber orientation affords little resistance to shear, but, if the insulation board is cut into strips and these strips are glued on edge between the sandwich panel faces, a large proportion of the fibers lie in planes at right angles to the faces of the panel. When a panel having such an end-grain core is subjected to bending, it is necessary to tear through many fibers to produce shear failure, so that superior strength would be expected when the fibers are oriented in this manner.

End-grain sandwich-type constructions are generally two to six times stronger and stiffer in bending than corresponding flat-grain panels. The highest test values are obtained with faces of relatively high tensile strength and high modulus of elasticity such as steel. Intermediate values are obtained with faces of moderate tensile strength such as wood. The lowest values are obtained when relatively weak face materials are used, such as gypsum board and chip board. Such panels tested in bending fail in tension or compression of the faces, and the difference in strength between panels having end-grain cores and similar panels having flat-grain cores is not so marked.

A special adhesive is required to bond the metal faces to the end-grain insulation-board core, but any cold hydraulic press or hot-plate press of adequate size as used for the manufacture of plywood can be employed. Some of the suitable adhesives set at room temperatures, whereas others require elevated temperatures.

3 DESIGN CRITERIA

Design criteria for sandwich construction differ from those associated with homogeneous sheet material or sheet with stiffened-rib construction. Much of the original development of sandwich design criteria was based on the work performed on plywood and other wood structures. Sandwich design, oversimplified, may be considered to consist of determining the thicknesses of the facings and core necessary to carry the moments, shears, and axial stresses induced by the loads on the sandwich.[1]

As previously noted, a sandwich panel is analogous to an I beam with facings and core corresponding to the flanges and web. The facings carry axial compressive and tensile forces. The core sustains shear stresses and because of its nearly continuous support of the faces it increases their ability to resist wrinkling or buckling under axial compressive load. When thick facings are used the amount of shear carried by them may be significant. The adhesive is subjected to shear and to transverse tensile stresses caused by the tendency of facings to buckle under axial compression.

Design is based upon sandwich theory and the properties of the materials, which allow the engineer and designer extensive latitude in the location and type of material.[21] He can use extremely high-strength thin faces on a thick plate without excessive weight. He can approach the structural problem with considerable freedom to select whatever combinations will provide the solution. There must be a strong and rigid bond between skins and core materials, plus the necessary inserts, attachment details, edgings and similar items. The bond must force the parts to act as a single unit under deflection by providing enough strength and stiffness to prevent separation and relative motion between the components, and have sufficient strength to force failure in either the core or the facings at ultimate load.

The core material may be of any density or strength compatible with loading and the general environment or nature of the part. Although core materials carry small direct stresses compared to those carried by the facings, these loads are readily determined and must be anticipated before choosing the core material.

Sandwich beams are generally stiffer than conventional types for a given weight. This is due to the more efficient geometric distribution and support of the main load-carrying material (facings) and results in a higher stiffness factor EI. The deflection of any beam varies inversely as the bending stiffness, that is, the deflection of the beam or plate is going to be proportionally less by virtue of the higher value of EI. Under edgewise compression loads, the geometry of the sandwich increases the value of EI.[8]

In aluminum honeycomb sandwich paneling, a high section modulus (I/c) is made possible with minimum weight. As compared to conventional sheet design, the more complete and efficient stabilization of the facings by the core permits development of stresses up to a high percentage of the face yield strengths. Designs have shown weight savings of 12 per cent to 50 per cent when compared with conventional designs of equivalent strength.

Stress distribution in a beam under load is such that bending stresses increase from zero at the neutral axis to maximum values at the outer edges. It follows, therefore, that the ideal sandwich core would be one with strength properties tailored to fit this distribution of stresses. In an effort to provide such a core Peter Hoppe and co-workers at Farbenfabriken Bayer A. G. in West Germany have developed a technique that involves the preparation of urethane sandwich cores having nonuniform density.[3,22] By the use of fibrous reinforcing materials such as glass mats and electrostatically flocked acrylic fibers, a density gradient is produced which ranges from the lowest values at the neutral axis to the highest values near the facings. Strength properties are directly proportional to density so it would appear that this technique fulfills one of the requirements for tailoring properties to match the distribution of stresses. It must be kept in mind that shear stresses are maximum at the neutral axis, and the material must be capable of withstanding shear.

The rigid urethane foam system typically employed in this technique is formulated for a free-rise density of approximately 1.5 pcf. It is then introduced into the cavity with a packing factor or overcharge of 200 per cent to 300 per cent. This overcharge is necessary to ensure the complete filling of the void and to encourage the complete penetration of the fiber layers by the rising foam. This deep penetration or mechanical bonding to the fiber layers is one of the key factors in preparing this structural panel.

The resultant core has a density ranging from 2 pcf at the neutral axis to upwards of 18 to 20 pcf in the outer areas. Average core density is 4 pcf, a function of the total material charged into the cavity.

3.1 Stiffness and strength

Sandwich design is discussed in Chapter 1 where the general mechanical properties are set forth. For sandwiches in which the facings are of equal thickness and have the same mechanical properties certain simplified formulas may be employed, as discussed by Kuenzi.[23] (The following discussion is adapted from Ref. 23.)

Sandwich stiffness

The stiffness D of a rectangular beam having thickness h, width b, and modulus of elasticity E is given by the formula

$$D = Ebh^3/12. \qquad (1)$$

If a sandwich is constructed with the same over-all dimensions but with a core of a different material of thickness c, the stiffness is given by

$$D = [E_f b(h^3 - c^3) + E_c bc^3]/12 \qquad (2)$$

where E_f is the modulus of elasticity of the facings and E_c is the modulus of elasticity of the core.

Since the core of the sandwich is to be extremely light weight compared to the facings, it will probably be a weak, limber material whose stiffness can be neglected in considering the sandwich stiffness. Thus, the last term of the preceding formula may be omitted, and the sandwich stiffness given simply by

$$D = E_f b(h^3 - c^3)/12 \qquad (3)$$

which can be rewritten as

$$D = \frac{E_f bh^3}{12}\left(1 - \frac{c^3}{h^3}\right). \qquad (4)$$

Thus far the design discussion has been limited to determining the flexural stiffness D or in the more usual terminology the EI of the sandwich. The deflection of a sandwich beam is not given entirely by the flexural stiffness. It must also include the deflection due to shearing deformations, because the core material may have a low shear modulus.

The general expression for deflection of a sandwich beam or panel is

$$y = k_B P a^3/D + k_s Pa/N, \qquad (5)$$

where y is deflection, P is total load, a is span, D is flexural stiffness given by formula 3, N is shear stiffness given by

$$N = (h + c)bG_c/2, \qquad (6)$$

where G_c is the core shear modulus, and k_B and k_s are constants dependent on the beam loading. Formula 5 gives deflection in terms of the usual bending deflection (first term) and the shear deflection (second term). For a long span, the first term of formula 5 will be of more consequence than the second term. This is also true in dealing with other materials where it is usually necessary to consider shearing deflections only if the span is very short. Values for the constants k_B and k_s for several beam loadings are given below:

sandwich thickness, c is the core thickness, and b is the sandwich width. The shear stress in the core is given by

$$S = 2V/(h + c)b, \qquad (8)$$

where S is the core shear stress, and V is the shear load on the sandwich.

Formulas 7 and 8 are approximate and can be used for most sandwiches with thin facings and moderately rigid and thick cores.

Sandwich construction may also be used effectively for carrying edge loads, such as might be required if it were used for bearing walls in buildings. Such design must be dependent upon the buckling resistance of a sandwich column or upon its ability to resist direct compression in the facings, whichever is the lesser.

Compressive stresses in the facings are given by

$$S = P/2tb. \qquad (9)$$

If the panel is simply supported at its ends, the column buckling load is given by

$$P = \frac{\pi^2 D}{a^2(1 + \pi^2 D/a^2 N)}, \qquad (10)$$

where P is total load, t is facing thickness, b is column width, a is column length, D is defined by formula 3, and N is defined by formula 6. The second term in the denominator of formula 10 accounts for possible shearing deformation in the core.

If the load-bearing wall panel is held in line at its vertical edges, the buckling load of the panel is given approximately by

$$P = \frac{4\pi^2 D}{b^2(1 + \pi^2 D/b^2 N)^2} \qquad (11)$$

Values of k_B and k_s for several sandwich beam loadings

Loading	Beam Ends	Deflections at	k_B	k_s
Uniformly distributed	Both simply supported	Midspan	5/384	⅛
Uniformly distributed	Both clamped	Midspan	1/384	⅛
Concentrated at midspan	Both simply supported	Midspan	1/48	¼
Concentrated at midspan	Both clamped	Midspan	1/192	¼
Concentrated at outer quarter points	Both simply supported	Midspan	11/768	⅛
Concentrated at outer quarter points	Both simply supported	Load point	1/96	⅛
Uniformly distributed	Cantilever, 1 free, 1 clamped	Free end	1/8	½
Concentrated at free end	Cantilever, 1 free, 1 clamped	Free end	1/3	1

Strength of sandwich

The strength of a sandwich beam under bending and shear loads is determined by the ability of the facings to resist compression or tension and that of the core and adhesive bond to resist shear. The stresses produced in the facings by bending moment applied to the sandwich are given by the formula,

$$F = 2M/t(h + c)b, \qquad (7)$$

where F is the mean compressive or tensile stress, M is the bending moment, t is the thickness of one facing, h is the total

for panels that are at least as long as they are wide and for which the second term in the bracket of the denominator is less than or equal to unity. In formula 11, P is total load, b is panel width, D is given by formula 3, and N is given by formula 6.

The preceding design criteria for sandwich stiffness and strength are suitable for sandwiches with thin isotropic facings and isotropic cores and are approximate for orthotropic materials.

3.2 Minimum weight

The concept of sandwich construction combining thin, strong facings on light-weight, thick cores immediately suggests possibilities of deriving constructions so proportioned that minimum weight for a given stiffness or loading capability is achieved. As explained in the work of E. W. Kuenzi of Forest Products Laboratory,[24] it is important to realize that the minimum-weight construction derived may not be practical because of unusually thin facings which are not available, or some other detail such as unusual light-weight core of great thickness. Since it is possible by theory to arrive at an impractical design, various minimum-weight analyses should be used with caution for comparing sandwich with other constructions unless the sandwich proportions are examined. Analyses of the efficiency of panels of various sandwich constructions of certain materials have been reported. The following sections are adapted from Kuenzi's analysis of minimum-weight sandwich (Ref. 24), considering stiffness, bending moment, and edge-loading capacity.

Bending stiffness

Since the primary purpose of a structural sandwich is to provide stiffness (hence low deflection under transverse load and high resistance to buckling under edgewise load), the analysis of minimum-weight sandwich to provide a specified bending stiffness is considered first.

Sandwich bending stiffness per unit width can be derived by elementary mechanics and is given by the following formula for a sandwich with thin facings and a core of negligible bending stiffness:

$$D = \frac{\dfrac{E_1 t_1}{\lambda_1} \dfrac{E_2 t_2}{\lambda_2} h^2}{\dfrac{E_1 t_1}{\lambda_1} + \dfrac{E_2 t_2}{\lambda_2}}, \qquad (12)$$

where D is bending stiffness; subscripts 1 and 2 denote facings 1 and 2; E is modulus of elasticity; λ is one minus the product of two Poisson's ratios; t is thickness; and h is distance between facing centroids.

After setting

$$\beta = \frac{E_2 t_2 \lambda_1}{E_1 t_1 \lambda_2}, \qquad (13)$$

formula 1 can be rewritten as

$$D = \frac{E_1 t_1}{\lambda_1} h^2 \frac{\beta}{1+\beta}. \qquad (14)$$

The weight of a sandwich with thin facings is given by the formula

$$W = w_1 t_1 + w_2 t_2 + w_c h + W_B \qquad (15)$$

where w is density (pci); W_B is total weight of bond (adhesive or braze) between facings and core (psi); and W is sandwich weight (psi). If we assume that bond weight is the same for all sandwiches of the type considered, then

$$(W - W_B) = w_1 t_1 + w_2 t_2 + w_c h. \qquad (16)$$

Now from expression 13

$$t_2 = \beta t_1 \frac{E_1 \lambda_2}{E_2 \lambda_1}.$$

Substituting this into 16 and minimizing with respect to h, we obtain

$$h^3 = \frac{2D\lambda_1}{E_1 w_c}\left(w_1 + \beta \frac{E_1 \lambda_2}{E_2 \lambda_1} w_2\right)\frac{1+\beta}{\beta}. \qquad (17)$$

The configuration of this minimum weight sandwich can be examined by substituting 14 for D in formula 17 and obtaining finally

$$\frac{t_1}{h} = \frac{w_c}{2\left(w_1 + \beta \dfrac{E_1 \lambda_2}{E_2 \lambda_1} w_2\right)}. \qquad (18)$$

Rewriting 18 and substituting 13 for β, we find

$$\frac{w_c h}{2(w_1 t_1 + w_2 t_2)} = 1; \qquad (19)$$

but the core weight is $W_c = w_c h$ and the facing weight is $W_F = w_1 t_1 + w_2 t_2$ and substitution of these into 9 produces $W_c = 2W_F$. Then the sandwich weight $(W - W_B) = W_c + W_F = 3W_F$, and therefore

$$\frac{W_c}{W - W_B} = \frac{2}{3}, \qquad (20)$$

and thus for minimum-weight sandwich of a specified stiffness the core weight must be two-thirds the weight of the sandwich minus bond weight.

Formula 17 can be used to find h and formula 18 to find t_1. There remains the determination of t_2 from 13 after β is found. The value of β after minimizing as above with respect to β is

$$\beta = \sqrt{\frac{w_1 E_2 \lambda_1}{w_2 E_1 \lambda_2}}, \qquad (21)$$

and substituting 21 into 13 we obtain

$$t_2 = t_1 \sqrt{\frac{w_1 E_1 \lambda_2}{w_2 E_2 \lambda_1}}, \qquad (22)$$

For many combinations of facing materials and cores the value of β can be assumed to be unity, thus resulting in $E_1 t_1/\lambda_1 = E_2 t_2/\lambda_2$.

The foregoing has not considered high stresses that might have to be carried by very thin facings. Because of stress limitations, availability of facings in proper thicknesses, and availability of cores of low density, it will often be found that the minimum-weight sandwich cannot be realized. It is also important to review other discussions of the minimum-weight sandwich to be sure that inherently impossible combinations of unusually thin facings on light-weight cores are not being examined.

Bending moment capacity

The bending moment resistance of a sandwich with thin, equal facings on a core of negligible bending stiffness is given by the formula

$$M = Fth, \qquad (23)$$

where M is bending moment per unit width, F is design facing stress, t is facing thickness, and h is distance between facing centroids. Following the same procedure as for bending stiffness, formula 23 is solved for t substitution of this in the weight formula $(W - W_B) = 2wt + w_c h$ and minimizing results in

$$h^2 = 2\frac{w}{w_c}\frac{M}{F}. \tag{24}$$

The configuration of this minimum-weight sandwich can be examined by substituting 23 for M in formula 24 and obtaining, finally

$$\frac{t}{h} = \frac{w_c}{2w}; \tag{25}$$

and further substitution of 25 into the weight formula leads to

$$W_c = \tfrac{1}{2}(W - W_B). \tag{26}$$

Thus the core weight for a minimum-weight sandwich of specified moment capacity must be one-half the weight of the sandwich minus bond weight.

The moment capacity was based on a facing stress F which may be an allowable stress or failing stress, and so on. If the possibility of local wrinkling or dimpling of sandwich facings exists, then moment capacity should be based on that stress at which wrinkling or dimpling of facings occurs. The ratio t/h remains as given in formula 25.

The dimpling stress of sandwich facings on honeycomb or corrugated core is dependent upon facing properties and unsupported width of facing. The dimpling stress is given by the formula

$$F_D = kEt^2/s^2, \tag{27}$$

where F_D is dimpling stress of facings; k is a theoretical or empirical buckling coefficient; E is effective elastic modulus of facing; t is facing thickness; and s is honeycomb core cell size or spacing between points of corrugated core supports for the facings. If it is assumed that the core density is related to the facing density and is inversely proportional to s, then

$$s = k_1 w/w_c,$$

and 27 becomes

$$F_D = KEt^2(w_c/w)^2 \tag{28}$$

where $K = k/k_1^2$. Proceeding as before and minimizing, it is found that

$$W_c = \tfrac{1}{4}(W - W_B). \tag{29}$$

Buckling under compressive edge load

The edge load capacity of a sandwich panel, precluding local facing failures by wrinkling, dimpling, or facing compression failure, is dependent on the buckling of the entire sandwich. This buckling is determined not only by the sandwich bending stiffness D but also by the shear stiffness.

The buckling load, per unit width, of a simply supported flat sandwich panel with isotropic facings and core, and having a length not less than its loaded width is given by the formula[25]

$$P_e = K\frac{\pi^2}{b^2}D \tag{30}$$

where

$$D = \frac{Eth^2}{2\lambda}, \ K = \frac{4}{(1 + C_e)^2}, \ C_c = \frac{\pi^2 Eth}{2\lambda b^2 G_c}, \ \lambda = 1 - \nu^2,$$

and D_e is buckling load per unit panel width; b is panel width (loaded edge); E is facing elastic modulus; t is facing thickness; h is distance between facing centroids; G_c is core shear modulus; and ν is facing Poisson's ratio. After substituting values for D, K, and V in equation 30 and minimization, we find

$$\frac{W_c}{(W - W_B)} = \frac{2}{3 - C_e}. \tag{31}$$

Thus if the core shear modulus is large and $C_e = 0$, the core weight is two-thirds of the sandwich weight minus bond weight. This also was the result obtained when prescribed bending stiffness was analyzed and was to be expected since for $C_e = 0$ buckling depends on bending stiffness (see 30). The effect of a $C_e \neq 0$ reduces the core weight relative to the sandwich weight.

Buckling of sandwich cylinders under axial compressive load

The axial compressive load capacity of a circular cylinder with walls of sandwich construction, precluding local facing failures by wrinkling, dimpling, or facing compression failure, is dependent on the buckling of the sandwich walls.

The buckling load, per unit circumference, of a sandwich-walled cylinder in axial compression is given by the formula[26,27]

$$P_c = KEth/r, \tag{32}$$

where $K = k_1(1 - k_2 C_c)$ for $C_c < \tfrac{1}{2}$ (approximately) and $C_c = Et/rG_c$ (approximately); P_c is buckling load per unit cylinder circumference; E is elastic modulus of facing; t is facing thickness; h is distance between facing centroids; r is mean radius of cylinder; k_1 is a coefficient dependent upon whether buckling is governed by small or large deflection theory; k_2 is a coefficient depending upon whether isotropic or orthotropic core is used and also upon small or large deflection theory; and G_c is core shear modulus associated with shear distortion axially.

Substituting, solving, and minimizing, we obtain

$$\frac{w_c h}{(W - W_B)} = \frac{W_c}{(W - W_B)} = \frac{1 - k_2 C_c}{2 - 3k_2 C_c}. \tag{33}$$

Thus, the core weight, W_c, is determined to be about one-half the sandwich weight for a sandwich proportioned to give minimum weight for a given cylinder buckling load. This was to be expected because for $C_e = 0$ the buckling load is dependent upon the product th as was the case when the prescribed moment resistance was analyzed for which $W_c = \tfrac{1}{2}(W - W_B)$.

3.3 Optimum material selection

It has been possible to establish simple formulas to determine the most economical relationship between the skins and the cores. Materials selection is particularly acute with composite panels because of the large number of available materials and the wide range of permissible sizes and thicknesses.

Hexagonal honeycombs display pronounced directional shear properties. Designers use this characteristic to advantage by orienting the core so that the major shear stresses are carried in the L (width, longitudinal, ribbon) direction. In the case of welded skins or nonflowing braze material, the node bonds are absent after exposure to high temperature. The shear strength of panels without node bonds is reported to be from 70 to 98 per cent in the L direction, and from 68 to 75 per cent in the W direction of the strengths obtained from the completely bonded nodes.

The deflection of thin, low-density composite panels results from compression and extension of the face panels and shear deformation of the core. Resistance to deflection for a given load and span can be improved by making the faces thicker; using a face material with a higher modulus of elasticity; and using a thicker core, increasing the shear modulus of the core, or both. There is an optimum combination of factors that will produce the most economical construction. Rib construction can also be evaluated (Figure 8.7).[28] One material selection procedure takes into account that any material to be used does not pass its rupture point. It is also assumed that the panels are uniformly loaded and are flat, not deeply corrugated. Other conditions are that the modulus of elasticity of the core is much lower than that of the facings. This leads to the common assumption that compared to facing material, the resistance of the core to compression and extension can be neglected. Thus, the core contributes to the stiffness of the panels only by resisting shear. Also the cost per unit weight of the core and the cost per unit volume of the faces are constant. The insulation value of the core can be ignored because it will normally exceed design requirements. Environmental and aesthetic requirements are not involved.

This analytical approach can be used to develop a more accurate method of producing the most structurally efficient panel at the lowest cost. It can be used in applications where pure guesswork is generally used in selecting materials. The basic formulas are as follows:

$$P_{ct} = \frac{2.38 \times 10^8 (c_1 + F_r c_2)\left(1 + F_r \frac{E_2}{E_1}\right) E_s^3}{E_2 F_r c_c p_c} \times \left(\frac{Y}{LW}\right)^2 \quad (34)$$

$$F_r = \frac{t_2}{t_1} = \sqrt{\frac{E_1 c_1}{E_2 c_2}}, \quad (35)$$

$$t_1 = \frac{5.05 \times 10^{-7} L c_c p_c}{E_s (c_1 + F_r c_2)} \times \frac{LW}{Y} \times \frac{(1 - N_{ct})}{N_{ct}} \times \frac{(1 - 2N_{ct})}{(2 - 3N_{ct})}, \quad (36)$$

$$t_c = \frac{\frac{8.7 \times 10^{-4} L}{E_s} \pm \sqrt{\frac{75.7 \times 10^{-8} L^2}{E_s^2} + \frac{Y}{LW} \times \frac{36 \times 10^{-5} L^3 \left(1 + \frac{E_2}{E_1} F_r\right)}{E_2 t_2}}}{\frac{2Y}{LW}} \quad (37)$$

where

P_{ct} = core cost factor,
c_1 = cost of one facing, ¢/cu in.,
c_2 = cost of other facing, ¢/cu in.,
c_c = cost of core, ¢/lb,
F_r = optimum ratio of thickness of the faces,
E_1 = modulus of elasticity for one facing, psi,

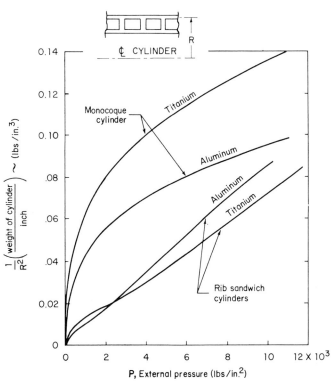

Figure 8.7   Comparison of Weight Between Optimum Rib-sandwich Cylinders and Monocoque Cylinders under External Pressure (Ref. 39)

E_2 = modulus of elasticity for other facing, psi,

E_s = shear modulus of core, psi,

p_c = core density, lbs/cu ft, pcf,

Y = allowable deflection, in.,

L = panel length, in.,

W = panel width, in.,

t_1 = thickness of one facing, in.,

t_2 = thickness of other facing, in.,

N_{ct} = optimum cost of core material in per cent of the total cost of core and facings,

t_c = core thickness, in.

These formulas take into consideration three rules concerning core density, core cost, and face thickness. Rule 1 states that a core material always should be selected with the lowest possible density. The optimum core material to use in composite panels is the material with the lowest density that will provide adequate shear strength and durability. It must also retain its form during end-use under handling loads.

Rule 2 states that the core should always cost at least twice as much as the two facings. This rule applies regardless of the span, load, properties, or the unit cost of any of the materials. In the case where the shear modulus of the core is so great that shear deformation is negligible, the core should cost just twice as much as the faces. Also, the optimum relative cost of the core, compared to the faces, will increase if lower shear modulus materials are selected for the core.

The optimum cost of the core materials (N_{ct}) can be determined by first calculating a core cost factor (P_{ct}) from equation 34. The point at which (P_{ct}) intersects the curve in Figure 8.8 gives the optimum cost of the core material.

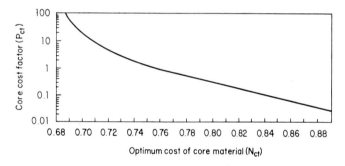

Figure 8.8   Optimum cost of core material is determined by first calculating core cost factor from Equation 34 and then extending cost factor to curve; intersection gives optimum core cost (Ref. 28)

Rule 3 states that the ratio of the thicknesses of faces should be equal to the inverse ratio of the product of either modulus of elasticity multiplied by the product of their costs per unit volume. This rule is based on equations 35 and 36. As shown in equation 35, the optimum ratio of the thickness of the faces (F_r) of the different face materials is equal to the square root of the inverse ratio of the products of their moduli of elasticity times their costs per unit volume. The thickness of either face panel (t_1) can be determined from equation 36. The thickness of the other face is then determined from the thickness ratio of the faces determined in equation 35. The thickness of the core (t_c) can be calculated from equation 37.

The following example will apply these equations to a composite panel made up of aluminum and steel faces over a

urethane foam core. The panel measures 240 in. long (L) by 48 in. wide (W) and the allowable deflection (y) is 1 in. The aluminum (face 1) has a modulus of elasticity (E_1) of 10^7 psi, and its cost (c_1) is 40¢/lb × 0.099 lb/cu in., or 3.96¢/cu in. Steel (face 2) has a modulus (E_2) of 3×10^7 psi, and its cost (c_2) is 11¢/lb × 0.283 lb/cu in. or 3.11¢/cu in. The polyurethane core material has a shear modulus (E_s) of 500 psi, a density (p_c) of 3 pcf and a cost (c_c) of 45¢/lb.

By substituting these figures in the equations, values obtained are

1. Relative thickness of the faces: The thickness ratio of the steel to aluminum face as calculated from equation 35 is 0.65.

2. Relative cost of the core: The core cost factor calculated from equation 34 is 5.6. Using this value in conjunction with the appropriate curve, the optimum cost of the core material (N_{ct}'— or the cost of the core divided by the combined cost of the core and faces) is 0.72. This means that the cost of the core should constitute 72 per cent of the total cost of core and facing materials.

3. Thickness of the faces: The thickness of the aluminum face as calculated from N_{ct} and equation 36 is 0.035 in. Then, from equation 35, the steel face is 0.023 in.

4. Thickness of the core: From equation 37, the thickness of the core is 7.2 in.

Filament-wound technology

This review has so far concerned itself with flat or relatively flat sandwich structures. Sandwich cylindrical or even box-shaped structures are sometimes designed to take internal and/or external pressures, as well as to meet stiffness requirements.[29]

As is pointed out above, the stiffness of a structure is a function of the modulus of elasticity and moment of inertia. Since the modulus of elasticity can only remain constant for a given material, a unique solution to the problem is to develop a sandwich construction.

Filament-wound facing structures for sandwiches have been developed having the highest strength-to-weight ratio of any known structural material. Design parameters have been developed based on experimental and production units. The design criterion is concerned with fiber orientation. The basic design method is to orient the fibers in the direction of the principal stresses and to proportion the number of fibers with respect to the sizes of the principal stresses.

In structures where it is geometrically impossible to orient the fibers precisely in the direction of the principal stresses they are oriented at some angle with them. A balanced structure can be achieved by proportionately locating fibers at two basically different winding angles (low helicals and circumferentials). A balanced structure is one in which the fibers oriented in any direction in the structure have equal stress applied to them under load.

Two major considerations in the design of deep-submergence vehicles are material strength and resistance to buckling. The traditional approach to design for resistance to buckling is to install ribs of some type inside the shell of the vessel. Ribs, however, prevent full utilization of the compressive strength of filamentary materials. The composite sandwich is one means of designing without ribs. The technical reason for the advantages offered by composite sandwich design may best be understood by examining the stress set up in ribbed structures (Figure 8.9) as contrasted with those in composite sandwich structures.[30]

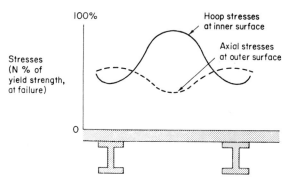

Figure 8.9 Use of stiffener ribs inside filament wound cylinder causes shear between RP and rib which increases fiber load (Ref. 30)

In composite sandwich design, the amount of material required for compressive strength is distributed between two concentric cylinders which are separated from one another by a light-weight core material in accordance with buckling requirements. Stresses are uniform throughout the filament-wound facings. The core need only sustain a radial pressure of about one-half the hydrostatic pressure, and there is only modest shear at the inner facing.[31]

The U.S. Naval Underwater Ordnance Station, Newport, R.I., undertook a program to evaluate structural performance of sandwich composites. Cylindrical models were built and pressure tested utilizing filament-wound reinforced-plastic (RP) facings and a syntactic core of hollow glass microspheres in a matrix of epoxy resin.

As the models were small and the RP facings necessarily thin, a 1C:1A filament orientation was used — alternate layers in the circumferential (hoop) direction and the axial direction. A 2C:1A orientation (two circumferential layers alternating with one axial layer) is used in larger vessels.

The test models were for research rather than prototype. They were designed to push the facings to compressive failure in one case and to collapse by buckling in the other. Results were as follows:

Failure Mode	External Hydrostatic Pressure
Compressive stress	
Design	4000 psi
Test	4200 psi
Collapse	
Design	2080 psi
Test	2200 psi

The stresses, measured by strain gauges, were uniform throughout the facings, and the type of failure was as predicted.

There are factors that make the task of torpedo hull design for great depths somewhat less forbidding: (1) torpedoes do not involve support of human life; (2) torpedoes are small enough to be easily tested; and (3) the number of pressure cycle excursions (cyclic loading) is limited. However, the ratio of hull weight to vehicle displacement increases sharply with depth. This creates space, weight, and buoyancy problems that are particularly acute for small vehicles because of the substantial weight and volume requirements of propulsion machinery, other internal hardware, and payload.

4 FABRICATION

Fabricating techniques will always be important in producing quality-controlled and structurally efficient parts. In practically all methods precise procedures are required in order to obtain maximum efficiency, as, for example, in aircraft to ensure meeting allowable specification requirements. Different procedures are used in fabrication. Brazing methods are used with steel; polymeric adhesives are used with plastics or metal-plastic composites; other methods are also used.

With metals such as aluminum, the surface must be carefully cleaned and prepared. After cleaning, careful handling is required. Even touching the surface with the fingers in handling has been found to result in open bonds.

4.1 Plastics

The fabrication of sandwich panels with glass-fabric plastic resin laminate facings requires a somewhat different technique than other materials. These panels can be prepared by a wet-laminating process in which the laminate is formed from individual layers of fabric and bonded to the core in a single operation. In this case a single resinous formulation serves both as the impregnating resin for the fabric and the adhesive for bonding to the core. It is obvious that this process requires a rather unusual fabricating process, or at least more quality control than bonding a precured laminate to the core. Certain polyester resins of the chemically unsaturated type have been developed which become thermosetting formulations by the incorporation of another unsaturated liquid monomer that serves first as the solvent to permit the necessary thinning of the resin for application and impregnation of the fabric, and then for obtaining the necessary adhesion to the core.

Unlike many materials which are first manufactured as stock sheets, bars, or profiles, and then fabricated into final form, parts made of fibrous composites are customarily fabricated in a two-step operation or directly from the basic constituents. Properties of the article are strongly influenced by the fabrication process.[2] The commonly employed glass-fiber facing composites, for example, are generally fabricated by the following methods; hand lay-up, preform, spray-up, and filament winding.

In hand lay-up layers of random fiber mat or of woven fabric are cut to size and laid on forms. Liquid resin is sprayed, brushed, or otherwise applied to the successive layers.

Complex three-dimensional shapes can be made by the preform method. They are first approximated by depositing chopped fibers on a screen of approximate shape and then given their final form by transferring this preform to a mold in a press. The screen deposits can be made by different techniques such as dry or wet lay-up.

In spray-up, bundles of continuous filaments, or rovings, are fed through a chopper. The chopped fibers are blown unto a mold simultaneously with a spray of liquid resin. The composite of resin and fiber is then usually compacted by rubber rolls.

In filament winding continuous filaments are wound onto a mandrel after passing through a bath of liquid resin.[29] The pattern of winding is controlled to conform to the requirements of the part. Tension in the filaments is also controlled. This method provides the greatest strength or the maximum efficient structure when compared on the weight basis. This process also provides for ease of quality-controlled fabrication of parts in fabricating a few parts, but more so in mass production.

After the reinforced plastics are in place on the form they may be allowed to stand until the matrix hardens by polymerization, with or without the addition of heat. They can be placed in a flexible bag (such as autoclave or hydroclave) which is then evacuated to draw out entrapped air and to provide some external air or water pressure. They can also be placed in a compression press with matched metal molds or flat platens.

4.2 Honeycomb cores

Fabrication methods are of course different from the riveting operations for conventional sheet stringer panel production. Flat panels are the simplest to fabricate, but contoured circular shapes are possible. As an example, if the radius involves only mild curves, constant section compound curves, when slight, can be fabricated in aluminum honeycomb sandwich structure with preformed facing sheets and the honeycomb core formed in the bonding fixture, or with preformed facing sheets and preformed core. The thinner the foil gage of the core, the easier the bending. For $\frac{1}{8}$-in. hexagonal cell size and 0.002-in. foil thickness, the minimum bend radius (measured to the inside of the curve) is ten times the core thickness for longitudinal bending and five times the core thickness for transverse bending.[8]

There are two basic methods of manufacturing any type (metal, plastic, paper, and so on) of honeycomb cores: the corrugation method and the expansion method. In the former method, the sheet material is first corrugated into halfhexagons. The corrugated sheets are then stacked and bonded. In the expansion method, adhesive lines are put on flat sheets which are then stacked, bonded to each other, and expanded after bonding to form the honeycomb pattern.

With fiber-reinforced-plastic cores the material fed into the stacking machine is a resin preimpregnated fibrous sheet. The most popular construction uses phenolic, polyester, or modified epoxy resins with glass, nylon, or cotton woven fabrics. Nonwoven-fiber-reinforced mats are also used. At present more work is being conducted with the new heat-resistant resins using high-strength fibrous structures of graphite, carbon, or boron. These types of composites are required to increase the heat-resistant properties of nometallic cores.

Honeycomb materials are normally supplied as flat panels, having standard length and width dimensions and cut to a constant specific thickness. In most sandwich applications it is necessary for core material to be cut from standard panels to the required length and width. Three methods are used depending upon the density and thickness of the material and quantity.[32] Where the density, thickness, and number of parts required are not too great, the least expensive method is hand-cutting with a serrated-edge bread knife.

Combinations of heavy and thick core may be cut by bandsaws. This method is also used where large panels or fairly large quantities of small panels are to be cut. A steel rule die is used in cutting all types of core where the high volume justifies cost of die.

Most carving techniques require that the honeycomb be held firmly during cutting. Honeycomb, as is, is difficult to hold in conventional fixtures. However a number of special methods are in use. The most satisfactory is to freeze the core piece to its chuck with polyethylene glycol. This material melts at 120°F to 140°F and is water soluble, nontoxic, and relatively inert. Its application can be facilitated by using a chuck with a manifold to circulate hot and cold water. After heating the chuck, the white flake material is applied and melted to form a layer $\frac{1}{32}$ to $\frac{1}{16}$ in. thick. The core is positioned in the melted plastic and held down by weights, vacuum, or a pressure pillow while the plastic sets. To speed up setting, cold water can be passed through the manifold chuck. Other methods are used to apply this plastic.

Limited machining may be accomplished with various taping techniques to hold the core. Double-back tape with pressure-sensitive adhesive on both sides can be used. Vacuum chucking is another method. This process requires a temporary skin to be attached on one side.

If unexpanded honeycomb can be used, problems of machining can be reduced or eliminated. Simple curves and tapers can be inexpensively produced by performing machining operations on the unexpanded billet before it is opened. Complex shapes can be carved in a milling machine.

4.3 Cellular cores

Cellular plastics, also called foamed or expanded plastics, are becoming materials of increasing commercial importance.[33] Foams may be classified according to structure into two general types: closed-cell foams in which each individual cell is completely closed in by a thin membrane of polymeric material, or open-cell foams in which the individual cells are open to each other.

Virtually every type of thermoplastic or thermosetting resin can be foamed. In general, the basic properties of the respective plastics are present in the foamed products, except those that are changed by the conversion process.

Cellular plastics can be commercially produced to form slabs, blocks, boards, sheets, molded shapes, extrusions, and sprayed coatings. In addition several types of expanded plastics can be foamed in place[34] in an existing cavity. A cellular plastic can be developed by several methods:

Air or gas is whipped into a suspension or solution of the plastic, which is then hardened by heat or catalytic action or both.

A gas is dissolved in a resin and expands when pressure is reduced.

A liquid component of the mix is volatilized by heat.

Water produced in an exothermic chemical reaction is volatilized within the mass by the heat of the reaction.

Carbon dioxide gas is produced within the mass by chemical reaction.

A gas, such as nitrogen, is liberated within the mass by

thermal decomposition of a chemical blowing agent.

Tiny beads of thermosetting resins, hollow or expandable by heat, are incorporated in a plastic mix.[35]

Among the commercially important types of cellular plastics are cellulose acetate, epoxy, phenol-formaldehyde, polyethylene, polystyrene, silicone, urea-formaldehyde, urethane and vinyl chloride.

By varying the ratio of ingredients, changes in foaming characteristics and/or foam properties can be achieved. In terms of the basic ingredients for most foams these may be categorized as follows:

Surfactant. Use of the recommended quantity of surfactant results in a uniform, fine cell size distribution. A three-fourths reduction of the recommended quantity results in a coarse, thick, cell-wall structure. An increase of up to 50 times the normal amount has a plasticizing effect during foaming and produces a tough, somewhat resilient foam. Small variations are not significant.

Blowing agent. The use of the recommended quantity of blowing agent results in optimum foam efficiency (lowest density for a given foam resin reactivity) and cell structure. Minor variations of up to 10 per cent are insignificant. With one-quarter the required quantity, a rapid increase in foam density occurs.

Catalyst. The type and amount of catalyst used is important in obtaining optimum foam expansion and minimum foam shrinkage. If less than the recommended amount is used, an insufficient rate of resin cure occurs which in turn causes serious foam shrinkage. If more than the desired amount of catalyst is used the foam sets before complete vaporization of water and blowing agent can take place. This results in increased foam density.

4.4 Metals

It is important that facing-to-core contact surfaces be accurately located in order to eliminate mismatch that would result in poor structural bonds.[36]

Brazing

Preliminary design calculations of future missiles and modes of transportation show the need for a constructional material capable of withstanding service temperatures in the range of 1600°F to 2200°F.

Molybdenum, a metal that is quite plentiful in the earth's crust, is 30 per cent more dense than iron and appears to be usable in this range. It retains an appreciable amount of its useful strength and stiffness up to its recrystallization temperature, which is quoted as 2400°F for a 0.5 per cent titanium molybdenum alloy. An example of brazing characteristics of small tee specimens in both argon and hydrogen atmospheres at various times and temperatures is as follows:[37]

A honeycomb sandwich structure consisting of 6061 aluminum alloy skins has been brazed to 3003 aluminum alloy core.[16]

Other test programs have involved finding usable brazing alloys. One project evaluated eight commercially available alloys for fillet formation, flow and strength when joining beryllium to itself, titanium, and 17-7 PH steel.[17] Metallographic examinations were made to determine the extent of diffusion. The brazing alloys selected from the study were

1. Beryllium to steel: Ag-Cu-Li braze alloy;
2. Beryllium to titanium: Ag-5A1-0.2 Mn;
3. Beryllium to beryllium: Ag-5A1-0.2 Mn.

Lap shear type tests of beryllium to titanium developed strengths in excess of 16,000 psi. Another important aspect of this study shows that silver brazing alloy density was so high that the strength-to-weight advantage of beryllium was lost. Preliminary tests were conducted on lower density alloys with some degree of success.

Different brazed structures are now being used. Design tests and service tests have been conducted on numerous primary structural applications, such as the B-58 bomber with RS 140 titanium skins with 17-7 PH stainless steel.[38]

Welding

Welded sandwich structures are employed. The more conventional welding methods have been used; however, limitations on size and shape can readily develop, particularly when cost considerations are involved. One of the new fabricating methods is identified as "roll-welding."[39] This process makes it practical to realize the advantages of sandwich structures. The three primary advantages are fabrication of complex parts, reliable diffusion bond, and low cost. (See Table 8.5.)

Titanium and titanium alloys are particularly well suited; others include aluminum, columbium, molybdenum, and stainless steel. An especial attribute reportedly is that neither new machines nor unusual techniques are required. Cores can be varied to suit the individual applications. The standard truss-core type with 45° ribs is reported as very likely to meet requirements.

In this process welding occurs along a line parallel to the ribs. Therefore, the roll-welding process is effective only for sandwich structures having unidirectional ribs, or for panels small enough to be cross rolled on commercial mills.

5 TEST METHODS

Different destructive and nondestructive test methods are used to control the production of sandwich panels. They are used for control from base materials to the end item. The more conventional sheet material tests are used in the preliminary

Alloy	Best brazing temp. (°F)	Flow	Wetting	Fillets	Room temp. strength
Foil copper	2050	Fair	Unsatisfactory	Fair	Good
AMS-4775 Nicrobraze	2000	Good	Good	Excellent	Good
Coast #50	2000	Fair	Good	Fair	Good
Nicrobraze #45	2250	Fair	Fair	Fair	Unsatisfactory

Table 8.5. Results of Mechanical Tests on Ti-6A1-4V Alloy Sandwich Structures (Roll-Welding)[39]

Length	Width	Thicknesss	Ultimate Load (lb)	Mode of Failure
	Dimensions of specimens, (in.)			
Flatwise Compression Test for Sandwich Core (Reference ASTM C 365-57)				
4.56	3.0	0.251	8360	Ribs failed in shear adjacent to bond junctures.
4.56	3.0	0.251	7300	Ribs failed in shear adjacent to bond junctures.
4.56	3.0	0.251	7180	Ribs failed in shear adjacent to bond junctures.
		Avg.	7613	
Edgewise Compression Test (Reference ASTM C 364-61)				
2.94	2.50	0.251	8780	Facings crimped in areas between bond junctures. Bonds did not fail
2.94	2.50	0.251	8710	Facings crimped in areas between bond junctures. Bonds did not fail
2.94	2.50	0.251	6270	Facings crimped in areas between bond junctures. Bonds did not fail
		Avg.	7943	
Flatwise Flexure Test (Reference ASTM C 393-61T)				
5.75	1.50	0.251	720	Face sheets buckled. Bonds did not fail
5.75	1.50	0.251	1004	Face sheets buckled. Bonds did not fail
		Avg.	862	

tests. However the sandwich tests in most cases require a different type of analysis and understanding because of complex mathematical analyses required.

Many of the present approaches are at least partially empirical. Research and development continues in the area of providing more meaningful tests to add to those already being used extensively.[7,40–48]

Probably the most impressive and useful tests are nondestructive. They are required in more sophisticated applications such as aircraft and underwater sandwich structures.[49,50] These techniques are also required to reduce cost and permit fast mass production of parts in other fields of interest such as building and construction paneling. As a field, nondestructive evaluation of materials is on its way to becoming well established although rather expensive. Nearly every known source of energy (for example, infrared, sonic, radiation, and X ray) has been used to establish workable systems.

6 DATA

The proper use of sandwich structures is obviously related to the proper use of materials and their assembly during fabrication. In turn the proper choice of materials depends on knowledge of their properties under different environments.[51]

In a field as new as structural sandwiches much needs to be done to fill in the gaps in knowledge respecting their properties and applications. Research in specific aspects is widespread and information comes from many different sources.

In this section selections of pertinent, recently developed data are summarized and quoted together with information affecting the engineering evaluation of data. These supplement material already presented in this chapter. Other chapters of this book also discuss aspects of sandwich behavior.

6.1 Failure modes-edgewise compression (adapted from Ref. 52)

Contributions to the acceptance of fiberglass-reinforced plastic specifically as an aircraft structural material were made through verification of existing theoretical strength relationships by the fabrication and testing of sandwich panels. The four basic failure modes were investigated for sandwich plates and plate columns loaded in edgewise compression. These were general buckling, face wrinkling, shear crimping and face dimpling (Figure 8.10).

In the development of a suitable structural sandwich ... the effect of adhesive filleting on the core strength and the effect of laminate thickness on facing strength properties were also isolated (Figure 8.11).

The effect of core thickness upon core shear strength and core shear modulus was evaluated (Figures 8.12 and 8.13).

Of the general buckling tests performed, the highest degree of precision was achieved in the tests involving the hinged boundary condition. It was found that the theoretical analysis was conservative for most of the cases investigated. The face wrinkling tests revealed that the symmetrical wrinkle would not always occur in sandwich constructions utilizing honeycomb cores as suggested by theory. A greater failure stress was generally realized when the load was applied parallel to the core ribbon direction than when applied perpendicular. The limited number of comparisons made showed a greater accuracy in predicting failure stress than for the general buckling mode of failure.

The limited study of shear crimping showed that such failure will not be a problem for honeycomb-core sandwich except for thin panels employing cores of very low shear modulus. The tests on intracellular buckling indicate that this mode will not be important for core cell sizes less than $\frac{1}{2}$ inch in combination with 3-ply, or thinner, reinforced plastics facings; however, a more thorough theoretical analysis is needed for the intracellular mode of failure.

6.2 Buckling coefficients (adapted from Ref. 53)

Design curves for the critical external radial pressure of circular cylindrical shells with sandwich walls, calculated

according to the formulas developed at the Forest Products Laboratory, are presented in Reference 53. The sandwich cylinder walls have isotropic facings of equal or unequal thickness, and of the same or different materials, and orthotropic or isotropic cores. It is assumed that Poisson's ratio is the same for both facings. The natural axes of the orthotropic cores are axial, tangential, and radial.[53]

Much investigative work has been done on isotropic cylindrical shells subjected to external pressure. It was found that experiment sometimes yields critical loads that are less than those predicted. This has been attributed to two causes. First, the experimental cylinders contained imperfections that lowered the critical load. Second, energy levels associated with postbuckling configurations of the cylinder are lower than

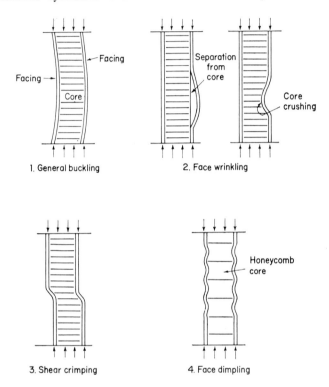

Figure 8.10   Modes of Failure of Sandwich Construction Under Edgewise Compressive Loads (Ref. 52)

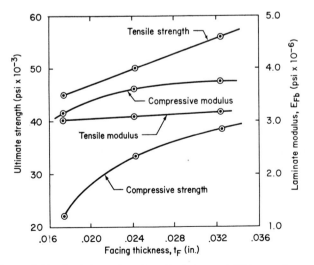

Figure 8.11   Variation of Facing Modulus and Ultimate Strength with Thickness (Ref. 52)

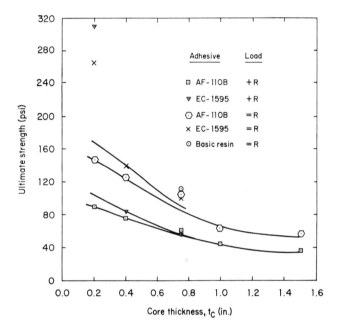

Figure 8.12   Relation Between Core Shear Strength of $\frac{3}{8}$-in.-cell, 5052-0.001P Aluminum Core in Sandwich Construction and Nominal Core Thickness (Ref. 52)

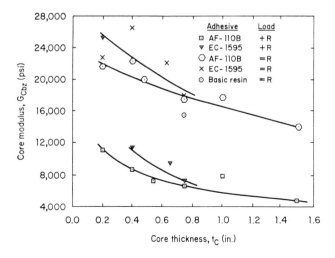

Figure 8.13   Relation Between Core Shear Modulus of $\frac{3}{8}$-in.-cell, 5052-0.001P Aluminum Core in Sandwich Construction and Nominal Core Thickness (Ref. 52)

those just at buckling. The energy levels associated with post-buckling may be reached without snap-through buckling. The energy necessary for snap-through buckling may be supplied by vibration or shocks. The curves in Reference 53 do not consider snap-through buckling or cylinders with imperfections. Sandwich cylinder walls, however, are much more perfect than their solid counterparts because they are thicker and the effect of an imperfection is in proportion to the ratio of its amplitude to the thickness of the cylinderical shell wall. Also, the curves neglect the stiffnesses of the individual facings. These stiffnesses add to the critical loads when the cylinders are short, and it is for short cylinders that snap-through is likely to occur.

The derivation of formulas for the buckling loads of rectangular sandwich panels subjected to edgewise compression is given in Reference 25. These formulas are derived for panels having dissimilar facings and orthotropic cores, the most general type of sandwich panel; and for several combinations of simply supported and clamped edges. The cores are assumed to be of such a nature that the stresses in them associated with strains in the plane of the panel may be neglected in comparison with the similar stresses in the facings and that the elastic modulus normal to the facings is so great that the related strain may be neglected.[54]

For honeycomb cores with hexagonal cells (a particular type of orthotropic core) it was found that the modulus of rigidity associated with the directions perpendicular to the ribbons of which honeycomb is made and the length of the cells is roughly 40 per cent of the modulus of rigidity associated with the directions parallel to those ribbons and the length of the cells. Making use of this fact, design curves for sandwich panels with simply supported and clamped edges and having isotropic facings and such honeycomb cores were published in Reference 55.

For the glass-fabric laminates currently used for facings, it was found that the numerical values of parameters involving the elastic properties of orthotropic facings that enter the formula for the buckling coefficient could be divided into three groups so that in each group the values do not vary greatly from one laminate to another. Using this fact, design curves for simply supported sandwich panels having similar orthotropic (glass-fabric laminate) facings were published in Reference 56.

Compressive buckling curves were also calculated for simply supported sandwich panels with honeycomb and isotropic cores and with one facing consisting of an orthotropic material (glass-fabric laminate) and the other facing of an isotropic material, and were presented in Reference 57.

This report also presents compressive buckling curves for sandwich panels with dissimilar, as well as similar, facings. The combinations of elastic properties of the orthotropic (glass-fabric laminate) facings may fall into any of the three principal groups. Curves are presented for four different combinations of panel edge support (simply supported or clamped).

Facing elastic properties (adapted from Ref. 54)

The various elastic properties of the facings can be combined into three convenient parameters for presentation of curves of buckling coefficients. These parameters are defined by the following:

$$\alpha_i = \sqrt{E_{bi}/E_{ai}},$$
$$\beta_i = \alpha_i \nu_{abi} + 2\gamma_i, \qquad\qquad (38)$$
$$\gamma_i = \frac{\lambda_i G_{bai}}{\sqrt{E_{ai} E_{bi}}},$$

where $\lambda_i = 1 - \nu_{abi}\nu_{bai}$; E_{ai} and E_{bi} are the moduli of elasticity of a facing in the a and b directions, respectively; G_{abi} is the facing shear modulus associated with shear distortion in the plane of the facing ($a - b$ plane); ν_{abi} is facing Poisson's ratio of contraction in the a direction to extension in the b direction due to a tensile stress in the b direction; ν_{bai} is facing Poisson's ratio of contraction in the b direction to extension in the a direction due to a tensile stress in the a direction; and $i = 1, 2$, denotes facing 1 or facing 2.

6.3 Fatigue (adapted from Ref. 58)

Dynamic methods, developed for determination of Young's modulus for honeycomb sandwich constructions, were successfully applied to obtain honeycomb core fatigue properties. *S-N* curves for extensional core fatigue were developed for representative samples of brazed, welded, and adhesive bonded honeycomb constructions. For tension-compression, the data show that core density is a dominant parameter in core fatigue. There are indications that core thickness up to and including 1 inch has little effect upon the data; more investigation is necessary to determine the effect of thickness greater than 1 inch. Feasibility of obtaining core fatigue using a two-specimen shear modulus approach was demonstrated, but more development is necessary in the design of testing systems to obtain dynamic balance and proper load levels.

The testing of honeycomb sandwich material has presented many problems in both the techniques used and interpretation of the final data. During sandwich testing it was found that the methods used for determining the extensional and shear moduli did not always give consistent results. Further investigation of testing techniques showed that dynamic methods could be used to obtain the extensional and shear moduli of brazed stainless steel honeycomb configurations.

This test method was investigated as an approach to fatigue testing of sandwich constructions. The investigation resulted in a program for the application of dynamic modulus testing approaches for comparing fatigue testing of sandwiches, including load levels, detection of cave fatigue, change in ratio of input-output, and so on.

The dynamic method originally was developed for the determination of honeycomb core moduli and was found to be consistent and reliable. The theory of the method is described as it relates core moduli to resonant frequency. Starting with a hypothetical mass and spring system, a transition to a real dynamic system is made in several steps to show the assumptions used and the related mathematical model.

For an undamped linear dynamic system with one degree of freedom, steady state simple harmonic motion is characteristic. The motion of indeterminate amplitude has a natural frequency, p, related to spring constant and mass by the following expression:

$$p^2 = K/m, \qquad (39)$$

where

$p =$ natural frequency, rad/sec.,
$K =$ spring constant, lb/in.,
$M =$ mass, lb/sec^2/in.,

The first step is to include the effects of positive damping and a harmonic force, Figures 8.22 and 8.23. As shown the resonant frequency differs slightly from the natural frequency (without damping) for small values of the damping coefficients. Low damping is experienced with core and thus the resonant frequency is used for the natural frequency in equation 39 with negligible error.

In the conventional treatment of dynamics problems the source of excitation is conveniently chosen as a hypothetical force applied to a mass restrained by a spring and damper. In this case it is more convenient to apply this force by the displacement of the spring and damper.

The final transition to the physical system used is made when the attachment and shaker head mass are taken into account in the system. Since the shaker is servo-controlled, the attachment and shaker head are effectively dummy masses, as a fixed acceleration can be maintained at the attachment thus giving a single mass system.

The compression-tension specimen and fixture has the free mass on top of the sandwich which acts as a spring and damper, and the dummy mass, which serves as a means of attachment to the head of the shaker.

For this system the spring constant of the sandwich can be expressed as a function of the modulus and planform area.

$$K = EA/c, \qquad (40)$$

where

$E =$ Young's Modulus of the core, psi,
$A =$ face area of specimen, in.^{2},
$c =$ core thickness, in.

Inserting equation 40 for K in equation 39 and solving for E gives

$$E = cmp^2/A, \qquad (41)$$

where m is the upper free mass.
The core stresses can be obtained from Newton's second law of motion

$$F = m\alpha \qquad (42)$$

or

$$f_b = BG/A_c, \qquad (43)$$

where

$\alpha =$ acceleration, in./sec^2,
$F =$ force, lb,
$f_b =$ core stress, psi,
$B =$ weight of free mass, lb,
$G =$ load factor of free mass,
$A_c =$ area of core, inc.^{2}.

The following discussion shows how this vibratory single mass system is utilized to obtain the extensional modulus and also

for fatigue testing of core. The specimen to be tested is bonded to a fixed mass fixture which is solidly mounted to the excitation source and to a free mass which is allowed to vibrate in the core tension-compression mode only. The fixtures are designed to give a minimum spring effect and the sizes are chosen so that they are compatible with both the frequency range of available electromechanical shakers and the core load level. The test setup has accelerometers installed on the electromechanical shaker. Both the fixed and free mass accelerators are mounted in the same vertical planes so that both input (fixed mass accelerometers) and the output (free mass accelerometers) can be monitored.

The initial frequency sweep of the system is made at low load output to obtain the resonant frequency of the core. This is used to determine the extensional modulus from equation 41. The fatigue testing is done at the resonant frequency and the core load level is monitored by the output accelerometers.

It is noted that two specimens are necessary to obtain symmetry of the shear test system. The analysis is similar, with the spring constant K_1 being a function of the core shear modulus. The method of testing is the same as described above for the extensional system.

$$K_1 = 2GA/C \qquad (44)$$

where

$G =$ shear modulus, psi.

6.4 Box beam structure (adapted from Ref. 59)

Analyses have been presented for the determination of the bending strengths and optimum efficiencies of box beam structures of conventional and advanced design. The methods are particularly applicable in preliminary design where the beam proportions for optimum strength-weight ratios are required. The theoretical analyses were supplemented by the design, fabrication and stuctural testing of fourteen box beam specimens of several configurations at both ambient and elevated temperatures.[59]

The analyses are primarily based on the assumptions of integral structure and exact geometry, thus resulting in theoretically optimum weight and strength values.

The methods presented for the preliminary design of box beam wing structures are useful in determining the theoretically optimum beam proportions which result in minimum weight for a given bending strength requirement. The methods are based, however, on the assumption of monolithic structure with the elements having perfect geometry. The degree to which the theoretical optimum can be achieved is dependent on the extent to which the component and attachments of the actual structure exhibit the assumed strength and stiffness characteristics.

Of the fourteen box beam specimens tested to determine the validity of the ultimate strength analyses, only five failed at, or above, their design ultimate moments.

The three sandwich plate cover panel, corrugated-web, steel box beams failed between 16 per cent and 35 per cent of their design ultimate moments, apparently because of weak riveting between the cover panels and the webs. The three 17-7PH stainless steel (Table 8.6), full depth honeycomb

Table 8.6. Mechanical Properties of 17-7 PH — RH 950 Stainless Steel (Properties from MIL-HDBK-5, "Strength of Metal Aircraft Elements"[60])

Temp (°F)	Ult. Tensile Strength (psi)	Tensile Yield Strength (psi)	Compressive Yield Strength (psi)
75	210,000	190,000	205,000
200	191,000	173,000	186,500
300	182,500	165,000	178,000
400	178,500	161,500	174,000
500	174,000	157,500	170,000
600	166,000	150,000	162,000
700	149,000	135,000	145,500
800	128,000	116,000	125,000
900	103,000	93,000	100,500
1000	73,500	66,500	71,700

beams failed between 10 per cent and 18 per cent of their design ultimate moments. The honeycomb cores of these beams were noticeably deviated from the assumed "perfect geometry," resulting in much smaller core property values than anticipated. The out-of-flatness of the large-sized honeycomb cell walls decreased the compressional and shear stiffnesses of the core, resulting in the premature failures of these beams.

It was apparent from the test results that designs which specify large, unsupported thin-gauged components will fail at lower than design loads because normal manufacturing tolerances invalidate the assumption of "perfect geometry" used in the derivation of the design methods.

The testing program was successful in substantiating the importance of the stiffness requirements of the supporting core media in box beam design. It also established a need for improved core-to-cover fastening techniques, especially in the cases of the more sophisticated designs.

6.5 Flutter (adapted from Ref. 61)

Flutter tests have been made on flat panels having a $\frac{1}{4}$-in. thick plastic-foam core (Figure 8.26) covered with thin fiberglass laminates. The testing was done in the Langley Unitary Plan wind tunnel at Mach numbers from 1.76 to 2.87. The flutter boundary for these panels was found to be near the flutter boundary of thin metal panels when compared on the basis of an equivalent panel stiffness.

The following conclusions can be drawn from the results of tests conducted on flat rectangular plastic panels at Mach numbers from 1.76 to 2.87.

1. Panel flutter coefficients for metal panels may be useful to predict the flutter of plastic panels by using an equivalent panel stiffness.

2. The depth of the cavity behind the panel has a strong influence on panel flutter. Reducing the cavity depth from $1\frac{1}{2}$ in. to $\frac{1}{2}$ in. reduced the dynamic pressure at flutter by 40 per cent, but when the cavity depth was further decreased to the minimum possible with the bottom of the cavity resting against the balsa strips on the backs of the panels, no flutter was obtained.

3. A small pressure differential across the panel is effective in increasing the flutter dynamic pressure.

4. The panel flutter coefficient is essentially independent of Mach number from 1.76 to 2.87 which indicates that it accounts for Mach number effects in this range.

6.6 Damping (adapted from Ref. 62,) Fatigue (adapted from Ref. 63)

A combined theoretical and experimental study was undertaken to develop an analytical approach for predicting the damping of sandwich configurations in flexure. The theory developed analyzes the various contributions to total damping, considering stress distribution and unit damping properties of skin and core, and employs a simple summation process to determine the damping of the composite. To confirm the theory a special test set-up was developed in which sandwich configurations were vibrated as free-free beams utilizing electromagnetic excitation. A series of tests were performed on several types of conventional sandwich beams. Damping predicted by the theory is in good agreement with that measured experimentally.[62]

The primary objective of this project was to develop a theoretical approach for predicting the damping of sandwich configurations in flexure, to devise methods for measuring experimentally the damping of sandwich materials, and to check theory and experiment. As part of the theoretical study the vibration characteristics and the shear and bending stress distribution along the beam were analyzed.

Based on the analysis made and the experiments performed the total internal damping of a sandwich beam in flexure may be predicted by the following procedure explained in detail in the report. From test data on the unit specific damping properties of the skin material and an analysis of the normal stress distribution in the skin, the contribution of the entire skin to the total damping of the sandwich may be calculated. From test data on the unit shear damping properties and a shear stress analysis of the core the contribution of the entire core to total damping may be determined. Total damping of the sandwich under the specified loading which produced the stress distribution considered is then calculated simply by adding the two contributions.

It is shown that damping of sandwich beams cannot be specified meaningfully by a simple single rating; the shear and normal stress distribution within the sandwich beam and the unit damping properties of the component materials must be considered in comparing the suitability of different sandwich configurations.

The second phase of this project was concerned with the fatigue of sandwich beams. The modes of failure in the sandwich beam are discussed in terms of localized stress and discontinuities.

The decrease in beam stiffness under sustained cyclic stress in the fatigue range is found to be gradual during the early stages of the fatigue but immediately preceding localized failure and rapid crack propagation the stiffness of the beam decreases rapidly.

In former years resonant vibrations under periodic excitation could often be avoided by properly separating the natural frequencies of a system and the forcing frequencies. However, this approach is no longer effective in many engineering situations, particularly where light-weight structures having a large number of natural frequencies are involved. Furthermore, random excitation, either of mechanical or of acoustical origin, is becoming more prevalent in service. Such excitation is generally capable of exciting structural resonances wherever they may lie within the broad frequency spectrum characteristic of random excitation. Experience in recent years has made it abundantly clear that resonant vibrations can no longer be avoided by clever design.

One of the most effective ways of controlling the deleterious effects of resonant vibrations is to minimize the resonant amplitude by maximizing the total damping in a structural configuration. Among the many contributors to total structural damping are (a) internal or hysteretic damping within the structural material itself, (b) slip at a structural interface, (c) shear in a viscoelastic layer, and (d) surface treatments such as coatings and damping tapes. A laminated beam, for example, may involve in addition to contribution (a) damping mechanisms such as (c) and (d), which can be particularly effective in dissipating energy in an optimized design. As a result various types of laminated configurations have been utilized during the past few years to produce high inherent damping in a structural configuration.

A particular type of laminated configuration which is being used in many new applications each year is the sandwich, consisting of a skin-core-skin combination. Significant improvement in acoustical fatigue life in many types of aircraft has, for example, been realized by replacing conventional aircraft skin panels by sandwich configurations. Yet, relatively little analytical study has been undertaken to provide improved understanding of the stress distribution, damping, stiffness, fatigue, and other properties of such laminates which critically determine their suitability for new applications. As a result

most sandwich applications have involved a trial-and-error approach followed by ad hoc testing to determine what improvement, if any, has resulted in acoustical fatigue life. Before a meaningful analytical approach can be developed, improved understanding of the damping, stiffness, and conventional fatigue properties of sandwich materials are required.

In another program (the following material is adapted from Ref. 63), work consisted of the determination of the basic dynamic properties of fiberglass-reinforced plastic sandwich structure suitable for use as a primary airframe structural material. The research program was carried out in two separate parts: (a) determining dynamic moduli and damping, and (b) determining fatigue behavior. In each part, two types of hexagonal-cell honeycomb core materials were investigated: 5052 aluminum foil and HRP (heat resistant phenolic) fiberglass (Table 8.7).[63]

In the dynamic moduli and damping experiment, beam strips were suspended at the nodes for the lowest symmetrical mode and excited acoustically at the frequency corresponding to this mode. Then the power was cut off and the decay in facing strain at the beam center line was recorded versus time. It was possible to use the static moduli (the modulus of elasticity of the facings and the shear modulus of the core) to predict the lowest natural frequency and corresponding node location. Over the frequency range covered (300 to 700 cps), the agreement between predicted and measured values was good, so that it was concluded that the dynamic moduli were the same as the corresponding static moduli.

There was no significant effect of stress level on damping (up to the maximum stress level covered, 1270 psi), but there was an effect of frequency, the peak occurring at approximately 500 cps.

In the fatigue experiment, a special specimen and an associated loading fixture were devised to subject a test length

Table 8.7. Properties of Sandwich Constituents

Constituent	Thickness[a] (in.)	Specific Weight[a] (psi)	Number Specimens	Load Orient.[b]	Type Loading	Strength[a] (psi)	Number Specimens	Modulus[a] (psi)	Number Specimens
Facing[c] (resin content: 35.7%)	0.0418 0.0339 0.0245	1.980×10^{-3} 1.950×10^{-3} 1.900×10^{-3}	12	(=W)	Tension	45,100 37,300 29,100	15	3.99×10^6 2.95×10^6 2.43×10^6	14
				(=W)	Compress.	44,900 40,500 30,400	13	4.34×10^6 3.72×10^6 3.11×10^6	11
Aluminum Core (5052-0.001P, $\frac{3}{16}$-inch cell)	$\frac{3}{4}$ (nominal)	1.372×10^{-3} 1.345×10^{-3} 1.323×10^{-3}	5	(=R)	Shear	233 200 216	35	47,000 40,700 34,200	35
HRP[d] Core (GF-11, $\frac{3}{16}$-in. cell)	$\frac{3}{4}$ (nominal)	1.991×10^{-3} 1.890×10^{-3} 1.815×10^{-3}	7	(+R)	Shear	200 181 154	27	7000 6200 5220	27
Adhesive	0.010	5.556×10^{-5}	—	—	—	—	—	—	—

[a] Values listed vertically in the order, high, average, and low.

[b] The symbols = and + are used to indicate whether the load was oriented parallel or perpendicular, respectively, to the fabric warp direction (W) or the core ribbon direction (R).

[c] The facing properties were obtained from one laminate only, except in the case of the specific weight measurements where the sampling represents 12 laminates.

[d] Heat-resistant phenolic.

of the specimen to a cyclic constant bending moment along the test length (Figure 8.14, Tables 8.8 and 8.9). Excitation was

Table 8.8. Fatigue Test Data—Aluminum Core[63]

Spec. No.	Fatigue Stress Level (psi)	Dynamic Deflection Imposed (in.)	Number of Cycles To Failure	Ratio of Fatigue Stress to Ult. Static Str.[a]
A1	50,999	Static	1	—
A2	48,816	Static	1	—
A3	53,270	Static	1	—
Average	51,028			
A4	25,000	0.357	16,380	0.50
A5	25,000	0.357	7,398	
A6	25,000	0.357	8,232	
Average			10,670	
A7	20,000	0.285	53,424	0.40
A8	20,000	0.285	176,956	
A9	20,000	0.285	49,732	
Average			93,704	
A10	17,000	0.240	223,800	0.34
A11	17,000	0.240	254,760	
A12	17,000	0.240	135,300	
Average			203,953	
A13	14,000	0.199	1,039,200	0.28
A14	14,000	0.199	1,802,400	
A15	14,000	0.199	2,033,300	
Average			3,934,966	
A16	12,000	0.171	3,143,000	0.24
A17	12,000	0.171	3,233,300	
A18	12,000	0.171	5,428,600	
Average			3,934,966	
A19	10,000	0.142	10,159,200[b]	0.20
A20	10,000	0.142	10,000,000[b]	
Average			10,000,000	
A21[c]	14,000	0.199	609,600	

[a] Based on average of six specimens: A1, A2, A3, F1, F2, and F3.
[b] Did not fail.
[c] Specimen had fabrication error introducing wave shape in fill direction.

Table 8.9. Fatigue Test Data—Fiberglass (HRP) Core[63]

Spec. No.	Fatigue Stress Level (psi)	Dynamic Deflection Imposed (in.)	Number of Cycles To Failure	Ratio of Fatigue Stress to Ult. Static Str.[a]
F1	50,982	Static	1	—
F2	48,185	Static	1	—
F3	47,153	Static	1	—
Average	48,773			
F4	25,000	0.353	5,743	0.50
F5	25,000	0.353	6,062	
F6	25,000	0.353	15,395	
Average			9,068	
F7	20,000	0.283	105,522	0.40
F8	20,000	0.283	117,236	
F9	20,000	0.283	52,236	
Average			91,665	
F10	17,000	0.240	112,839	0.34
F11	17,000	0.240	359,580	
F12	17,000	0.240	165,600	
Average			212,673	
F13	14,000	0.198	2,034,145	0.28
F14	14,000	0.198	878,600	
F15	14,000	0.198	571,400	
Average			1,161,384	
F16	12,000	0.170	3,397,430	0.24
F17	12,000	0.170	4,081,506	
F18	12,000	0.170	3,797,600	
Average			3,758,835	
F19	10,000	0.141	10,000,000[b]	0.20

[a] Based on average of six specimens: A1, A2, A3, F1, F2, and F3.
[b] Did not fail.

provided by an electromagnetic exciter with force control achieved by use of a piezoelectric force gauge. No significant difference was found between the fatigue behavior of the specimens with aluminum cores and those with fiberglass cores. There was no apparent endurance limit within 10 million cycles of load as often occurs in metals. However, design for various cyclic lives can be accomplished in terms of corresponding fatigue strength values. Thus, the life corresponding to a stress amplitude of 38 per cent of the ultimate static strength is approximately 10^5 cycles, while for a stress of 29 per cent the life is approximately 10^7 cycles.

6.7 Fabrication variables (adapted from Ref. 64)

A knowledge of the manner in which the strength properties of a material vary with changes in the basic fabrication process is necessary before it can be accepted as a primary structural element. This is particularly true of fiberglass-reinforced plastic materials because of the large number of fabrication process variables and the sensitivity of the material properties to each of these variables. Examples of material properties are shown in Figures 8.15 and 8.16 on plastics cores and Figure 8.17 on metal cores. The goal of a research program was to contribute to this body of knowledge in the realm of sandwich materials for aircraft structural use.

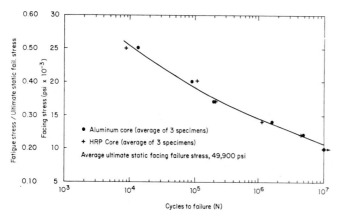

Figure 8.14   Fatigue *S-N* Curve for FRP Facing and Honeycomb Core Sandwich Structure Subjected to Pure Flexural Loading (Ref. 63)

The principal materials used were hexagonal-cell honeycomb core, electrical (E) glass 181-weave Volan A-finished fiberglass fabric, epoxy resin. Limited data were obtained for HTS finish 181 fiberglass fabric. Consideration was given to both the separately bonded and single-step fabricated thin facing sandwich. A program was conducted to detect the relation between the curing cycle variables and the mechanical properties of fiber-glass-reinforced plastics (FRP) facing sandwich fabricated by the single-step process and of bare

trends were expected to be established along with areas suitable for more detailed examination.

The program encompassed the two types of sandwich fabrication used industrially, the one-step process method and the method whereby prelaminated facings are bonded to the core in a separate step. Thin, 3-ply facings of these materials used in conjunction with honeycomb cores were the basic elements of the structural sandwich.

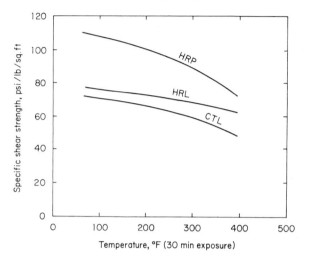

Figure 8.15   Specific Shear Strength of Heat-resistant Glass-fabric-reinforced-plastics honeycombs (L-direction) (Ref. 21)

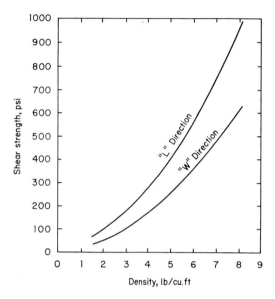

Figure 8.17   Shear Strength versus Density for 5052 Alloy Aluminum Honeycomb (Ref. 21)

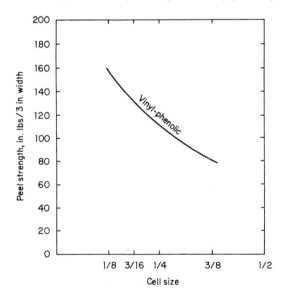

Figure 8.16   Peel Strength Versus Cell Size for Glass-fabric-reinforced-plastics Honeycomb (Ref. 21)

FRP facings prelaminated for use in the separately bonded-type sandwich. The program was carried out for the wet and the B-staged type of lay-up. For the separately bonded-type sandwich, an adhesive study was performed to indicate optimum bonding conditions for suitable adhesives and core materials. All testing was done at room temperature.

Though it was realized that no one area could be studied intensely because of the time and expense involved, valuable

The program included determining the effect of curing time, pressure, and temperature, as well as lay-up method, on the basic strength properties of the sandwich. Following the premise that the best bonded type sandwich would be the combination of the facings cured under the optimum conditions (as for the entire single-step type sandwich) and the optimum adhesive system and cure for a given core material, the facings and their bonding to the core were investigated separately. Some major conclusions drawn from the research are as follows:

1. The initial resin content of the facings of sandwich material and the uniformity and extent of voids in the impregnation have an important effect on its final strength properties. This is especially true when the facings are thin. The effect is also associated with the flow that takes place during the resin cure cycle.

2. The mechanical coating process produces more controllable and uniform resin impregnations of fiberglass fabric laminations than the hand squeegee method.

3. For the vacuum blanket technique of applying pressure to the epoxy resin impregnations employed, 20 inches of mercury is a safe lower vacuum pressure to prevent resin bubbling within the 160°F to 200°F temperature range.

4. Room temperature B-staged lay-ups of laminations employed are more convenient to work with and can be expected, in most cases, to produce slightly higher strengths.

5. A serious condition of resin starvation on the upper facing can be expected when the wet lay-up of the facings is

used directly in the fabrication of sandwich by the single-step method. This condition is not nearly so severe when the 10-hour room-temperature B-staged lay-up is used.

6. The separately bonded-type sandwich material investigated can be expected to develop a maximum of 30 per cent higher compressive strength than the single-step-type sandwich.

7. The optimum conditions of fabrication for tensile and compressive properties are not the same; hence further testing is required to establish flexural optimum strength properties.

6.8 Environmental effects—outdoors (adapted from Ref. 65)

Three constructions of glass-fabric plastic sandwich that were exposed at the seashore in south Florida for 3 years were evaluated in flatwise flexure and edgewise compression. The constructions had facings approximately 0.03-in. thick of glass-fabric laminate treated with polyester, epoxy, and heat-resistant phenolic resins. Cores were of glass-fabric honeycomb treated with the same kind of resin as the facings, except that the core for the sandwich with epoxy facings was treated with nylon-phenolic resin.

All constructions were tested at room temperature. Edgewise compression tests were conducted at 500°F for the heat-resistant sandwich. Polyester sandwich panels were evaluated both painted and unpainted. The unpainted panel deteriorated the most in 3 years' weathering; the edgewise compressive strength was reduced 40 per cent and the flexural strength 30 per cent. Painting the other polyester panel with a MIL-E-7729 enamel of the type normally used on exterior reinforced plastic aircraft parts reduced the loss in edgewise compressive strength to 22 per cent, but the reduction in flexural strength was still 30 per cent. In comparison, an unpainted polyester solid laminate panel $\frac{1}{8}$ in. thick had compressive strength reduced 28 per cent by 3 years' weathering.

At 500°F, the heat-resistant phenolic sandwich panel showed an 8 per cent increase in edgewise compressive strength, compared to the control value at 500°F.

6.9 Moisture content (adapted from Ref. 66)

Sandwich constructions, composed of strong, thin facings bonded to a thick, lightweight core, can be used to produce stiff, lightweight structural panels for use in aircraft and other flight vehicles. Resin-treated paper honeycomb cores have been used for bulkheads, cockpit floors, baggage racks, and other aircraft applications where minimum weight coupled with moderate strength and stiffness are required. The purpose of this study was to determine the compressive and shear properties of two commercially produced paper honeycomb cores.

The cores were of 60- and 125-lb kraft paper formed to $\frac{7}{16}$-in. hexagonal cells and then treated with alcohol-soluble phenolic resin to produce core of 1.7 and 3.7 pcf (pounds per cubic foot) density, respectively.

Honeycomb core shear and compressive strengths tend to decrease as core thickness increases. Criteria for buckling of the cell walls predict that core strength of a relatively thin core changes rapidly as core thickness changes; but strength for a core with a thickness greater than the cell size is relatively insensitive to changes in core thickness. Accordingly, a minimum thickness of $\frac{1}{4}$ in. was selected as the thinnest core likely to be of practical use, and additional thicknesses of $\frac{1}{2}$ and 1 in. were chosen to determine the shape of the curve showing strength as affected by thickness.

A maximum thickness of 2 in. was chosen to determine the lower boundary of the strength-thickness relations because strengths of thicker cores would not be expected to be significantly lower.

Aircraft sandwich panels in service may be exposed to a variety of humidity conditions affecting both strength and stiffness of paper honeycomb cores. Cores were evaluated after conditioning in rooms maintained at 73°F to 80°F and 30, 50, and 90 per cent relative humidity, producing core moisture contents ranging from 3 to 11 per cent. A more severe moisture exposure, occasionally encountered in service, is infiltration of liquid water into the interior of a sandwich panel. Accordingly, properties of cores immersed in water for 48 to 72 hours (60 to 80 per cent moisture content) were evaluated to establish a lower limit for the strength-moisture relations.

Compressive strength of $\frac{1}{4}$-in.-thick cores was 25 to 80 per cent greater than that of 2-in .thick cores, while strengths of 1-in.-thick cores were 10 per cent greater, at most. Compressive strength of cores at 10 per cent moisture content was about 50 to 70 per cent of that at 4 per cent moisture content, while cores at about 75 per cent moisture content were about 20 per cent as strong. Oven-dry cores were about 20 to 40 per cent stronger than cores at 4 per cent moisture content. The modulus of elasticity at 10 per cent moisture content was about half that of oven-dry core, while cores soaked in water (60 to 80 per cent moisture content) were only about 20 per cent as stiff.

Core shear strength of $\frac{1}{4}$-in.-thick cores was about two to four times as high as that for 2-in.-thick cores. Core shear strength and modulus of rigidity at 10 per cent moisture content were about 50 to 70 per cent of those at 4 per cent moisture content, while cores soaked in water had 20 to 40 per cent of the strength and stiffness of those at 4 per cent moisture content. Shear strength in the TW plane was about 55 per cent of that in the TL plane, while the shear modulus in the TW plane was about 35 per cent of that in the TL plane.

6.10 Elevated temperatures (adapted from Refs. 60 and 67)

Different tests have been conducted at elevated temperatures.

In one of the adhesive test programs,[60] the objective was to evaluate the elevated temperature characteristics of adhesive-bonded, 17-7 PH-RH 950 honeycomb sandwiches in edgewise column compression (see Tables 8.10 and 8.11 and Figure 8.18). Specimens were bonded with a modified polybenzimidazole (PBI) resin and a commercially available epoxy-phenolic adhesive.

The sandwich specimens bonded with the epoxy-phenolic adhesive exhibited approximately 5 per cent higher skin stresses, at room temperature, than the specimens bonded with the PBI adhesive. At elevated temperatures, the PBI-bonded specimens exhibited superior thermal stability and yielded skin stresses 10 to 30 per cent higher than the epoxy-phenolic bonded specimens.

Table 8.10. Test Results of Sandwich Column Compression at
Various Temperatures for 17-7 PH–RH 950
Stainless Steel Sandwiches Bonded with PBI Adhesive.
(Tested after 10-minutes Soak at Temperature)[60]

Specimen No.	Test Temp (°F)	Area (in.²)	Load (lb)	Skin Stress (psi)	Type of Failure
45–1	RT	0.1982	33,600	169,500	Buckling
45–2		0.1980	30,500	154,000	
45–3		0.1983	24,600	124,000	
				149,100	
44–1	365	0.1982	40,450	204,000	
44–2		0.1982	38,800	195,700	
44–3		0.1984	35,150	177,100	
				192,270	
43–1	500	0.1986	36,450	183,500	
43–2		0.1986	26,200	131,900	
43–3		0.1985	23,000	115,800	
				143,700	
46–1	600	0.1985	33,950	171,000	
46–2		0.1984	30,800	155,200	
46–3		0.1985	35,800	180,300	
				168,800	
51–1	700	0.1984	26,600	134,000	
51–2		0.1984	23,200	116,900	
51–3		0.1984	22,700	114,400	
				121,700	
52–1	800	0.1984	20,050	101,000	
52–2		0.1982	17,000	85,700	
52–3		0.1986	13,800	69,500	
				85,400	
53–1	900	0.1986	14,800	74,500	
53–2		0.1985	18,850	94,900	
53–3		0.1987	16,900	85,000	
				84,800	
54–1	1000	0.1984	8,850	44,600	
54–2		0.1984	11,150	56,200	
54–3		0.1986	11,800	59,400	
				53,400	↓

The PBI-bonded specimens developed 68 to 109 per cent of the allowable theoretical maximum skin stress of the sandwich specimens when tested at temperatures from 75°F to 1000°F for short periods. The PBI-bonded specimens also retained sufficient structural integrity, after extended aging at 600°F and 700°F, to be considered as a useful engineering material.

In other test programs[67] a heat-resistant honeycomb consisting of AF-R-100 polybenzimidazole (PBI) resin and glass-fabric reinforcement was fabricated to a nominal 6 pcf density. Stabilized flatwise compressive strength and modulus were determined in air at room temperature, at 500°F, 600°F, 700°F,

900°F and at 1000°F after $\frac{1}{3}$ hour and 1 hour at temperatures. Flatwise compressive strength and elastic modulus were also measured on the material at 600°F after heat aging for 1, 10, 50, 100, 150, and 200 hours at 600°F in air. In addition, flatwise tensile strength was evaluated at room temperature on specimens heat aged 1, 10, 100, and 192 hours at 500°F, 550°F, and 600°F, respectively. The outstanding heat resistance of this honeycomb was evident when compared with a commercially available heat-resistant phenolic honeycomb tested under similar conditions.

Polyaromatic resins, exhibiting outstanding resistance to oxidation at high temperatures (600°F to 700°F), hold substantial promise of useful application in honeycomb sandwich structures.

For this application a composite web material, consisting of E glass fabric (style 116) impregnated with the PBI resin,

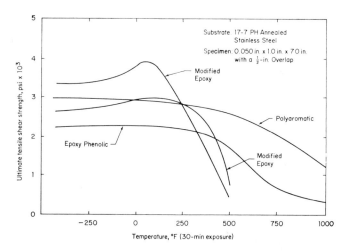

Figure 8.18   Effect of Elevated Temperature on the Different Classes of Structural Adhesive (Ref. 60)

Table 8.11. Test Results of Sandwich Column Compression at
700°F After Conditioning at 700°F for Various
Time Periods; Bonded with PBI Adhesive, 17-7 PH, RH 950
Stainless Steel[60]

Specimen No.	Hr @ 700°F	Area (in.²)	Load (lb)	Skin Stress (psi)	Type of Failure
56–1	1	0.1985	24,300	122,400	Buckling
56–2		0.1985	27,250	137,200	Buckling
56–3		0.1985	17,150	86,400	Adhesive
				115,300	
57–1	10	0.1986	16,600	83,580	Adhesive
57–2		0.1986	20,100	101,200	Adhesive
57–3		0.1987	28,800	144,900	Buckling
				109,900	
58–1	24	0.1986	9,250	46,500	Adhesive
58–2		0.1984	7,400	37,300	Adhesive
58–3		0.1985	14,050	70,700	Adhesive
				51,600	

Table 8.12. Compressive Strength and Modulus of PBI and Standard HRP-GF13 Honeycomb Controls[67]

Specimen Number	PBI Controls, Room Temperature Tests					Specimen Number	Controls at 500°F After ¼ Hour at 500°F				
	Comp. (psi)	Modulus (psi ×10³)	Utiliza-tion (%)	Resin (%)	Density (lb/ft³)		Comp. (psi)	Modulus (psi ×10³)	Utiliza-tion (%)	Resin (%)	Density (lb/ft³)
1-5-a	514	—	89.3	33.0	3.86	—	—	—	—	—	—
1-5-b	447	—	54.4	34.4	3.86	—	—	—	—	—	—
1-5-c	480	—	66.6	33.7	3.86	—	—	—	—	—	—
1-5-d	496	—	69.0	35.6	3.86	—	—	—	—	—	—
Average	484	—	69.8	34.2	3.86	—	—	—	—	—	—
2-2-c	541	85.3	73.1	22.3	3.80	2-2.a	399	60.9	66.6	21.3	3.80
2-2-d	539	77.6	68.5	19.6	3.80	2-2.b	431	69.5	86.0	21.2	3.80
Average	540	81.5	70.8	21.0	3.80	Average	415	65.2	76.3	21.3	3.80
Specimen	HRP-GF13 Controls, Room Temperature Tests					Specimen	Controls at 500°F After ¼ Hour at 500°F				
HRP-1-a	400	75.4	58.5	35.9	3.88	HRP-1-b	360	63.8	68.3	35.6	3.88
HRP-1-c	397	64.8	81.2	35.4	3.88	HRP-1-d	a	a	a	36.0	3.88
Average	398	70.1	69.9	35.7	3.88	Average	360	63.8	68.3	35.8	3.88
HRP-2-c	356ᵇ	70.9	79.2	22.2	3.80	HRP-2-a	313	53.6	58.8	20.6	3.80
HRP-2-d	472	72.2	87.8	22.3	3.80	HRP-2-b	339	60.5	84.2	21.3	3.80
Average	472	71.6	83.5	22.3	3.80	Average	326	57.1	71.5	21.0	3.80

ᵃ Test negated because of excessively high loading rate (4 in./minute).

ᵇ Excluded from average because zero indicator had not been reset prior to testing.

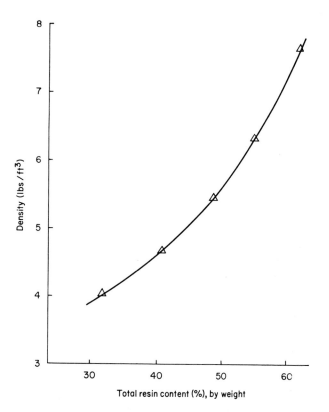

Figure 8.19   PBI Honeycomb Density versus Total Resin Content (Ref. 67)

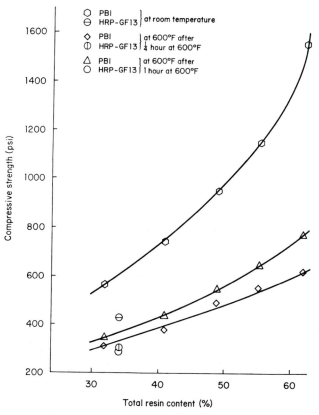

Figure 8.20   Compressive Strength versus Total Resin Content (Ref. 67)

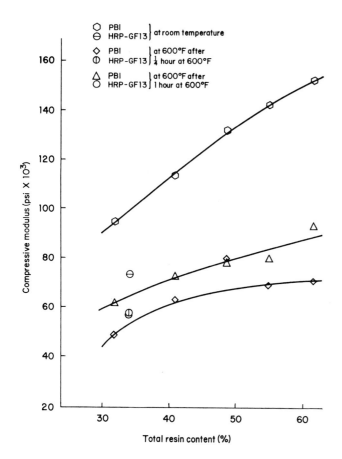

Figure 8.21   Compressive Modulus versus Total Resin Content
(Ref. 67)

was converted to honeycomb. This core material was tested in flatwise compression at elevated temperatures in air. Specimens heat-aged in a forced draft oven were also treated at elevated temperature. In addition, flatwise tensile tests were applied at room temperature to heat-aged honeycomb.

Strength and modulus values of this honeycomb were compared with similar properties of a standard commercially available heat-resistant honeycomb tested under similar conditions. These properties were correlated with resin content, temperature, and time at temperature for the various tests.

Results are shown in Table 8.12 and Figures 8.19, 8.20, and 8.21.

REFERENCES

1. L. J. MARKWARDT, "Developments and Trends in Light-weight Composite Construction," U.S. Dept. Agr. Forest Serv., Forest Prod. Rept., Article, Forest Products Laboratory, Madison, Wis., 1952.

2. A. G. H. DIETZ, "Composite Materials," 1965 Edgar Marburg Lecture, American Society for Testing and Materials, Philadelphia, Pa., 1965.

3. R. L. SIREN, "Recent Developments in Sandwich Construction with Rigid Urethane Foam," *Soc. Plastics Engrs. J.* **21**, 1290 (Nov. 1965).

4. D. V. ROSATO, "Plastics Foam Use Rises and Rises," *Plastics World* **23**, 67–73 (Oct. 1965).

5. "Sandwich Panel Design Criteria," Publ. 798, Building Research Institute, Washington, D.C., 1960.

6. V. GRINIUS, "Application of Sandwich Construction to Aerospace Structures," ASM Report D5-14.4, Air Force, Dayton, Ohio, p. 16, 1965.

7. *Composite Construction*, Military Handbook, MIL-HDBK-23, Armed Forces Supply Support Center, Washington, D.C., 1963.

8. J. K. KUNO, "Design Considerations for Sandwich Construction," Whittaker Corp., San Diego, Calif., 1964.

9. R. J. SEIDL, D. J. FAHEY, and A. W. VOSS, "Properties of Honeycomb Cores as Affected by Fiber Type, Fiber Orientation, Resin Type and Amount," NASA Tech. Note 2564, National Aeronautics and Space Administration, Washington, D.C., Nov. 1951.

10. R. MARK, "Balsa Cores for Reinforced Plastics Structures," Balsa Ecuador Lumber Corporation, New York, 1956.

11. Balsa Ecuador Lumber Corporation, New York, Data Sheet No. 5-C, Oct. 1962.

12. D. V. ROSATO, "Plastics Foam Use Rises and Rises," *Plastics World* **23**, 67–73. (Oct. 1965).

13. Foamalum Corporation, Status Report, Oct. 14, 1965.

14. J. BJORKSTEN, "Foamed Metal Low Density Core Material for Sandwich Construction," WADC TR 52-51, Part 3 (AD 29436), ASTIA, Springfield, Va., May 1954.

15. W. E. Jahnke and E. W. Kuenzi, "Performance of Brazed Stainless Steel Sandwich at High Temperatures," WADC TR 55-417 (ASTIA 202 492), ASTIA, Springfield, Va., Oct. 1958.

16. McDonnell Aircraft Corporation, Evaluation of Brazed Aluminum Honeycomb Panel," Rept. A 729 (ASTIA 441156), ASTIA, Springfield, Va., June 10, 1964.

17. "Development of Brazed Beryllium Sandwich Construction," Rept. ERR-FW-043, General Dynamics, Ft. Worth, Texas, Apr. 4, 1962.

18. Boeing Airplane Company, "Development and Test of Ceramic Bonded Stainless Steel Structural Components," BuWeapons Report D2-4961-5 (ASTIA 257516), ASTIA, Springfield, Va., Sept. 30, 1960.

19. D. V. ROSATO, "Plastic-based Adhesives," *Plastics World*, **23** (Aug. 1965).

20. A. G. H. DIETZ, *Engineering Laminates*, John Wiley & Sons, Inc., New York, 1949.

21. A. C. Marshall, Novel Design in Sandwich Structures, ASME Conference, Erie, Pa., Oct. 9–12, 1960.

22. P. HOPPE, "Sandwich Laminates by In-situ Polyurethane Foaming," *Brit. Plastics*, p. 73 (Jan. 1965).

23. E. W. KUENZI, "Structural Sandwich Design Criteria in Sandwich Panel Design Criteria," Publ. 798, National Academy of Sciences–National Research Council, Washington, D.C., 1960.

24. E. W. KUENZI, "Minimum Weight Structural Sandwich," U.S. Dept. Agr. Forest Serv., Forest Prod. Lab. Rept. 086, Forest Products Laboratory, Madison, Wis., Jan. 1965.

25. W. S. ERICKSEN and H. W. MARCH, "Compressive Buckling of Sandwich Panels Having Dissimilar Facings of Unequal Thickness," U.S. Dept. Agr. Forest Serv., Forest Prod. Lab. Rept. 1583-B, Forest Products Laboratory, Madison, Wis., Rev. 1958.

26. H. W. MARCH and E. W. KUENZI, "Buckling of Cylinders of Sandwich Construction in Axial Compression," U.S. Dept. Agr. Forest Serv., Forest Prod. Lab. Rept. 18,30 Forest Products Laboratory, Madison, Wis., Rev. Dec. 1957.

27. J. J. ZAHN and E. W. KUENZI, "Classical Buckling of Cylinders Sandwich Construction in Axial Compression — Orthotropic Cores," U.S. Dept. Agr. Forest Serv., Forest Prod. Lab. Rept. 018, Forest Products Laboratory, Madison, Wis., Nov. 1963.

28. J. A. HARTSOCK, "Material Selection," Olin Mathieson Chemical Corporation, New York, Mar. 1966.

29. D. V. ROSATO and C. S. GROVE JR., *Filament Winding: Its Development, Manufacture, Applications and Design,* John Wiley & Sons, Inc., New York, 1964.

30. J. W. JOHNSTON, "The Composite Sandwich: An Appraisal," *Reinforced Plastics* **4**, 26–27 (Sept.–Oct. 1965).

31. G. A. ROSSI and J. H. JOHNSON, "Composite Sandwich for Small, Unmanned Deep-Submergence Vehicles," ASME Publ. 65-UNT-2, American Society of Mechanical Engineers, New York, May 1965.

32. "Carving and Forming Honeycomb Materials," Bulletin TBS-117, San Francisco, Calif., May 1963.

33. I. N. EINHORN, "A Review of Cellular Plastics," Wayne State University Conference, May 24–28, 1965.

34. E. DUPLAGO and B. D. RAFFEL, "Foamed-in-place Sandwich Construction — Metal Faced," WADC TR 53-72 (ASTIA 15928), ASTIA, Springfield, Va., July 1953.

35. A. D. GOLLADAY, "Development of Plastic Expanded Pellet-Type Core Material for Sandwich Construction," WADC TR 53-327 (ASTIA 24746), ASTIA, Springfield, Va., Dec. 1953.

36. R. F. COX, "Mismatch Metal Sandwich Construction," Review, Boeing Co., Wichita, Kan., Jan. 1966.

37. General Dynamics, "Sandwich-Brazed Titanium Molybdenum — Construction Methods for — Development of," Rept. MR-57-9 (ASTIA AD 272096), ASTIA, Springfield, Va., Jan. 30, 1962.

38. General Dynamics, "Material-Design Study of Brazed R. S. 140 Titanium Sandwich For B-58," Report FGT-2783, ASTIA, Springfield, Va., Jan. 15, 1962.

39. "Roll-Welded Sandwich Structures," Rept. 4534, Douglas Aircraft Co., Inc. and Battelle Memorial Institute, Nov. 1963.

40. "Sandwich Constructions and Core Materials; General Test Methods," Military Standard, MIL-STD-401, July 1952, revised-updated.

41. Plastics for Aerospace Vehicles," Military Handbook, MIL-HDBK-17, Armed Forces Supply Support Center, Washington, D.C., 1961.

42. "Metals," Military Handbook, MIL-HDBK-5, Armed Forces Supply Support Center, Washington, D.C., 1965.

43. J. H. CUNNINGHAM and M. J. JACOBSON," Design and Testing of Honeycomb Sandwich Cylinders Under Axial Compression," NASA Symposium, Oct. 24, 1962.

44. "Design-Testing-Data," Advance Excerpts Brochure D, Hexcel Products Inc., San Francisco, Calif., 1956.

45. B. G. HEEBINK, E. W. KUENZI, and W. S. ERICKSEN, "Evaluation of a Vacuum-Induced Concentrated-Load Sandwich Tester," U.S. Dept. Agr. Forest Serv., Forest Prod. Lab. Rept. 1832-B, Forest Products Laboratory, Madison, Wis., Feb. 1953.

46. "Study of Ultrasonic Techniques for the Nondestructive Measurement of Residual Stress," WADD TR 61-42, Part III, ASTIA, Springfield, Va., May 1963.

47. "Industrial Laminated Thermosetting Products," NEMA Publ. No. L1 1-1965, National Electrical Manufacturers Association, New York, 1965.

48. "Meaningful Testing," Society of Plastics Engineers, RETEC, Stamford, Conn., Dec. 2–3, 1965.

49. "Nondestructive Evaluation of Materials, NBS OTR-124, U.S. Government Printing Office, Washington, D.C., Aug. 1965.

50. D. V. ROSATO, "Plastics In Oceanography," *Plastics World* **23**, 24 (Sept. 1965); "Plastics and Ocean Environment," *Plastics World* **24**, 26 (Apr. 1966).

51. D. V. ROSATO and R. T. SCHWARTZ, *Environmental Effects on Polymeric Materials,* John Wiley & Sons, Inc., New York, 1967.

52. G. M. NORDBY and W. C. CRISMAN, "Strength Properties and Relationships Associated with Various Types of Fiberglass-Reinforced Facing Sandwich Structure," Army Ft. Eustis Report 65-15 (AD 621522), ASTIA, Springfield, Va., Aug. 1965.

53. E. W. KUENZI, B. BOHANNAN, and G. H. STEVENS, "Buckling Coefficients for Sandwich Cylinders of Finite Length Under Uniform External Lateral Pressure," U.S. Dept. Agr. Forest Serv., Forest Prod. Lab. Rept. 0104, Forest Products Laboratory, Madison, Wis., Sept. 1965.

54. E. W. KUENZI, C. B. NORRIS, and P. M. JENKINSON, "Buckling Coefficients For Simply Supported and Clamped Flat, Rectangular Sandwich Panels Under Edgewise Compression," U.S. Dept. Agr. Forest Serv., Forest Prod. Lab. Rept. (ASTIA AD 610 683), Forest Products Laboratory, Madison, Wis., Dec. 1964.

55. C. B. NORRIS, "Compression Buckling Curves for Sandwich Panels with Isotropic Facings and Isotropic or Orthotropic Cores," U.S. Dept. Agr. Forest Serv., Forest Prod. Lab. Rept. 1854, Forest Products Laboratory, Madison, Wis., Jan. 1958.

56. C. B. NORRIS, "Compression Buckling Curves for Simply Supported Sandwich Panels with Glass Fabric Laminate Facings and Honeycomb Cores," U.S. Dept. Agr. Forest Serv., Forest Prod. Lab. Rept. 1867, Forest Products Laboratory, Madison, Wis., Dec. 1958.

57. C. B. NORRIS, "Compression Buckling Curves for Flat Sandwich Panels with Dissimilar Facings," U.S. Dept. Agr. Forest Serv., Forest Prod. Lab. Rept. 1875, Forest Products Laboratory, Madison, Wis., Sept. 1960.

58. Northrop Corp., "Fatigue Testing of Honeycomb Sandwich Construction," ASD TR 61-388 (ASTIA AD 277 799), ASTIA, Springfield, Va., Apr. 1962.

59. L. BERKE and P. VERGAMINI, "Design Optimization Procedures and Experimental Program for Box Beam Structures," WADD TR 60-149 (ASTIA AD 247 125), ASTIA, Springfield, Va., June 1960.

60. R. REED, "Polybenzimidazole and Other Polyaromatics for High Temperature Structural Laminates and Adhesives," AFML TR 64-365, ASTIA, Springfield, Va., Part 1, Vol. 1, Nov. 1964.

61. W. J. TUOVILA and J. G. PRESNELL JR., "Supersonic Panel Flutter Test Results for Flat Fiber-Glass Sandwich Panels with Foamed Cores," NASA Tech. Rept., National Aeronautics and Space Administration, Washington, D.C., D-827 (ASTIA AD 257 853), National Aeronautics and Space Administration, Washington, D.C., June 1961.

62. L. KEER and B. J. LAZAN, "Damping and Fatigue Properties of Sandwich Configurations in Flexure," ASD TR 61-646 (ASTIA AD 272 016), ASTIA, Springfield, Va., Nov. 1961.

63. G. M. NORDBY, W. C. CRISMAN, and C. W. BERT, "Dynamic Elastic, Damping and Fatigue Characteristics of RP Sandwich Structure," Army Ft. Eustis Rept. 65-60, ASTIA, Springfield, Va., Oct. 1965.

64. University of Oklahoma, "Research in the Field of RP Sandwich Structure for Airframe Use," Army Ft. Eustis Rept. 64-37, ASTIA, Springfield, Va., July 1964.

65. P. M. JENKINSON and E. W. KUENZI, "Effect of Environment on Mechanical Properties of Glass-Fabric Plastic Sandwich," WADC TR 60-202, Sup. 1 (ASTIA AD 282-010), ASTIA, Springfield, Va., May 1962.

66. P. M. JENKINSON, "Effect of Core Thickness and Moisture Content on Mechanical Properties of Two Resin-Treated Paper Honeycomb Cores," U.S. Dept. Agr. Forest Serv., Forest Prod. Lab. Rept. 35, Forest Products Laboratory, Madison, Wis., Sept. 1965.

67. W. T. JACKSON and B. R. GARRETT, "Heat Resistant Plastic Honeycomb for Sandwich Construction Based on Polyaromatic Resin," AFML TR 65-157, May 1965.

Chapter 9

Molding Laminates and Sandwich Materials

Charles B. Hemming

1 FOREWORD

In the first edition of *Engineering Laminates* which carries the date of 1949, this chapter described what was then a new industry, half art, half science, forced into adolescence by the exigencies of war. In the period between 1939 and 1941, there were embryonic attempts to do molding and laminating in which the emphases were on large areas, simple and compound curves, and the use of cores that could not possibly withstand the pressures then common for what might properly have been classed, at that time, as engineering laminates.

There were straws in the wind. The plywood industry was entering a stage of growth, later to prove spectacular, using laminating pressures of 300 psi and under. There were some attempts to make stiff panels, light in weight, by attaching metal skins to plywood panels and early workers even went so far as to slice balsa wood into uniform sections and attach faces to make the first elementary sandwich panels.

Figure 9.1 Portable Gallery Walls of Modular Sandwich Panels

CHARLES B. HEMMING is Director, Research and Development, United States Plywood Corporation, New York, New York.

The Second World War changed all this and accelerated the development of these techniques, forcing them into production and telescoping, into a few years, time that might have been measured in decades.

This chapter, in the earlier edition, documented all of this.

In reviewing what has happened since 1949, one discovers that this new industry, which has no generic name but is often called the field of structural plastics, was found to have languished except in certain isolated fields and is only now beginning to be industrially significant.

Progress has hardly been in the field of basic discoveries in either materials or techniques but, rather, it has been a period of consolidation and refinement principally of equipment with some improvement in methods and efficiency.

Figure 9.1 illustrates one interesting use of panels. The portable gallery walls for the new Whitney Museum of American Art are based on a two-foot modular concept and feature floor-to-ceiling particle-board panels. When not arranged as divider walls, panels can be stacked in gallery corners or placed around the room.

2 INTRODUCTION

2.1 Definitions, pressure ranges

There has been some confusion of terms, but molding is properly reserved for forming parts or structures with curved surfaces, whereas laminating, without further description, implies that a flat sheet is the finished product. Naturally, such a distinction is arbitrary, but it helps simplify discussion and thinking and is more or less supported by practices that have existed since this young industry came into being. There have been some changes that tend to confuse definition. For example, flat sheets rarely were formed from reinforcing agents that were not first made into some prefabricated form such as paper, fabric, mats, veneers, or metal sheets. This is no longer strictly true and some of the advances have involved simultaneous formation of the material that subsequently becomes a flat sheet. These advances will be detailed later.

Looking at molding and laminating broadly, we find a large pressure range, although it is substantially less wide than formerly since some of the laminates requiring extremely high pressures have fallen by the wayside in favor of superior ones made at lower pressures. An arbitrary but generally accepted division of the art as it now exists can be made according to Table 9.1.

Tooling or dies range from strong steel platens or matching male and female die parts through combinations of hard rubber pads or plugs and steel platens or molds to wood and rubber

195

Table 9.1.   Laminating Pressures

0 to 25 psi		Vacuum pressures and contact-pressure molding (a variation of the latter is dead weight molding or laminating)
25 to 100 psi		Low-pressure molding (sometimes uses positive dies)
100 to 500 psi		Low-pressure laminating: the "plywood" range
500 to 750 psi		Intermediate-pressure and positive-die molding
750 to 1500 psi		High-pressure molding and laminating

bags. With reference to Table 9.1 on the arbitrary division of pressure ranges, the following are accepted tooling practices for each range.

Tooling of the 0 to 25 psi range, in the case of flat laminate production, is generally a single light platen that can be used with or without a flexible bag. The principal improvement here has been advancement of what are known as half-bag techniques rather than totally enclosing envelopes. Cast epoxy resins figure prominently as molds in this range and are a logical development of the older attempts to use cast mineral compositions as dies. For molded work, patterns or forms such as will be described in the 25 to 100 psi range may also be used. Either mating, which requires considerable accuracy, or some version of flexible bag may be used in some instances. In this range, if surface tension forces are all that is required to make a molding, no bag is necessary because no pressure is necessary. The burden is on the resin which must, peculiarly, be sticky when the lay-up is made or become sticky with moderate heat and not lose its tack until cure is sufficiently far advanced that the molding cannot self-delaminate. There is a balance here between the tackiness of the resin and the stiffness of the fabric or reinforcing agent which is being molded. For light positive pressure, some form of envelope, which may be rubber, synthetic rubberized fabric, polyvinyl alcohol in film form, or some of the newer, very strong, flexible films such as the polyesters or polyvinyl fluoride may be used. Advancements here have come from improved tooling which involves membranes that may be clamped over the part after lay-up on the

mold. This type of tooling has increased the speed of production, lowered cost, and utilized fewer expensive membranes. An autoclave for application of fluid pressure is shown in Figure 9.2.

The range of 25 to 100 psi requires but light platens, generally with flexible pads, preferably rubber, for flat work. For molding, fluid pressure acting on membranes or flexible bags which, in turn, bear against patterns, either male or female, are required. Various castable and easily machinable alloys are used, the and new cast epoxies are particularly suited. Plaster is rarely used anymore except for break-away molds where the part does not have what is called a molding exit. Wood is often used for prototypes or short runs, but if accuracy is required wood patterns are made of phenol-resin-stabilized wood known as impreg. This latter technique is a new adaption of an old art and has solved some important problems in the automotive industry. Since the fluid pressure must, in some way, be confined, either an autoclave or a unit-type pressure vessel is required except in the special case of the female mold or pattern which is fitted with a cover and makes use of an inflatable rubber bag similar to the bladder in football. For substantial runs, male and female mating molds can be used, but because of the accuracy required, the molds are costly and not justified for a few parts. It is in this range that molding plywood is done. Since this technique involves actually gluing to contour rather than true molding, it also embraces the procedures used in the manufacture of a variety of compound-curved wooden surfaces for decorative and utilitarian purposes. This is also the range used for the production of radomes which are as important as ever not only to our fighter planes but to just about all of our modern passenger-carrying aircraft.

In the very important 100 to 500 psi range, both flat laminates and molded objects are produced. This is known as the plywood range. Hydraulic presses for flat laminates are generally specially adapted plywood hot presses. A multiple-opening platen press is shown in Figure 9.3. Platens and press

Figure 9.3   Multiple-opening Platen Press for Manufacture of Plywood

Figure 9.2   Autoclave for Application of Pressure to Flat or Curved Molded Parts

frames are of medium weight and strength compared to the more massive high-pressure presses. So-called "stress-frame" presses are popular and often homemade.

Very large platen areas and large openings or daylights can be accomplished inexpensively.

Molded sections in this range are made by two methods. The one more frequently used requires light but sturdy mating metal dies and hydraulic presses of modest pressure capability but large daylights. Machining must be of highest quality although accuracy of mating can be substantially lower than in the case of true high-pressure molding. This process is a form of die molding known as positive-die molding and borrows from both high- and low-pressure procedures. The other method takes advantage of the fact that rubber is a liquid and therefore incompressible. A positive female mold forms the outer contour. A rubber plug held captive by a back-up male plug forms the inner contour. As the pressure is applied, the rubber "flows" and compresses the material to be formed. Very substantial pressures for molding can thus be obtained with the correct choice of rubber durometer. A parting sheet is used to protect the rubber plug from contact with the molding resin.

Decorative laminates and the industrial equivalents known as phenol-paper have been called molded sheets. To a certain extent, this term is justified because of the high pressures used. However, there is a counterpart in the 100 to 500 psi range, which is a low-pressure lamination process better described as gluing to contour than as a molding; it can be used for either flat or simple- and compound-curved surfaces. It came into being to satisfy certain needs for thin high-accuracy phenol-paper laminates. The generic term is papreg, a development of the U.S. Forest Products Laboratory. Little of it is used in that form in the present day but a modern development of it is used in very large quantities for the surfacing of plywood for industrial purposes. The so-called medium- and high-density overlaid plywoods follow this principle.

The 500 to 750 psi range for flat laminates is practically extinct commercially. It was originally a hybrid between papreg and high-pressure laminates with the advantages of neither. However, some positive-die molding is done in this range.

The 750 to 1500 psi range makes use of heavy steel platens in massive hydraulic presses for laminating flat sheets. Heavy accurately machined and highly finished mating steel dies are necessary for molding. These, in turn, are placed in a massive hydraulic press which is arranged for quick closing. It is in this range that our decorative and industrial high-pressure laminates known generically as phenol-fiber, phenol-paper, phenol-fabric, and the newer glass-fiber combinations are produced. Developments in this field have included the expanded use of melamine resins and other resins specially formulated for required physical properties. Details of advances of the art in this range are discussed elsewhere in the book and need not be repeated here.

2.2 Temperatures

Temperatures required throughout the range depend principally on the type of resin involved rather than the pressure. This is obvious when one considers that molding or laminating almost always involves the act of first fusing the resin, at which time it takes the proper shape or plastic deformation and proceeds, in the case of thermoset materials, to the cured stage. In the case of thermoplastic materials, of course, cooling below the softening point is necessary before pressure is released.

Since these are viscosity phenomena, they are time and temperature functions. Pressure only bears on the temperature requirements insofar as it affects rate of flow at the chosen temperature. General practice is to use temperatures required by the several resins as outlined in Table 9.2

It should be remembered that the temperatures listed in the table are to be considered as those existing within the laminate or in the molding rather than ambient temperatures. It is obvious that within certain limits resins from a lower temperature class can be used at higher temperatures with resulting shorter cures. It is likewise obvious that very high-speed processes can make use of temperatures possibly as high as 1000°F if the effective temperature in the laminate is correct and the part is such that heat flow is no problem. As cures are made shorter and shorter by the use of higher and higher

Table 9.2. Temperature Requirements for Several Resins

70°F	Room-temperature-setting urea-formaldehyde, resorcinol-formaldehyde, and certain epoxy resins
75°–120°F	Resorcinol-formaldehyde, phenol-resorcinol-formaldehyde, acid-catalyzed casting-type phenol-formaldehyde, and certain polyesters in which catalyst-promoter combinations are used.
180°–235°F	Melamine-formaldehyde, urea-formaldehyde, "fortified" urea-formaldehyde, certain acid-catalyzed phenol-formaldehyde resins, polyester resins using promoters, and the hot-setting epoxy resins.
180° and 290°F	This is a special combination of temperatures particularly for polyester resins in which the resin is first jelled at a modest temperature and then hard-cured for durability and strength at the higher temperature. Certain special phenol-resorcinol resins also respond best to this procedure, and it is sometimes used for epoxy resins.
250°–290°F	Alkali-catalyzed phenol-formaldehyde resins of the resol[a] type and two-stage neutral phenol-formaldehyde resins of the novolac[b] type. Melamine formaldehyde resins can be cured at this high temperature although it is unnecessary and control must be quite accurate.
310°–325°F	Cross-linked thermoplastics, phenol-acetal resin compositions, and some of the more easily cured silicone resins.
450°–600°F	Certain silicone resins for bonds that will see high-temperature service and butadiene acrylo-nitrile-phenol-formaldehyde resin combinations otherwise difficult to cross-link.

[a] A resol, in simplest terms, is a complete phenol-formaldehyde resin in uncured form. It needs no further addition of any chemical to turn it into its final insoluble, unfusible form. Time and temperature alone accomplish the change from a resin to a thermoset plastic.

[b] A novolac, in simplest terms, is an incomplete phenol-formaldehyde resin. It will not cure or become a thermoset plastic unless an additional ingredient is added. This is generally hexa-methylene-tetramine. Note that this is not a catalyst but a true ingredient of the final cured plastic. Once hexa-methylene-tetramine has been added, time and temperature can accomplish the transformation into a plastic.

Figure 9.4   Preform Machine for Molding Reinforced Plastics

temperatures, it is found that, all things being equal, each resin has a limit beyond which it cannot be pushed without experiencing frothing, burning, or excessive cross-linking with loss of strength. If the heat flow is sufficiently rapid, extremely high temperatures may cause curing to occur so fast that the consolidation of the laminate and its compliance with the desired contour may not take place.

2.3 Time

As has been suggested, time is the other of the two vital factors controlling the cure or the flow of the mass being laminated or molded. Time varies, depending on the following existing conditions:

The amount of mass to be heated.
The rate of application of heat and whether heat is applied from one or both sides of the part.
Rate of heat conduction of the mass or assembly.
Time necessary to complete the flow and chemical action, if any.

Periods involved may be as short as a minute or as long as an hour or more. Special compositions involving aggressive catalysts may be expected ultimately to be useful when proper equipment is developed. High-energy particles are being employed to promote polymerization and cross-linking. This kind of application may be expected to play a substantial part in the very near future and may change the entire field of molding and laminating insofar as the time factor is concerned.

With this picture of the field of molding or laminating before us, we will now deal specifically with the techniques of molding

of laminates and sandwich materials with particular emphasis on those types engineered to produce predetermined sets of controllable physical properties.

2.4 Summary of improvements since 1949

The principal changes that have occurred in the intervening years have been ones of fabrication techniques, and this chapter has been rewritten with emphasis on fabrication.

The industry that has grown the fastest is the reinforced-plastics industry, now conveniently known as RP.

In the field of simple laminates, the biggest changes that have occurred have involved a wider range of grades and finishes with variations in thickness for various uses. Backing or balancing sheets are of a more economical construction. Perhaps the most interesting development has been the creation of an oil-rubbed appearance which is nonfingerprinting and of low glare. Improvements in construction which have led to thinner, more economical sheets while still minimizing telegraphing of the cores have been important contributions.

The so-called true cigarette-proof grade has largely disappeared from the market for reasons of cost, but reasonable National Electrical Manufacturers' Association Standards of Quality continue to be met.

In the field of sandwich construction, techniques are pretty much the same, but a greater variety of adhesives meeting a wider range of specifications has led to configurations that avoid both over and under design, again, with an eye to economy.

The greatest change has come about as a result of learning to use foam cores produced by a variety of methods and made from a variety of plastic resins.

In the reinforced-plastics field we now have — in addition to the earlier method involving lay-up of resin-saturated glass-fiber cloth — spray methods, a form of hat-molding, roving and filament-wound techniques, and positive-die molding.

While other reinforcing agents are used to a limited extent, the greatest progress has been made using glass fiber in various forms.

There are available chopped-strand mat, continuous-strand mat, straight roving, chopped-strand unmatted, and spun roving. Woven roving also continues to be used because of progress in lowering its cost. Hand lay-up is used for custom work and small runs, but wherever possible automatic or semi-automatic processes have come into their own. A preform machine for producing reinforced-plastics parts is shown in Figure 9.4.

The various winding techniques are, of course, used for cylindrical vessels and tubing. Matched-die molding is used for configurations that cannot be wound and here the hat process and matched-die techniques are preferred.

Taking a leaf from the older technique of surfacing wooden boat hulls with reinforced plastics, the most recent development involves stock for containerization in which the core is made of plywood for lightness and stiffness, whereas the overlay of reinforced plastic contributes its own unique properties to the surface.

While much of the containerization stock is still done by improved hand lay-up techniques and bag molding, special guns have been developed allowing spray coating. These guns chop the mat, coat it with plastic resin, and deposit the mixture in the necessary thickness on the surface (Figures 9.5 and 9.6).

Improved methods of getting adhesion to the wood and improved resins and catalysts are contributing to better durability, better appearance, and better physical properties.

For structures requiring the utmost in beauty and smoothness, gel-coating techniques have been developed so that surfaced-plastic bathtub assemblies now have the beauty and performance of porcelain-enameled iron. Furthermore, these one-piece bathtub and shower units are of lighter weight and avoid the coldness of the porcelain.

The molded bathtubs meet industry's standards which are set up by the National Association of Home Builders.

In the offing are revived forms of prepreg allowing dry lay-up.

New resins are promised which are expected to reduce the cure time to the order of one-half minute and which will require but a small amount of pressure other than simple contact. Orthochlorostyrene-based resins are an example and require 15- to 20-psi molding pressure.

Phenolic resins are by no means out of the picture where their durability properties are required, and acid-cured novolacs are showing encouraging results. While the wound-strand and roving techniques are particularly suited for maximum-strength requirements, the spray-applied strand is the most versatile and does not prohibit orientation for control of strength.

"AF fibers," a form of glass fiber developed for heat insulation and acoustic purposes, can now be furnished shot-free and are of considerable promise where the lower physical properties permit. The AF fibers are durable outdoors, providing the binder is adequate for the purpose. They have from one-third to two-fifths of textile fiber strength.

The AF glass-fiber wool is furnished in the form of an uncured pelt. The resin is a "B" stage novolac phenol-formaldehyde composition. One of the new uses for this product is a glass-fiber hardboard containing 85 per cent glass. It requires 300-psi molding pressure. This is not true low pressure but well inside the so-called plywood range.

In nonwound constructions, bag molding still furnishes the maximum physical properties.

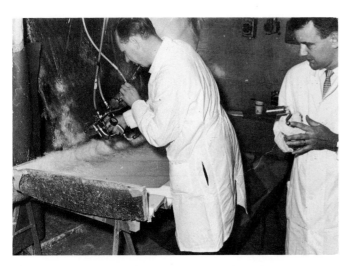

Figure 9.5   Chopped Roving Deposition with a Special Gun which Chops Roving and Deposits it along with the Necessary Resin

Figure 9.6   Compacting the Deposited Chopped-roving and Resin Mixture

Figure 9.7   Gel-coating a Section of a Reinforced Plastic Molding

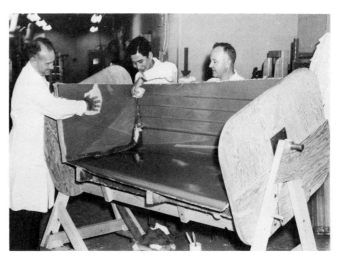

Figure 9.8   Mold Preparation for Reinforced Plastic Molding

A superior new mat is expected to be available about the time this book is published, with eight to ten times the strength of the older surfacing mats. However, from a fabrication viewpoint its principal interest lies in the fact that it will come in rolls, which will result in additional flexibility in manufacture similar to that of woven fabrics but without their high cost. There will be a choice of binders.

Glass-fiber scrims are still available and have reached new low prices. They are of interest where their physical properties are adequate.

One of the ways of obtaining exterior durability in reinforced plastic is, of course, pigmentation. Pigmentation has its principal effect on the binder or plastic resin, but it must be remembered that it is part of the filler material and since it offers very little strength as compared to that offered by the glass fiber, the physical properties must be balanced against the durability. In the case of pigmentation, great care must be exercised in picking inert substances that do contribute to durability and do not act simply as another filler. The choice of these pigments is very critical.

Figure 9.7 shows the application of a pigmented gel-coat overlay for a reinforced-plastic part. In Figure 9.8 a mold surface is being prepared to produce a fine finish.

In the manufacture of both laminates and overlays, the practice has been to put on $\frac{1}{16}$-in increments with wet-out rolling in-between to ensure uniform distribution of the plastic-resin binder.

2.5 Reinforcing agents

In the low-pressure molding and laminating field, and particularly in that part of the plastics industry which now classes itself as RP, glass fiber in one of its several forms has been almost universally used as the reinforcing agent. As a result of this, we have forgotten the lessons originally learned from the high-pressure plastics industry, only to relearn them relatively recently.

There is no question but that glass fiber is an extremely important reinforcing agent, but even the most avid enthusiast would be willing to admit that there are other reinforcing agents and that they may possibly have advantages for specific uses. It would be an unwise product manufacturer and molder who would ignore other materials that can have a superior group of properties for specific product end uses.

Originally a reinforcing agent was thought of as a tough, flexible material that gave added physical properties to relatively brittle resins used as binders. More recently the reinforcing agents have been called upon for special properties.

Since this is essentially a chapter on technology, we point out that even in the reinforced-plastics market there are some twenty different materials that are suitable. The technique of using these agents depends on their nature. For example, paper and cotton fabric can be handled very much like fiber-glass cloth. Chrysotile asbestos, after suitable preparation, can be handled very much like mat or chopped roving.

Polyester fiber and other synthetic materials may be had in the chopped form, as a mat or paper, and as a fabric, and are handled accordingly. The same is true of nylon, which is available as fabric, yarn, and monofils. Monofils can be utilized in filament-winding techniques.

Any fibrous material that can be pulped can also be used as a reinforcing agent using forming techniques and resin application similar to the time-honored procedures for molding paper pulp.

Moldings and laminates are now being designed in which the reinforcing agent is a composite or a combination of several different materials, which even include layers of metal fabric or wire cloth. These composites introduce no particular technological problems beyond the fact that each form must be suitably clean and, if sizing coats are necessary, the sizing must be applied ahead of time. Cleanliness and surface treatment to ensure adhesion of the binder is fundamental and varies with the reinforcing agent. However, all workers in the field of glass fiber know its importance. Perhaps the only thing unique in composites involving metal is that the metal must be chemically clean and frequently surface-etched by suitable chemicals using treatments familiar to sandwich panel maunfacturers who utilize metal faces or skins.

2.6 Filament-winding techniques in reinforced-plastics molding

In this new, fast-growing, and sophisticated fabricating technique known as filament winding, we will again concern ourselves for practical reasons primarily with glass fiber, but it is to be remembered that with suitable variations metal wires and tapes can be similarly used, either alone or in combination with glass.

Pretreatment, of course, depends upon the classic steps of cleaning and sizing, the latter often actually employing coupling agents to ensure adhesion between the binder and the filament.

Glass in particular, with its high coefficient of friction and abrasive properties, will tend to destroy itself and must carry a lubricant up to the point of utilization. This necessitates cleaning so the coupling agent or sizing can bond adhesively to the glass.

Most of these coupling agents are well known although certain new derivatives of the older ones are now in use. Common ones are methacrylatochromic chloride, allyl trichloro-silane-resorcinol, vinyl dischlorosilane, amino triethyl oxysilane, and, newest of all, a series of epoxy resins specially made for sizing.

One must be careful to ensure that moisture is removed from the fibers before application of the epoxy resin, and because of this necessity heat-cleaning procedures are often used.

Naturally, epoxy resin size brings to mind epoxy resins for molding, and these are being used extensively. There are also epoxy-polyester hybrids in which the lower cost of the polyesters is optimized against the superior stamina of the epoxies.

In a modern production unit, the commercial yarn, roving, or fabric tape proceeds in sequence to a cleaning bath or a heat cleaner, a sizing bath to add the coupling agent, and a suitable oven for drying. While it is not necessary to apply resin at this point, it can be done if the resin is of the so-called wet type most commonly used or if it is unusually sticky. The treated, coated glass should go right to the winding equipment rather than being rewound for stock.

Some of the very latest developments involve resins that are dry after application to the glass. These can be stored in any reasonable configuration, to be used later in the winding operation.

A variant of either of these procedures is one in which the cleaned, sized, and dry glass is wound simultaneously in the formation of the part while being impregnated with the appropriate resin.

The winding equipment, as might be expected, has a lathe-like application head that controls the uniformity of wall thickness. The winding tension is adjusted in order to give consolidation and to assure proper physical conditions on an after-cured basis. The essentials of coating rovings for filament winding are shown in Figure 9.9.

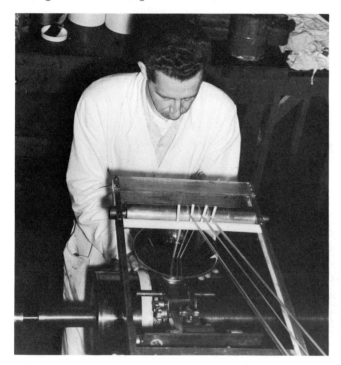

Figure 9.9   A Version of Filament Winding-molding Showing the Use of Roving to Make a Cylinder

It should be mentioned in passing that while we ordinarily speak of E glass as the source of our glass fiber for reinforced plastics, even this is changing. There are new high-strength glass fibers and modifications of E glass which markedly increase the physical properties.

Other fibers that are being studied and used in certain special-purpose laminates are boron and metal fibers. There is presently only limited information available on the technology of utilization of these fibers in reinforced plastics, but it is not expected that production techniques will be particularly more complex than those for glass fibers except in special, extremely critical end-use requirements.

When a wire becomes a fiber rather than an extremely fine wire, is a matter for arbitrary definition. Technically, wire is a material that is produced by drawing rods through dies or by rolling as in tape, rather than by casting or forging. Of course, extremely fine wire as small as 3 mils in diameter requires more careful handling in winding and treating than wire in the normal ranges, but this obvious difference has nothing to do with the fundamental requirements of the

process. What is more specific is the alloy that must be chosen according to the end use, and that is the subject of Chapter 6. The point at which the technology takes a hand has to do with laying down the filaments, and it can be easily visualized that precise control of the winding pattern and direction of the filaments is required for maximum strength. This, in turn, can only be achieved with controlled machine operation. Uniformity of wall thickness, as already mentioned, is also best controlled with machine operation, and, finally, the all-important winding tension must be adjusted so that maximum strength is achieved when all major stresses are carried by the filaments in tension.

With these problems confronting the industry it is natural to expect that suitable filament-winding machinery would be developed and made available. This is definitely what has happened.

2.7 Filament-wound reinforced-plastic tanks

The detailed description of the manufacture of tanks is chosen because it illustrates the new techniques of filament winding to produce reinforced-plastic parts; but it also includes other steps since filament winding is usually limited to cylindrical sections and a tank must have a top, a bottom, and often a skirt.

If the tank rests on a skirt, which must be an integral part, of course, the filament-winding technique can be used to produce the skirt. The skirt is connected by hand lay-up to an appropriate part on the tank bottom after it has been formed.

The dished ends are molded separately and are generally prepared with a hand lay-up and bag molding for maximum strength. The premolded ends are then lap bonded to the cylindrical shell and the resulting joint overlaid with an appropriate resin, usually polyester.

Filament winding has come to be a general term and filaments or groups of filaments may be the appropriate form, or bias tape may be used if strength and absolute uniformity of wall thickness and strength are critical.

Whether a filament or a tape is used, appropriate cleaning and sizing are necessary. This is done continuously by passing the material through the necessary treatments and through a drying oven and, finally, rewinding.

A steel male mandrel of appropriate inside diameter is made ready and a tape, preimpregnated with the desired resin, commonly epoxy or polyester but also phenolic in some cases, is fed to the winding jig, usually at a 15° angle.

After each layer of tape is wound on, longitudinal strength can be obtained by use of longitudinal plies.

While bias tape can be wound to almost any angle to the center line, an increase in angle decreases the longitudinal tensile strength of the part. At 90° orientation, it is easy to see, therefore, that only the resin between the fabric plies holds the structure in longitudinal tension. The 15° angle is an engineering compromise to ensure fail-safe characteristics in the tape-wound part, and this angle is easy to wrap. It has good longitudinal strength in itself and also resists cracking during cure.

When the tape wrapping is completed, the entire lay-up is encased in a rubber bag equipped with the usual parting layer and bleeder strips. The assembly is placed in an autoclave and a vacuum of at least 28 in. of mercury is drawn. The cure time and temperature depend, of course, on the resin used, but can be of the order of one and one-half hours at a temperature of 200°F.

A newer refinement of this technique involves the use of a hydroclave instead of an autoclave. The hydraulic fluid used in place of the gas (steam) or steam-air mixtures gives better molding of complex shapes and variable contours. A wider range of pressures is also safely available.

Longitudinal laminae, if needed, are generally composed of unidirectional glass fabric impregnated with the same resin.

For high-pressure services the tubing can have a final circumferential winding to give additional hoop strength.

Of course, if the shape permits, the hydraulic press can be used as well as the hydroclave and autoclave, but this requires matched dies. A modification of filament winding has been developed which requires no auxiliary pressure vessels. In this case the mandrel is capable of being heated or cooled with appropriate media, and the impregnated tape is tension wound, curing by preheating with infrared lamps positioned immediately prior to wrapping. The pressure is minimal because one ply is compressed at a time. Wrapping speeds are controlled so that adequate heating for cure is obtained.

3 METHODS OF LOW-PRESSURE FORMING OF LAMINATES AND SANDWICH MATERIALS

3.1 Definitions

The language of the art has changed little though there is more tendency to use the term "structural laminates" and "reinforced plastic laminates" than the older term "low-pressure laminates." Nevertheless, the basic criterion is that the molding pressures in this field are in the range of 0 to 500 psi. Roller laminating or nip roll bonding is a new and special case and it has been argued that high pressures are involved even though they occur along a line. Nevertheless, consideration of the lack of mass in the equipment adequate for the work and the fact that relatively easily crushed cores can be overlaid with these techniques permits us to include them in low-pressure forming of laminates and sandwich materials. Here the special requirement is more that of being able to furnish the face stock in roll form as opposed to sheet form for flat press or compound-curved piece molding.

Forming is an obvious term which includes the making of both flat sheets and curved parts. Laminating may or may not be involved in the strict sense, but it generally is. Laminating, often used interchangeably with molding, indicates that the unit is formed from a series of laminae or layers which are welded into one piece by the action of heat and pressure on the impregnating resin contained in the laminae or furnished to the laminae at the instant that consolidation takes place. Laminating generally carries the interesting connotation that a flat piece is being made. Molding similarly indicates that a curved part is being made although neither term is used strictly as indicated.

Considerable study has already gone into the definition of a sandwich construction or sandwich material, as it is obvious that there are many types of sandwiches in common use with which we are not concerned. In this industry, a sandwich material is generally considered to be composed of relatively

dense, high-strength skins which are relatively thin and permanently affixed to either side of a central core material which is relatively thick and of relatively low density. It has, generally, much lower strength than the skins since its sole requirement is to transfer sufficient shear forces to develop the strength of the skins. Additional strength within the core is wasteful and generally incurs a weight penalty. A little reflection will show that a sandwich representing the reverse of these requirements could be fabricated; although easily made, it would have little structural value though it might, conceivably, be decorative or could be of value in acoustical applications. In sandwich constructions under consideration, both the skins and the core are generally fabricated in advance and then combined to produce the sandwich. A notable exception to this is the three-layer particle board and the modern plastic-laminate overlaid plywood; both are made in a one-shot operation for reasons of economy of production. Laminates are composed of layers or laminae generally uniform and identical to each other with respect to thickness, density, and strength. The individual laminae or plies are generally relatively strong so that the resulting composite is merely a combination of the individual properties. Laminates can, of course, become the faces of structural sandwiches simply by being attached to either side of an appropriate core.

3.2 Choice of pressures

There are a number of important reasons justifying the use of low pressure in producing laminates and sandwich constructions. These constructions are almost invariably for strength or structural purposes. In solid laminates and in the skins of sandwiches the filler or reinforcing fabric furnishes the strength characteristics, whereas the resin or the adhesive holds the construction together. The requirements of these constructions are that they be of uniform strength, stress-free and flat or of accurate contour. There must, therefore, not be any crushing, displacement, or deformation due to the forming pressure. A disproportionately sharp loss in stability of shape or flatness and strength performance occurs due to initial eccentricity with even relatively small amounts of crushing displacement or deformation.

Sandwiches in particular, owing to the low density of the core materials, must be assembled with low pressure not only to meet the foregoing requirements but also so that the internal construction of the core is not buckled or distorted or prestressed. Recovery from elastic displacement can further complicate the stability problem and cause disfiguring telegraphing of the internal parts of the structure through the faces.

Finally, the large areas of a unit or a panel normally required would necessitate excessively heavy and costly equipment if high pressures were to be used. In discussing the actual methods consideration will first be given to producing flat laminates, then curved laminates, followed by flat sandwich construction and, finally, curved sandwich construction.

3.3 Preparation of materials

Since flat laminates are, as was noted in the definition, the result of the joining of separate layers called laminae or plies, attention must first be given to the preparation of the stock of which these layers are composed. Each layer comprises a reinforcing sheet, fabric, or batt which has been appropriately treated with a resin that will later secure it firmly to its neighbors and hold each individual fiber in place. The reinforcing sheet must, therefore, be treated by some means that attaches the necessary resin to it in a convenient manner for subsequent handling. This procedure can involve simple saturation of the web or reinforcing sheet by passing it through a bath, or coating of the outside layers without penetration. For special purposes, true impregnation in addition to superficial saturation or coating may be required, and the simple deposition of dry powdered resin followed by sintering to make it adhere has been used successfully.

Since the stock so prepared is generally treated with a resin solution, it is necessary that it be dried before further use. This drying must dispel as much volatile solvent as is possible without causing cure of the resin. If the volatile solvent is not removed, the roll or pack of treated stock may flow together and form a solid block which is quite useless. This phenomenon known as "blocking" must not occur under any reasonable storage conditions.

During the drying operation it is often desirable to initiate the advancement of the resin toward the cured condition. An unadvanced resin may flow excessively under the conditions of lamination or may volatilize from the heat with the result that the finished laminate is starved and does not have sufficient binder to resist delamination under stress.

Finally, the resin must be chosen and treated so that the impregnated reinforcing sheet must be stable for the necessary period of storage until it can be used.

4 PRODUCTION

4.1 Assembly

The first step in the actual production of the laminate is the lay-up or assembly of the laminae or plies of prepared reinforcing fabric or fiber. For flat rectangular sheets this merely means cutting to length and width, slightly oversize to allow for accurate trimming to size, and stacking the proper number of layers to give the required final thickness. If the laminae are anisotropic, it is often desirable to cross-band the assembly. This means that alternate layers are cut so that they can be turned 90° with respect to the other layers in order to obtain balanced panel with equal strength on both axes.

If the panel is of an odd shape, it may be more economical to lay up templated sheets than to cut finished parts from a rectangular laminate. Such templating can be done by means of steel-rule dies and can be made a very economical operation.

If the panels are larger in any dimension than available sheet stock can produce, it becomes necessary to decide whether each layer will be pieced by means of a lap or a butt. Both have their advantages and disadvantages, so that individual requirements should dictate the chosen practice. Generally laps give greater strength but less accuracy or uniformity in thickness. Butt joints are just the reverse.

In order to obtain maximum strength it is important to avoid wrinkles or slack in each layer. Ideally each layer should be stretched taut on both axes until the moment that lamination or resin cure actually begins. However, if the stock has been

properly handled up to the point of lay-up, adequate performance will be obtained from the laminate as long as ordinary care is used in stacking the individual sheets.

4.2 Pressing

After the lay-up is completed, it is necessary to press or form the laminate. The curing cycle, with respect to both time and temperature, must be chosen with regard to the requirements of the resin used to impregnate the sheet. Sources of pressure were given earlier in this discussion and should not need repetition here except to indicate that the hydraulic press is the most common source for large volume production, although certain of the newer laminates are now being made in a continuous-laminating process in which rolls furnish the pressure. The heat is furnished by means of steam, hot oil, electrical-resistance heating, radio-frequency electrostatic heating, and occasionally by means of radiant-heat (infrared) lamps.

Precuring and bridging must be avoided if a sound laminate is to result. Precuring most commonly results from too much exposure to heat before the pressure is applied, and the remedy is obviously to shorten the time required to load the press. For a multi-opening press and a fast-curing laminate, it may also be necessary to place the assemblies for the bottom few openings on insulating cauls in order to retard the application of heat. Bridging, which is a form of precuring, is a condition that occurs in thick laminates wherein the outer layers become hard and inflexible before the center has heated to the fusing point of the resin. Bridging has the effect of removing the pressure from the central layers so that they are not well pressed together and in extreme cases may actually be loose. The remedy is to reduce the thickness of the laminate or to make use of radio-frequency electrostatic heating in which the problem of heat flow through the laminate does not exist.

It is important that the assembly being cured not adhere to or contaminate the pressing platens with resin. It is general practice to include metal cauls above and below each assembly which are, in turn, lubricated with parting agents. These parting agents include metallic soaps, certain petroleum-oil or vegetable-oil fractions, and silicone greases. It is possible to apply the parting agent directly to the pressing platens, but, since the platens are generally heated continuously, the parting agents tend to burn off, and recoating is simpler in the case of the detached caul. In some cases, the resin manufacturer formulates the parting agent into the resin so that little or no additional material need be added to the cauls or press-platen surfaces once these surfaces are "broken in."

Blisters are a common fault, particularly in the formation of flat laminates, and they are most frequently caused by residual volatile solvent not properly dried out at the time of impregnation or by too high a proportion of volatile resin component. The only real remedy is, of course, to treat the impregnated web properly.

If susceptibility to blistering is expected, certain laminates can be made successfully by the expedient known as "breathing the press." This operation involves reduction of press pressure for a brief period after the charge is thoroughly heated but before curing has set in or all flow has ceased. The aim is to release the gases formed by the heat from the excessive quantity of volatile material but at a rate low enough to avoid explosive release and rupture of the partly formed sheet. For some laminates pressure is only reduced to the point where it is in equilibrium with the gas pressure of the volatile involved at the pressing temperature. For still others the pressure is just reduced to zero, but any opening of the press is avoided. For very porous plies the press can actually be momentarily opened but must be reclosed as soon as possible.

Starvation has been described earlier. This fault is caused by excessive flow of the resin under pressure and results in a weak laminate just as surely as would an under-impregnated stock. The resin may squeeze out or disappear into the plies if the latter are of the surface-coated type. Lowering press pressure below normal is a suitable expedient, but proper preadvancement of the resin after impregnation is the correct cure. For resins that cannot be preadvanced a two-step curing temperature cycle is necessary.

4.3 Discharging and stacking

Discharge from the press may be hot or cold, depending on the requirement of the resin used. Thermoplastic resins must generally be discharged cold for the obvious reason that during lamination they are above their melting point and, if they are discharged hot, they will either delaminate or deform during handling. Certain thermosetting laminates, particularly of the decorative type, require some cooling before discharge if the best possible finish and perfect flatness are required.

When discharging hot from the press, it is well to adopt one or two preferred procedures in order to assure reasonable flatness. One of these commonly used is known as "dead piling." This, as the name implies, simply means piling the hot panels one on top of the other when they come from the press until a stack is built up. This stack when properly covered cools very slowly, and stresses are thereby minimized in a kind of annealing action. A certain small percentage of the curing time required for the resin can be subtracted from the press time and completed during dead piling as there is often enough sensible heat remaining in the pile to complete the cure. Obviously it is wise to be sure that dependence on the heat in the pile is not carried too far.

An alternate method advisable where rapid cooling is necessary is known as "finger racking." As they come from the press the panels are supported in such a manner that the room-temperature air has equal access to the whole area on both sides of the panel. This results in rapid, uniform cooling and if the construction is reasonably well-balanced, it produces satisfactory flatness.

4.4 Finishing

Final steps in the production of the product are trimming and inspection. Inspection procedures are more or less obvious and are hardly within the scope of this discussion, but a word or two on suitable tools for trimming may be in order. High-grade woodworking tools perform reasonably well when used on plastics in which the reinforcing agents are cellulosic in nature, but they have a rather short life. Better production economy and less maintenance results from the use of at least nonferrous-metal working tools or preferably high-speed steel tools. For plastics reinforced with glass fibers, asbestos, or other mineral reinforcing agents, high-speed steel tools are suitable, but better life is obtained with tungsten carbide tipped tools or diamond-tipped tools and saws. The rate of production, the quantities handled, and the amount of maintenance required, must all be balanced against each other in determining the economical type of operation.

4.5 Radio-frequency heating

Radio-frequency energy as a source of heat, as noted previously, is indicated particularly where thick assemblies are involved or where it is desirable to avoid any delay having to do with heat transfer. Radio-frequency energy in an electrostatic field generates heat uniformly throughout most plastics. Furthermore, it does so very rapidly in all cases except those few plastics that are extremely efficient insulators, such as polystyrene, polytetrafluoroethylene, and polyethylene. Radio-frequency energy can be used to cure thin laminates by placing multiple assemblies suitably separated in the radio-frequency field. This procedure is likely to be of doubtful economic value if it must compete with quick-acting hydraulic presses or continuous roll-type laminators. Thin laminates of relatively large surface area are very difficult to cure by means of radio-frequency energy because of the hazard of flashover from one electrode to the other which results in erratic operation of the equipment and burning of the laminate.

A desirable use for radio-frequency energy in the curing of thick masses or thick sections is as a means of preheating. The sections ready for curing are placed in the field and brought to curing temperature, advantage being taken of the rapid action of radio-frequency energy. The specimens can then be transferred to a hot press or an oven in order to maintain the temperature since all the hot press or oven has to do is take care of heat losses. Often this procedure is easier and more reliable in production than completing the cure by means of radio-frequency energy. Since the amount of power required to bring the specimen up to temperature is very high compared to the amount of power required to keep it warm once the temperature is reached, the radio-frequency requirement and therefore energy costs are substantially lowered if excessive rates of preheating are avoided. It is not always possible to take advantage of continuous application of high power, allowing the temperature to rise far above normal curing temperature and thus effecting extremely short preheating or curing periods, because charring in the center is very apt to take place, and indeed even explosions may occur, owing to the generation of high gas pressures in the center caused by the development of excessively high temperatures in thick masses. This occurs because most resin-curing reactions are exothermic, and thick sections are good heat insulators. Thus the curing reaction may run away with itself.

In investigating the economics of radio-frequency heating it is well to bear in mind another point. In any curing procedure where there is an appreciable percentage of water, say as much as 10 per cent, it is advisable to use resins that cure below the boiling point of water, for, if it becomes necessary to pass through the boiling point, it is found that the energy required simply to get rid of the water is as much as four times (and sometimes more) the total energy required to raise the temperature of the mass without the water from room temperature to a curing temperature of 290°F.

It is generally safe to say that once an operation involving radio-frequency electrostatic heating has been properly set up any reasonably skilled help can operate the equipment day in and day out and obtain satisfactory results. On the other hand, the individual electrode problem must not be underestimated, and it is wise to have access to the advice of a competent radio engineer with experience in this field in order to assist in deciding whether or not a proposed operation will be economical and to help in the solution of the electrode problem, without which there simply can not be worthwhile results.

4.6 Curved laminates

In considering the production of curved laminates it is convenient to divide the discussion, dealing first with the production of simple curvatures and finally with the production of compound-curved parts, for although the technique for making compound curvatures is applicable to simple curves, there are certain simplified procedures suitable only for making simple curves.

4.7 Postforming

Of continuing popularity is the technique of producing simple curves by means of postforming. Wood has been bent for years by means of heat and moisture, and certain time-honored molding processes for dealing with thermoplastics could be classified as postforming. Flat sheets based on thermosetting resins can be postformed. There are two classes of thermosetting resins particularly suited to this process. One of these makes use of the two-stage phenolic resins which are applied to the web for the production of the stock flat laminate from alcohol solution. Any low-molecular-weight or water-soluble material is carefully avoided. The laminates are truly heat-hardened and generally considered completely cured, although minimum temperatures are to be preferred, and occasionally judicious undercuring is helpful. These laminates, if then subjected to an additional 25°F to 50°F over their original curing temperature, can readily be bent to curvatures following surprisingly small radii and, within certain limits, can be drawn. As might be expected, the nature of the reinforcing fabric has some effect on the success of the operation, particularly as far as drawing is concerned. For best results the forming fixtures should be well engineered, and it is desirable that the clamping action of the hold-down member of the fixture be adjusted so that it is applied at the proper time and in the proper position to prevent extreme stressing of the fibers at the outer or tension side of the bend. Application of tension to the extreme ends of the stock while drawing around the form is of considerable assistance in relieving undue stresses on the fibers. Of course, the form around which the stock is bent must be heated to the desired additional temperature even though the stock may be preheated. Occasionally it is necessary to heat the hold-down as well.

Partial and occasionally complete cooling in the fixture is necessary to avoid spring-back and loss of shape while hot. An excellent compromise that avoids tying up the bending equipment during cooling is to transfer the hot part to a fixture that holds it to the required curvature while it cools, but which can be a much simpler device since the bending operation has already been accomplished.

Another type of resin that is suitable for postforming is the group known as compound copolymers or polyesters. These resins in passing from the liquid to the solid stage pass through a gel condition which is sufficiently stable to allow some handling. As a result, the laminated stock can be brought to the gel stage but not completely cured. It can then be reheated and bent to the desired curvature and cross linkage can be completed which will render the resin hard and substantially infusible. Since not all the resins of this family have a very pronounced gel stage, some care must be used in choosing the best resin for the purpose.

The disadvantages of postforming are that with the exception of modest draws it is limited substantially to simple curves, and the finished part is not stress-free. There are all too frequently incipient fractures in the laminate and torn reinforcing fibers which result in weakness in the structure. Whether or not these weaknesses are serious depends on the use to which the finished part is to be put.

4.8 Compound curves

Parts of substantial compound curvature must be molded. As in the case of the flat laminates, once the coated or impregnated stock is properly prepared, the first step is the lay-up. Since we are concerned with low-pressure molding which we wish to be stress-free and to be an assembly of laminae of known critical performance held together so that each can cooperate with the other, we must remember that care in the lay-up is extremely important. If the operation as a whole is considered one of gluing laminae to contour, it is easier to visualize why care is important. Thus we must template properly, cut the necessary gussets, decide on and properly locate and distribute the types of joints, and avoid wrinkles or slack as much as possible. The preparation of the actual stock, including the preparation, drying, and handling, is identical with that for the production of flat laminates.

It is comparatively simple to lay up the laminae for a thin section tightly and free from wrinkles or slack, but for a thick section serious problems arise because the outer layers do not occupy the same position after lay-up and before compression that they do when molding is complete. In the case of a male mold, the outer layers cover less area after compression than before, and there is a great tendency for wrinkles to occur. In the case of the female mold or cavity, the reverse is true, with the result that tearing is apt to occur. In order to minimize these defects, very thick sections must be molded in steps. For example, if $\frac{1}{2}$ in. of laminate is required, it may be necessary to mold this in two, three, or four steps, in which case it may also be found necessary to include as an outer layer a sheet of reinforcing fiber or fabric that is unimpregnated, which serves as a "gluing layer" so that the subsequent molding is thoroughly bonded.

In molding thick sections, radio-frequency electrostatic heating has often been proposed, and no doubt it can be made to prove its worth, provided it is remembered that radio-frequency equipment furnishes heat only and not pressure. Thus, there is the triple problem of properly locating the electrodes with respect to the piece to be cured so that application of pressure is not interfered with, so that arc-over and puncture do not occur, and finally, so that at least one electrode is reasonably well insulated from the charge in order that heating by conduction currents can be avoided.

The forms or dies required almost always call for a definite and an indefinite member. The definite member is a male or female pattern or form, and the indefinite member is a flexible membrane or bag or, in one case to be discussed later, a semi-hard rubber plug. The definite member can be wood, plaster, laminated plastic, low-melting alloy, sheet metal, or cast metal. Wood is built into a pattern in the usual fashion. Plaster similarly needs no special treatment beyond possibly wire or coarse-fiber reinforcement and occasionally phenolic-resin surface impregnation. Low-melting alloys can be used in two ways. They can be cast to the approximate dimensions and machined. However, a simpler and more interesting use is to force into a pot of low-melting alloy a suitable wood or plaster pattern and allow the alloy to freeze. On removal of the pattern a little finishing is all that is necessary to put the cavity to use. The pot in which the alloy was originally melted can be jacketed for oil heating. By lowering the temperature of the heating oil a safe distance below the melting point of the alloy, heat necessary for the cure of the resin can be introduced. After the mold has served its purpose, it is necessary simply to remelt it to produce another shape. Sheet-metal forms can be spun or hammered, and both these techniques have proven valuable in the quick production of low-cost tooling. Generally, supporting rims or frames are necessary to protect sheet-metal forms. Cast-metal forms of bronze, steel, or aluminum are the most permanent and the most costly and can only be justified for large runs.

The indefinite member can be a vented bag as used in connection with the autoclave or a simple vacuum or pressure bag which is used in the same manner as a football bladder. It can also be a rubber plug or an incompatible liquid. Most of these devices are obvious from their name, but the rubber plug needs some additional identification. The aircraft industry forms aluminum and magnesium in huge presses called "hydropresses" using a metal shape and a plug of semihard rubber. Advantage is taken of the fact that rubber acts like a fluid when it is completely confined. It has the additional advantage that until the moment it is completely confined it exerts pressure only at high spots which can thus be chosen so that wrinkles and tearing can be avoided. This procedure is applicable to low-pressure molding using softer rubber and less sturdy molds and lighter presses. Incompatible liquid molding is of value in some instances. The cavity with the uncured laminae in place is filled with a liquid that can be heated and that is not compatible with and does not interfere with the curing of the resin. If the relative surface tensions of the resin and the incompatible liquid are correct and if the reinforcing material or fabric is sufficiently fine-grained in construction that the incompatible liquid will not be forced through the pores, very creditable moldings of complex shapes can be accomplished. Since the pressures applicable in this process are sharply limited, the so-called contact resins should be used. Obviously, certain of the very low-melting alloys are indicated for this procedure.

A hybrid low-pressure molding process borrowed from high-pressure compression technique is casting molding. It takes two forms. The more important of these consists of laying up unimpregnated fabric on or in an appropriate form, pouring resin monomer, resin syrup, or varnish over the fabric, taking care to have a slight excess, and then lightly forcing a mating form into position. This last step squeezes the resin through the fabric, displaces the air, and floods the molding space between the two dies. After cure is completed, a highly satisfactory and well finished piece of work results.

In designing the tooling for any of these specific techniques it is important that there be a molding exit and that the form have sufficient draft so that it is not excessively difficult to remove the finished part. Certain complex parts that do not permit a normal molding exit must be handled by means of collapsible mandrels or split female mandrels that can be disassembled for part removal. Finally, some thought must be given to thermal expansion and contraction of the mold material as compared to the plastic that will be molded, since it is quite possible to develop a shrink fit between the mold and the part which precludes removal of the part. This is particularly true if the draft must be very small.

4.9 Sources of heat and pressure

Sources of heat are saturated steam, hot oil, radiant energy (also called infrared-ray heating), hot-air oven heating, and radio-frequency electrostatic heating. Sources of pressure are steam and mixtures of steam and air, simple vacuum, hydropress, or simply hydraulic presses where rubber is used as the fluid or where casting molding is used and it is desired to have a positive force to close the dies. Finally, there is the dead-weight method of producing pressure which includes the special case of low-pressure molding wherein surface tension of the resin binder holds the uncured laminate in position until cure is completed.

Cure, trimming, finishing, and inspecting are generally similar in requirements to those of flat work and need not be repeated here.

5 PRODUCTION OF SANDWICH MATERIALS

Progress in the production of sandwich materials has followed two diverse paths naturally following the leadership set by uses for these structural components. On the one hand, uses have become much more sophisticated. On the other, concentrated attempts have worked toward simplification and economical adaptation to extend the use. Shaped or curved constructions have characterized the sophisticated uses but by far the largest part of the more economical applications has dealt with flat panels. Thus, in this discussion most emphasis is placed on the methods of producing flat assemblies, although for the sake of completeness curved assemblies are included in the final section. In considering sandwich constructions in general, it is well to review the purpose of each component.

A sandwich, in the field under discussion, consists of relatively thin, high-strength, relatively high-density skins on either side of the relatively thick, relatively low-density, weak core. In addition to the two main components of a sandwich, there is the very necessary adhesive required to attach the skins to the core, the auxiliary adhesive necessary in other parts of the assembly which will be described later, and the rails or edge banding. Note that in some constructions, notably foam cores created *in situ*, the core itself is also the primary adhesive.

The edge banding is often necessary for protection or for joining of one or more sections to produce a large panel or for attachment. It is quite generally understood that the core performs the simple function of keeping the skins separated a chosen distance and in a fixed position relative to each other, at the same time desirably contributing as little weight as possible to the finished assembly. The core is thus said to stabilize the skins. The skins, in turn, furnish the strength and, to a large extent, the inherent engineering characteristics that can be expected. A properly designed sandwich always develops the strength of the skins when stressed.

5.1 Skins for sandwiches

Skins are generally made of metal (most often aluminum), stainless steel, high-strength laminates (made by processes already described), and combinations of these materials.

5.2 Cores of sandwiches

Currently popular core materials are foamed-in-place thermosetting resins, prefoamed core stock of which there are several variations, honeycomb construction in a variety of materials, and, for certain uses, end-grain balsa wood. Rails or edge banding include metal, both solid and in structural forms, laminates, plywood (often on edge), and solid wood or dimension stock.

Incidental to the production of sandwich panels are the inserts or blocks which are generally similar in construction to the edge banding or rails. These blocks are necessary for the attachment of fittings since it is obvious that one cannot bolt, screw, or rivet to a sandwich panel as one would to a piece of sheet steel or wooden planking.

5.3 Adhesives

The primary adhesives are those required to attach the skins to the core. The secondary adhesives are those that may be required in the assembly of the core, the assembly of the rails, and in the attachment of the core to the rails. The secondary adhesive is also used to attach inserts to the core if these are used. These adhesives vary in chemical nature from simple thermosetting resins used ordinarily in wood bonding to polyfunctional adhesives that adhere to metals and nonmetals equally well. In between are bonding processes discussed later in which both types are used in a procedure known as two-step bonding. Although not strictly adhesives, it must be remembered that some cores formed by foaming in place are powerfully adhesive in their own right so that no additional adhesives are necessary.

5.4 Production procedures

The important types of sandwich construction are at present metal-faced, honeycomb-cored panels, plywood-faced honeycomb-cored panels, plywood faces with balsa core, metal faces with balsa core, and both metal and plywood faces over various cellular cores, some formed in advance, some formed *in situ* by various foaming processes or by expansion of especially prepared plastic beads. Other combinations are obviously possible, but if one understands the production methods required for popular forms, the necessary procedure for special combinations is easily visualized. There is one possible exception to this statement in which the core is prepared by foaming an appropriate resin in place. It is necessary in this procedure to locate the skins properly with respect to each other in a fixture that ensures maintenance of their positions until the foaming is completed. This process has the particular advantage that it can often be accomplished in a single operation and can be arranged so that an attaching adhesive is unnecessary. It is in this field that the greatest advances have been made and what was originally a difficult technique has become as reliable as any other method.

An extremely economical combination, not basically new, involves hardboard skins over a variety of cores. Where economy is more important than the strength-weight ratio, these sandwiches are worthy of consideration.

5.5 Foam-core sandwich constructions

The reinforced-plastic-laminate skin with cellular core is a sandwich construction that continues to be of value in the production of radomes and microwave windows. While foam-in-place techniques can be used either for flat panels or for compound curved surfaces, the older assembly method of preforming of the core and the skins is presently relegated to flat panel work.

Except for possible improvements in jigs and fixtures, the techniques are standard and can be applied not only to reinforced-plastic laminates but to metal and plywood skins. Since a preformed core generally forms a skin, this must first be removed in the preparation of the core panel. Removing the skin also removes any parting agent that might interfere with adhesion. In this operation, adjustment to accurate thickness can also be accomplished. After the skin is removed and the core is cut to size, assembly can proceed.

Lay-up for the two-step method includes lamination and curing of the two skins in advance and cutting and fitting of the cellular core. At the moment of assembly, one of the faces with adhesive on its inner side is placed on the lay-up table and a light frame, or jig, is laid over the skin for guidance in fitting the core. The jig is not absolutely necessary if the finished panel is to have edge rails or banding. If there is to be edge banding, it should be located properly on the lower skin. The sections of core, preferably coated with adhesive on both sides, are fitted in place, and adjacent edges are coated with adhesive pressed together to ensure a bond that transfers shear. When all the core is in place, the upper skin, coated on its underside with adhesive, is laid over the assembly. The completed lay-up is then put under heat and pressure depending upon the requirements of the adhesive used. Stops may be necessary unless the edge banding can act as stops or unless the press is accurately controllable in the very low-pressure ranges. Excess pressure or inaccurate adjustment of core and edge-band thickness results in telegraphing which renders a panel unsightly and may be cause for rejection.

A variant of the two-step process involves the use of contact cements wherein the actual laminating, after the contact cement has reached the proper stage of dryness, is accomplished by means of nip rolls or a quick-closing–quick-opening press.

The one-step process requires greater care but is faster and under certain circumstances, as in radomes, may be necessary for electrical reasons. In this procedure prepreg, or layers of impregnated fabric, are laid up in such a manner as to produce the lower skin. If the resin used in the skin-formation technique is adequately adhesive, the core is laid directly over the impregnated fabric or the prepreg. Otherwise, additional adhesive is applied to the underside of the core pieces. When all the core is in place after it has been ascertained that the edges are firmly forced together and adequately coated with adhesive, the layers of impregnated fabric comprising the upper skin are laid over the core. The entire assembly is then placed in the hot press or in bag molding equipment and all uncured adhesives and impregnating resins are simultaneously cured. This step forms the upper and lower skins, attaches them to the respective sides of the core, and cures the adhesive used in edge bonding the core. An edge band can be laid up and scarfed into the core and also cured and attached at the same time. Very careful choice of adhesives is necessary, but this procedure is entirely practical and fast and produces an excellent sandwich.

Foaming in place to produce a sandwich, from flat to compound-curved surfaces, requires a mold with considerable rigidity and strength so that foaming pressure does not displace the male and female molding surfaces. Either the one-step or the two-step process can be used although greater flexibility of choice of materials and greater accuracy of the sandwich dimensions is obtained in the two-step process.

In the two-step process, the skins are premolded to fit accurately the male and female forms. In fact, this can be done in or on the forms that will ultimately be used in making the sandwich, and so they will not need to be removed. It may be required to give a light sanding of the faying surfaces to make sure that subsequent, adequate adhesion is developed. When all is in readiness, the foaming resin with proper catalysts and foaming agents is introduced into the space between the two skins, one located in the female mold and the other located on the male mold. Timing, temperature, and quality of mix control the density and uniformity of the foam. When cure is complete, at which time the bond should also be cured, the male mold is withdrawn and then the finished sandwich is taken from the female mold, trimmed and made ready for inspection.

A variant of this involves coating the skins with a special adhesive, introducing a proper quantity of expandable plastic beads, and introducing steam into the core space. The heat expands the beads to form the foam of the required density, causes them to adhere to each other to make a continuous core, and causes the core to adhere by means of the special adhesive to the previously prepared skins.

Flat structures are in no wise different except that platens serve as the equivalent of the male and female molds. In reality the upper and lower platens must be rigid so foaming pressure does not spring them and spoil the uniformity of thickness of the sandwich.

5.6 Honeycomb core, metal skin

The most important sandwich at the present time is the metal-faced honeycomb-cored sandwich of which perhaps the most frequently used modification is one in which the skins are aluminum and the honeycomb is formed from phenolic-resin-impregnated fabric or paper. There are two generally accepted methods of making such a sandwich, and they center around the method used for forming the honeycomb. In one case the impregnated fabric or paper is corrugated by means of meshing belts carrying suitably shaped and spaced bars or gears, and the corrugations are glued together at their crests. The other is formed by striping the flat web of paper or fabric at intervals with adhesive, laying alternate pieces of the web so that even-numbered layers have their stripes centered between those on the odd-numbered layers. After the adhesive in the stripe is activated and cured, the pad is sliced into sections of the proper thickness, and the honeycomb is expanded in the same manner as is used in opening a paper Christmas bell.

In the formation of honeycomb by corrugation, impregnated paper or fabric of the same general quality as that used in making laminates by low-pressure means is fed into the corrugating machinery. This equipment must be adjusted either to cure the impregnant in the web after which the corrugations are postformed into the web, or so that during the period of time that the web is between the corrugating mechanism the resin is fused and then cured. Either of these procedures is satisfactory, but it must be remembered that not all resins postform. Generally, it will be found that the postforming procedure is faster than the other. Corrugated web cut into convenient lengths is next punched with registering holes, the ridges or crests are coated with an assembly adhesive, and the layers are dropped on on top of another in a restraining jig. After the desired thickness has been built up, the assembly is placed between cauls and put under light pressure to ensure contact at the crests until the adhesive is cured. For certain types of impregnants and certain types of combining adhesives it may be necessary to sand the crests of the corrugations before coating with adhesive. After the assembly adhesive is cured the honeycomb is ready for slicing and inspection. Figures 9.10–9.12 show several steps in honeycomb core preparation.

The Christmas bell or expanded type of honeycomb is easier to register accurately, as the assembly adhesive is applied in stripes across the web by means of a printing machine is such a fashion that its spacing cannot vary. It is then not a difficult problem to locate alternate sheets so that the stripes of the even-numbered sheets fall exactly between the stripes of the

Figure 9.10   Oven Curing of Assembly Bond of Corrugated-type Honeycomb Blocks

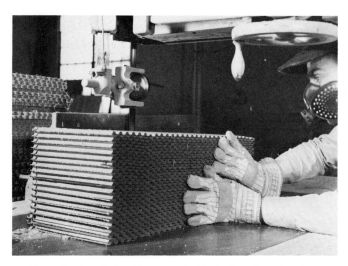

Figure 9.11   Slicing Honeycomb Blocks Into Core Sections

Figure 9.12   Honeycomb-core Coating Operation

odd-numbered sheets. As mentioned before, when a sufficient thickness of the assembly is built up, the whole can be placed under pressure (and heat if necessary), curing the stripe. Obviously a cold-setting adhesive can be used, in which case the assembly or lay-up should be made wet, and the pack should be put under pressure with due observance for the assembly time of the adhesive used. In the expanded-type honeycomb, one can impregnate the web before or after the formation of the honeycomb. If the impregnation is done after expansion, the procedure is obviously one of lowering the sliced expanded honeycomb on tray-shaped screens into the impregnating bath. This is followed by removal, draining, drying, and curing, which sets the expansion. If the chosen procedure requires the impregnation of the paper or fabric in the web, care must be observed in choosing the impregnating resin and the amount used so that the web is not self-laminating under the conditions of temperature and pressure used to set the striping or assembly adhesive. The only alternative to this requirement is the tedious one of laying masking strips such as cellophane between each sheet, located so that they do not interfere with

attachment at the striped points. The final step in this alternative procedure for expanded honeycomb is expansion to the desired degree and curing of the impregnating resin in order to fix the expansion, followed by removal of the masking strips. strips.

Preparation of metal skins involves cutting to size, inspection for defects, and cleaning. Cleaning cannot be overemphasized. Adequate bonds simply will not be obtained if the metal is not chemically clean on the side to be attached to the core. If the metal is very greasy, it should be solvent-washed or vapor-degreased. If free grease is not present, degreasing may be omitted. In any event the metal must be cleaned in a bath of a type recommended by the adhesive supplier for the metal involved. This cleaning should be carried out for the time and at the temperature recommended. Cleaning is followed by a thorough clean-water rinse. Cleaning must be sufficient to prevent water films from breaking or creeping because of surface tension during draining. This test must always be made, and cleaning must be repeated if perfect water wetting is not obtained. Scrubbing while in the cleaning bath may occasionally be necessary.

For unusually severe services an alkaline bath and rinse may beneficially be followed by an acid bath which generally comprises an appropriate acid, aliphatic alcohol, and water. Here again the adhesive supplier's instructions should be followed. This is followed by a water rinse and another check for cleanliness and drying. The drying operation is preferably done in a forced circulation of clean, warm air. Roller drying and handling with bare hands are to be avoided. Wiping may redeposit grease films. The clean, dry metal should be coated immediately with adhesive to prevent collection of grease films from the air.

In either case the cured, expanded, and sliced honeycomb is, at this point, ready for use in sandwich construction. The lay-up procedures do not differ greatly from those used in producing end-grain balsa or preformed cellular or foam cores. For "one-step" bonding the appropriate adhesive is necessarily applied to the core and preferably also to the insides of the faces or skins. Likewise, it is necessary to apply the adhesive to the edges of the core sections that adjoin each other and that come in contact with any inserts or edge banding so that proper shear transfer will result. At these points there is a local increase in density which is unavoidable, but, if proper slicing and laminating or pressing techniques are followed, the denser areas do not leave marks, and the bond is uniform and adequate over the whole panel. There are two precautions to be observed. One of these involves choosing an adhesive for bonding the core to the faces which, owing to surface tension, builds up a little fillet or bead on the edges of the core. Since it is obvious in the case of honeycomb that the point of contact is a narrow line, this must be reinforced by a very strong adhesive which is shock-resistant, craze-free, and produces a broad bead increasing the area of contact at the point where the skin touches the core. The other precaution to be observed during lay-up is to crowd slightly the sections of honeycomb so that under pressure they tend to expand, making firm contact between sections and between filler blocks and edge banding.

On completion of the glue application and lay-up the assembly is bonded in a press. If edge rails and framing are adequate in strength and area to act as stops, pressures normal to plywood manufacture may be used. These are more than adequate to close the press to the rails. If there are no edge bands, the same procedure may be used by placing stops or machined metal bars in the press. An alternate procedure is to press at very low pressures (15 to 30 psi), using a press with special low-pressure controls. The hot press is generally used, as most of the currently available special adhesives adhere best and give the most creep-resistant bonds when cured under combined heat and pressure.

There is, however, a method known as two-step bonding which permits the use of the cold press. The metal skins after cleaning are coated with a polyfunctional adhesive just as though they were to be hot-press-bonded to the core. Instead, they are first baked in a circulating hot-air oven or under radiant-heat lamps until a normal cure is obtained. Then the "primed" skins are cold-bonded, using room-temperature-setting resorcinol-formaldehyde or phenol-resorcinol-formaldehyde resins. Presses in this case are generally screw, fire-hose jig, or cold-baling types.

Faults or defects are unbonded areas, buckled honeycomb, dimpled skins, and panel warping. Unbonded areas are most frequently caused by dirty metal or uneven core thickness. Buckled areas are almost always caused by a core that was cut too thick but may occasionally be caused by excessive pressure which compresses the stops or the edge bands and can result from undercured core. Warping is a fault not easy to analyze, but if identical faces are used it should not occur. The best insurance against warping is to make absolutely sure that, the moment the adhesive sets, both faces are at exactly the same temperature.

Honeycomb-cored sandwich panels are no longer limited to glass-fiber fabric, cotton fabric, or paper honeycomb but make use of aluminum and stainless steel. Such sophisticated cores are often needed for reasons of conductivity or heat resistance or special strength properties. Since stainless steel honeycomb cores are generally assembled by welding or brazing, they can hardly be considered part of a chapter on molding techniques, although preformed stainless steel cores can, of course, be introduced just like any other honeycomb core and can be adhesively bonded to skins. Aluminum cores frequently are formed by bonding precorrugated strips to form the honeycomb and then, in turn, bonded to the skins to form a sandwich. The general techniques do not differ from techniques used in the production of fabric or paper honeycomb cores except for the special adhesive requirements and for the cleaning of the aluminum before coating with adhesive.

5.7 Honeycomb core, plywood skin

Production techniques for this type of construction differ only in detail from those required in the production of reinforced plastic-faced or metal-faced sandwich constructions.

Practically any grade of plywood can be used for faces, ranging from the special aircraft grades, which are very thin, to decorative grades later to be appropriately finished. The cut plywood must be free of dust, loose particles, and grease but generally the cleaning problem for plywood skins is less serious than for metal. Either hot or cold bonding of the faces to the cores can be used, depending on the choice of the bonding resin. In either case, either exterior or interior durability adhesives

are available. For best results it is preferable to coat adhesive on the core and on the inner sides of the plywood faces. It is, however, wise to coat the plywood very lightly and apply most of the adhesive to the core both for economic reasons and in order to save weight. It is not so important to have the plywood skins at identical temperatures at the moment the resin sets, but it is extremely important that they both have the same moisture content. Plywood carelessly handled causes more trouble with warping than metal, but this trouble can be avoided if reasonable care is taken.

5.8 Curved sandwich materials

With the exception of radomes, compound-curved sandwich constructions have not become popular. This is not only for economic reasons but because other means have been found to produce structures in which only an outer skin is important and the support comes from a supporting grid such as in the geodesic dome.

Improved die-molded reinforced plastics are another reason for the abandonment of compound-curved sandwiches. Larger parts, as has been mentioned previously, are bag molded but the improved results and adequate designs have largely eliminated sandwich constructions from this field.

Two notable exceptions are worth mentioning. One is simple-curved surfaces such as sections of a cylinder where great stiffness is important along with light weight. The production technique of such curved constructions is a simple extension of the several production techniques for flat panels in which either mating dies or bag-molding techniques are effective. The other exception involves special aircraft wing constructions involving stainless steel honeycomb, but since this is hardly a molding operation in the sense with which we are here dealing, details of the assembly must be left to other sources.

5.9 Balsa core, plywood skin

One of the oldest sandwich panels still enjoying commercial production involves end-grain balsa core of selected density and plywood skins of the proper configuration. In production of this construction the core can either be prepared in advance or laid up on one face after which the second face is affixed.

As the core is assembled, and in order to have it transmit shear forces, the sections of the balsa must be edge-glued to each other. This is most often done with a cold- or warm-setting resorcinol-formaldehyde resin. The completed core section is set aside at room temperature or in a low-temperature oven to cure after which it is carefully sanded or sliced to thickness and is ready for faces. The skins, whether they be plywood, metal, or hardboard, must be clean, dry, and fit for the adhesive operation. Again, either warm-setting or room-temperature adhesives of suitable durability are available. Assembled panels can be pressed in a bale and set aside to cure, or they can be cured rapidly in a hot press. In the latter case, it is advisable to use stops in the press in order to ensure against crushing the core or exaggerated telegraphing of edge banding which may be present. The only special precautions that must be exercised is the use of an adhesive so formulated that it will not all be absorbed by the balsa core and result in starved glue lines. Priming or sealing of the balsa before application of the glue may be necessary.

Obviously, a wet assembly can also be used and, in a sense, this may be considered a one-step operation. It puts a greater burden on proper dimensioning of the individual balsa blocks, which must be done in order to ensure a uniformly thick panel of good appearance; however, it does save considerable construction time.

SUPPLEMENTARY READINGS

H. C. Engel, C. B. Hemming, and H. R. Merriman, *Structural Plastics*, McGraw-Hill Book Company, Inc., New York, 1950.

An important series of reports by Forest Products Laboratory on "Reinforced Plastics" and on "Sandwich Constructions." Forest Products Laboratory lists should be consulted. These reports date from 1950 to the present day and were issued in cooperation with the ANC-23 Panel of the Departments of the Air Force, Navy, and Commerce.

U.S. Forest Products Laboratory, "Composite Construction for Flight Vehicles," ANC-23 Bulletin, U.S. Government Printing Office, Washington, D.C., May, 1958.

D. V. Rosato and C. S. Grove, Jr., *Filament Winding: Its Development, Manufacture, Applications, and Design*, John Wiley & Sons, Inc., New York, 1964.

R. H. Sonneborn, A. G. H. Dietz, and A. S. Heyser, *Fiberglass Reinforced Plastics*, Reinhold Publishing Corporation, New York, 1959.

H. A. Perry, *Adhesive Bonding of Reinforced Plastics*, McGraw-Hill Book Company, Inc., New York, 1959.

Society of the Plastics Industry, *Plastics Engineering Handbook*, Third Edition, Reinhold Publishing Corporation, New York, 1960.

Society of the Plastics Industry, "Sixteenth Annual Technical and Management Conference, Reinforced Plastics Division," 1961.

"For Versatility in Building: Sandwich Panels," *Mod. Plastics* **40**, No. 9, 100 (May 1963).

"Making Truck Panels Automatically," *Mod. Plastics* **37**, No. 3, 89 (March 1964).

George J. Neh, "Deep Draw Moldings with Glass Fiber Mats," *Reinforced Plastics*, March-April, 1964.

"Filament-wound Reinforced Plastic Tanks for Paint Emulsion Storage," *Mod. Plastics* **41**, No. 9, 118 (May 1964).

"Dip Coating Honeycomb." *Reinforced Plastics* (May-June 1964.)

D. V. Rosato, "An Introduction to Filament Winding," *Reinforced Plastics* (May-June 1964).

"Filament Winding Machinery," *Reinforced Plastics* (May-June 1964).

D. V. Rosato and C. S. Grove, Jr., "Filament Winding Technology, Reinforcements, Part I," *Reinforced Plastics* (July-Aug. 1964).

Edward F. Barro, Sr., "Phenolics — More Versatile Than Ever," *Mod. Plastics* **42**, No. 2, 102 (Oct. 1964).

T. Walter Noble, "Picking the Reinforcement for your Plastic Product," *Plastics World* **23**, No. 3, 38 (Mar. 1965).

"For Fast Redesign-Filament Wound RP," *Mod. Plastics* **42**, No. 8, 110 (Apr. 1965).

"Filament Winding Without Auxiliary Pressure Vessels," *Mod. Plastics* **42**, No. 8, 117 (Apr. 1965).

Chapter 10

Thermostat Metals

Unto U. Savolainen
Raymond M. Sears

1 INTRODUCTION

The subject covered by this chapter is known variously as bimetal, thermometal, thermostatic bimetal, thermostatic metal, or thermostat metal. The first of these terms describes the material as being composed of two metals although as explained further there is no restriction regarding dual combinations. The plurality of components thus brings this type of material within the classification of engineering laminates. The other terms which contain the expression "thermostat" indicate the general field of application of these materials. Originally they were used as the temperature-sensitive elements of devices for controlling, regulating, or indicating temperature. "Thermostat metals"* have since been utilized in a great variety of ways, in any way in which a change in temperature could be used to control, regulate, compensate, indicate, and the like. A partial list of uses follows:

Air dryers
Air heaters
Air valves
Alarm devices
Altitude meters
Aquarium heaters
Automatic chokes
Automatic exhaust heat
  control
Blueprint machines
Bread-wrapping machines
Candy mixers
Carburetor temperature
  regulators
Chicken brooders
Cigar lighters
Circuit breakers
Clocks
Cord sets
Damper controls
Demand indicators

Dental furnaces
Dental sterilizers
Draft controls
Electric-light plants
Electric meters
Electric motors
Fans
Fire alarms
Fluorescent starters
Gasoline gauge indicators
Gas safety pilots
Glue pots
Hat stretchers
Heating pads
Hot beds
Humidifiers
Incubators
Instruments, electric
Instruments, testing
Ironing machines
Irons, electric

Laboratory ovens
Lamps, electric
Lamps, therapeutic
Light flashers
Lighting systems
Machine tools
Motor protection
Necktie pressers
Oil-burner
  control
Oil gauges
Oil purifiers
Ovens, electric and gas
Percolators
Photomounting machines
Popcorn machines
Portable electric tools
Radiator shutters
Ranges, electric and gas
Recording thermometers
Refrigerators
Relays, overload

Relays, signal
Room thermometers
Room thermostats
Scales
Shock absorbers
Signal devices
Sign flashers
Soldering irons
Stack controls
Starting devices
Steam radiators
Steam traps
Time switches
Toasters, electric
Transformer temperature
  indicators
Type-metal pots
Voltage regulators
Waffle irons
Water heaters, electric and
  gas
Windshield defrosters

* "Thermostat metal" as a standard descriptive term for these materials has been accepted by The American Society for Testing and Materials and is used throughout this section.

UNTO U. SAVOLAINEN is Chief Engineer, Industrial Metals, Metals & Controls Inc., Corporate Division of Texas Instruments Inc., Attleboro, Mass.
RAYMOND M. SEARS is Senior Applications Engineer, Thermostat Metals Department, Metals & Controls Inc., Corporate Division of Texas Instruments Inc., Attleboro, Mass.

2 DEFINITION

Thermostat metal is a composite material, usually in the form of sheet or strip, comprising two or more materials of any appropriate nature, metallic or otherwise, which, by virtue of the differing expansivities of the components, tends to alter its curvature when its temperature is changed.

3 HISTORY

Thermostat metal has been in use for a long time — as early as 1766 it was suggested for use for temperature compensation in chronometers. Breguet's thermoscope (a combination of gold, silver, and platinum) invented in 1817, Ure's thermostat of brass and iron patented in 1831, Wilson's U.S. patent 24,896 of 1858 (brass and steel) are early recorded instances of its use. Villarceau in 1863 published an analysis of its characteristics in the *Annales de l'Observatoire Impérial de Paris*. After that time references occurred with increasing frequency until in 1897 C. E. Guillaume and C. Dumas applied for a patent on an anomalous alloy of 36 per cent nickel and 64 per cent iron which later became known as Invar. The discovery of this alloy with its nearly zero coefficient of thermal expansion sparked the development of the large variety of present-day commercial thermostat metals which use it or variations for their efficient characteristics.

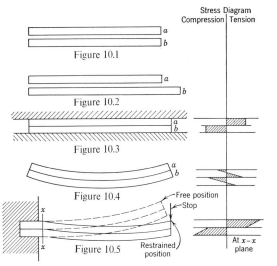

Figures 10.1 to 10.5 Brief Diagrammatic Explanation of Thermostatic Action and Stress Patterns under Various Conditions

4 PRINCIPLES UNDERLYING THERMOSTAT METAL ACTION

Before the various types of thermostat metals are described the principles underlying the bending of a thermostat metal strip with a change in temperature will be discussed. To illustrate the bending, consider two metal strips a and b (Fig. 10.1) having low and high coefficients of thermal expansion (henceforth called expansivities), respectively, and having identical dimensions. When the temperature is raised, their relative lengths are as in Figure 10.2. If the two are bonded together, clamped as in Figure 10.3, and then the temperature is raised, the one with the high expansivity is under uniform compression, and the one with the low expansivity is under uniform tension. These forces produce a bending moment, and, when the clamps are released, the free element assumes a uniform arc (Fig. 10.4). If the combination is straight or has an initial uniform curvature, the resulting curvature for uniform temperature change is uniform, that is, a true arc of constant radius. This follows since the stresses producing the bending moment are uniform for any cross section.

5 GENERAL EQUATION FOR BENDING OF THERMOSTAT METAL STRIP

The bending of a two-component thermostat metal strip when uniformly heated is mathematically described by the following equation:[1]

$$\frac{1}{\rho} = \frac{6(\alpha_2 - \alpha_1)(T_1 - T_0)(1 + m)^2}{t[3(1 + m)^2 + (1 + mn)(m^2 + 1/mn)]}, \quad (1)$$

where

α_1 and α_2 = temperature coefficients of expansion (expansivities),
E_1 and E_2 = moduli of elasticity,
t_1 and t_2 = thicknesses of components,
t = thickness of strip,
ρ = radius of curvature of strip,
T_1 and T_0 = temperatures,
$t_1/t_2 = m$, $E_1/E_2 = n.$ (2)

If the thicknesses of both metals are the same,

$$t_1 = t_2, \qquad m = 1,$$

and

$$\frac{1}{\rho} = \frac{24(\alpha_2 - \alpha_1)(T_1 - T_0)}{t(14 + n + 1/n)}. \quad (3)$$

Further, if the moduli of elasticity are the same,

$$E_1 = E_2, \qquad n = 1,$$

and

$$\frac{1}{\rho} = \frac{3}{2} \frac{(\alpha_2 - \alpha_1)(T_1 - T_0)}{t}. \quad (4)$$

The radius of curvature is thus directly proportional to the difference in the expansivities and the temperature change and inversely proportional to the thickness of strip. The radius of curvature is affected by the ratio of moduli of elasticity of the components. Which component has the larger modulus is immaterial as long as the ratio is the same. The radius of curvature also is affected by the ratio of thicknesses, $t_1/t_2 = 1$ being the best.

A discussion of these factors affecting thermostat strip bending or radius of curvature is illuminating. The relationship between the radius of curvature and the ratio of thicknesses of components varied in either direction is shown in Figure 10.6, where the moduli of elasticity are assumed to be equal. It is interesting to note that the ratio can be widely varied, for instance, from 0.5 to 2.0, and still retain a high percentage, 88 per cent, of the maximum thermal activity.

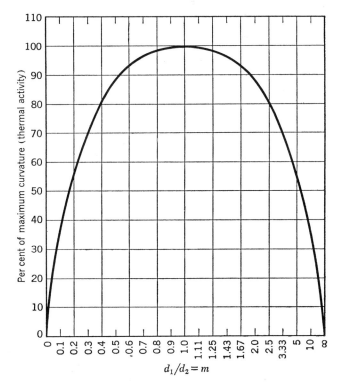

Figure 10.6 Effect of Ratio of Thicknesses on Radius of Curvature of Strip

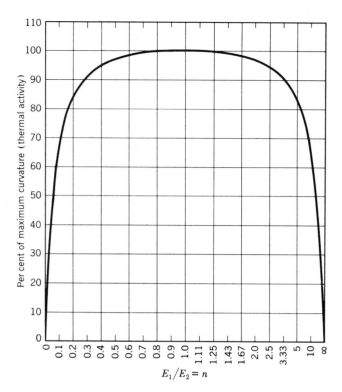

Figure 10.7   Effect of Modulus of Elasticity on Radius of Curvature

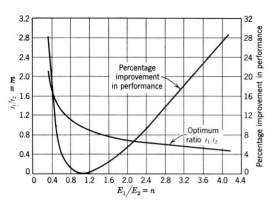

Figure 10.8   Best Ratios of Thicknesses of Components to Obtain Maximum Output for Various Ratios of Moduli of Elasticity

ment the thicknesses of the components should be inversely proportional to the square roots of the moduli of elasticity. Figure 10.8 shows the optimum ratios of t_1/t_2 and E_1/E_2 and the percentage improvement in performance. With reference to Figure 10.8, if the ratio of moduli n is 3.0, the optimum ratio of thickness m is 0.6, and the percentage improvement in performance is 16 per cent over a ratio of thickness of 1.0. Obviously, when n is 1.0, m also is 1.0. This is the optimum ratio, and no improvement in performance is possible by varying m.

Many thermostat metals incorporate a third component in order to obtain a particular electric resistivity as noted on p. 216. Because of their importance, the formula for the thermal deflection of a three-component thermostat metal is given:

$$\frac{1}{\rho} = \frac{12E_1t_1E_2t_2E_3t_3\left[\left(\frac{t_1+2t_2+t_3}{2E_2t_2}\right)(\alpha_3-\alpha_1)+\left(\frac{t_2+t_3}{2E_1t_1}\right)(\alpha_3-\alpha_2)+\left(\frac{t_1+t_2}{2E_3t_3}\right)(\alpha_2-\alpha_1)\right]T_1-T_0}{E_1^2t_1^4+E_2^2t_2^4+E_3^2t_3^4+2E_2t_2E_3t_3(2t_2^2+3t_2t_3+2t_3^2)+2E_1t_1E_3t_3(2t_1^2+2t_3^2+6t_2^2+6t_2t_1+6t_3t_2+3t_1t_3)}$$
$$+2E_1t_1E_2t_2(2t_1^2+3t_1t_2+2t_2^2) \quad (6)$$

where

α_1, α_2, and α_3 = temperature coefficients of expansion,
E_1, E_2, and E_3 = moduli of elasticity,
t_1, t_2, and t_3 = thicknesses of components,
ρ = radius of curvature of strip,
T_1 and T_0 = temperatures.

The modulus of elasticity of a thermostat metal may also be calculated from the moduli of elasticity and the thicknesses of its components. The formulas for both two- and three-component thermostat metals follow:

Two-component:

$$E = \frac{4E_1\{(t_1-c_1)^3+c_1^3+(E_2/E_1[(t_2+t_1-c_1)^3-(t_1-c_1)^3])\}}{t^3},$$
$$(7)$$

where

$$c_1 = \frac{E_1t_1^2+E_2t_2(2t_1+t_2)}{2(E_1t_1+E_2t_2)}; \quad (8)$$

Three-component:

The modulus of elasticity will be the same for both components only by coincidence. The effect of unequal moduli is illustrated in Figure 10.7 it being assumed that the component thicknesses are equal. Figure 10.7 surprisingly shows that the ratio of moduli of the two components can vary from 0.3 to 3.3 and still show 91 per cent of the maximum thermal activity. Combinations having too great a difference in moduli, such as Invar–cadmium and Invar–hard rubber, show no thermal activity whatsoever. Work that a thermostat metal element can do is proportional to the square of the difference of expansivities and proportional to the modulus of elasticity. The modulus of elasticity, or stiffness or rigidity, therefore is considered in the following. The general equation for stiffness or rigidity can be written as follows:

$$S = \frac{1}{3}(E_1\alpha_1^3+E_2\alpha_2^3)-\frac{1}{4}\frac{(E_1\alpha_1^2-E_2\alpha_2^2)^2}{(E_1\alpha_1+E_2\alpha_2)}, \quad (5)$$

where S = stiffness.

The stiffness is therefore affected by the ratio of the moduli and also by the ratio of thicknesses of components. Analyzing the effect would be too lengthy for this article; therefore it is left out, but the conclusions of such an analysis are given in the following. To obtain the maximum work from a thermal ele-

$$E = \frac{4E_1\{(t_1 - c_1)^3 + c_1^3 + (E_2/E_1)[(t_2 + t_1 - c_1)^3 - (t_1 - c_1)^3] + (E_3/E_1)[(t_3 + t_2 + t_1 - c_1)^3 - (t_2 + t_1 - c_1)^3]\}}{t^3}, \qquad (9)$$

where

$$c_1 = \frac{E_1 t_1^2 + E_2 t_2(2t_1 + t_2) + E_3 t_3(2t_1 + 2t_2 + t_3)}{2(E_1 t_1 + E_2 t_2 + E_3 t_3)}, \qquad (10)$$

E_1, E_2, and $E_3 =$ moduli of elasticity of components,
t_1, t_2, and $t_3 =$ thicknesses of components,
$E =$ modulus of elasticity of thermostat metal.

6 STRESSES

It is difficult to analyze properly the stresses in thermostat metal since there are so many factors that are difficult to evaluate. Stresses that result from factors created during the functioning of a finished element are caused by (1) thermal changes, (2) mechanical loading, and factors originating from the manufacturing operations, (3) bonding, hot rolling, cold rolling, slitting, flattening, fabricating, and heat treating. The effect of (1) and (2) can be worked out fairly simply but (3) is almost impossible. Theoretical thermal and mechanical stresses are discussed in the following.

With reference to Figure 10.4, the stress at the bond of a heated strip which is free to move is

$$\text{Stress} = \tfrac{1}{2}E(\alpha_2 - \alpha_1)(T_1 - T_0), \qquad (11)$$

with the high-expansive component in compression and the low-expansive in tension. The outer fiber stresses for both components are one half of the bond stress with the high in tension and the low in compression. Zero stresses occur one sixth of the total thickness in from the outer surfaces. With reference to Fig. 10.3, the stress resulting from uniformly heating and uniformly restraining the strip is

$$\text{Stress} = \tfrac{1}{2}E(\alpha_2 - \alpha_1)(T_1 - T_0), \qquad (12)$$

with the high-expansive component in compression and the low-expansive component in tension, the stress being uniform throughout the thickness of each. Note that these stresses are the same as the bond stresses of a free strip. If a straight cantilever strip of thermostat metal is heated with the free end unrestrained, the free end moves up as shown by the dotted lines in Figure 10.5. If the free end is restrained in its original position and the strip heated, the stresses due to heating and the mechanical restraint are maximum at or near the point of clamping (plane x–x, Fig. 10.5). The maximum stresses are at the outer fibers and are given by the following equation:

$$\text{Stress} = \tfrac{7}{8}E(\alpha_2 - \alpha_1)(T_1 - T_0). \qquad (13)$$

The bond stresses, however, remain the same as the bond stresses in a free or uniformly restrained strip.

With reference to the cantilever strip previously mentioned, obviously this strip of uniform cross section is not efficient since only at the clamped end is it worked at full capacity. A tapered beam of uniform strength illustrated in Figure 10.9 gives, for the same volume of thermostat metal, 33 per cent greater contact force and 43 per cent less stress. The thickness alone can also be tapered[2] as well as both width and thickness. These forms give better form economy than strip of uniform cross section, because of stressing the length uniformly. Lessening of the tendency to vibrate is another desirable feature of these strips tapered in both width and thickness.

Stresses due to mechanical loading can be calculated if the element is considered not as laminated but as a single material, and usual formulas are employed. Stresses due to bonding, hot rolling, and the like are difficult to determine; therefore, after calculation of the simple thermal and mechanical stresses high safety factors must be used to arrive at allowable stresses.

The stress that a thermostat metal element can withstand without taking a permanent set (allowable stress) depends on the temperature and previous history such as the rolling, forming, and heat treatment. For most thermostat metals having for both components ferrous alloys, the following allowable working stresses can be used:

Temperature °F	Allowable Working Stress, f
75	25,000
300	23,000
500	18,000
700	13,000
900	5,000
1,000	0–1,000

The following equation can be used to determine the allowable working loads for cantilever elements and helix- and spiral-coil elements:

$$P = fwt^2/6L \text{ or } R), \qquad (14)$$

where

$P =$ allowable load, lb,
$f =$ allowable working stress, psi,
$w =$ width, in.,
$t =$ thickness, in.,
$L =$ length of cantilever, in.,
$R =$ radius arm of helix or spiral coil.

The allowable stresses listed are approximate only and should be used with caution. As an illustration of the effect of previous history, experiments to determine allowable stresses on helix coils coiled in various ways (high-expanding component inside or outside) were made by loading so as to unwind and also to wind the coils. Allowable stresses calculated from the data varied from 20,000 to 100,000 psi.

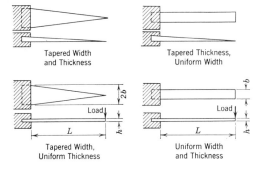

Figure 10.9 Cantilever Elements of Various Designs

7   ELECTRICAL RESISTIVITY

An increasing number of products are made, such as thermal relays, circuit breakers, and motor overload protectors, in which heat generated by passage of an electric current through a thermostat metal element operates the device. These applications require a variety of thermostat metals covering a wide range of electrical resistivities. A number were available for these applications several years ago, but the over-all variation in electrical resistivity was not sufficiently wide, and the resistivities were not properly spaced.

It is possible to make circuit breakers and similar equipment with widely different ratings using one type of thermostat metal by varying the size and shape of the element. But circuit-breaker manufacturers desired to make a line of circuit breakers all of the same physical size but varying in circuit interrupting power in uniform steps from low to high. The thermostat metal producers designed a line of thermostat metals to fill this need.

The heating effect of an electric current is

$$H = ri^2\theta/4.18, \tag{15}$$

where

$H =$ heat, calories,
$r =$ resistance, ohms,
$i =$ current, amperes,
$\theta =$ time, sec.

The temperature rise (disregarding heat losses) of a resistor is given by

$$\Delta T = 0.43ri^2\theta/MS, \tag{16}$$

where

$\Delta T =$ temperature rise, °F,
$M =$ mass, grams,
$S =$ specific heat.

This formula may also be rewritten

$$\Delta T = 1.73 \times 10^{-9}i^2R\theta/(wt)^2, \tag{17}$$

where

$R =$ electrical resistivity per circular-mil-foot,
$\Delta T =$ temperature change, °F,
$w =$ width, in.,
$t =$ total thickness, in.

The heating effect is proportional to i^2r in which i^2 is analogous to the current rating of a breaker and r is analogous to the electrical resistivity of the thermostat metal. Throughout a line of circuit breakers, the heating effect which is proportional to i^2r must be a constant for uniform tripping time. Therefore, the series of thermostat metals must vary in electrical resistivity as the square of the rating. The following equation is useful in laying out a line of breakers since, after the type of material has been experimentally determined for one rating, the others can be approximately calculated.

$$I^2R = \text{constant},$$

where

$I =$ current rating of breaker,
$R =$ resistivity of thermostat metal.

The equation is also useful in determining whether the design of the line of breakers is such as to make best use of the range of resistivity materials available. The highest rating should use the lowest resistivity material or the lowest rating the highest resistivity material.

The range of electrical resistivities available before the advent of special electrical-resistivity series is illustrated by the thermostat metals listed in Table 10.1. Thermostat metals are not yet standardized, and, therefore, the types mentioned in the table and later in this chapter are types manufactured by Metals and Controls, Inc., Texas Instruments, Incorporated.

Table 10.1

Thermostat Metal	Electrical Resistivity at 75°F, ohms per circular-mil-ft
A1	90
N1	95
E5	345
E1	500

The special thermostat metals developed to meet the requirements set forth above were based on standard thermostat metals similar to type E1 in Table 10.1. The resistivity was varied by incorporating a third layer of variable thickness of a relatively low-resistivity metal. This filled in the range from 100 to 470 ohms per circular-mil-foot. A series from 15 to 100 ohms per circular-mil-foot utilized high-conductivity copper alloys for the third layer.

A recent innovation was the introduction of a new series of special three-component thermostat metals utilizing a high-manganese alloy. Due to the high coefficient of expansion of the manganese alloy, a series of electrical resistivity materials were designed and made, each of which has a higher flexivity for a given electrical resistivity than previously available. This series ranges from 30 to 600 ohms per circular-mil-foot.

The resistivity can be calculated easily if the three laminations are considered as a parallel circuit:

$$\frac{x}{r_1} + \frac{y}{r_2} + \frac{z}{r_3} = \frac{100}{R}, \tag{18}$$

where

x, y, and $z =$ thicknesses of components x, y, and z in per cent of the total thickness of the three,
r_1, r_2, and $r_3 =$ resistivities of the three components,
$R =$ resistivity in ohms per circular-mil-foot of the combination.

The recent introduction of a high-manganese alloy of 1050 ohms per circular-mil-foot as a thermostat metal component has increased the top limit to 850 ohms per circular-mil-foot. Electrical resistivities above 500 ohms per circular-mil-foot have been obtained using nickel-chromium resistance alloys as well as nickel-chromium–aluminum–iron and nickel–manganese–aluminum–iron alloys.

Figure 10.10 gives the resistivity versus temperature curves of a series of electrical-resistivity thermostat metals whose re-

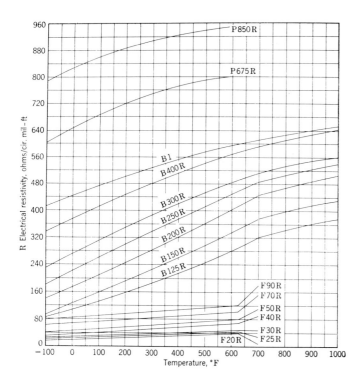

Figure 10.10   Electrical Resistivity versus Temperature Curves of a Group of Thermostat Metals Designed for Use in Circuit Breakers and Electric Overload Devices

sistivities at 75°F cover the range from 20 to 850 ohms per circular-mil-foot. The resistance versus temperature information is important in design work for obvious reasons. The resistance at the operating temperature and the rate of increase with temperature rise (coefficient of resistance) is equal in importance to that at room temperature.

8   CHOICE OF ALLOYS

In choosing components to use in thermostat metals numerous factors must be kept in mind to produce an efficient, economical, and accurate finished product. First, since the rate of thermal activity is proportional to difference in expansivities of the two components (see equation 1) materials must be chosen to have as low and as high expansivities as possible. Also the rate of change of the expansivities must give the required linearity or required deviation from linearity of thermal activity over the required operating range. Usually constancy of expansivities is desirable although for special applications varying expansivities are used to give a desirable activity rate. Some types of characteristic activity curves are as follows: (1) relatively low activity until a predetermined temperature is reached, (2) accelerating activity, (3) decelerating activity, (4) decelerating activity above a predetermined temperature, (5) no activity above a predetermined temperature, and (6) reversed activity above a predetermined temperature. Depending on the temperature range encompassed, several of the types mentioned may occur in the characteristic activity curve of one thermostat metal. High or low expansivities are useless if the expansivity is not reversible, since the resulting hysteresis would destroy the reproducibility and reliability of operation.

Therefore, either elemental metals or solid-solution alloys are preferred to those containing compounds, exhibiting phase changes, or changing solubility of one constituent with temperature.

The ability to do work is related to the modulus of elasticity as previously mentioned, and therefore, high modulus materials are preferred. Materials with high strengths, proportional limits, and creep strengths are needed to withstand the high thermal and mechanical stresses created in elements at high temperature and when heavily loaded.

For low-expansive components alloys of the Invar group (nickel–iron alloys) are generally chosen. The expansivities versus temperature curves of a group are shown in Figure 10.11.

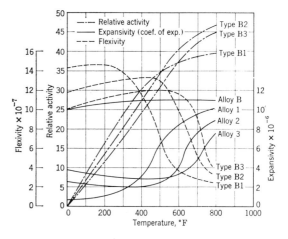

Figure 10.11   Curves Illustrating Relation among Expansivities (Coefficients of Expansion), Relative Activities, and Flexivities of High and Low Expansivity Alloys and Thermostat Metals Composed of Them

Salient features as the nickel content increases above 36 per cent are as follows:

1. The inflection temperature increases.
2. The temperature at which the minimum expansivity occurs increases.
3. The temperature range corresponding to a relatively low expansivity increases.

Therefore, when these nickel–iron alloys are laminated to a given high-expansive component with an expansivity that does not vary significantly with temperature, increase in the nickel content:

1. Raises the temperature at which maximum activity occurs.
2. Decreases the magnitude of the maximum activity.
3. Increases the temperature range of relatively uniform activity.

These effects are shown in Figure 10.11. This chart gives considerable information. As already noted, the expansivity versus temperature curves for three nickel–iron alloys (increasing in nickel content from 1 to 3) are given as well as that for an alloy having a high expansivity. The flexivities of the three types of thermostat metals composed of the alloy with the high expansivity bonded to the three alloys having low expansivities, respectively, are given. Flexivity, a fundamental activity constant, is defined in the section on Standard Test and Purchase Specifications, page 220. These calculated flexivities come surprisingly close to the flexivities obtained from actual activity data, as may be seen by comparing these flexivity curves with curves on Figure 10.17. The relative-activity curves of the three types of thermostat metals are included in the chart. These are the characteristic activity curves of movement versus temperature in inches or angular degrees, or the like, depending on the form of element.

The flexivity curve is the curve of the first derivative of the relative-activity curve. Any point on the flexivity curve denotes the slope of the corresponding point on the relative-activity curve. The calculation of the flexivity from the expansivity data is relatively simple, since from the general equation of the curvature of thermostat metal strip the following relation can be derived:

$$\text{Flexivity, } F = \tfrac{3}{2}(\alpha_2 - \alpha_1) \tag{19}$$

The Invar-type low-expansive alloys in some cases have been modified by the addition of chromium, molybdenum, and cobalt for high-temperature strength and scaling resistance.

High-expansive components are of a greater variety than low. Brass was one of the original ones and is still used because of its high expansivity. Disadvantages are its low maximum temperature of use, 300 to 350°F and the difficulty of processing the soft brass in combination with hard Invar. Monel was an improvement followed by austenitic nickel–chromium–iron and nickel–manganese–iron alloys. Pure nickel and nickel–molybdenum–iron alloys also are used.

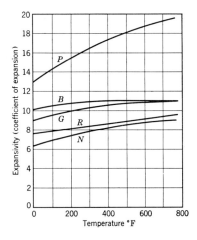

Figure 10.12 Expansivities (Coefficients of Expansion) of a Few High-Expansivity Alloys

The introduction of the high-manganese nickel–copper alloys about 1941 was the most significant advance in high-expansion alloys since the advent of thermostat metals. These alloys have approximately 50 per cent greater expansivity than previously used alloys. In addition to high expansivity they have very high electrical resistivity, as mentioned on page 216 in the section on electrical resistivity. The expansivity versus temperature curves of some high-expansivity alloys are given in Figure 10.12.

9 HEAT TREATMENT

As mentioned in the section on stresses, stresses of unknown magnitude due to cold rolling, roller flattening, slitting, piercing, forming, and so on are present in an element. If assembled into a thermostat or other thermally operated device, relief of these stresses and subsequent permanent deformation and upsetting of calibration may take place. Stress relief or, more correctly, stress redistribution is obtained by appropriate heat treatment. The heat treatment relieves the mechanical stresses of fabrication, and, also, the thermal stresses produced by heat treatment cause yielding in highly stressed points, and this yielding produces a redistribution of the stresses in such manner as to withstand normal operating stresses due to temperature and loading. The temperature of heat treatment should be approximately 100°F over the maximum temperature met with in use. Any forming or permanent mechanical change in a heat-treated element should be followed by another heat treatment. If the element is to be used for low-temperature work, subjecting it to lower temperatures than those met with in use imparts stability. Cycling several times from high to low temperature is efficient in stabilizing an element quickly.

In ordinary high-temperature thermostat metals, thermal activity constants decrease with increase of heat-treating temperature up to approximately 700–900°F and then increase again. The increase may be due to incipient recrystallization (annealing). The usual heat treatment does not affect the structure of the component alloys in any way or soften or harden them. Elements which in service are under restraint can be materially improved in resistance to permanent set by heat treating under similar restraint. The element must in service be subjected to restraint only in the same direction since restrained heat treating decreases the resistance to stresses opposite to those developed by the treatment.

10 HARDNESS

Since thermostat metals are composed of elemental metals or solid-solution alloys, hardness is obtained only by cold rolling. This cold rolling determines the hardness and elastic properties. High hardnesses usually indicate high elastic properties; therefore, high degree of hardness is desirable unless severe forming operations or sharp bends require a softer material. The elastic properties can be conveniently evaluated by the hardness, and hardness tests are routine in quality control. Most thermostat metal is used in relatively thin strip in which one component is usually one-half the total thickness. Therefore, hardness test methods are limited to those having very light loads and depths of penetration such as Vickers, Eberbach, microhardness, Tukon, or Knoop. Even with these instruments the hardness of strips having thicknesses of the order of 0.001 to 0.003 in. is difficult to determine accurately.

11 SURFACE CONDITION

As in any high-quality metal article the surface should be as free from defects and imperfections as possible. However, since thermostat elements operate by changes in temperature, the surface should be as receptive to absorption of heat or radiation of heat as possible. This calls for a dull matte finish rather than a bright polished finish. In most applications, rate of absorption or radiation of heat is not too critical, and normal cold-rolled finishes are used. The thinner the strip, the more important is the surface for its performance, since very thin strip consists essentially of two surfaces with very little material between. Surfaces approaching perfection are desired in this instance.

If a matte or black finish is desired it can be produced chemically or mechanically. One chemical process is to electroplate with copper and then oxidize by heat treatment or reaction with sulfides. A recently developed process of producing an adherent thin black surface finish on high-alloy ferrous alloys, consisting of heating in a molten sodium or potassium dichromate bath at 650–750°F, lends itself admirably to thermostat metal since the surface conditioning and heat treatment can be combined into one operation. This process is said to increase corrosion resistance. Fine abrasive blasting processes (Vapor blast, Microblast) are mechanical means of producing a matte finish. The finest meshes of abrasive should be used so that only the outer surface skin is affected. The detrimental effect of a coarse rough surface is illustrated by a 15 per cent decrease in thermal activity observed on a 0.020-in.-thick element blasted with 16-mesh metallic grit. Carburizing, nitriding, calorizing, and other treatments that chemically affect the surface should be avoided.

12 CORROSION RESISTANCE AND PROTECTION

Although most alloys used in thermostat metals are inherently corrosion-resistant, when they are combined in a thermostat metal galvanic couples may be produced. In corrosive media such as water and steam, careful choice of type of thermostat metal must be made to prevent rapid and disastrous failures. Brass–Invar and silicon–bronze–Invar combinations in some hard waters may disintegrate sufficiently in a few weeks to become inoperative. Combinations must be chosen in which the components individually are corrosion-resistant and which in combination do not produce a destructive galvanic couple. Types which are considered corrosion resistant are listed in Table 10.2 along with some of their mechanical and physical properties. The best of the group, type J7, possesses the best corrosion resistance of any type in the series. It has been used as an element in water-mixing valves for over 19 years, in all types of water, without failure.

Of course, the mounting brackets and the like must be of material that will not react with the thermostat element. Tin and lead coatings by dipping are used in some thermostat metal elements with good results. Zinc, cadmium, nickel, and chromium electroplating is also common practice to protect the surface with usually, however, indifferent results.

13 THERMAL CONDUCTIVITY

The rate of transfer of heat from its surface into the body of a thermostat element is dependent on its thermal conductivity. Therefore, the thermal conductivity may be, depending on the application, of considerable importance.

Heat can be transferred to an element by radiation, convection, or conduction. The heat thus transferred is absorbed into the body of the element by conduction. Heat supplied by a heater near the element is an example of the first way, and heat supplied by circulating air is an example of the second way. Since in most cases thermostat metal elements are made of relatively thin strips or sheets, the surface condition is more important for heat transfer by convection and radiation than the thermal conductivity. This follows because the mass is small compared to the surface area.

In numerous applications, such as iron thermostats, aquastats, and fluorescent starters, heat transfer to the element is mainly by conduction through the mounting bracket. In these cases the thermal conductivity is of importance.

The thermal conductivities of most metals and alloys are roughly proportional to their electrical conductivities. This fact can be used to get a rough approximation of the thermal conductivity:

Thermal conductivity, cal per sec per sq cm per °C per cm
$$= 12/\text{electrical resistivity, ohms per circular-mil-foot} \quad (20)$$

Thermostat metals are made up of two or more components which may greatly differ in thermal conductivity. Naturally, the longitudinal passage of heat parallel to the laminae is not the same as that perpendicular to them. If one knows the thermal conductivities of the components, it is simple to calculate the longitudinal and/or perpendicular conductivity of the whole. The flow of heat parallel or perpendicular to the laminae follows the same laws as flow of electricity through parallel and series resistances. The thermal conductivity can be changed to thermal resistance (the reciprocal), and thus formulas analogous to electrical-resistance formulas can be used.

Table 10.2. Mechanical and Physical Properties of Corrosion-Resistant TRUFLEX Thermostat Metals

TRUFLEX Type	ASTM Flexivity (F) 50–200°F × 10⁻⁷	Maximum Sensitivity Temperature Range °F	Useful Deflection Temperature Range °F	Recommended Maximum Temperature °F	Modulus of Elasticity (E) lb per sq. in. × 10⁶	Electrical Resistivity (R) at 75°F		Recommended Stabilizing Heat Treatment One hour at °F
						ohms per c m f	ohms per s m f	
J7	56	0–625	−100 to 500	625	22.0	96	75	500
G7	61	0–800	−100 to 1000	1000	27.5	450	353	700
J8	75	0–400	−100 to 500	625	19.4	135	106	500
GA12	104	0–300	−100 to 1000	1000	24.4	500	392	700
GB14	98	0–300	−100 to 1000	1000	25.8	511	401	700
MB18	141	0–300	−100 to 1000	1000	24.0	470	369	700
M8	65	0–400	−100 to 1000	1000	24.0	457	359	700
T14	110	0–300	−100 to 350	350	19.0	76	60	300

14 SPECIFIC HEAT

The specific heat determines the amount of heat energy necessary to raise the temperature of a thermostat metal element any given amount. Usually the amount of heat available is so great that this factor is not too important except possibly in the case of the heating by passage of electric current (see page 216). There is, however, little choice among the various thermostat metals available, since all their specific heats are very close to 0.12 calorie per gram per degree centigrade (the value is the same for British thermal units per pound per degree Fahrenheit).

15 STANDARD TEST AND PURCHASE SPECIFICATIONS

The properties and attributes that affect performance of thermostat metal are logical items that are covered by specifications. These include chemical analysis, quality of bond, quality of surface, thickness, width, length, cross, lengthwise and edgewise curvatures, hardness, thermal activity, electrical resistivity, temperature coefficient of electrical resistivity, stiffness or torque, and marking or identification of type and/or high- or low-component side.

Because thermal activity is the most important it has in the past been specified in a large variety of ways using numerous test specimen sizes and shapes. The American Society for Testing and Materials has developed method B106 for flexivity, which is a great improvement over previous methods since it gives a fundamental value of activity independent of size or shape. Flexivity is defined as "the change of curvature of the longitudinal center line of the specimen per unit temperature change for unit thickness" and is given by the following formula (see Fig. 10.13):

$$F = \frac{(1/R_2 - 1/R_1)t}{T_2 - T_1},\qquad (21)$$

where

$$\frac{1}{R} = \frac{8B}{L^2 + 4Bt + 4B^2},\qquad (22)$$

F = flexivity,
R_2 and R_1 = radii of curvature, in.,
T_2 and T_1 = temperature, °F,
t = thickness, in.,
B = movement, in.

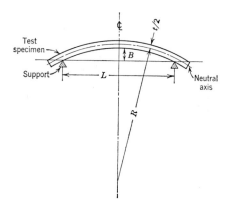

Figure 10.13   Schematic Diagram of Test for Flexivity

Other ASTM test methods are as follows:

1. Resistivity of Metallically Conducting Resistance and Contact Materials, B63;
2. Change of Resistance with Temperature of Metallic Materials for Electrical Heating, B70;
3. Modulus of Elasticity of Thermostat Metals (Cantilever Beam Method), B223;
4. Maximum Loading Stress at Temperature of Thermostat Metals (Cantilever Beam Method), B305;
5. Mechanical Torque Rate of Spiral Coils of Thermostat Metal, B362;
6. Specification for Thermostat Metal Sheet and Strip, B388;
7. Thermal Deflection Rate of Spiral and Helical Coils of Thermostat Metals, B389;
8. Mean Specific Heat of Thermal Insulation, C351;
9. Diamond Pyramid Hardness of Metallic Materials, E92.

Thermostat metals in the trade are classified into two main groups: low-temperature and high-temperature thermostat metals. This classification, unfortunately, has little to do with the temperature range of uniform or linear activity but only defines the types as to maximum temperatures to which they can be subjected without damage. Brass–Invar and silicon-bronze–Invar, and nickel–iron–chromium-alloy–Invar thermostat metals are examples of low- and high-temperature types with maximum temperatures of use of 350 to 600°F and 1000°F, respectively. Usually low-temperature types have for their high-expansion components nonferrous copper-base alloys. The high-temperature types are mainly composed of components both of which are ferrous alloys.

Thermostat metals as yet are not standardized, and, therefore, each producer has his own proprietary designation for each type.

16 BOND

Four main processes are used to create the bond between components. The process used depends on the types of materials and in some degree on the end use.

1. Casting the lower-melting alloy on the solid higher-melting one.
2. Joining the components directly by heat and static pressure in a press.
3. Joining the components directly by heat and the dynamic pressure of a hot-rolling mill.
4. Joining the components directly by dynamic pressure alone at room temperature.

Note that none of these processes involves the use of a brazing or soldering material, and, with the exception of method 1, all are done in the solid phase.

Method 1 can be used only where the melting points differ considerably and is limited to low-temperature metals such as brass on Invar.

Method 2 is used for only a small number of combinations which are not amenable to hot rolling.

Method 3 is used to bend the majority of thermostat metals and is the most widely used for high-quality, high-temperature metals.

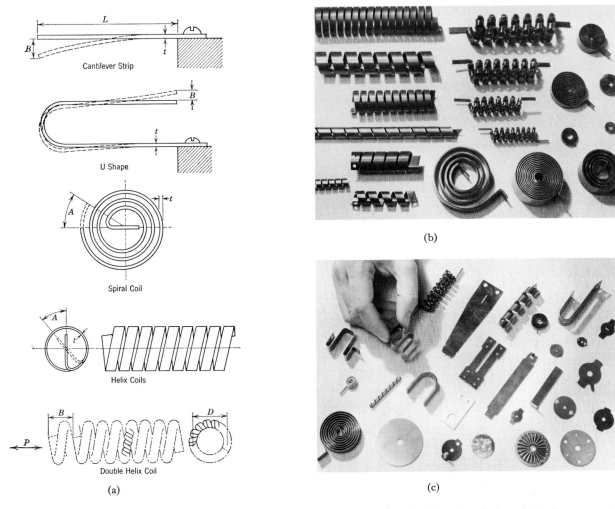

Figure 10.14 (a) Fundamental Types of Thermostat Metal Elements. (b) Helix and Spiral Coils. (c) Variety of Thermostat Metal Elements in Common Use.

Method 4 is the second-largest method in respect to quantity bonded and utilizes continuous lengths of materials.

17 TYPES OF ELEMENTS

The varieties of thermostat-metal element shapes possible are almost without number, but they can be classified into the following general group shapes:

1. Straight strips — the fundamental form. Straight strips can be used as cantilever beams (clamped at one end, other end free) or as simple beams (supported at both ends but ends not clamped since this would restrict movement).
2. U shapes. A U shape is an element that is a combination of a half circle and two straight strips.
3. Spiral coils — sometimes called clock-spring-type coils.
4. Helix coils.
5. Double helix coils.

These five shapes are illustrated in Figure 10.14(a). Many shapes are possible by combining these shapes or portions of them [Fig. 10.14(b), (c)].

18 FORMULAS

Formulas have been developed relating the dimensions, temperature change, and activity and torque or force constants of the various shapes of elements, and these are listed in Table 10.3. These formulas are all directly developed from the general equation of thermal bending (equation 1) and from standard deflection–load formulas. These are, of course, simplified equations. Insignificant terms have been dropped, and, where several constants are involved, they have been combined into one. Although not theoretically exact, they are useful in general work for calculating various types of elements for specific uses.

19 APPLICATION OR USE

In designing a thermostat element into a thermometer, thermostat, or other thermally operated device, the problem is one of choosing the most economical type of material and shape of element to produce the required movement and/or force. The type having the highest thermal activity in the working temperature range is usually chosen, although this is not always true. If a wide range of calibration or adjustment in a

Table 10.3. Summary of Formulas

	Thermal Deflection	Mechanical Force	Thermal Force
Cantilever	$B = \dfrac{0.53F(T_2 - T_1)L^2}{t}$	$P = \dfrac{4EBwt^3}{L^3}$	$P = \dfrac{2.12EF(T_2 - T_1)wt^2}{L}$
Simple beam	$B = \dfrac{0.133F(T_2 - T_1)L^2}{t}$	$P = \dfrac{64EBwt^3}{L^3}$	$P = \dfrac{8.51EF(T_2 - T_1)wt^2}{L}$
U shape (for small radius at bend)	$B = \dfrac{0.265F(T_2 - T_1)L^2}{t}$	$P = \dfrac{16EBwt^3}{L^3}$	$P = \dfrac{4.24EF(T_2 - T_1)wt^2}{L}$
Spiral and helix	$A = \dfrac{67F(T_2 - T_1)L}{t}$	$P = \dfrac{0.0232EAwt^3}{Lr}$	$P = \dfrac{1.55EF(T_2 - T_1)wt^2}{r}$
Double helix	$B = \dfrac{0.46F(T_2 - T_1)LD}{t}$	$P = \dfrac{5.2EBwt^3}{LD^2}$	$P = \dfrac{2.39EF(T_2 - T_1)wt^2}{D}$

Flexivity

$$F = \frac{(1/R_2 - 1/R_1)t}{T_2 - T_1}$$ (See also formula 21)

where $\dfrac{1}{R} = \dfrac{8B}{L^2 + 4Bt + 4B^2}$ (See also formula 22)

General Laws Governing Thermostat Metals

Deflection on temperature change varies:

For Cantilever, Simple Beam, and U Shape	For Spiral and Helix	For Double Helix
Directly as temperature change	Directly as temperature change	Directly as temperature change
Directly as length squared	Directly as length	Directly as length
Inversely as thickness	Inversely as thickness	Directly as diameter
		Inversely as thickness

Force exerted on temperature change varies:

For Cantilever, U Shape, and Simple Beam	For Spiral and Helix	For Double Helix
Directly as temperature change	Directly as temperature change	Directly as temperature change
Directly as width	Directly as width	Directly as width
Directly as thickness squared	Directly as thickness squared	Directly as thickness squared
Inversely as length	Inversely as radius	Inversely as diameter

t = thickness, in.
w = width, in.
L = active length of strip, in.
B = deflection, strip, in.
A = angular rotation, coils, degrees
P = force, ounces
r = radius (at point load is applied), in.
$T_2 - T_1$ = temperature change, °F
E = modulus of elasticity
F = flexivity
R = radius of curvature

small space is required, a low-thermal-activity material may be optimum. Also where the length of element and movement for a given temperature range are specified, a thin low-thermal-activity material element can be used to advantage. An element of high thermal activity may require increased thickness to bring down the movement to that specified. Besides the thermal activity, other factors such as corrosion resistance, thermal conductivity, and electrical resistivity influence the choice.

There are three conditions of operation that must be considered:

1. Thermal deflection.
2. Thermal force.
3. Thermal deflection and force.

19.1 Thermal deflection

In this case, the free deflection is utilized without the development of any force. The thermal deflection equations for a particular design of element give a ratio of thickness to length with a resultant infinite variety of combinations. The final size is determined by factors other than the free-deflection specification, such as rigidity, resistance to vibration, space limitations and practicality for production.

19.2 Thermal force

In this case, the basic free deflection is completely prevented, and thus only force is developed. The thermal force equation gives a ratio of width, thickness, and length with a resultant infinite variety of combinations. As in the case of thermal deflection the final size is determined by the factors other than thermal force, plus the factor of allowable stress.

19.3 Thermal deflection and force

Here, part of the potential thermal deflection is prevented and transformed into force. The element thus deveolps in response to a temperature change a combination of deflection and force. This is the most common application of thermostat metal. A minimum volume of material is required if one-half the available temperature change is used for developing thermal deflection and the other half is used for developing thermal force. When one of the variables of thickness, width, or length is selected, the other two are automatically fixed, resulting in only one specific size of a given type of material to satisfy the conditions.

It may be desirable to deviate from this optimum division of the temperature change because of dimensional, fabricating, assembly, or allowable stress conditions. An infinite number of size variations can be calculated by varying the temperature changes allotted to each property. However, there are a few limiting points of interest. If one third of the total available temperature change is used for developing force, the formulas give the shortest element for a given width for these types: cantilever, simple beam, U shape, spiral, or helix coils. If two thirds of the total available temperature change is used for developing force, the thinnest element for a given width is obtained for cantilevers, simple beams, and U shapes. For spiral and helix coils, the strip can be made as thin as is practical by using most of the available temperature change to produce force. This requires increasing the length of the strip proportionally.

After choosing the type of material, the shape of the element is chosen so that it moves thermally in the correct way to perform the required cycle of operation and so that it fits into the available space. The straight strip gives essentially linear movement for small temperature changes. The U shape similarly gives linear motion with a more compact design. The coils, spiral, and helix give rotating motion, which is uniquely adapted to thermometers and other dial-type devices. The spiral coil is also the most compact type of element.

The double helix coil has the special property of expanding or contracting (depending on the way of coiling — high-expansive component coiled inside or outside in the primary helix, primary helix coiled left- or right-handed, and secondary helix coiled left- or right-handed) with temperature rise. If a slight angular rotation is disregarded, the movement is true linear movement. Another feature is the small cross-sectional area and long length of the strip making up the coil. This gives a relatively high electrical resistance and high thermal-energy release on passage of electric current.

20 PHYSICAL CONSTANTS

In Table 10.4 is listed a large group of thermostat metals and their physical constants. These include all the constants that apply to the formulas given in Table 10.3 and also other pertinent physical data of use to design engineers. Further comment on these would be superfluous.

The relative thermal-activity curves of some of the thermostat metals listed in Table 10.4 are given in Figure 10.15. These characteristic curves give the relative movement of any type of

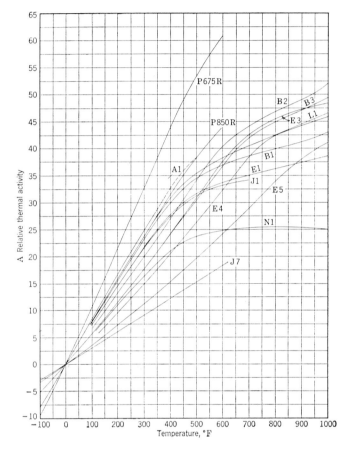

Figure 10.15 Relative Thermal Activity Curves of a Group of Thermostat Metals.

element with temperature. These curves give the following information: Straight portions of the curves indicate uniform linear thermal movement, curved portions indicate nonuniform or nonlinear movement, and the steepness or slope of the curves indicates the magnitude of the movement. These curves are valuable in choosing the type of material for a particular use. For instance, for an application calling for relatively straight-line operation between 300 and 750°F the type E4 would be the logical choice.

Since the relative thermal-activity curves are not linear, the thermal activity constants vary with the particular temperature range chosen. In Figure 10.16 are given the thermal-activity-constant versus temperature curves. The flexivity, strip-activity constant, and coil-activity constant can be expressed by the

Table 10.4   Mechanical and Physical Properties of Thermostat Metals

Type	ASTM Flexivity F × 10⁻⁷ 50 to 200°F Temperature Range	Maximum Sensitivity Temperature Range	Maximum Sensitivity Temperature Range °F	Useful Deflection Temperature Range °F	Recommended Maximum Temperature °F	Modulus of Elasticity E lb/sq. in. × 10⁶	Electrical Resistivity at 75°F ohms/cmf	Electrical Resistivity at 75°F ohms/smf	Recommended Stabilizing Heat Treatment One Hour at °F	Density lb/cu. in.	Type
A1	150	148	0 to 300	−100 to 350	350	18.0	74	58	300	0.30	A1
AG1	141	141	0 to 250	−100 to 300	300	15.0	20	15.7	250	0.34	AG1
B1	147	144	0 to 300	−100 to 700	1000	24.0	472	371	700	0.29	B1
B2	124	129	100 to 550	−100 to 1000	1000	24.0	424	334	700	0.29	B2
B3	112	122	200 to 600	−100 to 1000	1000	24.5	410	322	700	0.29	B3
B11	140	145	150 to 450	−100 to 1000	1000	24.0	447	350	700	0.30	B11
BN	46	45	0 to 400	−100 to 1000	1000	28.5	95	75	700	0.31	BN
B100R	105	104	0 to 300	−100 to 700	1000	24.5	100	78.5	700	0.30	B100R
B125R	125	124	0 to 300	−100 to 700	1000	24.5	125	98	700	0.30	B125R
B150R	133	131	0 to 300	−100 to 700	1000	24.5	150	118	700	0.30	B150R
B175R	136	135	0 to 300	−100 to 700	1000	24.5	175	137	700	0.30	B175R
B200R	141	139	0 to 300	−100 to 700	1000	24.5	200	157	700	0.29	B200R
B250R	145	143	0 to 300	−100 to 700	1000	24.5	250	196	700	0.29	B250R
B300R	146	144	0 to 300	−100 to 700	1000	24.5	300	236	700	0.29	B300R
B350R	147	145	0 to 300	−100 to 700	1000	24.5	350	275	700	0.29	B350R
B400R	147	145	0 to 300	−100 to 700	1000	24.5	400	314	700	0.29	B400R
C1	150	147	0 to 300	−100 to 700	1000	24.5	476	374	700	0.29	C1
C3	117	132	200 to 600	−100 to 800	1000	24.5	415	326	700	0.29	C3
C11	140	146	150 to 450	−100 to 900	1000	24.0	450	353	700	0.29	C11
D560R	140	138	0 to 300	−100 to 700	1000	24.0	560	440	700	0.28	D560R
E1	133	131	0 to 300	−100 to 700	1000	24.0	497	390	700	0.29	E1
E3	100	118	200 to 600	−100 to 1000	1000	24.5	435	342	700	0.29	E3
E4	80	103	250 to 700	−100 to 1000	1000	25.0	395	310	700	0.29	E4
E5	60	79	300 to 800	−100 to 1000	1000	25.0	360	282	700	0.29	E5
EN5	60	79	300 to 800	−100 to 1000	1000	25.0	282	222	700	0.29	EN5
F1	127	124	0 to 300	−100 to 500	700	20.0	20	15.7	500	0.31	F1
F15R	67	66	0 to 300	−100 to 500	700	20.0	15	11.8	500	0.32	F15R
F20R	127	125	0 to 300	−100 to 500	700	20.0	20	15.7	500	0.31	F20R
F25R	130	127	0 to 300	−100 to 500	700	22.0	25	16.0	500	0.31	F25R
F30R	134	131	0 to 300	−100 to 500	700	23.0	30	23.6	500	0.30	F30R
F35R	137	134	0 to 300	−100 to 500	700	23.5	35	27.5	500	0.30	F35R
F40R	140	137	0 to 300	−100 to 500	700	24.0	40	31.4	500	0.30	F40R
F50R	141	138	0 to 300	−100 to 500	700	24.0	50	39.3	500	0.30	F50R
F60R	143	140	0 to 300	−100 to 500	700	24.0	60	47.1	500	0.30	F60R
F70R	144	141	0 to 300	−100 to 500	700	24.0	70	55.0	500	0.30	F70R
F90R	145	142	0 to 300	−100 to 500	700	24.0	90	70.7	500	0.30	F90R
F100R	145	142	0 to 300	−100 to 500	700	24.0	100	78.5	500	0.30	F100R
G1	139	137	0 to 300	−100 to 700	1000	24.5	472	370	700	0.29	G1
G3	103	117	200 to 600	−100 to 1000	1000	24.5	420	330	700	0.29	G3
G7	61	61	0 to 800	−100 to 1000	1000	27.5	450	353	700	0.28	G7

Type	Density lb/cu. in.	Recommended Stabilizing Heat Treatment One Hour at °F	Electrical Resistivity at 75°F ohms/smf	Electrical Resistivity at 75°F ohms/cmf	Modulus of Elasticity E lb/sq. in. ×10⁶	Recommended Maximum Temperature °F	Useful Deflection Temperature Range °F	Maximum Sensitivity Temperature Range °F	ASTM Flexivity F ×10⁻⁷ 50 to 200°F Temperature Range	ASTM Flexivity F ×10⁻⁷ Maximum Sensitivity Temperature Range	Type
GA12	0.29	700	392	500	24.4	1000	−100 to 1000	0 to 300	104	103	GA12
GB14	0.29	700	401	511	25.8	1000	−100 to 1000	0 to 300	98	97	GB14
H1	0.29	700	365	465	24.0	1000	−100 to 700	0 to 300	156	154	H1
J1	0.31	500	74	94	19.0	625	−100 to 500	0 to 300	132	131	J1
J7	0.30	500	75	96	22.0	625	−100 to 500	0 to 625	56	56	J7
J8	0.31	500	106	135	19.4	625	−100 to 500	0 to 400	75	75	J8
K1	0.29	700	118	150	24.4	1000	−100 to 500	0 to 250	93	93	K1
L1	0.29	700	382	487	24.0	1000	−100 to 700	0 to 300	147	145	L1
M7	0.29	700	342	435	27.2	1000	−100 to 1000	0 to 800	40	39	M7
M8	0.29	700	359	457	24.0	1000	−100 to 1000	0 to 400	65	64	M8
MB18	0.29	700	369	470	24.0	1000	−100 to 1000	0 to 300	141	140	MB18
N1	0.31	700	75	95	26.0	1000	−100 to 500	0 to 300	101	100	N1
NF4	0.30	500	23.6	30	22.0	700	−100 to 700	250 to 700	42	63	NF4
P3	0.28	500	432	550	20.0	600	−100 to 600	200 to 600	180	210	P3
P30R	0.30	500	23.6	30	19.0	600	−100 to 500	0 to 400	175	174	P30R
P35R	0.30	500	27.5	35	19.0	600	−100 to 500	0 to 400	184	183	P35R
P40R	0.30	500	31.4	40	19.0	600	−100 to 500	0 to 400	190	189	P40R
P50R	0.30	500	39.3	50	19.0	600	−100 to 500	0 to 400	196	195	P50R
P60R	0.29	500	47.1	60	19.0	600	−100 to 500	0 to 400	200	199	P60R
P70R	0.29	500	55.0	70	19.0	600	−100 to 500	0 to 400	202	200	P70R
P90R	0.29	500	70.7	90	19.0	600	−100 to 500	0 to 400	204	202	P90R
P100R	0.29	500	78.5	100	19.0	600	−100 to 500	0 to 400	205	203	P100R
P125R	0.29	500	98.0	125	19.0	600	−100 to 500	0 to 400	206	205	P125R
P150R	0.29	500	118	150	19.0	600	−100 to 500	0 to 400	207	206	P150R
P175R	0.29	500	137	175	19.0	600	−100 to 500	0 to 400	207	206	P175R
P200R	0.29	500	157	200	19.0	600	−100 to 500	0 to 400	207	206	P200R
P250R	0.29	500	196	250	19.0	600	−100 to 500	0 to 400	208	207	P250R
P300R	0.29	500	236	300	19.0	600	−100 to 500	0 to 400	205	203	P300R
P350R	0.29	500	275	350	19.0	600	−100 to 500	0 to 400	207	206	P350R
P400R	0.29	500	314	400	19.0	600	−100 to 500	0 to 400	207	206	P400R
P450R	0.29	500	353	450	19.0	600	−100 to 500	0 to 400	208	207	P450R
P500R	0.29	500	392	500	19.0	600	−100 to 500	0 to 400	208	207	P500R
P550R	0.29	500	432	550	19.0	600	−100 to 500	0 to 400	204	202	P550R
P600R	0.29	500	471	600	19.0	600	−100 to 500	0 to 400	207	206	P600R
P675R	0.28	500	530	675	19.0	600	−100 to 500	0 to 400	209	208	P675R
P850R	0.27	500	668	850	19.0	600	−100 to 500	0 to 400	150	147	P850R
PJ	0.30	500	96	122	17.0	625	−100 to 625	0 to 600	75	81	PJ
R2	0.31	700	275	350	24.0	800	−100 to 700	250 to 550	90	104	R2
1513	0.29	700	310	395	22.8	1000	225 to 1000	500 to 800	−21	81	1513
7110	0.29	700	471	600	24.0	1000	−100 to 400	0 to 350	86	82	7110

Table 10.5.   Tolerances

Thickness, in.	Tolerance, in.	Width, in.	Tolerance, in.	Length, ft.	Tolerance, in.
Under 0.005	± 0.0003	Up to $\frac{1}{2}$ incl.	± 0.003	Up to 1 incl.	$\pm \frac{1}{32}$
0.005 to 0.0099 incl.	± 0.00035	Over $\frac{1}{2}$ to 1 incl.	± 0.004	Over 1 to 4 incl.	$\pm \frac{1}{16}$
0.010 to 0.0149 incl.	± 0.0004	Over 1 to 3 incl.	± 0.008	Over 4 to 12 incl.	$+\frac{1}{2} \; -\frac{1}{16}$
0.015 to 0.0199 incl.	± 0.0005	Over 3	± 0.010	Edgewise camber	$\frac{9}{32}$ inch maximum
0.020 and over	$\pm 2\frac{1}{2}\%$				in three feet

Flexivity $\pm 4\%$ within the maximum sensitivity range for those metals having a flexivity of 100×10^{-7} and greater. Electrical resistivity $\pm 3\%$ to 10% depending on type of material.

same curve, using different ordinates, since they all express the same function (the first derivative of the relative thermal-activity curve) for various shapes of elements. These curves, therefore, give the instantaneous values of the constants.

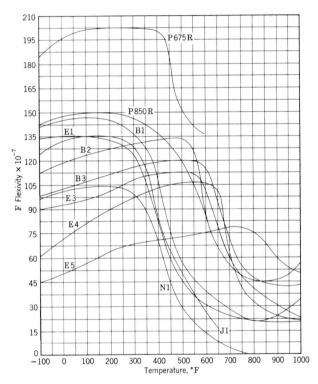

Figure 10.16   Thermal Activity Constant versus Temperature

21   STANDARD TOLERANCES

Standard tolerances for physical dimensions and some properties are given in Table 10.5.

Throughout this paper, an attempt has been made to concentrate on thermostat metal itself rather than its uses and applications. A discussion of the large variety of applications would take several times the length of this article. The reader is referred to the extensive literature in current journals and in patents for this information.

REFERENCES

For a more complete list of references, refer to the *Bibliography and Abstracts on Thermostat Metals 1806–1959;* Special Technical Publication No. 288; supplements No. 288-A, 1960–

1961, and No. 288-B, 1962–1965. These are published by the American Society for Testing and Materials, Philadelphia, Pa.

1. S. Timoshenko. "Analysis of Bimetal Thermostats," *J. Opt. Soc. Am.* **2**, 233–255 (Sept. 1925).
2. William B. Elmer, "Tapered-Thickness Bimetal," *AIEE Trans.* **64**, 661–664 (Sept. 1945).
3. H. E. Cobb, "Stress Analysis and Derivation of Formulae," *Elec. J.* (*London*) **25**, 288–291 (June 1928).
4. William Crawford Hirsch, "Historial Résumé of Stages of Development of Bimetal and General Discussion of Its Uses," *Elec. Mfg.* **2**, 181–182 (Oct. 1928).
5. Howard Scott, "Characteristics of Low-Expansion Nickel Steels," *Trans. ASST*, 829–847 (May 1928).
6. W. Rohn, "Bimetal," *Z. Metallk.* **21**, 259–264 (Aug. 1929).
7. K. Becker, "Bending with Temperature of Bimetallic Strips of Any Curved Shape," in *Research and Technique*, W. Peterson, Editor, Julius Springer, Berlin, 1930, pp. 340–348.
8. Herbert Buchholz, "Behavior of Bimetallic Strips When Heated," *Z. Tech. Physik.* **11**, 273–275 (Nov. 7, 1930).
9. H. Scott, "Thermostatic Metal — or Bimetal; Nature and Utility," *Metal Progress* **22**, 29–33 (Nov. 1932).
10. Robert Edler, "Coiled Bimetallic Strips," *Elektrotech. Maschinenbau* **21**, 546 (Oct. 8, 1933).
11. H. D. Matthews, "Mechanical Design of Thermostatic Bimetal Elements," *Prod. Eng.* **4**, 420–422 (Nov. 1933).
12. T. A. Rich, "Thermo-Mechanics of Bimetal," *Gen. Elec. Rev.* **37**, 102–105 (Feb. 1934).
13. S. Timoshenko, "Buckling of Flat Curved Bars and Slightly Curved Plates. Includes the Case of the Bimetallic Strip," *J. Appl. Mech.* **2**, A17–A20 (Mar. 1935).
14. Enrico Erni, "Maximum Mechanical, Thermal, and Electrical Load-Carrying Capacity of Bimetals," *Assoc. suisse des Elec. Bull.* **27**, 732–737 (Nov. 25, 1936).
15. G. Keinath, "Bimetals," *Arch. tech. Messen*, **5**, T55–T56, Z972 (Apr. 1936).
16. Hakar Masumoto, "On the Thermal Expansion of the Alloys of Iron, Nickel, and Cobalt, and the Cause of the Small Expansibility of Alloys of the Invar Type," reprinted from *Tohoku Imp. Univ. Science Repts. Ser.* I, **20**, 1.
17. S. R. Hood, "Thermostat Metal," *ASTM Bull.* (Jan. 1942).
18. F. R. Hensel and J. W. Wiggs, "Special Metals in the Electrical Industry. Bimetals," *Elec. Eng.* **62**, 297–299 (July 1943).
19. P. R. Lee, "Bimetal Performance at 800°F," *Metals & Alloys* **20**, 346–349 (Aug. 1944).

Chapter 11

Aluminum Alloy Laminates: Alclad and Clad Aluminum Alloy Products

Robert H. Brown

Laminates composed of two aluminum alloys of different compositions metallurgically bonded have been produced in large quantities since 1928. Most of the production has been, and still is, in sheet products, although there also has been some production in wire and tubular products.

The core alloy is chosen so as to have the required mechanical characteristics. Many cores are used because of their relatively high strength, but some are chosen because of their fabricability, formability, or ductility.

The coating alloys vary in thickness on one or both sides of the core alloy from $1\frac{1}{2}$ to 15 per cent of the total thickness of the product. The coating alloys originally were limited to those compositions that had electrode potentials in most natural environments anodic to those of the core alloy. Such a product was given the name "Alclad" by the Aluminum Company of America who registered the term as trademark. In the interest of allowing industry-wide use of the name, the company subsequently relinquished the trademark rights. Although for an interval the Reynolds Metal Company used the words "Clad" and "Pureclad," the term "Alclad" has been adopted by the Aluminum Association to designate an aluminum alloy product with a coating that will electrochemically protect the core. Alclad products are unique in that they are products having "built-in" cathodic protection of long life.

Three other types of aluminum alloy laminated products have found wide use. A coating alloy that has a melting point appreciably lower than that of the core, and good fluidity in the liquid state, is used as cladding on brazing sheet. For specific applications requiring resistance per se against specific chemical solutions, a corrosion resistant coating is selected as a cladding. Finally, a group of clad sheet products is widely used because of their important characteristics, having a pleasing uniform appearance after etching and finished anodic coatings or transparent organic coatings. This latter group of sheet products includes those that may be described as "self-clad" because the core and coating are of the same composition. The high degree of total reduction imparted to the coating layer results in improved etching and anodizing characteristics. For these three types of clad products, the coating alloy compositions are selected for special characteristics other than solution potential. The term adopted by the Aluminum Association for such laminates is clad products.

1 ORIGINS OF ALCLAD

Mr. E. H. Dix, Jr., the originator of what was later to be known as "Alclad," had his interests kindled sometime around

ROBERT H. BROWN is Assistant Director of Research, Alcoa Research Laboratories, New Kensington, Pa.

1925 by the experiments at the Bureau of Standards. These experiments concerned the corrosion protection given 2017-T4 (then known as 17S-T) by spraying with liquid aluminum (commercially pure). The corrosion protection was excellent but the solidified metal was porous and brittle. Even with ductile coating, metal spraying was not considered practical for handling relatively large surface areas. Dix pursued the idea of using wrought commercial aluminum to clad both sides of 1017-T4 sheet or plate to obtain mechanical protection. Three practices for applying the commercial aluminum to 2017-T4 were developed.[1] By 1927 substantial quantities of sheet were being produced, and by 1928 Dix published articles on the evaluation of various corrosion tests of Alclad 2017-T4.[2,3] A contemporary paper[4] by his colleagues further stimulated Dix's belief in Alclad 2017-T4. Data in this paper showed that commercial purity aluminum was anodic to 2017-T4 and, hence, the aluminum coating not only would afford mechanical protection in corrosive environments, but also would electrochemically protect exposed areas of 2017-T4, such as edges. The further dividend to be obtained by the use of alclad was

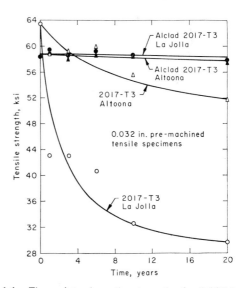

Figure 11.1   These data show the strength of a 0.032-in. sheet of 2017-T3 and Alclad 2017-T3 after various exposures up to 20 years in two corrosive atmospheres — one at Altoona, Pa., and the other at La Jolla, Calif. The atmosphere at Altoona is laden with soot, dirt, and acid fumes. The atmosphere at La Jolla is marine and heavily laden with salt spray which deposits on the specimens and is not washed off because of the low rainfall. Of the seven geographical locations where Subcommittee VI of ASTM Committee B3 exposed specimens, these two were the most corrosive. It will be observed that the Alclad 2017-T3 had essentially the same tensile strength after 20 years' exposure in either atmosphere, whereas there was a substantial decrease in tensile strength of the 2017-T3 in both atmospheres

that the attack only penetrated into the sheet the thickness of the coating and then spread out with coating electrochemically protecting the exposed core (or diffusion zone).

The great utility of this concept, of course, is best demonstrated by the widespread use of this type of sheet product for many years in aircraft without weakening of the sheet by corrosion.

Figure 11.1 is a graph showing the strength of 0.032-in.-thick sheet of both 2017-T3 and Alclad 2017-T3 after exposures as long as 20 years in the corrosive atmospheres of Altoona, Pa., and LaJolla, Calif., locations used by Subcommittee VI of ASTM Committee B-3 on Corrosion of Non-Ferrous Metals and Alloys.[5]

The data show dramatically that in environments which caused substantial corrosion of the 0.032-in.-thick 2017-T3 sheet, the strength of Alclad 2017-T3 was virtually unchanged after 20 years, even though exposed without maintenance. In the other environments, corrosive effect on the 2017-T3 was less pronounced and, of course, the strength of the Alclad 2017-T3 was not changed.

2 EARLY ALCLAD SHEET

The first laminated product to be produced in relatively large quantities was Alclad 2017-T3, originally known as Alclad 17S-T. In the latter half of the 1920's, Alclad 2017-T3 skin sheet was used on a number of aircraft such as the Tri-Motored Ford Monoplane, the Hamilton Metalplane, and the Sikorsky S-38. An interesting application of this period was the use of 0.0095-in.-thick Alclad 2017-T3 for the skin of the metal airship ZMC-2.

3 CURRENT HEAT-TREATABLE ALCLAD SHEET

Alclad 2017-T3 sheet is now obsolete and has been supplanted by a number of other alclad sheet products. The first successor was Alclad 2024-T3, which also is composed of commercial purity aluminum coating and a core of an Al-Cu-Mg-Mn alloy. The 2024-T3 is appreciably stronger than 2017-T3 (also an Al-Cu-Mg-Mn alloy). Figure 11.2 is a photomicrograph (125×) of the cross section of 0.032-in.-sheet of Alclad 2024-T3.[6] The two coating layers of 1230 alloy and the diffusion zone in these layers are clearly discernible from the Al-Cu-Mg-Mn alloy core of 2024-T3.

Panels of 0.064-in. alclad sheet with the coating removed by machining for a distance of 1 in. were exposed to Miami tidewater for two years A panel of 2024-T3 was machined in a similar manner to the same depth as the Alclad 2024-T3 and exposed with the latter panel. A comparison of Figures 11.3 and 11.4 (85×) demonstrates the electrochemical protection afforded the exposed core of the Alclad 2024-T3 after two years in seawater off the eastern coast of Florida. Both the machined and unmachined areas of 2024-T3 (nonclad) are pitted, whereas the coating of the Alclad 2024-T3 adjacent to the 2024-T3 core exposed by machining shows the beneficial effect of electrolytic action required to prevent corrosive attack of the core.

The ability of Alclad 2024-T3 to resist the ravages of

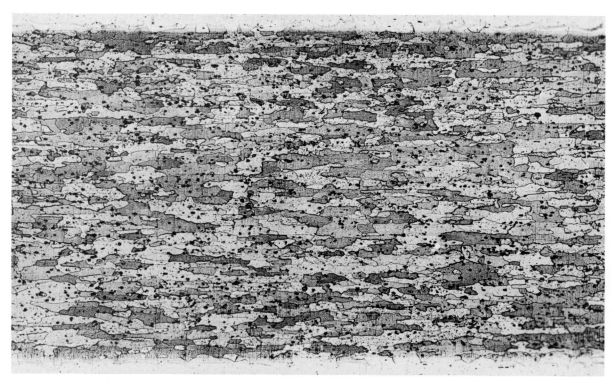

Figure 11.2　Full cross section of 0.032-in. Alclad 2024-T3 sheet. Etching differentiates between surface layers of 1230 aluminum and 2024-T3 core. Diffusion zone resulting from migration of copper into the coating layers is evident. Etch: HF-HCl-HNO₃ solution 15 sec. Magnification 125×

Figure 11.3   This photomicrograph shows part of a cross section of an area of Alclad 2024-T3 exposed for two years to seawater off the eastern coast of Florida between high and low tide. Prior to exposure, the coating was removed by machining. In this operation the machining was carried down into the core. The dotted white line shows the original contour of the exposed surface. This photomicrograph should be compared with that in Figure 11.4. Note that although some of the alclad coating has been consumed, the exposed core of 2024-T3 is virtually unattacked. This is an example of the built-in cathodic protection that is present in alclad products. The tool marks in the machined area were clearly visible after exposure. Etch: HF-HCl-HNO$_3$ solution 15 sec. Magnification 85×

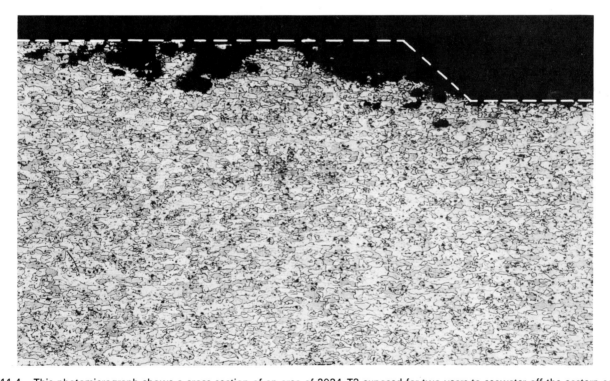

Figure 11.4   This photomicrograph shows a cross section of an area of 2024-T3 exposed for two years to seawater off the eastern coast of Florida between high and low tide. Prior to the exposure, the 2024-T3 was also machined to give the same surface contour as that of the Alclad 2024-T3 in Figure 11.3. Again, as in Figure 11.3, the dotted line shows the original contour of the surface. It is noteworthy that not only is there attack in the unmachined area, but the tool marks in the machined area did not show after exposure. Etch: HF-HCl-HNO$_3$ solution 15 sec. Magnification: 85×

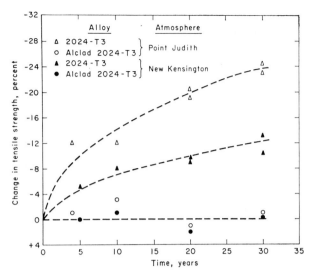

Figure 11.5   These data show that even after 30 years' exposure to the marine atmosphere at Pt. Judith, R.I., or the industrial atmosphere at New Kensington, Pa., 0.064-in.-thick sheet of Alclad 2024-T3 shows practically no loss in tensile strength. The degree of protection rendered by the alclad coating is evident by the much higher losses in tensile strength of 0.064-in.-thick sheet of 2024-T3 exposed for the same periods of time in these same environments

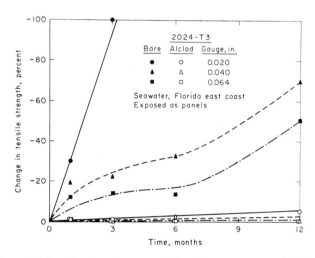

Figure 11.6   The low losses resulting from exposure of Alclad 2024-T3 sheet, even as thin as 0.020 in., when exposed to a corrosive environment such as seawater, is demonstrated by the data in this graph. The degree of protection rendered by the alclad coating is evidenced by the much lower losses in the case of the alclad sheet when compared with the bare 2024-T3

Table 11.1.   Electrode Potentials of Some Commercial Aluminum Alloys

Alloy	Temper	Use	Type[a]	Potential,[b] Volts
2017	T4	Core	H.T.	−0.68
2024	T3	Core	H.T.	−0.68
2014	T6	Core	H.T.	−0.79
2024	T81	Core	H.T.	−0.80
7178	T6	Core	H.T.	−0.81
7075	T6	Core	H.T.	−0.81
2219	T81	Core	H.T.	−0.81
2219	T87	Core	H.T.	−0.81
6053	T6	Core	H.T.	−0.83
6061	T6	Core	H.T.	−0.83
6063	T5, T6	Core	H.T.	−0.83
7075	T73	Core	H.T.	−0.84
7079	T6	Core	H.T.	−0.87
3003	O, H18	Core	S.H.	−0.84
3004	O, H18	Core	S.H.	−0.84
3005	O, H18	Core	S.H.	−0.84
5050	O, H36	Core	S.H.	−0.84
5052	O, H34	Core	S.H.	−0.85
5154	O, H34	Core	S.H.	−0.86
5155	O, H38	Core	S.H.	−0.86
5056	H192	Core	S.H.	−0.87
6003		Coating		−0.83
1230		Coating		−0.84
7072		Coating		−0.96
6253		Coating		−0.99
7472		Coating		−1.01

[a] H.T. — Heat Treatable.
   S.H. — Strain Hardenable.
[b] (0.1 N Calomel Reference).
Typical measurement in a salt-peroxide solution (53 grams/liter NaCl and 3 grams/liter H_2O_2).

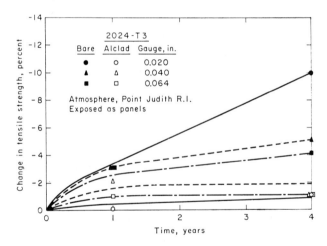

Figure 11.7   The data in this graph is for the exposure of the same material as in Figure 11.6, except that the exposure was made in a marine atmosphere at Pt. Judith, R.I. Again, it is evident that even in the case of thin alclad sheet, the protection rendered by the alclad coating is almost completely effective. The effectiveness of the alclad principle in combating corrosion is even more strikingly demonstrated by these thin sheets

corrosive environments of the marine atmosphere at Point Judith, R.I., and the highly industrial atmosphere of New Kensington, Pa., is shown in Figure 11.5.* This graph shows that after 30 years the Alclad 2024-T3 sheet 0.064-in.-thick did not lose tensile strength, whereas 2024-T3 sheet of the same thickness showed substantial loss in tensile strength.

* It must be emphasized that discussions of corrosion performance throughout this chapter are on products that were freely exposed for the entire period of exposure without any maintenance, protective organic coatings, or attention other than repair of the exposure fixtures.

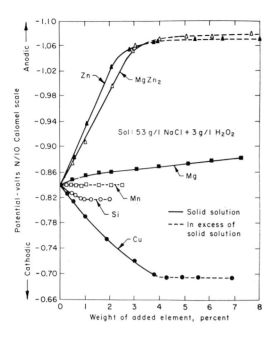

Figure 11.8   The effect of various alloying elements in solid solution on the electrode potential of aluminum alloys is shown in this graph. Any alloy with a potential more electronegative than another will afford some electrochemical protection of the other alloy. For example, a 2 per cent Zn alloy would afford some electro-chemical protection of a 1 per cent Zn alloy, of pure aluminum, of a 4 per cent Mg alloy, of a 1 per cent Mn alloy, of a 1 per cent Si alloy, or of a 2 per cent Cu alloy, if these elements were in solid solution. Similarly a 4 per cent Mg alloy would afford some electrochemical protection of a 1 per cent Mn alloy, which in turn would afford some electrochemical protection of a 3 per cent Cu alloy, again providing these alloying elements were in solid solution

By the use of alclad, protection from corrosion can be obtained for thin-sheet products in highly corrosive environments. Figure 11.6 shows the percentage change in tensile strength resulting from the exposure of 0.020, 0.040, and 0.064-in.-thick 2024-T3 and Alclad 2024-T3 sheet in seawater off the eastern coast of Florida. Even the thinnest (0.020 in. thick) Alclad 2024-T3 decreased less than 5 per cent in tensile strength, whereas a similar thickness of 2024-T3 sheet was completely destroyed in this exposure. Figure 11.7 is a graph showing the corrosion performance in the Point Judith, R.I., atmosphere (marine) for four years, as evaluated by percentage change in tensile strength of the same two sheet products over the range of thicknesses. Again the ability of alclad to retain its initial strength under highly corrosive conditions is clearly demonstrated.

4   EXPANDING THE ALCLAD CONCEPT

Since it had been established that copper in solid solution causes the electrode potential of an aluminum-copper alloy to become cathodic to aluminum, the first step in widening the application of alclad was to develop cladding alloys that would be anodic to aluminum so that core alloys equipotential or anodic to aluminum could be coated with an alloy that would electrochemically protect the core.[7,8]

Figure 11.8 shows the effect of various alloying elements in solid solution on the electrode potential of the alloys in chloride-saline environments. The solid portion of the curves indicates the potential of solid solutions and the dotted portion the potential of the solid solutions plus excess constituent. The alloys were given an elevated temperature solution heat treatment and quenched in cold water. Alloying elements such as Mg, Zn, or Zn and Mg in the ratio of $MgZn_2$ in solid solution causes the potential to become more anodic at room temperature than that of aluminum; whereas Cu and Si in solid solution cause the potential to be altered in the cathodic direction. Mn in solid solution does not materially alter the potential.

Table 11.1 shows potentials of some commercial aluminum alloys — both of heat-treatable and strain-hardenable types — especially those used for cores or coatings of alclad sheet, tube, and wire.[9]

The nominal compositions of alloys composing alclad and clad products are given in Table 11.2. Table 11.3 lists the commercially available clad and alclad products. This table lists the

Table 11.2.   Nominal Composition of Aluminum Alloys Used in Alclad and Clad Products

Alloy	Cu	Si	Mn	Mg	Zn	Cr	Si + Fe[a]	Al[b]	Others
1100	—	—	—	—	—	—	1.0	99.00	
1135	—	—	—	—	—	—	0.65	99.35	
1175	—	—	—	—	—	—	0.15	99.75	
1185	0.01[a]	—	—	—	—	—	0.15	99.85	
1230	0.10	—	0.05	—	—	—	0.70	99.30	
1260	—	—	—	—	—	—	0.40	99.60	
2014	4.4	0.8	0.8	0.4	—	—	—	—	
2017	4.0	—	0.5	0.5	—	—	—	—	
2024	4.5	—	0.6	1.5	—	—	—	—	
2219	6.3	—	0.3	—	—	—	—	—	0.1 V, 0.15 Zr
3003	—	—	1.2	—	—	—	—	—	
3004	—	—	1.2	1.0	—	—	—	—	
3005	—	—	1.2	0.4	—	—	—	—	
4343	—	7.5	—	—	—	—	—	—	
4045	—	10.0	—	—	—	—	—	—	
5050	—	—	—	1.4	—	—	—	—	
5052	—	—	—	2.5	—	0.25	—	—	
5154	—	—	—	3.5	—	0.25	—	—	
5155	—	—	0.4	4.2	—	0.15	—	—	
5056	—	—	0.1	5.2	—	0.10	—	—	
6003	—	0.70	—	1.2	—	—	—	—	
6951	0.25	0.35	—	0.65	—	—	—	—	
6053	—	0.70	—	1.3	—	0.25	—	—	
6253	—	—	—	1.25	2.0	0.25	—	—	
6061	0.25	0.6	—	1.0	—	0.20	—	—	
6063	—	0.4	—	0.7	—	—	—	—	
7072	—	—	—	—	1.0	—	—	—	
7472	—	—	—	1.2	1.6	—	—	—	
7075	1.6	—	—	2.5	5.6	0.30	—	—	
7178	2.0	—	—	2.7	6.8	0.30	—	—	
7079	0.6	—	0.2	3.3	4.3	0.20	—	—	

[a] Maximum.
[b] Minimum.

Table 11.3. Commercially Available Alclad and Clad Aluminum Alloy Products

Item	Sheet Composite	Sheet Gauge	Core Alloy	Coating Alloy	Coating Sides Coated	Coating Thickness, %
1	Alclad 2017[a]	All	2017	1230	2	5
2	Alclad 2014	0.010–0.039	2014	6003	2	10
		0.040 and thicker	2014	6003	2	5
3	Alclad 2024	0.010–0.062	2024	1230	2	5
		0.063 and thicker	2024	1230	2	2½
		0.188 and thicker	2024	1230	2	(b)
4	Alclad 2219	0.010–0.039	2219	7072	2	10
		0.040–0.099	2219	7072	2	5
		0.100 and thicker	2219	7072	2	2½
5	Alclad 6061	0.020 and thicker	6061	7072	2	5
		0.020 and thicker	6061	7072	1	5
6	Alclad 7075	0.025–0.062	7075	7072	2	4
		0.063–0.187	7075	7072	2	2½
		0.188 and thicker	7075	7072	2	1½
		0.025–0.062	7075	7072	1	4
		0.063–0.187	7075	7072	1	2½
		0.188 and thicker	7075	7072	1	1½
7	Alclad 7079	0.025–0.062	7079	7072	2	4
		0.063–0.187	7079	7072	2	2½
		0.188 and thicker	7079	7072	2	1½
8	Alclad 7178	0.025 and thicker	7178	7072	2	4
9	Alclad 3003	All	3003	7072	2	5
		All	3003	7072	1	5
10	Alclad 3004	0.010 and thicker	3004	7072	2	5
		0.020 and thicker	3004	7072	1	5
11	Alclad 3005	0.010–0.018	3005	7072	2	10
		0.019 and thicker	3005	7072	2	5
12	Alclad 5050	0.020 and thicker	5050	7072	2	5
		0.020 and thicker	5050	7072	1	5
13	Alclad 5052	0.020 and thicker	5052	7072	2	5
14	Alclad 5154	0.020 and thicker	5154	7072	1	5
15	Alclad 5155	0.020 and thicker	5155	7072	2	5
16	Clad 1135[c]	0.124 and thinner	1135	1135	2	10
		0.125 and thicker	1135	1135	2	5
17	Clad 1100[c]	0.124 and thinner	1100	1100	2	10
		0.125 and thicker	1100	1100	2	5
18	Clad 1100[d]	0.064 and thinner	1100	1175	2	15
		0.065 and thicker	1100	1175	2	7½
		0.064 and thinner	1100	1175	1	15
		0.065 and thicker	1100	1175	1	7½
19	Clad 3003[c]	0.124 and thinner	3003	3003	2	10
		0.154 and thicker	3003	3003	2	5
20	Clad 3003[d]	0.064 and thinner	3003	1175	2	15
		0.065 and thicker	3003	1175	2	7½
		0.064 and thinner	3003	1175	1	15
		0.065 and thicker	3003	1175	1	7½
21	Clad 6053	0.020 and thicker	6053	1185	2	5
		0.020 and thicker	6053	1185	1	5
22	Clad 3003[e]	0.063 and thinner	3003	4343	1	10
		0.064 and thicker	3003	4343	1	5
		All	3003	4343	1	15
23	Clad 3003[e]	0.063 and thinner	3003	4343	2	10
		0.064 and thicker	3003	4343	2	5
		All	3003	4343	2	15
24	Clad 6951[e]	0.090 and thinner	6951	4343	1	10
		0.091 and thicker	6951	4343	1	5
		All	6951	4343	1	15
25	Clad 6951[e]	0.090 and thinner	6951	4343	2	10
		0.091 and thicker	6951	4343	2	5
		All	6951	4343	2	15
26	Clad 6951[e]	0.090 and thinner	6951	4045	1	10
		0.091 and thicker	6951	4045	1	5
27	Clad 6951[e]	0.090 and thinner	6951	4045	2	10
		0.091 and thicker	6951	4045	2	5
		All	6951	4045	2	15

Item	Tube Composite	Tube Size	Core Alloy	Coating Alloy	Coating Location	Coating Thickness, %
28	Alclad 1100	All	1100	7072	Inside	10
		All	1100	7072	Outside	7
29	Alclad 3003	All	3003	7072	Inside	10
		All	3003	7072	Outside	7
30	Alclad 5050	All	5050	7472	Outside	8
31	Alclad 6061	All	6061	7072	Inside	10
		All	6061	7072	Outside	7
		All	6061	7072	Both Sides { in	10
					out	7
32	Alclad 6063	All	6063	7072	Inside	10
		All	6063	7072	Outside	7

Item	Wire and Round Rod Composite	Size	Core Alloy	Coating Alloy	Coating % Cross Sect. Area
33	Alclad 5056	All	5056	6253	20

[a] Obsolete, replaced by Alclad 2024.
[b] Standard coating thickness 2½%, available as special with 1½%.
[c] For anodized finishes.
[d] Reflector sheet.
[e] Brazing sheet.

alloy-coating thickness available in relation to thickness of the sheet or tube. In the case of rod and wire, the cladding is expressed in percentage of cross-sectional area.

In Table 11.4 are *typical* tensile properties for 0.064-in.-thick sheet of clad and alclad in more popular tempers. Tempers of intermediate hardness, such as H12, H14, and H16 and H32, H34, and H36 have typical properties intermediate between those of 0 and H18, or 0 and H38 tempers. The typical properties for the heat-treatable alloys are included only for solution-heat-treated and solution-heat-treated and artificially age-hardened conditions, these being the most widely used tempers.

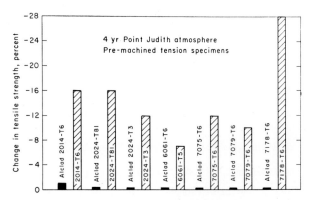

Figure 11.9 As shown in Figure 11.8, the potential of the coating alloy can be chosen so that it is anodic to practically any aluminum core alloy, with the result that a number of different alclad alloys have been commercialized. This figure shows the superior resistance to corrosion of the alclad sheets when compared with their nonclad counterparts. These data show that after four years' exposure of 0.064-in.-thick sheet to the Pt. Judith atmosphere, the losses in tensile strength were negligible for the alclad sheets, whereas they are appreciable for the nonclad sheets

Table 11.4.   Typical Tensile Properties of Alclad and
Clad Sheet Products

Alloy	Coating	T.S.	Y.S.	% El
Alclad 2017-T3	1230	59	38	18
Alclad 6063-T5[a]	7072	25	19	12
-T6[a]	7072	32	28	12
Alclad 6061-T6	7072	42	37	12
Alclad 2024-T3	1230	65	45	18
-T81	1230	65	60	6
Alclad 2219-T81	7072	60	46	10
-T87	7072	62	52	10
Alclad 2014-T6	6003	67	59	10
Alclad 7079-T6	7072	73	64	11
Alclad 7075-T6	7072	76	67	11
Alclad 7178-T6	7072	81	71	10
Alclad 3003-O	7072	16	6	30
-H18	7072	29	27	4
Alclad 3005-O	7072	20	8	25
-H18	7072	26	25	5
Alclad 3004-O	7072	26	10	20
-H36	7072	38	33	5
Alclad 5050-O	7072	16	6	24
-H36	7072	28	24	7
Alclad 5050-O[b]	7472	16	6	24
Alclad 5052-O	7072	27	12	25
-H34	7072	35	29	10
Alclad 5154-O	7072	34	17	27
-H34	7072	41	32	13
Alclad 5155-O	7072	36	17	22
-H38	7072	50	41	10
Alclad 5056-H192[c]	6253	66	—	—
-H392[c]	6253	62	—	—
Clad 1100-O	1100	13	5	35
-H18	1100	24	22	5
Clad 1100-O	1175	13	5	35
-H18	1175	24	22	5
Clad 1135-O	1135	12	4	35
-H18	1135	23	20	5
Clad 3003-O	3003	16	6	30
-H18	3003	22	21	8
Clad 3003-O	1175	15	5	30
-H14	1175	21	20	8
Clad 6053-O	1185	15	8	30
-T6	1185	34	29	12
Clad 3003-O	4343	16	6	30
-H14	4343	22	21	8
Clad 6951-O	4343	16	6	30
-T6	4343	39	33	13
Clad 6951-O	4045	16	6	30
-T6	4045	39	33	13

[a] Pipe.
[b] Tube.
[c] Wire.

5   ADDITIONAL HEAT-TREATABLE ALCLAD SHEET

The availability of alloys having potentials more anodic than the potential of commercially pure aluminum (1230) permitted the coating of a wider variety of cores. Alclad 7075-T6 and Alclad 7178-T6 are representative of the heat-treatable sheet of this class and have the highest static ultimate tensile strengths.[10] Alclad 2014-T6[11] and Alclad 2219 have been made available but their use has been limited. On the other hand, Alclad 6061-T6 has been used in many applications because of versatility of the 6061-T6 core.

The ability to choose coating alloys that will electrochemically protect a specific core, the selection having been made by screening alloys by means of potential measurement (Fig. 11.8 and Table 11.1), is demonstrated by exposures to Point Judith atmosphere. The fact that the cores at the edges of the premachined specimens were not attacked demonstrates that the coating alloys were electrochemically protecting the core alloys, as had been predicted by the potential measurements. Figure 11.9 is a bar graph comparing the performance of 0.064-in.-thick sheet of 2014-T6, 6061-T6, 7075-T6, 7079-T6, and 7178-T6 with their alclad counterparts after four years' exposure to the Point Judith atmosphere. It is evident that Alclad 2014-T6, Alclad 6061-T6, Alclad 7075-T6, Alclad 7079-T6, and Alclad 7178-T6 retain their tensile strengths as well as Alclad 2024-T3, which is used as a basis of comparison. Furthermore, the losses in tensile strength for the alclad sheets were markedly less than any of the nonclad sheets.

6   STRAIN-HARDENABLE ALCLAD SHEET

Alclad sheet with cores of strain-hardenable alloys is used when resistance to perforation is one of the foremost requirements, and a selection of tensile yield strengths over a range of 6–41 ksi is possible (see Table 11.4). Figure 11.10 shows the cross sections (30×) of 3003-H14 and Alclad 3003-H14, 0.032-in. thick, after 22 months' exposure to flowing seawater. Although the corrosive attack had penetrated the alclad-coating alloy, the corrosion only penetrated to the cladding-core interface, which was to a lesser depth than in the case of the nonclad 3003 sheet.

Many applications require sheet products that are resistant to perforation. For example, certain types of thermal insulation, when moist, cause sufficient corrosion to locally perforate sheet. The condition exists also when metals are buried in moist soils. Leaves, dirt, and other debris can cause a like condition to exist in gutters and downspouts of rain-carrying equipment. It has been found convenient to term this condition of exposure "poultice action."

Even though improvements in the manufacture of thermal insulation have resulted in the reduction of the corrosivity of the insulation, it is still believed advisable to use alclad strain-hardenable sheet when contacting many of the commercial thermal insulations. A bar chart in Figure 11.11 shows the depth of attack expected upon the exposure of nonclad and alclad strain-hardenable alloys under conditions of poultice action. The greater resistance of the alclad aluminum alloys to perforation is very marked. Figure 11.11 also shows the superior resistance to perforation of alclad strain-hardenable alloys when immersed in corrosive waters such as exist in certain types of cooling towers or in seawater. Under these conditions, the marked superiority of the alclad sheet is completely demonstrated.

7   LAMINATES OF ALUMINUM ALLOYS AVAILABLE IN TUBE FORM

These are of the alclad type because there is a coating alloy layer on either the inside or outside, which is anodic to the

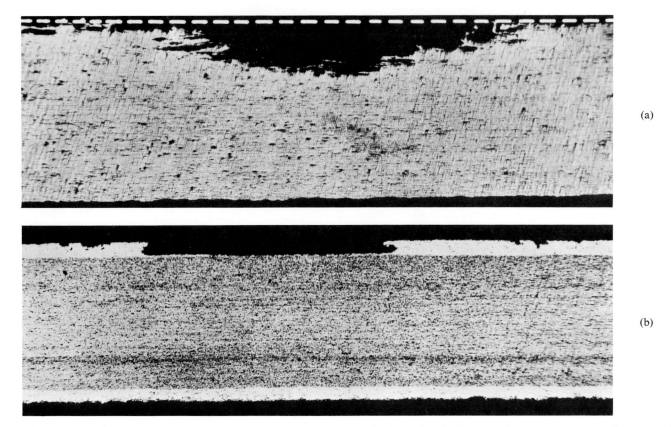

(a)

(b)

Figure 11.10   The ability of alclad products to minimize the depth of attack and resist perforation is shown in these two cross sections, one of Alclad 3003-H14 and the other of 3003-H14 exposed to flowing seawater for 22 months. The dotted white line shows the original surface of the 3003. Note the depth of attack is much deeper than in the case of the Alclad 3003 where the attack had only penetrated to the coating-core interface. Etch : HF-HCl-HNO₃ solution 15 sec. Magnification : 30×

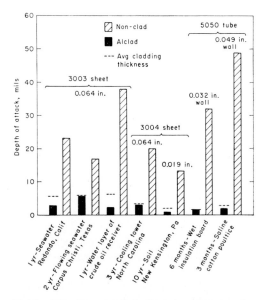

Figure 11.11   This bar graph further demonstrates the extreme resistance to perforation of alclad sheet and alclad tubing that had cores of strain-hardenable alloys. The depth of attack in the alclad products was never greater than that of the coating thickness and, in turn, was much less than that of the nonclad products. The effectiveness of the coating alloy in markedly retarding perforation was evident also for the exposure conditions that produce poultice-type action

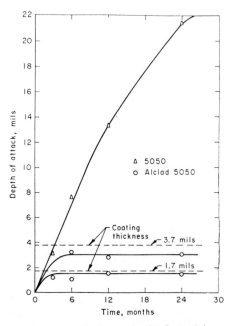

Figure 11.12   The markedly lower depth of attack in outside Alclad 5050 tubing, when compared with nonclad 5050 tubing, both being exposed to a saline environment, is shown graphically in this chart. In no case was the depth of attack in the outside Alclad 5050 tubing greater than that of the coating alloy thickness, whereas the depth of attack in the nonclad tubing was several fold greater

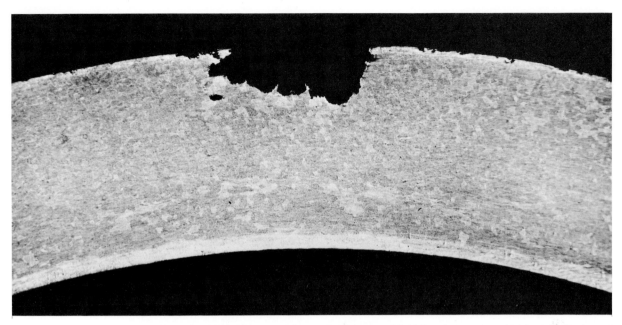

Figure 11.13 This shows a cross section of a 6063-T5 type with a wall thickness of 0.14 in. after seven years' burial in moist soil at New Kensington, Pa. Etch: HF-HCl-HNO$_3$ solution 15 sec. Magnification: 15×

core alloy of the tube. The tube products that are available in alclad form are listed in Table 11.3. This table lists the core alloy and the nominal thickness of the cladding alloy and whether it is on the outside or inside circumference of the tube.

The resistance to perforation of alclad tube is shown in Figure 11.12. In this graph the depths of attack encountered upon exposure to sodium chloride spray of 5050-0 and Alclad 5050-0 are compared. It should be noted that after exposures of 6 to 24 months, the measured depths of attack in the Alclad 5050-0 did not exceed the thickness of the 7472 alloy used to

coat the outside circumference of the Alclad 5050 tube, and the depth of attack was completely controlled by the thickness of the coating alloy. However, in the case of the nonclad 5050-0 tube, the depth of attack continued to increase after the 6-month period. The reduction in depth of attack and greater resistance to perforation resulting from the use of alclad pipe is illustrated in Figures 11.13 and 11.14. The cross sections of the alclad and nonclad pipe demonstrate the greater resistance to perforation of the Alclad 6063-T5 pipe when compared with the nonclad 6063-T5 pipe.

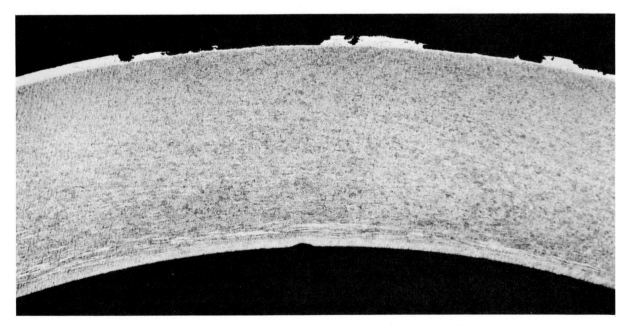

Figure 11.14 This shows a cross section of an outside Alclad 6063-T5 type with a wall thickness of 0.14 in. after seven years' exposure to moist soil at New Kensington, Pa. Note that the attack has only penetrated to the coating-core interface in the case of the outside alclad pipe, whereas the pitting penetrates to a much greater depth in the case of the 6063-T5 type shown in Figure 11.13. Etch: HF-HCl-HNO$_3$ solution 15 sec. Magnification: 15×

8   WIRE AND ROUND ROD

One type of alclad wire has been in regular production for a number of years — Alclad 5056 (Table 11.3). The main production is approximately 0.013-in. diameter and is used to fabricate woven screen cloth. The product has an alloy coating which usually accounts for about 20 per cent of the cross-sectional area of the wire. When consideration is taken of the small diameter of the wire, it is not surprising that a small amount of pitting could cause marked losses in tensile strength. Figure 11.15 compares the performance of 5056-H192 wire

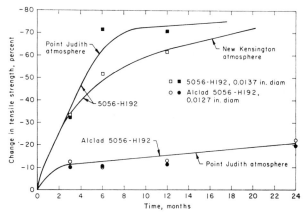

Figure 11.15   The data in this graph is for the exposure of 5056-H192 wire, 0.0137-in. diameter, and Alclad 5056-H192, 0.0127-in. diameter, to the marine atmosphere at Pt. Judith, R.I., and to the industrial atmosphere at New Kensington, Pa. These data strikingly emphasize the superior resistance to corrosion of small diameter alclad wires as compared with that of nonclad wires

having a diameter of 0.0137 in. with that of Alclad 5056-H192 having a diameter of 0.0127 in. after exposures to the marine atmosphere at Point Judith, R.I., and the industrial atmosphere at New Kensington, Pa. After an exposure of 12 months, the loss in tensile strength of the Alclad 5056-H192 wire was less than 15 per cent in both environments, whereas the loss for the nonclad 5056-H192 wire was four to fivefold greater. Even after 24 months' exposure to both environments, the loss in tensile strength of the Alclad 5056-H192 wire was only about 20 per cent, indicating that the attack had penetrated the 6253 alloy coating but had not penetrated the core. Microscopic examination of cross sections of the wire showed that the attack had reached the 5056 core and was spreading out along the wire.

9   CLAD-SHEET PRODUCTS

Previously, reference has been made in this article to laminates composed of two aluminum alloys chosen because of the mechanical characteristics of the core, with the coating alloy selected because of its specific finishing characteristics.

The Aluminum Association refers to two of these sheet products as clad reflector sheet. There are several of these clad-sheet products that are used because of their finishing characteristics, especially when suitably processed by anodic oxidation.[12] This group of clad-sheet products is listed in Table 11.3 and includes items 16, 17, 18, 19, and 20. It will

be noted that in some instances the cladding is the same alloy as the core (items 17 and 19). The term "self clad" has been applied to this product. The relatively high degree of total metalworking imparted to the surface layer results in such a finely divided metallurgical structure that when anodized, a highly uniform surface finish that has no visible pattern is produced.

Another group of clad-sheet products does not have as wide an application as the previous group but has been required for certain applications because of the chemical characteristics of the alloy used for the cladding, the core having been chosen because of its strength characteristics. Clad 6053 (Table 11.3, item 21) has been used largely because of the low copper content of the cladding alloy, enabling alkaline cleaners to be used.[13] Even though the cladding may be attacked to a small degree by the alkaline cleaner, harmful copper residues on the surface are reduced to a minimum.

Still a third group of clad sheet products are those termed brazing sheet (Table 11.3, items 22 to 27, inclusive). The coating alloy in the case of brazing sheet is chosen so that it will have a melting point lower than that of the core alloy and so that it will have sufficient fluidity to flow into recesses, crevices, and so on, and upon solidification, accomplish the joining.[12,14] The core in the case of these sheet products is chosen not only because of its mechanical characteristics but also because of its ability to resist the diffusion of the cladding alloy when in a liquid state and because it has a higher melting point than the cladding alloy. The core and cladding also must have composition relationships to each other so that any harmful effects of solid-state diffusion are minimized.

In Table 11.3, items 22, 23, 24, 25, 26, and 27 list the core and coating alloys that are available in brazing sheet along with the percentage thickness of coating that is associated with various thicknesses of brazing sheet. Table 11.4 lists the mechanical properties of brazing sheet. After brazing, clad 3003 which had been coated with 4343 would have the tensile properties approximating those of annealed 3003. In the case of brazing sheet with a core of 6951 alloy which is susceptible to heat treatment and artificial aging, the properties obtained after brazing depend upon the rate of cooling from the brazing temperature. The figures given in Table 11.4 for the T6 temper represent the properties that should be obtainable with a relatively rapid cool. For example, some actual properties obtained on 0.064-in.-thick clad 6951 that was brazed at 1120°F and air cooled to room temperature were 34 ksi tensile, 32.5 ksi yield, and 2.5 per cent elongation after aging for eight hours at 350°F.

Brazing is accomplished by employing an alloy that becomes at least partially liquid at a lower temperature than the two other pieces to be joined. In a number of instances, joining is accomplished by heating three pieces of aluminum alloys together, two of which have a higher melting point than the other. Upon placing them in the furnace, the alloy with the lower melting point becomes liquid and upon solidification, joins the other two pieces. Some of the items listed in Table 11.3 consist of brazing sheets composed of a core alloy and a brazing alloy metallurgically bonded. With such brazing sheet, more complex structures can be produced. Also, the number of pieces of sheet that have to be held in place in the brazing furnace is reduced.

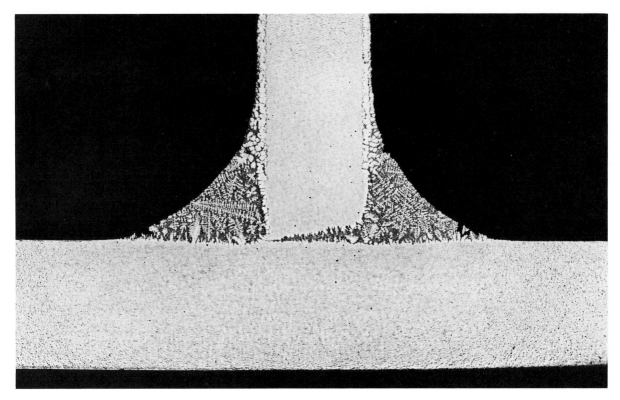

Figure 11.16 This photograph shows a tee joint made by brazing at a temperature of 1130°F. The horizontal piece was 0.064-in.-thick 3003 and the vertical piece was clad 3003 (No. 12 Brazing Sheet). The solidification of the fillet resulted after the flow of liquid Al-Si alloy (Si enriched compared with the unmelted coating) from the coating on the clad 3003. Etch: HF-HCl-HNO$_3$ solution 15 sec. Magnification: 20×

Figure 11.16 shows a cross section of a brazed joint composed of a horizontal piece of 0.064-in.-thick 3003 and a vertical piece of clad 3003 (No. 12 brazing sheet — see Table 11.3, item 23). This joint was made by brazing at 1130°F using a flux. The remainder of the cladding after melting and solidifying can be seen on the vertical portion of the clad sheet, and the cross section also reveals the fillet composed of the Al-Si alloy, much of which is the eutectic drained from the coating of the clad 3003.

10 APPLICATIONS

An important use of alclad sheet continues for air frames. Many planes over the years have used alclad sheet for skin sheet and structures formed from sheet. Experience has dictated the use of alclad sheet as part of the structure of aircraft, even to the present time. Comparison of Figure 11.17 (the Tri-Motor Ford of 1927) with Figure 11.18 (Lockheed C-141 transport of 1965) emphasizes the growth of the airplane in less than 50 years. Just as the basic principles involved

Figure 11.17 This trimotored aircraft built in 1927 and designated Ford Model 4-AT had a wing span of 74 feet, a gross weight of 10,130 pounds, and its three Wright Whirlwind reciprocating engines of 300 hp each enabled it to attain a cruising speed of approximately 100 miles per hour. Much of the sheet used in the construction of this plane was Alclad 2017-T3

Figure 11.18 This four-motored aircraft, known as the StarLifter, is one of the world's largest air freighters and is designated Lockheed Model C-141. This plane has a wing span of 159 ft $\frac{3}{4}$ in., a gross weight of 318,000 pounds, and its four turbofan Pratt & Whitney engines of 21,000 pounds thrust enable it to attain a cruising speed of approximately 500 miles per hour. Much of the sheet used in this plane was Alclad 7079-T6

in both planes are common, the basic concept of alclad sheets used in both planes is related. The fact that Alclad 2017-T3 was used in the Tri-Motor Ford and Alclad 7079-T6 in the Lockheed C-141 only serves to emphasize Dix's awareness of a need.

Alclad 6061-T6 plate has been used for tube sheets in tubular condensers and heat exchangers. The coating alloy of Alclad 6061-T6 affords some cathodic protection of the ends of the condenser tubes.

Sheet and plate of Alclad 6061-T6 are used to fabricate van containers for shipboard and highway transport. This sheet and plate product is also used for the construction of a variety of storage bins.

Figure 11.20   This photograph shows some examples of reflectors, after forming and etching, fabricated from clad 1100 sheet (see Table 11.3, Item 18)

Figure 11.19   Over 90,000 square feet of clad 1100 that were given the Duranodic 300 finish were used for the spandrels of this high-rise building in New York City. (The type of clad 1100 used is shown in Table 11.3, Item 17)

rain-carrying equipment is fabricated from Alclad 3003 because the accumulation of vegetation and atmospheric soil may cause "poultice action," as will certain types of materials of construction contacting the metal. For these reasons alclad sheet is chosen to provide an additional measure of insurance against localized perforation for such unexpected corrosion hazards.

Clad strain-hardenable sheet and plate used in the chemical industry are usually weldable. These products are used for various types of stationary and transport tankage.

Large quantities of strain-hardenable clad sheet are employed for items that are anodized, and they cover a wide range of uses, from reflectors to facings for a variety of buildings.

A very extensive application of clad sheet is shown in the 30-story high-rise building in New York City where clad 1100 is employed for the spandrels (see Table 11.3, item 17). A view of this building is shown in Figure 11.19. Figure 11.20 shows some examples of reflectors made of clad 1100 (see Table 11.3, item 18).

Alclad tubes used for heat exchangers generally employ cores of strain-hardenable alloys. The coating alloy is on the outside or inside of the tube, depending upon the design of the exchanger. If the cooling water is inside the tube, inside alclad tube is used, and the converse is true if the cooling water is on the outside. Alclad 3003 tube has been used in cooling systems for the gasoline engines of motor torpedo boats.

In recent years Alclad 7178 plate has been used as the principal material for the assembly of spheres for deep-sea buoyancy tanks.

Alclad sheets with strain-hardenable cores, such as 3003, 3004, and 3005, are used in large quantities for the highest quality industrial siding and roofing panels. The best quality

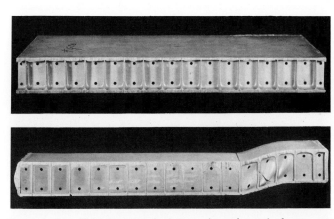

Figure 11.21   This photograph shows a brazed panel of egg-crate construction, extremely lightweight and possessing a high degree of stiffness. The panel is shown before and after testing to destruction

Figure 11.22 This is a photograph of the world's largest building (by volume) at Cape Kennedy. It will be used to assemble rockets for the U.S.A. aerospace program. Enclosing the walls are over a million square feet of prefabricated insulated aluminum panels. The exterior facing of these panels is Alclad 3004-H16

Brazing sheet was used in the construction of the extremely lightweight, high-strength and stiff panels shown in Figure 11.21. The interior of the panel is hollow except for the egg-crate type assembly which contributed to the stiffness.

Figure 11.22 shows the building at Cape Kennedy in which large multistage vehicles for space exploration can be assembled. This large building is entirely covered with Alclad 3004 sheet. This is probably the largest single application of clad or alclad sheet.

REFERENCES

1. EDGAR H. DIX, U.S. Patent 1,856,089, "Corrosion Resistant Aluminum Alloy Articles and Methods of Making Same," June 28, 1928 — filed Jan. 22, 1927.

2. E. H. DIX, "Alclad — A New Corrosion Resistant Aluminum Product," NACA Tech. Note 259, National Advisory Committee for Aeronautics, Washington, D.C., Aug. 1927.

3. E. H. DIX, "Application of Alclad Sheet to the Aircraft Industry," *Aviation* 25, 26 (Dec. 22, 1928).

4. J. D. EDWARDS and CYRIL S. TAYLOR, "Solution Potentials of Aluminum Alloys in Relation to Corrosion," *Trans. Am. Electrochem. Soc.* 56, 27 (1929).

5. E. H. DIX and R. B. MEARS, "The Resistance of Aluminum Base Alloys to Atmospheric Exposure," Symposium on Atmospheric Exposure Tests on Non-Ferrous Metals, American Society for Testing and Materials, Philadelphia, Pa., Feb. 1946.

6. F. KELLER and R. A. BOSSERT, "Revealing the Microstructure of 24S Alloy," Alcoa Research Lab. Tech. Paper 8, Aluminum Company of America, Alcoa Research Laboratories, New Kensington, Pa., 1942.

7. ROBERT H. BROWN, U.S. Patent 1,997,165, "Duplex Metal Article" (Apr. 9, 1935 — filed Oct. 20, 1933).

8. Robert H. Brown, U.S. Patent 1,979,166, "Duplex Metal Article," Apr. 9, 1935 — filed Oct. 20, 1933.

9. E. H. DIX, "Acceleration of the Rate of Corrosion by High Constant Stresses," *Trans. AIME, Inst. Metals Div.* 137, 11–40 (1940).

10. J. A. NOCK, JR., U.S. Patent 2,240,940, "Aluminum Alloy," May 6, 1944 — filed Sept. 28, 1940.

11. T. L. FRITZLEN and L. F. MONDOLFO, "R301—Reynolds New High Strength Aluminum Alloy," *Metals & Alloys* 20, No. 4, 926–933 (Oct. 1944).

12. "Standards for Aluminum Mill Products," First Edition, Aluminum Association, New York, June 1955, p. 11.1.1. "Standards for Aluminum Mill Products," Eighth Edition, Aluminum Association, New York, Sept. 1965, p. 20.

13. C. J. WALTON, R. V. VANDENBERG, and R. H. BROWN, "Alclad 53S-T6, The New Aluminum Alloy for Beer Barrels," Annual Report (1951) of the American Society of Brewing Chemists, pp. 130–140.

14. G. O. HOGLAND, U.S. Patent 2,312,039, "Duplex Metal Article," Feb. 23, 1943.

SUPPLEMENTARY READINGS

1. ROBERT H. BROWN, U.S. Patent 1,975,778, "Duplex Metal Article," Oct. 9, 1934 — filed Oct. 20, 1933.

2. ROBERT H. BROWN and LOWELL A. WILLEY, U.S. Patent 2,011,613, "Magnesium Duplex Metal," Aug. 20, 1935 — filed Oct. 6, 1934.

3. ARTHUR W. WINSTON, U.S. Patent 2,023,498, "Method of Producing Composite Wrought Forms of Magnesium Alloys," Dec. 10, 1935 — filed July 21, 1932.

4. E. H. DIX, "Superficial Corrosion Attack on the Surfaces of Alclad Sheets," discussion in Note 14, Civil Aeronaut. Authority, Tech. Development Div., Washington, D.C., Dec. 1938.

5. FREDERICK C. PYNE, "Ten Years' Service Experience with Alclad Materials in Aircraft," *J. Soc. Automotive Engrs.* 44, 221–228 (1939).

6. G. O. HOGLUND, "Brazing the Aluminum Alloys," *Am. Welding Soc. J.* 19, 123s–125s (Apr. 1940).

7. M. A. MILLER, "The Flow of Metal in Brazing Aluminum," *Am. Welding Soc. J.* 20, 472s–478s (1941).

8. G. O. HOGLUND, U.S. Patent 2,258,681, "Method of Joining," Oct. 14, 1941.

9. R. L. TEMPLIN, E. C. HARTMANN, and D. A. PAUL, "Typical Tensile and Compressive Stress Strain Curves for Aluminum Alloy 24S-T, Alclad 24S-T, 24S-RT, and Alclad 24S-RT Products," ARL Tech. Paper No. 6, Aluminum Company of America, Alcoa Research Laboratories, New Kensington, Pa., 1942.

10. M. A. MILLER, "Aluminum Brazing Sheet, Fundamentals of Metal Flow," *Am. Welding Soc. J.* 22, 596s–604s (1943).

11. E. H. DIX, "New Developments in High Strength Aluminum Alloy Products," *Trans. Am. Soc. Metals* 35, 130–155 (1944).

12. R. B. MEARS, R. H. BROWN, and E. H. DIX, "A Generalized Theory of the Stress Corrosion of Alloys," Symposium on Stress Corrosion Cracking of Metals, ASTM-AIME, American Society for Testing and Materials, Philadelphia, Pa., 1944, pp. 323–339.

13. F. KELLER and R. H. BROWN, "Diffusion in Alclad 24S-T Sheet," *Trans. AIME, Inst. Metals Div.* 156, 377–386 (1944).

14. J. A. NOCK, JR., "75S—Alcoa's New High Strength Alloy," *Metals & Alloys* 20, No. 4, 922–925 (Oct. 1944).

15. PUZANT W. BAKARIAN, U.S. Patent 2,366,168, "Bonding Magnesium Alloy Sheet," Jan. 2, 1945 — filed May 2, 1942.

16. HAROLD A. DIEHL and JOHN C. McDONALD, U.S. Patent 2,366,185, "Rolling Composite Magnesium Base Alloy Sheets," Jan. 2, 1945 — filed May 4, 1942.

17. M. A. MILLER, "New Developments in Aluminum Brazing," *Metal Progr.* 48, 477–483 (1945).

18. OWEN LEE MITCHELL, "Wrought Aluminum Alloys," Manual 12, *Materials & Methods* 23, 2 (Feb. 1946).

19. M. A. MILLER, U.S. Patent 2,602,413, "Aluminous Brazing Product and Method of Brazing," July 8, 1952.

20. M. A. MILLER, U.S. Patent 2,821,014, "Composite Aluminum Metal Article," Jan. 28, 1958.

21. R. V. VANDENBERG, "Finishes for Architectural Aluminum," *Products Finishing* (Cincinnati) 26, 36–54 (Feb. 1962).

22. J. R. TERRILL, "Diffusion of Silicon in Aluminum Brazing Sheet," *Welding J.* (N.Y.) 45, No. 5, 202s–209s (1966).

23. "Heat Treatment of Aluminum Alloys," Mil Spec. MIL-H-6088D, Government Printing Office, Washington, D.C., Mar. 24, 1965.

Chapter 12

Stainless-Steel-Clad Metals

Jack B. Morgan

The use of two metals that have been bonded together by suitable means dates back to antiquity, before the birth of Christ. Man through the ages has realized the advantages of combining the desirable properties of two metals into a composite that is superior to either of the components in one or more ways. With the development of so-called "stainless" steel in the early part of this century, a new worthwhile cladding material was made available. Stainless steel was developed because of the need for a material that would be more corrosion resistant and less likely to stain than the carbon or low alloy steels. The advantage of cladding mild steel with stainless was soon recognized as evidenced by the issuance of a patent shortly after stainless steel was developed.

1 STAINLESS-CLAD STEEL

1.1 Bonding techniques

Various methods of cladding carbon or low alloy steels with stainless have been used in the past. One of these consisted of placing two stainless steel plates that had been welded together in an ingot mold. Molten steel was then cast around the plates. The solidified mass was then hot rolled to a convenient thickness and the welded areas of the stainless plates cut away. This then provided two single clad plates, that is, two plates that consisted of stainless on one side and carbon or low alloy steel on the other. Double clad, that is stainless on either side of carbon or low alloy steel, was produced in a similar manner except that stainless plates were properly positioned in the ingot mold and molten steel poured between them.

Another method used was one in which a mixture for producing stainless steel was placed on a steel slab and the granulated materials melted and fused to the surface of the slab by electric arcs. The solidified composite was then reduced to proper thickness by rolling. A similar method used today employs electric arcs to melt the carbon steel slab face as well as the required ferroalloys and other metals to form the stainless steel. A mold is used around the upright carbon steel slab to retain the molten metal until solidification occurs. In this method, as well as the preceding one, a protective slag would normally be employed.

Perhaps the most commonly used method is one that employs a pack assembly. To produce single-clad material by this method two plates of stainless steel are placed together. A parting compound is placed between contacting faces of the plates to prevent bonding. The parting compound consists of some suitable inert material such as chromium oxide. Heavier

JACK B. MORGAN is Senior Supervising Metallurgist, Allegheny Ludlum Steel Corporation, Research Center, Brackenridge, Pa.

carbon steel plates are placed on the exposed sides of the stainless steel as can be seen in Figure 12.1. The edges of the pack are welded for two reasons. Welding tends to minimize oxidation of the surfaces to be bonded and keeps the various components in proper register during further processing. After

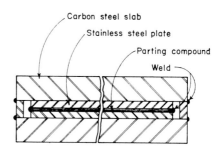

Figure 12.1 Sketch Showing Assembly Build-up for Producing Stainless-Clad Carbon Steel — Single Clad

welding, the assembly is heated and hot rolled sufficiently to achieve a bond. Once this has been done the welded edges can be removed and the resultant single-clad material further reduced to finish gage.

In the production of double-clad material by this method, stainless steel plates are placed on either side of a heavier carbon steel plate. The assembly is completely welded — usually by the use of stainless steel side bars around all four edges of the carbon steel plate — to fill the space between the stainless plates, as shown in Figure 12.2. The stainless plates are usually wider and longer than the carbon steel plate to

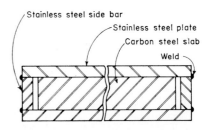

Figure 12.2 Sketch Showing Assembly Build-up for Producing Stainless-Clad Carbon Steel — Double Clad

allow space for expansion and movement during heating and rolling. Because of the relatively high chromium content of stainless steel, there is a strong tendency for chromium oxide to form on the surface. This is a severe deterrent to bonding. In an effort to prevent this chromium oxide formation, several techniques have been developed and described in patents. In one of these techniques the stainless surface to be bonded is plated or coated with iron. In another, nickel is used instead of iron. In still another, the pack is evacuated to a low residual pressure after welding to remove essentially all oxidizing atmosphere.

In order to achieve optimum bonding, it is important that all surfaces to be bonded be clean and as free from oxide as possible. Also, the assembly must be at the proper temperature for hot rolling. Sufficient reduction must be taken in the rolling step to bring the surfaces to be bonded in intimate contact and adequately break up any oxide film that may have formed.

Another technique for cladding stainless steel to carbon steel is the vacuum brazing technique. With this technique brazing alloy is placed between the stainless and carbon steel surfaces to be bonded. The assembly is welded all around the edges, evacuated and heated under vacuum to achieve bonding. The vacuum acts to create uniform contact between the cladding and base metals thus assuring continuous bonding of large areas. It also is helpful in removing any residual oxygen which may still remain or be evolved at the surfaces to be bonded. In preparing the assembly, the plates used are generally that of the finished thickness desired since no further reduction is required to achieve bonding.

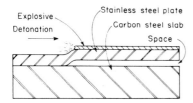

Figure 12.3   Sketch Showing Explosive Bonding Technique for Producing Stainless-Clad Carbon Steel

A different type of technique which is used to bond stainless steel to carbon steel and other metals and alloys employs explosives. Figure 12.3 is a sketch showing the arrangement used for this bonding technique. With this technique the clad metal is held at a controlled distance from the base metal and the explosive charge detonated. This brings the surfaces to be bonded into intimate contact and bonding is achieved. It is not necessary to heat the materials prior to bonding or to use any intermediate between the clad and base metals with this technique.

Numerous techniques for bonding stainless steel to carbon steel or other alloys have been briefly described. Several of these techniques are commercially employed in the production of stainless clad. No attempt has been made to closely outline the details of these techniques since this information can generally be obtained from the literature, and also the techniques are not necessarily limited to the production of stainless clad. They can be used for the production of various clad materials. This chapter is concerned only with stainless clad.

1.2   Product types

Stainless-clad steel is generally available as plates, sheets, and coils. Both single and double clad are available; however, in the case of plate single clad is more common. Double clad is more common for light gage products.

Stainless-clad steel plate is commercially available in many stainless grades. The composition of stainless steels generally used for cladding is given in Table 12.1. The backing or base steel is generally A-285 flange quality. However, other grades and quality levels are available. These are listed in Table 12.2.

The amount of cladding, expressed as a percentage of the total thickness of the plate, is available in thicknesses from 5 to 50 per cent. The most widely used thicknesses of the clad are 10 and 20 per cent.

The principal use of light-gage stainless-clad steel is for the manufacture of cookware. In this application a double-clad material is used that consists of approximately 20 per cent Type 304 stainless on either side of an SAE 1006 or 1008 carbon steel core.

Stainless-clad steel plate is available in essentially the same sizes as solid stainless. Thicknesses range from $\frac{3}{16}$ inch to several inches, widths range to 180 inches, and lengths range to 480 inches. Greater widths and lengths are possible.

1.3   Treatment and fabrication

In the handling of stainless-clad steel it is important to consider several factors. In pickling it is necessary to realize that the normal pickling solutions used for solid stainless can cause rapid attack on the carbon steel base. To prevent this, normal pickling techniques used for carbon steel can be employed; for example, a 10 per cent solution of sulfuric acid with inhibitor at a temperature of 150–180°F. A final 10 per cent nitric acid pickle can be used to whiten the stainless and passivate its surface. Sodium hydride as well as sandblasting are used for descaling.

Because of the difference in the coefficient of expansion of the two components of stainless-clad steel, distortion can occur as a result of heating and cooling. This is particularly noticeable in the case of single clad. This is not at all uncommon in bimetallic combinations. Its effect tends to be minimized in the case of double-clad material and where the material is formed into shapes such as circular vessels. The extent of this warping differs for the different stainless steels used as cladding material because the coefficient of expansion is different. For example, the coefficients of thermal expansion of the 400 series of stainless steels are much closer to carbon steel than those of the 300 series.

In annealing stainless-clad steel various temperatures are used depending primarily on the type of stainless. Typical annealing temperatures are given in Table 12.3.

The heat treatment may vary depending on (1) whether or not additional work or fabrication is to be performed, (2) the nature of any additional processing, and (3) the ultimate use of the product.

The heat treatment may be made to favor one material or the other making up the clad product in order to produce certain desired properties. Usually, the treatment is directed toward obtaining optimum corrosion-resistant and mechanical properties.

It is desirable to maintain the integrity of the stainless-steel-clad composition because of its benefits in regard to corrosion resistance and the like. Carbon migration can occur from the base metal to the clad itself; this is a function of time at an

Table 12.1. Chemical Composition of Stainless Steel Types Used as Cladding Materials

Type	Carbon,[a] max., %	Manganese, max., %	Phosphorus, max., %	Sulfur, max., %	Silicon, max., %	Chromium, %	Nickel, %	Other Elements
304	0.08	2.00	0.045	0.030	1.00	18.00 to 20.00	8.00 to 12.00	
304L	0.03	2.00	0.045	0.030	1.00	18.00 to 20.00	8.00 to 12.00	
309S	0.08	2.00	0.045	0.030	1.00	22.00 to 24.00	12.00 to 15.00	
310S	0.08	2.00	0.045	0.030	1.50	24.00 to 26.00	19.00 to 22.00	
316	0.08	2.00	0.045	0.030	1.00	16.00 to 18.00	10.00 to 14.00	Mo 2.00 to 3.00
316L	0.03	2.00	0.045	0.030	1.00	16.00 to 18.00	10.00 to 14.00	Mo 2.00 to 3.00
317	0.08	2.00	0.045	0.030	1.00	18.00 to 20.00	11.00 to 15.00	Mo 3.00 to 4.00
317L	0.03	2.00	0.045	0.030	1.00	18.00 to 20.00	11.00 to 15.00	Mo 3.00 to 4.00
321	0.08	2.00	0.045	0.030	1.00	17.00 to 19.00	9.00 to 12.00	Ti 5 × C min; 0.70 max
347	0.08	2.00	0.045	0.030	1.00	17.00 to 19.00	9.00 to 13.00	Cb + Ta 10 × C min; 1.10 max
405	0.08	1.00	0.040	0.030	1.00	11.50 to 14.50	0.60 max	Al 0.10 to 0.30
410S	0.08	1.00	0.040	0.030	1.00	11.50 to 13.50	0.60 max	
430A	0.12	1.00	0.040	0.030	1.00	14.00 to 16.00	0.75 max	

[a] The carbon analysis shall be reported to the nearest 0.01 per cent.

Table 12.2. Types of Backing Steels Used in Producing Stainless-Clad Steels

ASTM Specification		Tensile Strength, psi	Yield Point Min., psi	Elongation[a] in 8 in., Min. %	Carbon,[a] Max. % Plate to 2 in.	Mn,[a] Max. %	Ni	Mo	Cr
A-285 Flange	Grade C	55– 65,000	30,000	24	—	0.80	—	—	—
A-285 Firebox	Grade A	45– 55,000	24,000	29	0.17	0.80	—	—	—
	Grade B	50– 60,000	27,000	27	0.22	0.80	—	—	—
	Grade C	55– 65,000	30,000	25	0.30	0.80	—	—	—
A-201 Firebox	Grade A	55– 65,000	30,000	25	0.24	0.80	—	—	—
	Grade B	60– 72,000	32,000	22	0.27	0.80	—	—	—
A-212 Firebox	Grade A	65– 77,000	35,000	21	0.31	0.90	—	—	—
	Grade B	70– 85,000	38,000	19	0.33	0.90	—	—	—
A-515 Firebox	Grade 55	55– 65,000	30,000	25	0.22	0.90	—	—	—
(A516)[b]	Grade 60	60– 72,000	32,000	23	0.27	0.90	—	—	—
	Grade 65	65– 77,000	35,000	21	0.31	0.90	—	—	—
	Grade 70	70– 85,000	38,000	19	0.33	0.90	—	—	—
A-203 Firebox	Grade A	65– 77,000	37,000	21	0.17	0.80	2.00–2.75	—	—
	Grade B	70– 85,000	40,000	19	0.20	0.80	2.00–2.75	—	—
	Grade C	65– 90,000	43,000	18	0.25	0.80	2.00–2.75	—	—
A-204 Firebox	Grade A	65– 77,000	37,000	21	0.21	0.90	—	0.40–0.60	—
	Grade B	70– 85,000	40,000	19	0.23	0.90	—	0.40–0.60	—
	Grade C	75– 90,000	43,000	18	0.26	0.90	—	0.40–0.60	—
A-387 Firebox	Grade A	65– 82,000	40,000	20	0.21	0.84	—	0.40–0.70	0.46–0.79
	Grade B	60– 82,000	35,000	21	0.17	0.69	—	0.40–0.70	0.75–1.31
	Grade C	60– 85,000	35,000	21	0.17	0.69	—	0.40–0.70	0.94–1.56
A-302 Firebox	Grade A	75– 95,000	45,000	17	0.23	1.35	—	0.41–0.64	—
	Grade B	80–100,000	50,000	17	0.23	1.55	—	0.41–0.64	—

Source: Lukens Steel Company.

[a] These values can vary with gage; refer to specification for full details.

[b] Fine Grain.

Table 12.3.   Typical Mill Heat Treatments of Clad Steels[a]

Stainless Steel Type	Type of Backing Material	Heat Treatment	Procedure
405, 410	A-285 A-201 A-212 A-515 A-204 A-387	First Anneal	Heat to 1625–1675°F. Hold 1 hr per in. of thickness. Air quench.
		Second Anneal	Heat to 1250–1300°F. Hold 1 hr per in. of thickness. Air quench.
430	A-285 A-201 A-212 A-515 A-204 A-387	First Anneal	Heat to 1625–1675°F. Hold 1 hr per in. of thickness. Air quench.
		Second Anneal	Heat to 1425–1475°F. Hold 1 hr per in. of thickness. Air quench.
304, 304L, 309, 310, 316, 316Cb, 316L, 317, 321, 347	A-285 A-201 up to A-212 2 in. gage A-515	Anneal	Heat to 1950°F min. (Range 1950–2150°F). Air quench.
(See Ftn. b) 304L, 316L, 316Cb 317L, 321, 347	A-201 over 2 in. A-212 gage A-515	Anneal	Heat to 1950°F min. (Range 1950–2150°F). Air quench.
		Normalize	Heat to 1600–1650°F. Hold 1 hr per in. of thickness. Air quench.
304, 304L, 309, 310, 316, 316Cb, 316L, 317, 317L, 321, 347	A-204 up to A-302 2 in. gage	Anneal	Heat to 1950°F min. (Range 1950–2150°F). Air quench.
(See Ftn. b) 304L, 316L, 316Cb 317L, 321, 347	A-204 over 2 in. A-302 gage A-387 all gages	Anneal	Heat to 1950°F min. (Range 1950–2150°F). Air quench.
		Normalize	Heat to 1600–1650°F. Hold 1 hr per in. of thickness. Air quench.

Source: Lukens Steel Company.

[a] The heat treatments listed here are generally correct for the material combinations shown. To meet specific requirements, some deviations may be made.

[b] Stabilized or low-carbon types of stainless steel should be used when this type of heat treatment is involved.

elevated temperature. In normal processing to produce the clad steel itself, little migration occurs because of the short time during which the clad is in an elevated temperature range at which significant migration occurs. In describing the various techniques for manufacturing clad steels, it was mentioned that one technique used a nickel plate on the stainless steel face to prevent oxide formation. The nickel also can serve as a barrier to carbon migration.

When working with stainless-clad steels in which the cladding is composed of one of the 300 series of stainless steels, that is, the chromium-nickel types, it is important that the same precautions be taken as would be taken with the solid metal to assure maximum corrosion resistance. If the steels are heated in the 800 to 1500°F range, it is important that they be fully annealed. If not, rapid intergranular attack can occur under corrosive conditions because of the presence of carbides formed during heating in the above range.

Fabrication of stainless-clad steel can usually be performed in a manner similar to carbon steel, that is, it can be hot or cold worked or formed, and bent or cut much the same as carbon steel. In shearing single-clad plate, it is desirable to shear with the stainless side up. In oxygen flame cutting, the cut should start from the backing steel side. Oxygen pressure should be low.

In welding, stainless-clad steel is usually beveled or grooved and the carbon steel side welded first. The stainless side is then grooved and welded. It is common practice to use a welding rod richer in the alloying elements than the original clad. This is done so that if dilution tends to occur, the higher percentages of the elements in the filler metal serve to maintain the high alloy composition of the clad — thus assuring the corrosion resistance and other properties associated with stainless steel. Annealing after welding may be required for reasons given above, that is, carbide formation. Where annealing is not possible, stainless clad containing columbium or titanium should be used. The filler metal used in the weld likewise should contain one of these elements. Welding may be done by the gas shield-arc or submerged-arc processes.

1.4 Uses and properties

Stainless-clad carbon steel has found use for several reasons. Because of the relatively heavy alloying required to make a steel stainless, it is necessarily more expensive. It would then appear that it might be advantageous from a cost standpoint to use a clad product that consists of only 10 or 20 per cent stainless steel. Actually, the saving as compared to solid stainless is not as great as might be expected. This results from the additional cost involved because of the cladding process itself. Nevertheless, there are savings that could justify its use. The amount of savings depends on a number of factors. These include

1. Type of stainless used;
2. Type and quality of backing steel used;
3. Percentage thickness of stainless clad.

Of course, there are other factors to consider. One of these is that the choice of backing steel can provide a range of mechanical properties. This may permit a design capable of utilizing the higher strengths available.

A considerable quantity of stainless-clad carbon steel is used in the production of cooking utensils. It finds use in this area because it has heat-transfer characteristics which are superior to solid stainless. Double clad is used in this application since the corrosion resistance and good appearance offered by the stainless steel are desirable for both exposed surfaces. The carbon steel core acts to spread heat horizontally.

Stainless-clad steel plates find use in a number of major industries such as chemical processing, oil refining, and food processing. Stainless-clad steel may be used in pressure vessels where its use is certainly utilitarian, or it may be used for aesthetic value where the fine appearance of the stainless is seen from the outside.

Stainless-clad steels could conceivably be used wherever the advantages of stainless steel (for example, its good corrosion resistance) are important. However, one must carefully weigh the disadvantages and the advantages of stainless-clad steel over solid stainless steel. As previously pointed out, there are disadvantages in the use of stainless-clad steel.

Mechanical properties of stainless-clad steel are generally related to the properties of the individual components. The influence of the stainless steel with respect to strength, for example, depends on the thickness of the cladding as a percentage of the total thickness. It can generally be assumed that the minimum strength is represented by that of the core or backing steel.

Bond strength of commercial stainless-clad steel exceeds the 20,000 psi requirement of ASTM Specifications A-263 and A-264. It generally is approximately twice this value.

Well-bonded material can be bent with the stainless in tension or compression without adverse effect. The radius of the object over which the bend is made is generally either $1T$ or $2T$ ($T =$ thickness) depending on the thickness of the clad product.

2 STAINLESS-CLAD ALUMINUM

Stainless steel is used as a cladding for aluminum. This presents an interesting combination of metals. It combines the cleanability, stain-resistance, strength, and toughness of stainless steel with the lightness and excellent heat-transfer characteristics of aluminum.

2.1 Bonding technique

This composite is generally produced in plate, sheet, or coil form by (1) cleaning and scratch brushing the surface to be bonded, (2) heating, and (3) rolling. The thicknesses of the stainless steel and aluminum are chosen so that after the reduction by rolling, which is required to achieve bonding, they are at the desired final thickness. This is important since further rolling, once bonding has been achieved, results in equal coextension of the two metals with work hardening of the stainless. When this occurs, maximum softness and ductility cannot be restored because of the differences in the properties of the two metals making up the composite. The annealing temperature required for the stainless exceeds the melting point of the aluminum.

2.2 Treatment and fabrication

Stainless-clad aluminum can be deep drawn. It can also be roll formed, blanked, sheared, and sawed. Finishing the surfaces presents no unusual problems: Normal techniques for the solid metals (abrasive belts or wheels, buffing, and the like) can be used.

In subjecting the clad to heat, it is important to know that a brittle intermetallic compound forms if the heat is excessive. This intermetallic compound, $FeAl_3$, forms rapidly above 1000°F.

The coefficients of thermal expansion of stainless steel and aluminum are significantly different. Therefore, warping can occur as a result of heating and cooling. As was mentioned in the case of stainless-clad steel, this is more noticeable for single clad than double clad. Its effect tends to be minimized where the material is formed into shapes such as circular vessels.

If a heat tint is encountered on the stainless surface, it can be removed by use of a 10 per cent nitric acid solution heated to approximately 130°F. A light heat tint can also be removed by a buffing operation.

Welding of stainless-clad aluminum presents problems. In order to get a satisfactory weld in the stainless portion, it is important that no aluminum enter the molten steel pool. Underlying aluminum is removed before the stainless is welded. Once the stainless has been welded the remaining groove is filled to produce the aluminum weld. Because of the relatively high temperature required to weld the stainless and the formation of a brittle intermetallic compound above 1000°F, it is difficult, if not impossible, to prevent some bond deterioration from welding. For these reasons brazing or soldering are to be preferred because of the lower temperatures required.

2.3 Uses and properties

The principal use of stainless-clad aluminum is in the manufacture of cooking utensils. It lends itself well to this application because it can be successfully deep drawn, it offers a readily cleanable and stain-resistant surface and it has excellent heat-transfer properties. Figure 12.4 shows the heat-transfer characteristics of this material in relation to those of other

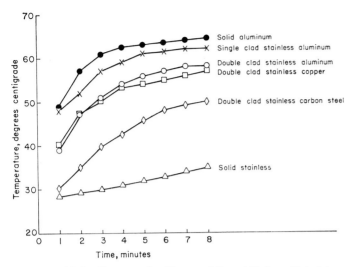

Figure 12.4   Heat-transfer Characteristics of Various Materials

materials. The data given in Figure 12.4 were obtained by placing a heat source in the center of a sheet specimen and measuring the temperature at a corner of the specimen as a function of time. All specimens were of the same nominal thickness. The thickness of the cladding (per side) varied from approximately 20 per cent in the case of the carbon steel core to 25 per cent in the case of the aluminum core and approximately 35 per cent in the case of the copper core. From the figure it can be seen that the stainless-clad aluminum exhibits the best heat-transfer properties of all the clads included in this work. Single clad is better than double clad in this respect. Stainless-clad copper is almost equal to the double-clad aluminum. It should be noted that the thickness of the stainless layer is greater in the case of the copper core. If the stainless-clad thickness were reduced and the amount of copper increased, improved properties could be expected. This reasoning has been substantiated in work done with clad materials having cores of various thicknesses. Also included in Figure 12.4 is stainless-clad steel, which showed properties significantly better than solid stainless.

The strength of stainless-clad aluminum varies as a function of the thickness of the stainless clad. Figure 12.5 shows that

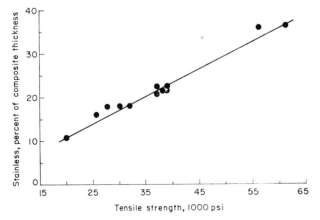

Figure 12.5   Graph Showing Tensile Strength of Stainless-Clad Aluminum as a Function of the Percentage of Stainless in the Composite

the tensile strength approaches that of aluminum as the thickness of the stainless clad is reduced. Yield strength follows a similar trend. Elongation values have been found to range from something less than normal for annealed aluminum to approximately that for annealed stainless. Elongations have been found to be greater for plate than for sheet. The effect of some work hardening which apparently occurred as a result of the bonding operation is evident. Elongation values measured from the stainless side are somewhat smaller than those measured from the aluminum side of the composite.

Elastic modulus values of stainless-clad aluminum are higher than that of solid aluminum and approach the value for solid stainless as the amount of stainless in the composite increases.

Bending characteristics of stainless-clad aluminum are generally good. It can be bent 180° around a $1T$ radius or less with either the stainless or aluminum on the outside. Material as thick as one inch has been subjected to such tests.

Results of impact tests made in the temperature range of −320°F to +320°F show the material to have fairly uniform properties in this range. Values of 60 foot-pounds have been obtained for $\frac{3}{8}$-in.-thick plate. Thinner material in which substandard specimens had to be used gave lower values. The values appear favorable for cryogenic application.

Figure 12.6   Photomicrograph Showing Stainless Steel–Aluminum Interface of Clad. Magnification: 500×

Figure 12.6 is a high-magnification photomicrograph of a section of stainless-clad aluminum showing the interface between the stainless steel and aluminum. It can be seen that it is essentially free of oxide. This is typical of well-bonded material.

In bond evaluation several tests have been employed. Peel tests in which 1-in.-wide specimens are used can be expected to give values of 100 pounds or more for the force required to cause separation. Shear tests across the bonded face of the composite yield values approaching that of the shear strength of solid aluminum. In conducting the test on a well-bonded specimen severe necking down of the aluminum occurs.

It would appear that this clad material might find future use in the electrical field. The good electrical properties of aluminum combined with the properties of stainless steel should provide a useful new product for this field.

3 STAINLESS-CLAD COPPER

Stainless-clad copper possesses several of the attributes of stainless-clad aluminum. The stainless offers cleanability, stain-resistance, strength, and toughness. The copper offers excellent heat transfer characteristics. Because of this, stainless-clad copper finds uses similar to stainless-clad aluminum.

3.1 Bonding techniques

Stainless-clad copper has been made by several techniques described previously when discussing the other stainless clads. One technique used is the "sandwich technique" in which stainless steel plates are placed on either side of a copper plate and the assembly welded together by use of stainless steel side bars around all four edges of the copper plate. Space between the side bars and copper plate is left to allow for expansion during heating and rolling. Experience shows that the copper must have a low oxygen content if satisfactory bonding is to be achieved. As previously noted, there is a strong tendency for chromium oxide to form on the stainless surface if oxygen is available. In addition to using a low-oxygen-content copper, the welded pack is evacuated to a low residual pressure.

If the stainless-clad copper is to be deep drawn, it is important that the grain size of the copper core be relatively fine. Otherwise, "orange peel," an undesirable surface irregularity, occurs on the drawn article. Severe orange peel may not be removable by conditioning, thus making the article useless. Proper control of grain size can be difficult because of the hot-rolling temperatures and annealing required in processing the composite. There are several elements that can be used to control grain size. Economic considerations exclude some of these.

Figure 12.7 Photomicrograph showing Stainless Steel–Copper Interface of Clad. Magnifications: 500 ×

Figure 12.7 is a photomicrograph of a section of stainless-clad copper showing the interface between the stainless steel and copper.

3.2 Treatment and fabrication

In discussing the other stainless clads, it was pointed out that distortion occurs upon heating as a result of the difference in the coefficients of thermal expansion of the two components. In the case of stainless-clad copper in which a chromium-nickel stainless is used, there is little difference in the coefficients so this is no problem.

Stainless-clad copper can be fabricated readily. It can be roll formed, deep drawn, sheared, and sawed. Welding presents problems. Inclusion of copper in the pool of molten steel results in an unsatisfactory and unsound weld. It is necessary to remove the copper for an appreciable distance away from the weld area.

3.3 Uses and properties

Stainless-clad copper is used in the manufacture of cooking utensils because of its heat-transfer properties as well as its easy-to-clean, stain-resistant characteristics. There has been interest in this product in the electrical field not only because of the good electrical properties of the copper core but also because of the heat-transfer characteristics.

Several stainless-clad materials have been described. They have found a useful place in our world of today, mainly because they possess unique properties obtained from the combination of two metals. It is expected that as our ingenuity is further tapped, new important uses for the clads described here will be found and new stainless clads will be developed in the future.

PATENTS

The following patents, expired and current, pertain to the production of stainless-clad steel, aluminum, or copper. They may cover other clad materials as well. It is not expected to be a complete listing.

Gillespie, 1,306,690 (1919)
Kelley, 1,365,499 (1921)
Andrus, 1,680, 276 (1928)
Johnson, 1,690,684 (1928)
Landgraf, 1,709,729 (1929)
Armstrong, 1,757,790 (1930)
Armstrong, 1,781,490 (1930)
Johnson, 1,824,898 (1931)
Armstrong, 4,826,860 (1931)
Andrus, 1,840,305 (1932)
Fifield, 1,864,590 (1932)
Ingersoll, 1,868,749 (1932)
Duff, 1,883,630 (1932)
Johnson, 1,886,615 (1932)
Maskrey, 1,896,411 (1933)
Case, 1,909, 952 (1933)
Swearingen, 1,920,534 (1933)
Ingersoll, 1,963,745 (1934)
Ingersoll, 1,967,754 (1934)
Ingersoll, 1,983,760 (1934)
Armstrong, 1,997,538 (1935)
Andrus, 2,015,173 (1935)
Howard, 2,056,673 (1936)

Andrus, 2,219,352 (1940)
Kerr, 2,219,957 (1940)
Huston, 2,225,868 (1940)
Hopkins, 2,226,403 (1940)
Chace, 2,226,695 (1940)
Chace, 2,235,200 (1941)
Armstrong, 2,241,572 (1941)
Medsker, 2,253,526 (1941)
Chace, 2,268,566 (1942)
Deutsch, 2,269,523 (1942)
Brown, 2,275,503 (1942)
Dodson & Newman, 2,288,184 (1942)
Young, 2,308,288 (1943)
Chace, 2,325,659 (1943)
British, 566,916 (1945)
Jones, British, 568,786 (1945)
Dyar, 2,414,511 (1947)
Liebowitz, 2,438,759 (1948)
Keene & Carlson, 2,468,206 (1949)
Kinney, 2,473,712 (1949)
Lynch, 2,539,248 (1951)

Johnson, 2,059,584 (1936)
Armstrong, 2,074,352 (1937)
Meier, 2,091,871 (1937)
Ingersoll, 2,094,538 (1937)
Hopkins, 2,107,943 (1938)
Gordon, 2,133,291 (1938)
Gordon, 2,133,292 (1938)
Gordon, 2,133,293 (1938)
Gordon, 2,133,294 (1938)
Huston *et al*, 2,147,407 (1939)
Hopkins, 2,151,914 (1939)
Moses, 2,164,074 (1939)
Merritt & Keller, 2,171,040 (1939)
Kinkead, 2,175,606 (1939)
Kinkead, 2,175,607 (1939)
Hopkins, 2,191,469 (1940)
Hopkins, 2,191,470 (1940)
Hopkins, 2,191,471 (1940)
Hopkins, 2,191,472 (1940)
Hopkins, 2,191,474 (1940)
Hopkins, 2,191,475 (1940)
Hopkins, 2,191,476 (1940)
Hopkins, 2,191,477 (1940)
Hopkins, 2,191,478 (1940)
Hopkins, 2,191,479 (1940)
Hopkins, 2,191,480 (1940)
Hopkins, 2,191,481 (1940)
Hopkins, 2,191,482 (1940)
Ostendorf, 2,191,321 (1940)
Chace, 2,211,922 (1940)
Trainer & Hodge, 2,214,002 (1940)

Kinney, 2,558,093 (1951)
Boessenkool & Durst, 2,691,815 (1954)
Hamilton, Pakhola & Smith, 2,704,883 (1955)
Brown, 2,713,196 (1955)
Ulam, 2,718,690 (1955)
Fromson, 2,728,136 (1955)
Kinney, 2,744,314 (1956)
Boessenkool *et al*, 2,753,623 (1956)
Chace, 2,757,444 (1956)
Ulam, 2,758,368 (1956)
Siegel, 2,767,467 (1956)
Campbell, 2,782,497 (1957)
Fayles, 2,786,266 (1957)
Pflumm & Rogers, 2,834,102 (1958)
Grovannucci *et al*, 2,845,698 (1958)
Boessenkool *et al*, 2,860,409 (1958)
Althouse, 2,932,866 (1960)
Chyle, 2,975,513 (1961)
Ulam, 3,050,834 (1962)
Gould *et al*, 3,078,563 (1963)
Allen, 3,133,346 (1964)
Cowan *et al*, 3,137,937 (1964)
Popoff, 3,140,537 (1964)
Holtzman, 3,140,539 (1964)

SUPPLEMENTARY READINGS

Articles

1. Z. W. B. KEELAR, "Welding and Cutting of Stainless Clad Steel," *Welding J.* **18**, 723–725 (Nov. 1939).
2. T. W. LIPPERT, "Armored Steel," *Iron Age* **147**, 35–45 (Mar. 6, 1941).
3. S. L. HOYT, "Stainless Clad Steels," *Metal Progr.* **41**, 51–53 (Jan. 1942).
4. A. B. KINZEL, "Solid Phase Welding," *Welding J.* **23**, 1124–1143 (Dec. 1944).
5. G. DURST, "A Few Observations on Solid Phase Bonding," *Metal Progr.* **51**, 97 (Jan. 1947).
6. L. W. TOWNSEND, "Diversified Applications of Stainless Clad Steel," *Steel* **123**, 95–96, 124, 126 (Sept. 6, 1948).
7. V. W. COOKE and A. LEVY, "Solid Phase Bonding of Aluminum Alloy to Steel," *J. Metals* **1**, 28–35 (Nov. 1949).
8. S. SIEGEL, "Modern Composite Metals," *J. Metals* **2**, 916–920 (July 1950).
9. W. G. THEISINGER, "Origin and Use of Clad Steel Plate," *Metal Progr.* **59**, 671–672, 708–716 (May 1951).
10. R. C. Bertossa, "High Strength Vacuum Brazing of Clad Steels," *Welding J.* (Research Suppl. **31**, 411s–447s (Oct. 1952).
11. "Proper Procedure Eases Welding of Stainless Clad Steels," *Industry and Welding* **25**, 77–78, 80, 82 (Dec. 1952).
12. T. T. WATSON, "The Manufacture and Properties of Clad Steel Plates, "*Blast Furnace and Steel Plant* **41**, 318–320, 326–327, 351, 354–355 (Mar. 1953).
13. J. M. PARKS, "Recrystallization Welding," *Welding J.* (Research Suppl.) **32**, 209S–222S (May 1953).
14. W. L. KEENS, "Clad Metals and Clad Metal Processing," *Steel Processing* **39**, 647–653 (Dec. 1953).
15. J. TEINDL, "Cladding of Thin Sheets with Stainless Steels" (in Czech), *Hutnik* **5**, No. 7, 204–205 (July 1955).
16. "Stainless Plated Plates" (in English), *Aciers Fins et Speciaux Francais*, No. 24, 73–75 (Dec. 1956).
17. J. HINDE, "The Welding of Clad Steels," *Welding and Metal Fabrication*, 59–62 (Feb. 1956).
18. A. FRANZEN, "Clad Metal Plate — Manufacturing and Use in Chemical Industry (in Norwegian), *Teknisk Ukeblad* **104**, 211–218 (Mar. 14, 1957).
19. W. H. FUNK, "Welding Clad Steel," *Iron and Steel Eng.* **33**, 104–107 (May 1956).
20. DAVID T. SMITH, "Savings in Stainless Clad Steel," *Chicago Purchasor*, January 1958, 63–67 (Jan. 1958).
21. G. DURST, "A New Development in Metal Cladding," *J. Metals* **8**, 328–333 (Mar. 1956).
22. M. ZIDEK, B. GLATZ, and K. GOTTWALDA (Czech), "Rolling Thick Steel Sheets Clad on One Side with Stainless Steels" (in Czech), *Hutnicke Listy* **13**, No. 8, 679–687 (1958).
23. W. RADEKER, "Metallurigcal Requirements for Welding Clad Steel Plates" (in German), *Schweissen und Schneiden*, **10**, 255–259 (July 1958).
24. L. CAPEL and C. NEDERVEEN, "Use and Application of Clad Steel" (in English), *Lastechniek* **24**, 136–149 (July 1958).
25. H. GERBEAUX, "Manufacture and Welding of Clad Steel Plates" (in French), *Soudage et Techniques Connexes* **12**, 441–459 (Nov.-Dec. 1958).
26. C. E. BOWMAN and T. J. DOLAN, T. & A. M. Report 164, Department of Theoretical and Applied Mechanics, University of Illinois, Urbana, Ill., May 1960.
27. "Clad Steels, Growing in Use, Offer Combined Benefits of Two Metals," *Steel Facts* **161**, 4 (June 1960).
28. K. BOIN, "Plated Sheets for High-Duty Applications" (in German), Paper from *Dechema Monographien* (Deutsche Gesellschaft für Chemisches Apparatewesen, E.V., Frankfurt am Main, Germany) **45**, No. 734–760, 65–76 (1962).
29. C. L. KOBRIN, "New Stainless-Clad Aluminum Aims for Varied Markets," *Iron Age* **189**, 97–99 (Mar. 29, 1962).
30. "Uses of Rolls Clad with Stainless Steel," *Engineering* **193**, 306 (Mar. 2, 1962).
31. W. G. CLARK, "Stainless Vessels; Solid Vs. Clad; Cost File," *Chem. Eng.* **69**, 188 (Apr. 16, 1962).
32. G. E. LINNERT and K. L. CROOKS, "Method; Key to Stainless Digester Overlay," *Welding Engr.* **47**, 62 (Nov. 1962).
33. W. M. ROGERSON and J. K. WAREHAM, "Working Stainless Steel Clad Aluminum Sheet," *American Machinist/Metalworking Manufacturing*, 80–84 (Nov. 25, 1963).
34. "Stainless Steel Cladding with Strip Electrode," *Engineer* **216**, 1039 (Dec. 29, 1963).
35. D. FISHLOCK, "Using Explosives to Bond the Incompatibles," *Metalworking Production* **108**, No. 11, 86–87 (Mar. 11, 1964).
36. "Weld Overlay Protects Vessels," *Iron Age* **194**, 82 (Aug. 13, 1964).
37. A. POCALYKO and C. P. WILLIAMS, "Clad Plate Products by Explosion Bonding," *Welding J.* **43**, 854–861 (Oct. 1964).
38. "Weld Overlay Solves Corrosion Problem," *Steel* **155**, 114 (Nov. 2, 1964).
39. E. C. VICARS, "An Efficient Joining Technique — Diffusion Bonding," *Metal Progr.* **87**, 125–130 (Apr. 1965).
40. J. J. OBRZUT, "Diffusion Welding: The Quest for Perfect Joints," *Iron Age* **195**, 95–102 (Apr. 15, 1965).
41. "Clad Metals, "*Welding Handbook*, Fourth Edition, American Welding Society, New York, Section 5, Chapter 95.

Specifications

1. Specification for Corrosion-Resisting Chromium Steel Clad Plate, Sheet and Strip, Specification A263–66, *Book of ASTM Standards*, American Society for Testing and Materials, Philadelphia, Pa., 1968, Part 3, pp. 231–242.
2. Specification for Stainless Chromium-Nickel Steel Clad Plate, Sheet and Strip, Specification A264–66, *Book of ASTM Standards*, American Society for Testing and Materials, Philadelphia, Pa., 1968, Part 3, 243–254.
3. Other ASTM Specifications referred to in this chapter can be found in Parts 3 and 4 of the 1968 *Book of ASTM Standards*, American Society for Testing and Materials, Philadelphia, Pa.

Chapter 13

Nonferrous-Clad Plate Steels

Albert Hoersch, Jr.

1 FOREWORD

The nonferrous-clad plate steels included in this chapter are

Nickel and low-carbon nickel clad,
Monel-clad,
Inconel-clad,
Copper-clad,
Cupro-Nickel clads (70/30 alloy and 90/10 alloy).

These materials are integral clad-steel plates, which means that the high-alloy component and backing steel are metallurgically bonded over their entire mating surfaces.

The first commercial integral clad-steel plate was nickel clad. Developed in 1930 by Lukens Steel Company, it was used in the construction of a tank car for the transportation of concentrated caustic. Production of other types of nonferrous-clad steels followed as markets for these products opened up.

Most of these clad steels are produced by hot rolling a specially prepared clad pack or "sandwich." The sandwich consists of a pair of backing steel slabs and a pair of alloy inserts (Fig. 13.1) in which the surfaces to be bonded are mechanically cleaned.

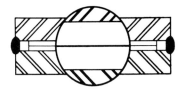

Figure 13.1   Cross Section of the Clad Pack

In addition, nickel plating is applied to the inserts, to the backing steel slabs, or to both. The inserts are made slightly smaller than the steel slabs to allow the placement of spacer bars. Further, the two surfaces of the inserts at the center of the pack must be coated with a parting compound to prevent them from bonding together. Welding around the circumference with a deeply penetrated submerged arc weld completes the preparation of the clad pack. From here, treated as an ingot, it is heated to about 2300°F and rolled to twice the thickness of an individual clad plate. After cooling, the sandwich is cut around the edges and separated into two clad-steel plates. Cleaning by blasting completes the process.

In the early 1950's, braze-bonded clad steels were developed to commercial importance. Pressure is required to hold the cladding and backing steel tightly together during the brazing

ALBERT HOERSCH, JR., is Technical Service Engineer, Lukens Steel Company, Coatesville, Pa.

operation. In some processes, the pressure is applied mechanically with a press or with weights. In another process, vacuum is applied to a clad pack and the resulting atmospheric pressure holds the components together.

The overlay welding method of cladding has recently gained commercial acceptance. Automatic welding methods are usually used to deposit two or more layers of cladding material. The development of modern welding processes that allow controlled dilution and welding materials with proper chemical composition is largely responsible for making this process feasible.

Explosion-bonded clad steels are the latest addition to this clad family. They are made by utilizing explosive energy to bond the cladding to the steel without the introduction of intermediate agents or the application of heat. The bond so produced is complete and metallurgical in nature.

Although the remainder of this chapter concerns hot roll bonded clads, the principles described apply also to clad steels made by other processes. All of these materials are integral clad steels, which means that they are metallurgically bonded over 100 per cent of the mating surfaces of the cladding and steel.

2 BOND STRENGTH

ASME and ASTM specifications are tested as shown in Figure 13.2. The shear strength of hot roll bonded nickel-clad, Monel-clad and Inconel-clad exceeds the 20,000-psi minimum required by the ASME specification and approaches the shear strength of the backing steel. Examination of the broken shear specimens of nickel, Monel, and Inconel usually show some backing steel sticking to the clad "nubbin." In copper-clad and cupro-nickel-clad, the fracture is through the copper or cupro-nickel. Typical shear strength values are shown in Table 13.1.

Figure 13.2   Arrangement for Measuring Shear Stress of Integrally Bonded Clad-Steel Plates According to the ASME Boiler Code and ASTM Specifications

Table 13.1. Typical Shear Strength Values of the Bond Obtained by Tests on Production Plates

Type	Gage	Percentage Clad	Shear Strength (psi)
Nickel	¾ in.	10	46,300
Monel	1 in.	10	48,000
Inconel	½ in.	20	48,700
Copper	½ in.	20	25,500

The strength of the bond can also be tested with bend tests with the cladding in compression or in tension, as shown in Figure 13.3.

3 BACKING STEELS

The nonferrous clads are commonly supplied with carbon steel backing plates conforming to an ASME or ASTM pressure-vessel or structural-steel specification. Low-alloy steels are available for nickel and nickel alloy clads for low-temperature or high-temperature service.

4 NICKEL AND HIGH NICKEL ALLOYS (TABLE 13.2)

Nickel, which is commercially 99 per cent pure, finds considerable use in clad form in the alkali, rayon, soap, and the process industries for equipment handling hot concentrated caustic. It has become practically the standard material of construction for evaporators used in the production of caustic. Other applications include pressure vessels and tanks for chemicals where product purity must be maintained. Nickel equipment is an excellent choice for the food and cosmetics industries, because nickel does not add any off-tastes or coloration to these products.

Low-carbon nickel (Table 13.2), a special grade of nickel in which the carbon content is less than 0.02 per cent, is employed in higher-temperature applications. Fused caustic reactors are often constructed of low-carbon nickel-clad steel. Pots for containing molten metallurgical heat-treating salts such as potassium-sodium, nitrate-nitrite mixtures are also made of this material.

Monel (Table 13.2) is often referred to as a seagoing metal because of its resistance to sodium chloride. This property

Figure 13.3 Bend Test Specimens in Tension (A) and In Compression (B)

makes Monel-clad steel useful for heat-transfer equipment used in coastal areas. Examples include the evaporators serving saline water plants and the tube sheets of heat exchangers used in power stations, chemical plants, and oil refineries. Monel-clad steel is also used to handle the brines from salt wells of inland caustic-chlorine plants. Many nonoxidizing mineral acids and their corresponding salts in some temperature and concentration ranges may be contained in Monel-clad steel equipment. Notable examples are dilute sulfuric acid, cold dilute hydrochloric acid, hydrofluoric acid, and phosphoric acid. Typical oxidizing corrosives to be avoided are nitric acid and ferric chloride.

Inconel (Table 13.2), a nickel-base alloy containing nominally 16 per cent chromium and 8 per cent iron, possesses the corrosion resistance of nickel and Monel plus properties

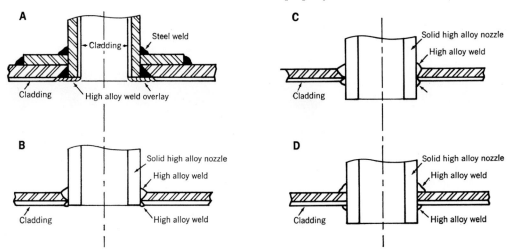

Figure 13.4 Typical Designs of Openings in Clad Vessels to Provide Continuity of Cladding

Table 13.2. Chemical Analysis Range of Cladding Materials

	Popular Name	ASTM Type	C max. %	Mn max. %	S max. %	Si max. %	Cr %	Ni[a] %	Cu max. %	Fe %
Nickel and nickel base alloys	Low Carbon Nickel Alloy 201	A 265 Nickel-Low Carbon B162	0.02	0.35	0.01	0.35	—	99.0 min.	0.25	0.40 max.
	Nickel Alloy 200	Nickel B162	0.15	0.35	0.01	0.35	—	99.0 min.	0.25	0.40 max.
	Monel Alloy 400	Nickel-Copper Alloy B127	0.3	1.25	0.024	0.5	—	63.0 to 70.0	Remainder	2.5 max.
	Inconel Alloy 600	Nickel-Chromium-Iron Alloy B168	0.15	1.0	0.015	0.5	14.0 to 17.0	72.0 min.	0.5	6.0 to 10.0

	Popular Name	ASTM Type	Copper[b] % min.	Nickel %	Manganese % max.	Lead % max.	Iron %	Zinc % max.
Cupro-Nickel alloys	70/30 Cupro-Nickel Copper Alloy No. 715 Copper Nickel 30%	B171 70/30 Copper-Nickel-Alloy	65.0	29.0 to 33.0	1.0	0.05	0.70 max.	1.0
	90/10 Cupro-Nickel Copper Alloy No. 706 Copper Nickel 10%	B171 90/10 Copper-Nickel-Alloy	86.5	9.0 to 11.0	1.0	0.05	0.5 to 2.0	1.0

	Popular Name	ASTM Type	Copper %	Phosphorus %
Copper	Oxygen-free without residual deoxidants	B152 102	99.92 min.[c]	—
	Phosphorized, high residual phosphorous	B152 122	99.90 min.[c]	0.015–0.040

[a] Cobalt counting as nickel.
[c] Includes small amount of silver.

[b] Copper plus named elements shall be 99.5% min.

similar to stainless steels. At high temperatures it resists steam, air, carbon dioxide, and sulfur compounds and is free from intergranular deterioration. In aqueous media, it resists dilute organic acids and strongly oxidizing solutions. Inconel-clad steel equipment is used in the chemical, petroleum, pharmaceutical, and food industries. Inconel is one of the few alloys that is useful in both oxidizing and reducing environments.

5 COPPER (TABLE 13.2)

Copper-clad steel is more than an economical substitute for solid copper. The strength imparted by the backing steel allows the corrosion resistant properties of copper to be utilized at temperatures and pressures far exceeding those where solid copper can be used.

Typical applications include stills, condensers, columns, kettles, heat exchangers, and storage tanks. Outstanding examples of chemicals that can be handled include alcohols, aldehydes, ketones, esters, organic acids, coal tar products, fluoride salts, sodium sulfate, sulfur dioxide, hydrofluoric acid, and hydrocyanic acid. The high thermal conductivity of copper in copper-clad equipment is particularly useful in processing heat sensitive materials such as varnishes and food products.

6 CUPRO-NICKEL CLAD (TABLE 13.2)

The two clad steel types, 70/30 cupro-nickel containing 70 per cent copper and 30 per cent nickel, and 90/10 cupro-nickel containing 90 per cent copper and 10 per cent nickel, are extensively used for tube sheets of condensers and heat exchangers where sea water or brackish water is the coolant. Their antifouling properties are particularly useful in seawater applications. The higher nickel content, 70/30 alloy, is often specified for more severe applications such as polluted seawater. Cupro-nickel metals possess useful corrosion resistance to the following chemicals under some environmental conditions: sulfuric acid, phosphoric acid, hydrofluoric acid, organic acids, fatty acids, and solutions of nonoxidizing salts similar to sodium chloride.

7 DESIGN

In designing nickel-clad, Monel-clad, and Inconel-clad vessels to ASME boiler code requirements, the cladding may be taken into account in calculating the vessel wall thickness. The bond strength must test to 20,000 psi minimum according to the SA265 specification.

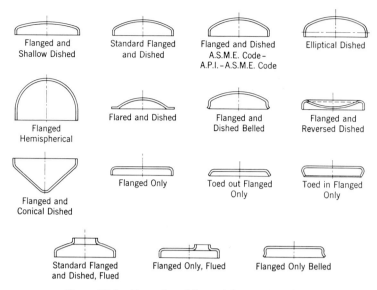

Figure 13.5   Examples of Several Styles of Formed Heads

Copper-clad and cupro-nickel-clad steel vessels are designed on the basis of the thickness of the backing steel only.

A continuous high-alloy surface must be maintained on the inside of clad-steel vessels to protect the backing steel from corrosive attack. Typical designs for nozzle openings in clad-steel vessels are shown in Figure 13.4.

8   FORMED HEADS

Almost any type of formed head can be made from clad steels. The bond of these hot-rolled nonferrous clad steels will withstand the rigors of hot spinning and hot or cold pressing. Examples of several styles of formed heads are shown in Figure 13.5.

9   FABRICATION

9.1   Handling

The general consideration in handling clad-steel plates is to avoid anything that might contaminate or mar the high-alloy cladding surfaces. (Protective plastic or paper coverings are often applied to cladding surfaces to protect them from travel dirt during shipment.) Wood-lath strips should be placed between the clad-steel plates when piling one on top of the other. Except for flame cutting and certain welding operations, clad plates should be handled with the cladding side up.

9.2   Bending and rolling

In performing these operations, the clad-steel plate and all working surfaces of the machine should be cleaned free of any loose scale, steel particles, or shop dirt to avoid embedding any foreign particles into the cladding. Bending rolls may be coated with paper to avoid contact with the cladding surface. The degree of bending or rolling which can be performed is limited only by the ductility of the cladding or the backing steel, depending upon which is in tension.

The condition of the plate edges is important for successful bending. Sharp corners should be rounded off. Flame-cut edges of hardenable backing steels should be softened or machined. Rough flame-cut edges also should be conditioned.

9.3   Hot forming

Certain precautions must be observed in heating operations to preserve physical properties and corrosion resistance of the cladding. Muffle furnaces operating with reducing atmospheres are desirable. Fuels such as coke, coal, crude oil, and others having sulfur content in excess of 0.5 per cent must be avoided as they can cause embrittlement of the cladding.

Before heating operations are performed, cladding surfaces should be cleaned thoroughly. Embrittlement or impaired corrosion resistance can result from diffusion of marking paints, oil, grease, and protective paper or plastic into the cladding. Marking paints containing zinc, lead, tin, or sulfur are especially harmful to nickel and nickel alloys.

9.4   Shearing and punching

Shearing and punching should be done with the clad side up so as to throw the burr on the backing-steel side. Shearing limits are practically the same as for carbon steel.

9.5   Cutting

Nickel-clad, Monel-clad and Inconel-clad steel in percentages to about 25 per cent can be flame cut without powder injection. Successful flame cutting with ordinary oxy-fuel gas equipment depends on the use of lower oxygen pressures and larger cutting tips than required to cut corresponding thicknesses of carbon steel. These clad steels also may be cut by the gas powder injection process or by the newer plasma-arc process.

Copper-clad steel cannot be cut with oxy-fuel gas methods and gas powder injection cutting does not produce satisfactory results. Therefore, it is necessary to use the plasma-arc process or to flame cut after chipping away the copper along the line of cut.

The cupro-nickel clad steels may be flame cut by the gas powder injection method or by the plasma-arc process.

9.6 Welding

Welds in clad steels, as well as mechanical joining, must provide a continuous high-alloy cladding surface. Another consideration with nonferrous clads is iron pickup in the cladding. Certain joint designs and welding techniques are especially useful to minimize this pickup.

The basic procedure for welding clad is a combination weld shown in Figure 13.6. Note that the steel side is welded first with steel welding materials.

Reverse-bevel combination welds and full alloy welds are commonly used variations of the basic procedure. Many shops find full alloy welds economical for thicknesses to about $\frac{1}{2}$ in. Low iron dilution is an added bonus for full alloy welds. (See Figure 13.7.)

9.7 Nickel-clad steel

Low iron content in welds of nickel-clad steel vessels is often required. Proper welding techniques and methods are as important as proper weld designs. Welding operators should use proper heat inputs, "stringer" beads, and small diameter electrodes. The inert-gas consumable-electrode welding process with its relatively low heat input is ideally suited for welding nickel cladding. Iron content also decreases with the number of weld layers. Two or more layers are preferable. It

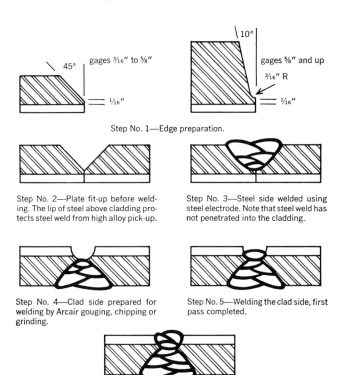

Step No. 1—Edge preparation.

Step No. 2—Plate fit-up before welding. The lip of steel above cladding protects steel weld from high alloy pick-up.

Step No. 3—Steel side welded using steel electrode. Note that steel weld has not penetrated into the cladding.

Step No. 4—Clad side prepared for welding by Arcair gouging, chipping or grinding.

Step No. 5—Welding the clad side, first pass completed.

Step No. 6—Completing the joint. Finish pass completed, clad side.

Figure 13.6 Basic Welding Procedure. Note that steel side is welded first

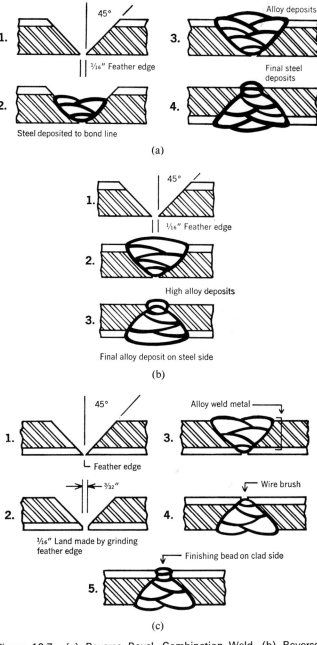

Figure 13.7 (a) Reverse Bevel, Combination Weld. (b) Reverse Bevel, Full Alloy Weld. (c) Standard Alloy Weld

may be necessary to remove about half of initially deposited weld beads before continuing with succeeding layers.

Low iron content is not necessary for many applications. For example, in caustic service, an important application of nickel-clad steel, weld metal diluted with as much as 24 per cent iron has been shown to possess adequate corrosion resistance.

9.8 Monel-clad steel

Monel-clad is welded by procedures similar to those used for nickel-clad. Control of iron dilution is important in many applications of this clad steel.

9.9 Inconel-clad steel

Inconel-clad is welded by procedures similar to those used for nickel-clad and Monel-clad. Iron pickup is not as critical, because Inconel itself contains iron. Therefore, welding materials containing no iron provide some compensation for iron dilution from the backing steel.

9.10 Copper-clad and cupro-nickel-clad steel

Copper-clad and the cupro-nickel-clads are welded using procedures basically similar to those for other clad steels. However, greater precautions are required in edge preparation, sequence of welding, and welding method to avoid pickup of copper by the steel weld or excessive dilution of the copper weld by steel. One method of achieving this is to deposit barrier layers of nickel, Monel, Inconel, or aluminum-bronze between the copper and steel.

Although under very carefully controlled conditions copper can be deposited directly on steel, it is preferable to deposit a barrier layer between the copper and steel. Full alloy welds also can be made using these barrier pass alloys for welding the steel. Commonly used welds are shown in Figure 13.8.

Electrodes and wires used for welding the cladding side of nickel-clad, low-carbon nickel-clad, Monel-clad, Inconel-clad, copper-clad and the cupro-nickel-clads are listed in Table 13.3.

9.11 Cleaning

Oil, grease, and ordinary shop dirt can be removed by washing with water and detergents or solvents. Sandblasting or other abrasive methods may be used for more persistent dirt or oxide scale.

The above methods plus an acid wash containing 10 per cent nitric acid to brighten the cladding surface after sand-blasting are the only cleaning methods suggested for copper-clad and cupro-nickel-clad. Chemical cleaning methods based on hydrochloric acid, however, may be used for nickel-clad, Monel-clad, and Inconel-clad.

A typical solution is composed of 1 gallon water, $\frac{1}{4}$ pint of hydrochloric acid (20°Be) and $1\frac{1}{2}$ ounces ferric chloride. Pastes made up with such an acid solution and lampblack or Fuller's earth are useful for holding reagents in contact with vertical or overhead surfaces.

Table 13.3. Electrodes and Wires for Welding the Cladding Side of Nickel-Clad, Monel-Clad, Inconel-Clad, Copper-Clad and Cupro-Nickel-Clad Steels

Cladding	For Covering Exposed Steel with First Pass or Layer		For Balance of Weld	
	Electrode	Wire	Electrode	Wire
Low Carbon Nickel	E4N11	ERN61	E4N11	ERN61
Nickel	E4N11	ERN61	E4N11	ERN61
Monel	E4N10	ERN60	E4N10	ERN60
Inconel	E4N12	ERN62	E4N12	ERN62
Copper	E4N10	ERN60	—	RCu
70/30 Cupro-Nickel	E4N10	ERN60	ECuNi	RCuNi
90/10 Cupro-Nickel	E4N10	ERN60	ECuNi	RCuNi

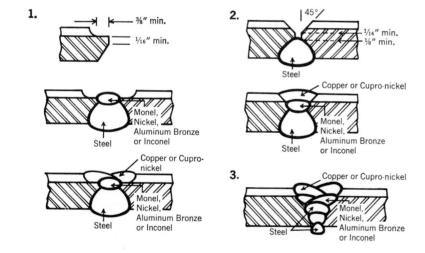

Figure 13.8 Commonly Used Weld Joints for Copper-Clad and Cupro-Nickel-Clad Steels

Chapter 14

Surfacing by Welding for Wear Resistance

Howard S. Avery

1 INTRODUCTION

The provision of a hard wear-resistant surface layer on metal parts by fusion welding is one of the most versatile expedients at the command of the engineer. Hard-facing alloys are available in the form of convenient welding rods, and sources of necessary heat, usually the oxyacetylene flame or the electric arc, are portable, relatively inexpensive, and familiar to engineers and mechanics.

It is the purpose of this chapter to provide information that will be helpful in the selection and application of hard-surfacing alloys. Welding techniques are adequately described in the American Welding Society's *Welding Handbook*[1] and consequently will receive only brief mention here. However, it may often be observed that surfacing alloys are used with a limited understanding of their characteristics.[2] Partially to remedy this and to present some salient aspects of the metallurgy of surfacing alloys as they affect engineering properties, the *materials* for hard-surfacing are featured in this discussion.

The selection of hard-facing as an engineering technique is strongly influenced by the need for armoring a part against severe service. In addition, the factor of protection in depth is generally required, though in some field repair work this may be less important than the ease of application.

If superficial protection alone is required, a manufacturer will probably select a case-hardening process, such as carburizing or nitriding,[3-5] which provides a hard zone a few hundredths of an inch thick to minimize wear; hard chromium plating[6] to discourage scratching; various electroplates or hot-dip coatings to combat corrosion; or perhaps metal spraying* to prevent surface deterioration or to provide heat resistance[7] (aluminum or chromium). All these processes are limited to the thickness of the protective layer that may be obtained practically or economically. If an armor $\frac{1}{32}$ to $\frac{3}{8}$ in. in thickness is required, a welded overlay is the preferred choice.

A surface overlay deposited by welding may be required for wear resistance, corrosion resistance, heat resistance, or perhaps all three. Other special properties, such as reflectivity, are occasionally necessary. However, most hard-surfacing is done to prevent wear, and subsequent discussion emphasizes this factor.

Wear has received considerable attention in engineering literature, but the optimum solution of a wear problem frequently is not clarified. Some reasons for this are the com-

plexity of wear as a phenomenon, the difficulty of identifying which of several wear factors dominates a specific situation, the variability of the materials used to combat wear, the poor reproducibility of most service tests used to evaluate materials, and the tendency of those who have studied wear intensively to specialize in relatively narrow fields. Despite these limitations hard-surfacing has been richly rewarding in many areas, as evidenced by the assortment of reports summarized below.

Hard-facing is reputed to increase the life of parts 2 to 25 times. Specific examples include 10 to 18 times the life of plain steel for hard-faced cement mill-grinder rings (with further economy from elimination of 9 replacement shutdowns and 15 to 20 per cent increase in production because of better machine efficiency); 7 to 10 times increase in life of hammer mill hammers; 3 to 15 times greater plowshare life; a tenfold increase for brick machine feeder shoe life after replacement of cast iron; fivefold improvement of vertical pug mill push shoes, in coke manufacture carbon scrapers were increased from a life of 3 days to 75 days by hard facing; coke pusher shoes had wear reduced to about $\frac{1}{80}$ that of steel; water-cooled pokers were improved from 3 months to 2 years to life; hard-faced power-shovel dipper teeth outwore an ordinary set 7 to 1; tungsten carbide insets increased the life of similar teeth before dulling from 6 to 120 hr, with improved efficiency; and armored dredge cutters outwore steel by 7 to 1.[8]

Steel mills have benefited by hard-facing rolling-mill guides, with a typical increase of 10 times in life. This application requires combined heat and abrasion resistance. In one case, a steel plant (in the Pittsburgh district) increased the life of the guides for a 15-in. rolling mill from an average of 600 billets for chilled iron to an average of 150,000 billets per set of guides hard-surfaced with a martensitic cast iron. Scratching of the billets was also eliminated. One steel company is said to have saved $10,000 per year on one item: hot-ingot grappling tong points. Wood-pulp shredder knives used in the paper industry were increased sixfold in life, with the annual cost per machine reduced from $2400 to $180 by associated benefits. Other machinery parts provide a multitude of opportunities for similar savings: tool-steel centers have been increased 6 times in life; hard-faced cams, 10 times. Engine valves are such an excellent application for hard facing that the procedure has become standard practice for many manufacturers. A PT-boat rocker arm, which had been giving trouble from wear, outlasted the engine after being hard-faced with a martensitic iron, it appearing in perfect shape when the engine was torn down after wearing out (Fig. 14.1).[2]

An aluminum-bronze overlay on a tool-steel base set of drawing and forming dies for stainless and mild steel gave a production life of 15,000 parts contrasted with an average life

---

* See Chapter 15 for discussion of metal spraying.

HOWARD S. AVERY is Research Metallurgist, Abex Corporation, Mahwah, N.J.

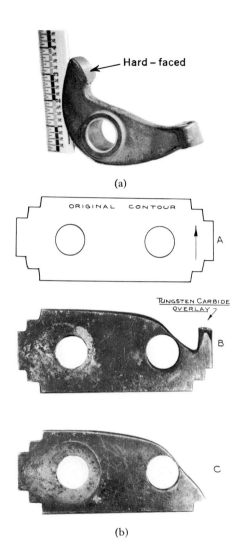

(a)

(b)

Figure 14.1   (a) PT Boat Rocker Arm, Hard-faced at the Critical Wearing Point with a Martensitic Cast Iron. The procedure changed the part from a problem to an item that outwore the motor (Refs. 2, 12). (b) Pulverizer Hammers, Showing (A) Original Contour, (B) Wear Pattern When Tip Is Protected with a Welded Overlay of Tungsten Carbide, and (C) Wear Pattern of Unprotected Hammer. The hard-facing procedure increased hammer life by 600 per cent in service pulverizing asphalt roofing trimmings. The arrow indicates direction of motion. This is a reversible hammer, which explains the two holes (Ref. 12)

of 3600 parts for the hardened and chrome-plated tool-steel dies.[9]

A great diversity is apparent in this list but it represents only a small sampling of the many applications where surfacing can benefit. Assuming that an engineer sees in this list an example of a situation with which he is concerned, an investigation of the original report (if it is accessible) is likely to be disappointing. A clear identification of the wear factors involved, a precise description of the service conditions, and an adequate definition of the materials (and their metallurgical status) are very rarely given. Without these, an attempt to translate a past experience into a future application can be treacherous. Recognizing this problem, this chapter will offer a number of useful generalizations that can aid engineering decisions. In using them it should be remembered that generalizations frequently sacrific precision. This may leave an element of uncertainty and an opportunity for individual experiment. However, unless such experiments are carefully carried out they may be misleading for the same reasons that reduce precision in the generalizations.

At the outset a decision about the applicability of welding is required. For a description of the variations, such as arc versus gas welding; flame or plasma spraying; manual, semiautomatic, or automatic methods; the surfacing chapter of the current A.W.S. *Welding Handbook*[10] is recommended. Other chapters of this same handbook provide details about the specialized equipment and methods that may be involved. A few highlights will be mentioned here: gas welding (usually oxyacetylene) is more precise and more expensive than electric arc welding, heating and cooling are usually slower with gas and thus thermal stresses are likely to be lower, and the equipment is very portable. Some metallurgical factors also favor gas welding. Arc welding is the most popular method, and reaches its maximum economy where large flat, cylindrical, or symmetrically curved surfaces can be surfaced with automatic equipment. Welding skill can be an important variable; precision of placement, control of thermal stresses, and avoidance of welding defects are very much dependent on the welder's skill. It is also helpful if he understands the materials, both base and filler metals, with which he is working.

Medium- and low-carbon steels are excellent base materials for surfacing, and most applications involve these. However, certain base metals require careful consideration. It is ordinarily impractical to overlay a high-melting-point filler metal on a lower-melting-point base. Also, cast irons and high-carbon alloy steels may be very sensitive to thermal stresses and make welding without cracking very difficult. Surfacing tool steels, for example, is an area where specialized advice on techniques may be essential. If the base metal is an air-hardening steel, that is, it can transform to martensite as it cools from welding heat, underbead cracking from the effect of hydrogen is a liability that should be minimized by using welding electrodes with "low-hydrogen" coating. These in turn require careful storage and handling to prevent moisture pickup, and frequently should be oven dried before using.

Granted that welding has been judged appropriate, seems better than use of a solid part of wear-resistant alloy, and the base is adaptable, the engineer is faced with selection of the surfacing alloy. To aid him there are a few technical articles describing certain alloy types, many brief articles and application notes that inadequately describe the service conditions and the alloys used, handbook chapters, a Specification for surfacing filler metals (AWS A5.13-56T),[11] and a variety of articles about wear. These together with research in this field serve as the background for this chapter.

2   A CLASSIFICATION OF WEAR

Wear has been defined as "deterioration due to use." This is very broad, and convention has tended to limit consideration to wear as a surface phenomenon. Even so, there are some who would exclude liquid corrosion and high-temperature oxidation from consideration. This trend is diminishing as it becomes clear that the same units of surface metal lost can be used to describe them as well as the more conventional mechanical and abrasive wear. Moreover, these factors can combine with mechanical wear and require consideration in selection of some surfacing materials. For this reason they will receive some attention here.

Six major wear factors are

Impact,
Abrasion,
Friction,
Heat,
Corrosion, and
Vibration.

Each of these has various ramifications. The first two are most important in many hard-facing situations and require consideration together because the materials tend to become less effective for one as they become better for the other. Thus engineering selection requires judgment as the two are weighed, with a compromise the usual result. Both factors are basically mechanical and the underlying cause of deterioration is stress.

3 IMPACT AS A WEAR FACTOR

Impact is a mechanical stress producer. Though these stresses may be brief, since impact is characteristically momentary, they may be astonishingly high. The ease with which a center-punch mark is made with a light hammer blow, contrasted with the difficulty the same person would have in making the same impression with steady pressure, illustrates the considerable mechanical advantage of impact.

Under the focus of an impact blow the stress is compressive. To the sides of the point of impact the stress is tensile, and between these two areas is a zone of shear stress. If the point of impact is close to the edge of brittle material, failure through the shear zone is likely. To minimize this it is helpful to chamfer the edges of such structures. What happens in the compressive zone depends on the magnitude of the stresses. If they do not exceed the elastic limit of the material the energy of the blow is returned as a bounce of the striking object and no damage should be suffered. If the elastic limit is exceeded deformation of the stressed zone is expected. This is lateral as well as vertical under the blow. The lateral flow tends to stress the surface in tension.

The ductility of the material struck becomes important as impact-deformation occurs. Brittle metals such as the hard irons can withstand some compressive flow without failure but they are characteristically unable to suffer more than a very tiny amount of tensile flow without cracking. Thus one severe impact blow or a succession of lighter blows can cause failure. Since such brittle materials may be the preferred choice for abrasion resistance the salient judgment is aimed at getting as much abrasion resistance as possible without having the part break under the impact conditions associated with the abrasion.

Impact stresses can be calculated if the conditions are known precisely.[12] In dealing with stresses below the elastic limit, compressive stress depends on the velocity of the blow, the velocity of the stress wave in the body struck, and the elastic modulus of its material. These are combined in the nomograph of Figure 14.2,[12] which will be useful in estimating

whether an impact blow will exceed the elastic strength of the metal struck.

It is convenient to type the conditions where the velocity of the blow does not exceed material elasticity as *light impact* and to accept the premise that light impact involves no hazard to the material.

The other extreme, *heavy impact*, involves so much kinetic energy in the blows that the engineering material cannot resist deformation, and appreciable flow is expected from each blow.

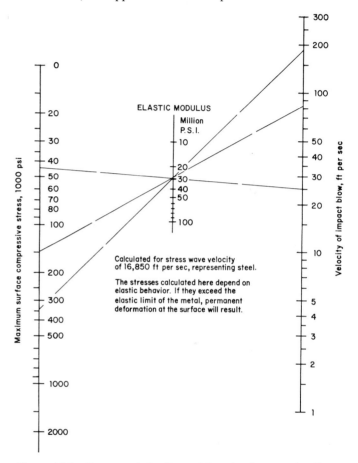

Figure 14.2 Nomograph for Determining the Compressive Stress That Will Be Produced on a Surface By Impact Blows of Different Velocities (Ref. 12)

Typical of such conditions is railway trackwork in frogs and crossings. These batter down under continual pounding of high-speed heavily loaded trains until the worn surface is considerably below the original level. These battered zones are then built up with welded overlays. The need for great toughness is obvious here and austenitic manganese steel is the preferred material, both for base metal and overlay. This is a case where impact is the dominant wear factor; abrasion and corrosion are negligible, while fatigue, though possible, is usually minor.

Between these extremes is the less well defined realm of *intermediate* impact. Hardenable steels are usually selected for such overlays, with consideration being given to the need for relative dimensional stability (high yield strength) balanced against freedom from the likelihood of cracking.

Figure 14.3   Gas-weld Overlay of Martensitic Iron, $\frac{1}{16}$-in.-Thick, on a 1-$\frac{1}{16}$in.-Thick Base of 1020 Steel; Cracked by One 500-ft-lb Impact Blow Delivered on a $\frac{7}{8}$-in.-Diameter Area. Cracking of the hard facing is due primarily to plastic deformation of the soft base (Ref. 12)

Design can play an important part in using impact resistant composites. Figure 14.3,[12] representing a hard-iron overlay on a soft 1020-steel base, has suffered one impact blow. Flow of the soft base underneath the overlay has stressed the latter in tension and numerous cracks resulted. Here the base is not strong enough to properly support the hard facing. In contrast, Figure 14.4 shows the same overlay material supported by an air-hardening steel base that is strong enough to withstand the transmitted stress. This composite took many blows without damage. Note that the specimen has the edges protected by chamfering and that the overlay is placed in a pocket of the base. Placement of hard facing in grooves and pockets is a very useful technique, as it minimizes the vulnerability to the undesirable tensile stresses.

4   ABRASION AS A WEAR FACTOR

Abrasion is a kind of mechanical wear in which the stresses are imposed through the medium of hard particles or fragments. Such abrasives are usually nonmetallic and most of them are either natural or artificial minerals. Since there are hundreds of known minerals, with many variations in hardness and fracture characteristics; and since they can occur in a great range of sizes, shapes, angularities, and types of support; and since they can impinge with a considerable range of velocities; the ramifications of abrasion are almost endless.

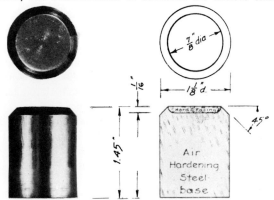

Figure 14.4   Gas-weld Overlay of the Same Martensitic Iron as in Figure 14.3, $\frac{1}{8}$ in. Thick, on a 1$\frac{5}{16}$ in. High Cylinder of Air-hardening Steel. This specimen took one hundred 500-ft-lb blows on its $\frac{7}{8}$-in.-diameter top surface with no sign of damage (Ref. 12)

Much simplification is necessary if an engineering judgment is to be conveniently applied.

Intuitively, it is expected that the harder minerals will be most abrasive and this is borne out considerably by experiment. Fortunately a few common rock-forming minerals are found to dominate most mining, quarrying, and earth-moving situations. Of these, quartz is by far the most common, and lacking more precise information the assumption that a given seriously abrasive environment is due to quartz or related hard silicates is likely to be true. Quartz is the preferred abrasive for laboratory testing because of such widespread occurrence.

If quartz is not the major cause of abrasion in a specific field situation, the mineral responsible very probably appears in Table 14.1, which is a list that combines the common rock-forming minerals with a few others needed to complete Mohs' mineralogical scale of hardness. The table contains supplementary information that will aid identification and evaluation of the minerals.

Rocks are aggregates of one or more minerals. Soils are relatively fine-grained aggregates of the decomposition products of rocks. In the evaluation and reporting of earth-moving abrasion problems it is helpful if the rock or soil type can be described with some precision. *Igneous rocks* result from the solidification of molten material; they exist in a great variety of both composition and texture. Their chief constituents are usually hard silicates with or without quartz. *Sedimentary rocks* are deposited in layers in water, such as stream beds, beaches, deltas, and shallow seas. The three common types are limestone, sandstone, and shale, with possibilities of various mixtures of these or transition types. *Limestone*, composed of the minerals calcite and dolomite, is a common sedimentary rock that may appear hard and compact. However, it is definitely softer than most sandstones and thus poses a much less severe abrasion problem. *Sandstone* is consolidated sand, usually from beach or delta deposits. As a rock its abrasiveness may depend upon how well the sand grains are cemented together. When broken down by crushing, grinding, or weathering, its abrasiveness depends much on particle angularity. *Shale* is consolidated mud, ordinarily composed largely of the clay mineral, kaolin. Kaolin is so soft that its abrasiveness is negligible. The abrasiveness of shales increases with their sand content until with high levels of quartz they merge into sandstones. *Metamorphic rocks* are those that have been altered by heat, pressure, and other geologic factors. The changes can be profound. Limestone can recrystallize into *marble*, shale can be hardened into *slate*, and sandstone can become *quartzite*, which is one of the most abrasive of rocks because the quartz grains are so well cemented together. The hybrid sedimentary rocks such as sandy shale can recrystalize with the formation of different and more abrasive minerals; thus soft kaolin may be changed into hard silicates. Silicification, with the addition of quartz or silicates usually in tough, fine-grained form, can occur. Petrified wood is an example. Silicification usually markedly increases the abrasiveness of a rock. Thus a silicified limestone can be highly abrasive, whereas ordinary limestone presents only mild abrasion.

Even with the above simplification there is a great range in character of rocks, ores, and soils. Some geological background coupled with close observation is very helpful in dealing with them.

Table 14.1. Chart of Natural Abrasive and Rock Forming Minerals

Name	Chemical Composition	Habit	Mohs' Hardness	Specific Gravity	Cleavage	Color	Occurrence
Diamond	Carbon	Diamond occurs as octahedral crystals, in darker rounded forms and fragments, or compact, granular or massive black or grayish-black "carbonado." Rare, it is listed here because it is the hardest substance known. (No. 10 on Mohs' scale)	10	3.516–3.525	Perfect octahedral	Colorless, various pale shades, black	
Corundum	Al_2O_3	In hexagonal crystals, often barrel-shaped. Rudely crystallized, massive, or granular. Emery is fine-grained corundum mixed with other minerals, chiefly magnetite. (No. 9 on Mohs' scale)	9	3.95–4.10	Sometimes nearly rectangular pseudo-cleavage	Blue, red, yellow, brown, grey, nearly white	As a minor mineral in metamorphic rocks, such as marble, mica-schist, gneiss, etc.
Topaz	$(AlF)_2SiO_4$	Orthorhombic prismatic crystals, crystalline masses, and granular. (No. 8 on Mohs' scale)	8	3.4–3.6	Highly perfect — Basal	Straw yellow, white, greyish, greenish, bluish, reddish	Pegmatite veins
Quartz	SiO_2	Hexagonal crystals, prismatic or pyramidal. In massive forms of great variety, as beach sand, in coarsely crystalline veins, as flint masses, and as microscopically crystalline aggregates such as chalcedony, chert, agate, jasper, and flint. (No. 7 on Mohs' scale)	7	2.65	None	Colorless to black, white, rose, amethyst, red, green, grey	Very widespread. In veins, sandstones, conglomerates, granites, gneisses, schists, rhyolite, pegmatites, stream and ocean deposits; and in most soils
Garnet	$3(Ca, Mg, Fe, Mn)O·(Al, Fe)_2O_3·3SiO_2$	Dodecahedral or trapezohedral crystals, rounded grains, coarse or fine massive granular	6.5–7.5	3.1–4.3	Dodecahedral parting	Red, brown, green	Common as an accessory mineral in metamorphic and igneous rocks; gneisses, mica-schists, hornblende schists, pegmatites and granites
Olivene	$2(Fe, Mg)O·SiO_2$	Rarely as orthorhombic crystals; usually as embedded grains or granular massives	6.5–7.0	3.3 ±	Distinct — Prismatic	Olive green	Occurs chiefly in dark igneous rocks such as gabbro, periodite and basalt. "Dunite" is made up almost entirely of olivine
Epidote	$4CaO·3(Al, Fe)_2O_3·6SiO_2·H_2O$	Sometimes as monoclinic crystals or fibers, but usually coarse to fine granular masses	6–7	3.4	Perfect — Basal	Pistachio green	Metamorphic rocks such as gneiss, amphibolite and various schists; frequently from metamorphism of impure limestone. Often associated with chlorite
Nepheline	$Na_2O·Al_2O_3·2SiO_2$	Usually massive, compact, or as imbedded grains. Rarely as small, prismatic hexagonal crystals. May have a greasy luster	5.5–7.0	2.6 ±	Distinct — Prismatic and Imperfect — Basal	White or yellowish; when massive, dark green, grey, bluish-grey, brownish-red, brick red	In igneous rocks such as phonolite nephelite-syenite, nephelite-basalt; and almost never associated with free quartz

Mineral	Composition	Form	Hardness	Sp. Gr.	Cleavage	Color	Occurrence
Plagioclase Feldspar	$Na_2O \cdot Al_2O_3 \cdot 6SiO_2$ to $CaO \cdot Al_2O_3 \cdot 2SiO_2$	Massive, cleavable masses, or compact. Frequently finely striated, and sometimes with a beautiful play of colors on cleavage faces	6.0-6.5	2.62-2.76	Perfect — Basal and Imperfect — Prismatic	White greyish reddish	Particularly in igneous rocks, such as granites, syenites, porphyries, felsite, while the darker varieties occur in diorite, gabbronorite, basalt, andesite, and amphibolite
Pyrite	FeS_2	Often in cubic or pyritohedral crystals, but frequently massive, granular, globular, kidney-shaped, or stalactitic forms	6.0-6.5	5.0	None	(Black streak) brass yellow	As an accessory mineral in many rocks. In veins and ore deposits. May be associated with coal; copper, lead, zinc and gold ores; and may also form large massive deposits
Orthoclase Feldspar	$K_2O \cdot Al_2O_3 \cdot 6SiO_2$	Prismatic, monoclinic crystals; usually crystallized or coarsely cleavable to granular. Sometimes fine-grained or massive	6	2.57	Perfect — Basal and Imperfect — Prismatic	White, colorless, pink, straw, flesh	A very common mineral. In granites, pegmatites, syenites, porphyries and gneisses; and sometimes in certain sandstones and conglomerates
Magnetite	$FeO \cdot Fe_2O_3$	Usually granular massive, coarse or fine; as grains in sand, and sometimes as octahedral crystals	5.5-6.5	5.2	Octahedral parting	Black, metallic	Forms large ore bodies; also occurs as an accessory mineral in many rocks from granites to gabbro. Black seashore sands. Associated with corundum to form "emery"
Hematite	Fe_2O_3	Usually earthy or in rounded shapes with a radiating crystalline structure. Sometimes micaceous or crystallized in hexagonal or rhombohedral form	5.5-6.5	5.1	Parting	(Red streak) red-black or bluish	As large ore bodies, as an accessory mineral in granites, as the cement on red sandstones, in metamorphic rocks, and as sedimentary beds
Pyroxene	$CaO \cdot (MgO \cdot FeO) \cdot 2SiO_2$	Coarse to fine granular, or as 4- to 8-sided square prisms	5-6	2.8±	Perfect — Prismatic interrupted; 87° or 93°	Green to black	Common in dark-colored igneous, and is the dominant constituent in gabbros and some basalts. In some gneisses and recrystallized impure dolomitic limestones. Rarely associated with free quartz
Hornblende (Amphibole)	$(Ca, Mg, K, Fe, Al)O \cdot SiO_2$	Crystals or crystalline masses; coarse to fine granular	5-6	2.9-3.4	Prismatic Perfect	Black	Diorites; other igneous and metamorphic rocks. Gneisses and schists, granites and syenites
Tremolite (Amphibole)	$CaO \cdot 3MgO \cdot 4SiO_2$	Slender crystals or fibers (asbestos)	5-6	3.0	Prismatic Perfect at angles of 56° and 124°	White	Metamorphosed dolomitic limestones
Actinolite (Amphibole)	$CaO \cdot 3(Mg, Fe)O \cdot 4SiO_2$	Crystalline, acicular; compact (jade)	5-6	3.0-3.2	Prismatic Perfect at angles of 56° and 124°	Green	Crystalline schists
Hypersthene	$(MgO \cdot FeO) \cdot SiO_2$	Foliated massive	5-6	3.2-3.5	Prismatic Perfect	Yellow-brown green to black	Peridotite, gabbro. Dark colored igneous rocks like peridotite or gabbro
Leucite	$K_2O \cdot Al_2O_3 \cdot 4SiO_2$	Trapezohedrons or grains	5.5-6	2.5	—	Grey	Igneous rocks that contain no quartz; recent lavas
Ilmenite	$FeO \cdot TiO_2$	Massive; compact, thick tabular crystals	5-6	4.5-5	—	Black	Frequently with magnetite; as beds or lens-shaped bodies in metamorphic rocks; in veins or masses
Limonite	$2Fe_2O_3 \cdot 3H_2O$	Earthy; massive; stalactitic	5-5.5	3.6-4	None	Brown to yellow	From other iron minerals by solution and deposition; masses in clay; beds in marshes

Table 14.1—continued

Name	Chemical Composition	Habit	Mohs' Hardness	Specific Gravity	Cleavage	Color	Occurrence
Apatite	$Ca(F, Cl) \cdot Ca_4(PO_4)_3$	Hexagonal crystals; granular massive to compact	5	3.2	Basal Imperfect	Green or brown; blue to colorless	In all kinds of rocks; beds; veins; Phosphate rock
Zeolites	Ca, Al, alkaline silicates with H_2O	As crystals in steam blow-holes	4±	2.2±	Various	White or light colored	In cavities of basalts
Fluorite	CaF_2	Octahedrons, cubes and crystalline masses	4	3.1±	Octahedral Perfect	Various; purple, white to colorless; green, blue	In-veins; as gangue in metallic ores; in limestones
Siderite	$FeCO_3$	Rhombohedrons; cleavable masses; earthy; globular concretions	3.5–4	3.8	Rhombohedral Perfect	Brown	As beds with clay or slate rocks and coal deposits; ore veins; in limestones
Dolomite	$(Ca, Mg)CO_3$	Rhombohedrons; cleavable masses; compact	3.5–4	2.8	Rhombohedral Perfect	Pinkish, grey; colorless or white to black	As dolomitic limestone in beds; in marble; in ore veins
Serpentine	$3MgO \cdot 2SiO_2 \cdot 2H_2O$	Massive or fibrous. Chrysotile is the asbestos form	2.5–4	2.6±	Prismatic Perfect in Crysotile	Green	Igneous and metamorphic rocks. Forms by alteration of other magnesium silicates. May make up almost entire rock mass
Calcite	$CaCO_3$	Massive; cleavable rhombohedrons; compact; earthy; stalactitic	3	2.7	Rhombohedral Perfect	Colorless, white, or light shades	Very common. As beds: limestones, chalks, marls, etc. Marbles. In veins and as gangue in ore deposits
Biotite (Mica)	$(HK)_2(Mg, Fe)_2Al_2 (SiO_4)_3$	Irregular foliated masses; disseminated scales or aggregates	2.5–3	3±	Basal Very Perfect	Dark green and brown to black	Igneous rocks, such as granite and syenite; Metamorphic rocks such as gneisses and schists; in pegmatite veins; in some lavas
Muscovite (Mica)	$K_2O \cdot 3Al_2O_3 \cdot 6SiO_2 \cdot 2H_2O$	Foliated, in sheets — large, small, or scales	2–2.5	2.8±	Basal Very Perfect	Transparent; light brown, green or yellow	Widespread. Igneous rocks: granites, syenites, pegmatite veins; gneisses and schists. Chief constituent of mica-schists
Chlorite	Hydrous Fe, Mg, Al silicates	Schistose; foliated massive	2–2.5	2.7±	Basal Highly Perfect	Green	Metamorphic rocks, from alteration of pyroxene, amphibole, biotite, garnet, and other magnesium aluminum silicates
Kaolin	$Al_2O_3 \cdot 2SiO_2 \cdot 2H_2O$	Earthy; claylike masses; compact or friable	2–2.5	2.6	—	White or light shades	Widespread. The chief constituent of clay and shales. From alteration of feldspars
Gypsum	$CaSO_4 \cdot 2H_2O$	Cleavable massive; granular massive; foliated; fibrous; tabular crystals	2	2.3+	Tabular Perfect	Colorless, white or grey	As thick beds; interstratified with limestones and shales; under beds of rock salt; as gangue in ore veins; as wind blown sand
Talc	$3MgO \cdot 4SiO_2 \cdot H_2O$	Foliated massive; compact. Soapy feel	1–1.5	2.7–2.8	Basal Perfect	White Green	Chiefly in metamorphic rocks as talc-schist or as soapstone rock, or in igneous rocks from alteration of magnesium silicates

Excluding impact, which has been treated as a separate factor, there are several different kinds of abrasion which can be recognized. They are sufficiently distinctive to sometimes reverse the merits of alloys that are being ranked. These types are low-stress scratching abrasion or erosion, high-stress grinding abrasion, and gouging abrasion. The gross stresses involved tend to increase in the order listed. *Erosion* usually involves low velocities and weak support of the abrasive. Wear of a plowshare in sandy soil is a typical occurrence. The machine-shop example is polishing with abrasive on a soft cloth. Thus the energy focused by the abrasive is low and impact is usually absent. However, velocities can be high, as in a sandblast nozzle; the severity of this abrasion is expected to increase as the third power of the velocity. In the laboratory erosion can be produced by the impingement of a fluid stream or by an abrasive-fed rubber wheel rubbing on a brake-shoe-type specimen.

Grinding abrasion involves the fragmentation of the abrasive, usually as initially small pieces, between two strong faces. It occurs in machine-shop lapping and in ball- or rod-mill grinding. The gross stresses may be moderate but on a local or microscopic scale the stress must be high enough to exceed the compressive strength of the abrasive. It is best recognized by a metal-sandwich situation with the abrasive between. Where impact is low or moderate, as in most ball mills, hard facing is not appropriate because the wearing parts can be made of the abrasion-resistant alloys in massive form, as balls, mill liners, and so on. In other industrial situations, especially where a tough base is required for protection against occasional impact, hard-facing alloys can provide valuable protection. The severity of this abrasion depends considerably on the toughness and strength of the abrasive; in one experiment the substitution of one quartz abrasive for another with only

one-third the compressive strength caused a threefold increase in wear rate.[13] Abrasive hardness has also been proved important. Not only do wear rates increase but the spread in wear resistance between good and poor materials tends to diminish as abrasive hardness increases (Fig. 14.5).

Gouging abrasion can be recognized by the prominent grooves or gouges that it leaves on wearing faces. In a machine shop it is simulated by abrasive-wheel action or by machine-tool cutting. In jaw and gyratory crushers it is associated with great compressive stresses. In power-shovel dipper teeth, especially with rock digging, it may involve heavy impact also. The typical abrasive is a sizable piece of rock or slag that is massive enough to provide considerable support to the cutting edge or point that does the gouging. Equipment subjected to gouging abrasion usually features replaceable parts and hard-facing is frequently used to extend their life or to rejuvenate them.

5 FRICTION AS A WEAR FACTOR

Frictional contact is the situation usually visualized as mechanical wear. In the absence of an abrasive and lubricant the basic mechanism is usually welding. It is recognized that the forces of interatomic attraction are very powerful at close range. If the atoms of two separate pieces of metal are brought together, that is to a separation near the crystal lattice spacing of the metal, they cling together from this attraction. If enough atoms are so bonded, the faces weld together. This phenomenon is the secret of "cold welding." It does not require heat, though heat facilitates the action by softening the metals so flow and more intimate contact result from the forces pushing the faces into contact. Soft metals thus weld more easily because they flow into contact more readily. Cleanliness is essential; films of dirt, oil, grease, or even adsorbed gases may keep the faces apart by more than the critical distance that prevents the atomic bonding. Since frictional movement may serve to wipe away such separating materials, close-fitting moving faces are more likely to weld and seize than are stationary objects.

Where the seizure just described occurs on a few minute areas it is evidenced as measurable friction. Presumably the forming and breaking of the atomic bonds explains the increased difficulty of movement associated with friction. The engineering solution of this problem is usually lubrication, which prevents the intimate metal contact while at the same time permitting close dimensional fit of moving parts. When lubrication fails and some seizure occurs, the forces of the moving machine may break atomic bonds to permit continued motion. Since the bonds may not be those just formed but may be within the original metal face materials, the torn out zone may leave a depression and appear on the opposite face as a welded-on projection. With motion this in turn may gouge or score the face from which it came, in the fashion of a dull tool. Seizing, scuffing, scoring, and galling are all terms used to describe this undesirable group of events. Where this occurs with normally well-lubricated parts they may require replacement. Fretting is a term that suggests that the debris of such events is making an added contribution as an abrasive, perhaps in the form of oxidized metal particles. The movement that causes fretting may be the undesired consequence of vibration.

Hard-facing is used less in the frictional area than it is

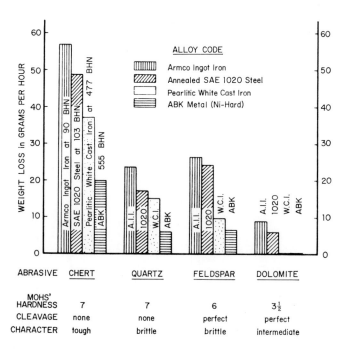

Figure 14.5 The Relative Abrasiveness of Common Mineral Abrasives As Determined By the ABEX Corporation Laboratory Test for Grinding (High Stress) Abrasion (Ref. 13)

for abrasion. However, it is used for some situations where lubrication is not feasible. Its logic is based on either surface films, such as aluminum oxide, that prevent easy metal-to-metal contact, or on provision of a contact face that has very high elastic strength to resist the flow that encourages intimate contact. Aluminum bronzes are examples of alloys that form effective films quite readily. The martensitic cast-iron surfacing alloys provide high elastic compressive strength.

It will be recognized that an uninvited abrasive in the lubricant may change the picture. If its dimensions are such that it can bridge across the normal lubrication film spacing it can act as a lapping agent and cause metal loss. This kind of wear is grinding abrasion, and it can shorten the life of normally well-lubricated parts. An example is the cylinder wear of gasoline engines that inhale dusty air. Nitrided cylinder liners are sometimes used as a form of protection but an air filter is more common. Hard-facing is not likely to be practical; the case is described here to show how one kind of wear can be complicated with another wear factor.

6 HEAT AS A WEAR FACTOR

Heat can cause steady or progressive deterioration of metals in a variety of ways. This may be through hot-gas corrosion or oxidation, the softening of hardened structures by the mechanism of tempering, the basic loss of strength as temperatures rise, or the more insidious deterioration associated with slow plastic flow (also called creep).

Oxidation is controlled primarily with chromium. It forms oxides readily, but these are usually tightly adherent, relatively impervious to further penetration by oxygen, and thus outstandingly protective. About 25 per cent chromium satisfactorily protects iron-base alloys up to 2000°F, holding surface metal loss in air to less than 0.1 in. per year. Some added protection is given by nickel or cobalt. If these elements are used to substitute for chromium they must be employed in larger amounts, being roughly one third as effective.

Hot-gas corrosion in other atmospheres or with certain other elements in the environment may be less easy to control. While oxidizing flue gases, even with considerable sulfur present as sulfur dioxide, behave much like air, the reducing flue gases that contain hydrogen sulfide can be quite destructive to nickel-containing alloys. Nickel sulfide tends to form on the surface and flux away the protective oxides. The elements lead, sodium, and vanadium form oxides that also can form low-melting surface compounds and thus are also detrimental to oxidation resistance.

For most surfacing alloys oxidation is unlikely to be much of a problem below 800°F. Between 800 and 2000°F chromium can be proportioned to the maximum temperature expected. There are available many suitable alloys in the chromium-iron, Cr-Ni-Fe, nickel-base and cobalt-base grades. Some of these will serve up to perhaps 2300°F.

The steels and irons that harden by the transformation of austenite during cooling are usually softened and toughened by reheating. For tool steels and structural engineering steels this thermal sequence may involve heating to a hardening temperature (to form austenite); rapidly cooling, usually by liquid quenching (to form hard martensite); and then tempering between 300 and 1200°F to toughen the brittle martensite. Carbon makes an important contribution to the maximum hardness thus attainable and the tempering temperature controls the final hardness, as shown in Figure 14.6.[14] The amount of alloying elements other than iron contribute chiefly through their control of the cooling rates that are effective for hardening; with little alloy as in simple carbon steels the cooling must be very fast (for example, water quenching); with significant amounts of such elements as chromium, nickel, and molybdenum present hardening can occur even with air cooling.

Many hard-facing alloys are of the latter air-hardening type and they thus harden as they cool from the heat of welding. A deliberately applied heat treatment is rare unless the overlays are tool steels used to rebuild a tool-steel base. Thus overlays are not often tempered. However, if service conditions involve heat such tempering can occur and softening may be expected.

In addition to the special case of softening by the tempering mechanism, all metals tend to soften as temperatures are

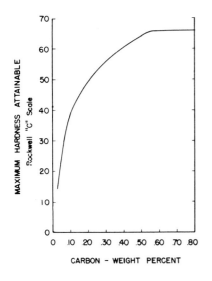
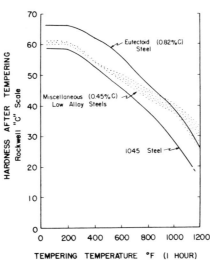

Figure 14.6 The Influence of Carbon Content and Tempering Temperature on the Hardness of Steels That Have Been Effectively Hardened By the Martensite Reaction. (Ref. 14)

raised. This means that they lose strength in proportion to the temperature of service. They differ considerably in the amount of such weakening, and the quantitative picture must be obtained individually for each alloy. Such softening is reversible; hardness is regained upon cooling, in contrast to the softening from reheating martensite.

Besides the effect of temperature there is a time effect. This takes the form of slow plastic flow or *creep*, which must be considered in the design of engineering parts that are to operate at elevated temperatures. In the field of overlays creep should be considered because of its rather potent effect on apparent hardness. It can be demonstrated and evaluated by adding a time factor to conventional hardness testing. Figure 14.7[14] shows the effect of creep on the apparent hardness of several alloys that may be used as overlays.

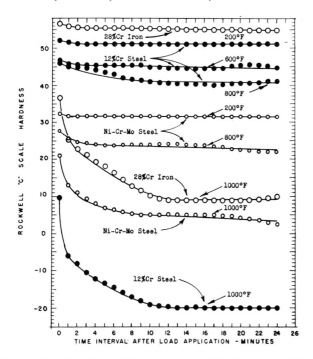

Figure 14.7   The Effect of Time on Apparent Hot Hardness of Three Iron-base Alloys Showing Creep Phenomena At the Higher Temperatures (Ref. 14)

If creep continues long enough it can cause rupture. Specialized creep-rupture tests serve to provide the information needed for engineering design, but such data are rare for hard-facing alloys. The best available criterion is hot hardness, but it should be recognized that it is an unvalidated wear test in most cases. This is a pertinent concern because hot-wear resistance is sought in most cases where overlays are exposed to heat. Fortunately there are several alloys for which a reasonable amount of experience provides guidance in selecting materials for hot-wear resistance.

7 CORROSION AS A WEAR FACTOR

The last decade has seen considerable growth in the use of overlays for protection against liquid corrosion. It would seem feasible to make weld deposits of most corrosion-resistant alloys but such utilization has been much slower than for abrasion-resistant alloys, even though such surface deterioration can occur as readily. Reluctance to accept corrosion as a form of wear has perhaps been inhibiting. However, there are other adverse factors. Among these are the need for greater integrity or surface quality, since liquid corrosives can search out imperfections more effectively than does abrasion. Large areas may require protection and the economics thus favors arc welding, usually by automatic methods. The high temperatures of arc welding cause considerable dilution of the overlay with the base metal, thus changing the composition of the filler metal. Unless the amount of such dilution is known and considered, the nominal corrosion resistance of a selected filler metal may be quite different from the behavior of the diluted weld deposit.

As a mechanism of deterioration corrosion can cause surface metal loss by *solution*, expressed in the same units as abrasive wear (weight loss per unit area per unit time or perhaps inches per year dimensional change). It can also cause a more insidious *intergranular corrosion*, which can lead to failure with little or no superficial damage. Inhibition of corrosion by additives to the liquid agent or electrical protection that reverses the trend where electrolytic attack is likely are added factors to be considered. The field of corrosion is too complex for effective treatment here and the rather generalized advice that the overlay should be of the final composition desired and that it should be free of penetrating imperfections must suffice.

8 VIBRATION AS A WEAR FACTOR

Alternating stresses, especially when they cause apparent deformation on each cycle, are an obvious cause of surface deterioration. The end result can be surface cracking and then failure. Even when deformation is not detectable such surface cracks can follow enough reversals of stress. Vibration is a common cause of this *fatigue*, but any normal flexing or bending in service can lead to fatigue failure if the cycles are repeated often enough and if the metal is weak enough to be vulnerable. Decarburized and thus weakened surface zones are particularly susceptible. Moreover, notches and similar surface imperfections or reentrant angles can encourage fatigue cracks. Besides the matter of good design a way to combat fatigue is to make the surface material stronger.

Overlay of a weak base metal with a stronger alloy is a means of adding strength where it is most needed for fatigue resistance. However, welding is rarely employed for this purpose. Possibly more will be done in the future. Such overlays should be made of strong and relatively tough alloys and the quality of the deposit should be high. Also, if possible the residual stress from welding should be compressive on the surface. If residual tensile stress is present it may be advisable to heat-treat the whole part before use.

9 THE MEASUREMENT OF IMPORTANT PROPERTIES

Comparison and selection of alloys requires knowledge of the properties that are pertinent to performance. It is desirable that this knowledge be quantitative. In the surfacing field some commonly recognized properties (for which well-standardized tests are available) are involved, but there are other less clearly defined properties that are quite important. Table 14.2 provides a list of properties that may be involved. Of these, hardness, impact resistance, abrasion resistance, oxidation resistance, and compressive properties merit discussion here.

Table 14.2.   Important Properties of Surfacing Alloys

A. Hardness
1. *Indentation hardness* — a measure of compressive strength
2. *Scratch hardness* — an indication of shear strength
3. *Microhardness* — as determined on individual single phase areas
4. *Macrohardness* — appropriate especially for single phase alloys, but providing only an average indication for complex multi-phase alloys
5. *Hot hardness*
   a. Instantaneous values — valid for transient loading
   b. Time dependent values — more valid for sustained loads and reflects the creep behavior
   c. The effect of temperature — an important variable

B. Abrasion Resistance
1. To low-stress scratching abrasion or erosion
   a. Effect of velocity of the abrasive
   b. Effect of angularity of the abrasive
   c. Effect of hardness of the abrasive
2. To high-stress grinding abrasion
   a. Effect of hardness of the abrasive
   b. The level of impact associated
3. To gouging abrasion
   a. Effect of toughness of the abrasive
   b. Effect of hardness of the abrasive
   c. Angularity of the abrasive
   d. The amount of associated impact

C. Impact Resistance
1. Velocity of the impact blows
2. Resistance to flow under repeated impact (This is related to yield strength and to work-hardening capacity)
3. Resistance to cracking or rupture under impact
   a. Cleavage tendencies
   b. Compressive yield strength — a quality that resists flow
   c. Compressive ductility — consists of an elastic component and a plastic component that determines the flow that is tolerable before fracture

D. Heat Resistance
1. Hot gas corrosion resistance
   a. Under oxidizing conditions
   b. Under reducing conditions
   c. In the presence of sulfur or other detrimental elements
2. Resistance to tempering of hardened structures, e.g. martensite or pearlite
3. Retention of strength at elevated temperatures
4. Creep resistance — resistance to flow under sustained loads
5. Rupture resistance — ability to creep without cracking
6. Resistance to thermal fatigue — a complex of properties to which thermal expansion, hot and cold yield strength, cold ductility, and modulus of elasticity all contribute

E. Frictional Properties
1. Friction coefficient
2. Yield strength
3. Surface film formation
4. Internal lubrication, e.g. graphite flakes or lead droplets
5. Plasticity — ability to flow and conform mating surfaces, as in bearings
6. Welding tendency — the ability of mating surfaces to weld or "seize," which is expected with similar simple metals and crystal structures

F. Corrosion Resistance
1. Chemical inertness
2. Presence of protective film
3. Presence of inhibitors
4. Electrochemical effects
5. Effect of alloy composition
6. Effect of the corrosive agent — including composition, temperature, pressure, and possible erosive agents in association

G. Miscellaneous Engineering Properties
1. Thermal expansion coefficient
2. Heat conductivity
3. Specific heat
4. Reflectivity
5. Magnetic properties
6. Optical properties, e.g. reflectivity
7. Surface smoothness
8. Melting points and ranges
9. Wetting action (surface tension)

Measurements of wear resistance, even if they are competently made, are not universally accepted. This is partly due to the fact that there is no universal wear test, so there must be a variety of methods for evaluating the different aspects of wear. Even so, a specific type of evaluation may lack recognition because the apparatus and techniques are not widely and precisely duplicated. This is the case for some of the criteria described here. However, it is essential that test results be used if quantitative and dependable comparisons are to be made.

Before using wear-test results certain aspects should be clarified and used to judge the quality of the data available. The reliability, the ranking ability, and the validity of a given test should be established and properly reported before confidence in the test is justified. Reliability serves to define the amount of experimental error involved. When it is large, poor reproducibility is shown by repeated tests. This in turn means that more tests are required to establish a dependable estimate of average performance. Reliability should be shown in statistical terms such as standard deviations, confidence limits, and expected scatter band ranges. Failure to use such engineering tools properly can lead to misleading results and erroneous conclusions.[13] If a test has proven reliability it can then be used to rank a variety of interesting materials. This ranking should show considerable spread if the test is to be useful. It is usually possible to include a few materials for which relatively valid opinions exist. If these differ considerably, the ranking by the test must indicate similar differences, or else the utility of the test is questionable until its validity can be established by pertinent supplementary tests. The final validation of a test is the demonstration that it can predict service performance. This is the most difficult step because it requires a correlation between the laboratory test results and those from a service test. A good service test must also have proven reliability and ranking ability. These attributes are quite rare. If a laboratory test were to show the great variability and the lack of statistical reproducibility that characterize most service results it would not be seriously considered for use. The pressure for production, the day-to-day changes in operating conditions, variability of feed, and frequently inadequate records, militate against good service evaluations. This is a major handicap in the evaluation of wear resistance and abrasion-resistant materials.

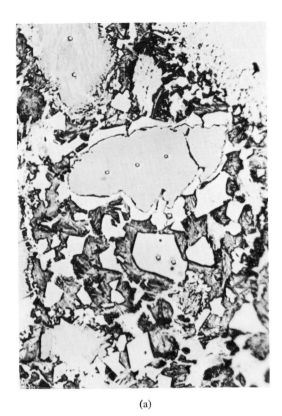

(a)   (b)

Figure 14.8   (a) Vickers Diamond Pyramid Hardness Impressions in Cast Tungsten Carbide (WC + W_2C) Granule (Upper Center) and Eta Carbide (Fe, W)$_6$C Crystal (Lower Center). The primary carbide granules have a VPN of about 2400, while the eta carbide that results from dissolving some of the tungsten carbide by the molten steel has a VPN of about 1745. A very light load (25 grams) is required for these tests. The darker matrix is martensite; its hardness is below 700 VPN. This is an oxyacetylene weld deposit from a composite tube rod containing 32% finer than 140 mesh and 68% 40/140 mesh granules. Magnification: 500×. (b) Higher Magnification Photomicrographs of a Gas-weld Containing Granules Sized 68% to 40/140 Mesh and 32% as Fines Below 140 Mesh, and Illustrating Formation of the Eta Double Carbide from a Granule of Primary Tungsten Carbide (a) As Well as the Transformed Austenite in the Background (b). Magnification: 1000× (Ref. 24)

10 HARDNESS

Probably the most common basis for approximately predicting wear resistance is a measurement of hardness. However, hardness can be misleading and should be relied upon only in the absence of better evidence. This is frequently the case. The engineer should be especially cautious in comparing two materials of quite different nature on the basis of hardness.

The reliability and the ranking ability of laboratory tests for hardness are well established and because of extensive treatment in engineering literature and standards they will not be questioned here. However, certain aspects of hardness itself merit discussion.

The term hardness applies generally to the properties that confer resistance to indentation, deformation, compression, cutting, scratching, or abrasion. From these, a correlation with wear resistance would be expected. However, the quality of toughness is also involved; in some cases it may be paramount, as in those where rubber outwears steel. In other cases, an approximate relationship exists, but, before hardness is used as an index of wear resistance the correlation should be established.

Minerals are commonly reported in terms of scratch hardness, whereas metals are generally tested by indentation methods. Data for comparing the two are therefore useful,

though it should be understood that they do not represent identical properties. With this qualification, several tables are included to facilitate approximate conversion from one system to another. As an engineer using these data may be familiar with some but not with the others, a very brief description of each is included.

Mohs' scale of hardness is a method that can be applied in the field, as it involves relative scratch hardness, with ten minerals as standards. The steps are not equal, unfortunately; the difference between No. 9 (corundum) and No. 10 (diamond) is probably as great as that between Nos. 1 and 9.

Brinell hardness for steels is obtained from the diameter of the indentation made by a 10-mm ball (hardened steel or cemented tungsten carbide) under a load of 3000 kg for 30 sec. It is reported as a Brinell hardness number, abbreviated here as BHN. It is a standard method for testing steels and other metals, but it cannot be used for brittle minerals such as those in Mohs' scale, because they fracture rather than yield to the indenting ball. Any correlation with Mohs' system must be based on another technique that is applicable to both types of material.

Rockwell hardness is another indentation method widely used for metals. It employs several types of indenters, the most useful of which is a diamond cone or "Brale." Each combination of load and indenter is designated by a letter. Thus the diamond Brale with a 150-kg load provides C scale values, here abbreviated as Rc. The Rockwell C scale is most applicable to hardened steels and hard-surfacing alloys. Brittle minerals fracture under the indenter, providing invalid results similar to the Brinell test behavior.

A diamond may be substituted for the round ball of the Brinell instrument. If very light loads are applied, an indentation may be obtained without fracture in brittle materials as well as in metals. The small size of such an indentation, which is measured with a microscope, permits reporting of the indentation hardness of the microconstituents of alloys (Fig. 14.8), provided very light loads are employed, whereas the Rockwell and Brinell indentations are so large that only the average hardness of a fine heterogeneous structure can be obtained. The Vickers instrument features a square-base diamond-pyramid indenter, the results of which are reported as Vickers pyramid numbers,* abbreviated as VPN. The Tukon tester also accommodates a Vickers indenter.

The Knoop indenter is a diamond pyramid differing from the Vickers type in proportions and depth of impression. It has a length-to-width ratio of 7.11 : 1.00, the long dimension of an indentation being measured microscopically in practice. It was developed to measure the hardness of very brittle materials and of microscopic areas. The apparent hardness values, reported as Knoop numbers,* are sensitive to the applied load and tend to rise sharply for very hard materials as the load is decreased. The Tukon tester is commonly employed for Knoop hardness determinations.

The Bierbaum "Microcharacter" is a scratch-hardness instrument for precision measurements. It employs a diamond point, cut as the corner of a cube, that is drawn across the surface to be measured under a load of 3 or 9 grams. The width of the V-shaped scratch is measured in microns under a calibrated microscope and converted to hardness numbers that are reported as K values. This type of hardness seems most

---

* Both the Vickers and Knoop hardness values reported herein are based on an applied load of 25 grams.

Table 14.3.   Knoop Hardness of Mohs' Minerals and Abrasive Materials[a]

Samples	Knoop Numbers
Gypsum	32
Calcite	135
Fluorite	163
Apatite parallel to axis	360
Apatite perpendicular to axis	430
Albite	490
Orthoclase	560
Crystalline quartz parallel to axis	710
Crystalline quartz perpendicular to axis	790
Topaz	1250
Carboloy	1050 to 1500
Alundum	1620 to 1680
Silicon Carbide	2050 to 2150
Boron Carbide	2250 to 2260
Diamond	5500 to 6950

[a] V. E. Lysaght, "The Knoop Indenter as Applied to Testing Non-metallic Materials Ranging from Plastics to Diamonds," *ASTM Bull.* **138**, 39 (Jan. 1946).

Table 14.4.   Relations between Mohs' Scale and Microhardness Numbers[a]. Variations in microhardness of the materials are due to variations in texture, impurities, and anisotropy.

Materials	Mohs' Scale	Microhardness	Average by H. C. Hodge and J. H. McKay
Talc	1	0.8– 21.5	1
Selenite	2	10.2– 56.6	11
Calcite	3	126 – 135	129
Fluorite	4	138 – 145	143
Apatite	5	870 –1740	
Orthoclase	6	2100 –2500	
Quartz	7	2066 –3906	2700
Topaz	8	2770 –4444	3420
Corundum	9	3906 –8264	5300

[a] As given in Williams, *Hardness and Hardness Measurements*, ASM, 1942, p. 133. Originally published in *Am. Mineral.* **19**, 4 (Apr. 1934).

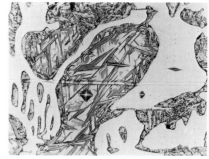

Constituent	Vickers	Knoop	Microcharacter	Rc*	BHN*	Wet-sand Abrasion Factor*
Martensite	362	555	1700± }	62	600	0.27
Carbide	1125	1130	5000± }			

*Rockwell and Brinell hardness and abrasion factor, when associated in this form, were determined, on the same specimen.

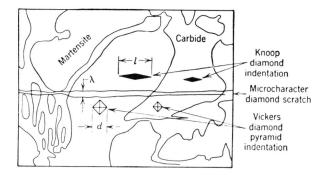

	Michrocharacter	Vickers	Knoop
Definition	$K = \dfrac{L-p}{\lambda^2 - q}$	$VPN = \dfrac{L}{Ar}$	$I = \dfrac{L}{Ap}$
Working Formula	$K = \dfrac{10}{\lambda}$	$VPN = \dfrac{1.8544L}{d^2}$	$I = \dfrac{L}{l^2 Cp}$

L = load in Kg
λ = width of scratch in microns (for 3 gram load.)
p and q = constants of diamond
d = diagonal in mm

l = long diagonal in mm
Cp = constant relating to projected area
Ap = projected (unrecovered) area
Ar = recovered area

Figure 14.9 Micro-hardness Methods Demonstrated on NI-hard Cast iron. Photomicrograph magnification : 300×

closely related to abrasive wear.[15] Unfortunately, the instrument is a research tool and is not adapted to shop or field use. The measurement of the impressions is an exacting and difficult operation. Like the Knoop and diamond-pyramid methods, it can deal with either brittle or ductile materials.

In Table 14.3 Knoop hardness and in Table 14.4 microcharacter values are reported in terms of Mohs' scale. These will permit evaluation of the microconstituents of hard-surfacing alloys in terms of the generally available data on minerals. In Figure 14.9, Knoop, diamond-pyramid and microcharacter impressions are shown. Table 14.5 contains values obtained on some of the constituents of hard-weld deposits. Table 14.6 is included for convenience in comparing Brinell and Rockwell C hardness value for ferrous metals.

Table 14.5. Hardness of Some Microconstituents in Hard-Surfacing Weld Deposits

Constituent	Microhardness		
	VPN,[a] 25-g Load	Knoop I, 25-g Load	Micro-character, K, 9-g Load
Ferrite			
In 0.20% C steel	104	154	
Pearlite			
In 1.6% Cr white cast iron	250– 355		
In 1.6% Ni:1.5% Cr white cast iron	314	453	
In 2.1% Ni:2.4% Cr white cast iron			1100–1525
In Ni-Cr white cast iron			772– 865
In 7.6% Cr white cast iron			550
Austenite			
3.9% Mn:3.6% Ni:2.1% Cr white cast iron			370– 590
1.5% C:11.8% Cr steel	353	412	
In Ni-Resist cast iron	165– 243		
Martensite			
In Ni-Hard cast iron	390– 640	506±	1200–2500
In Cr–Mo cast iron	473	682	
Cementite			
Iron carbide in gray cast iron	940		
In 1.6% Cr white cast iron	855–1080		
In Ni-Hard cast iron	1052–1600	1108+	2660–9000
In pearlitic white cast iron	1128–1270	1720	
Chromium carbides			
24% Cr white cast iron			5350–7400
1.5% C:11.8% Cr steel	1960		
Cr–Co–W	1672–2520	1720–2220	5730–7570
Chromium boride			
Ni–Cr–B–C	2668	2912	

[a] The Vickers pyramid numbers are roughly equivalent to Brinell hardness numbers in the range around 200 BHN, with heavy loads. At higher levels, the values diverge (695 VPN = about 614 BHN). This note is included for the benefit of engineers who are most familiar with Brinell hardness, with the caution that there is no precise equivalency, especially at low loads, although the mathematical bases are comparable.

Table 14.6. Approximate Hardness Conversions. Rockwell "C" Scale vs. Brinell and Vickers Pyramid Numbers

Rockwell "C" (conical diamond Brale)	Hard Steel Relationships		
	VPN[a] (diamond pyramid)	BHN[a] (steel ball)	BHN[b] (tungsten carbide ball)
40	389	375	368
41	404	388	375
42	420	401	392
43	429	408	401
44	437	415	411
45	454	429	425
46	472	444	436
47	494	461	450
48	505	470	465
49	515	477	478
50	540	495	492
51	553	505	508
52	567	514	523
53	598	534	544
54	616	544	565
55	633	555	586
56	654	566	593
57	675	578	610
58	717	601	623
59	741	614	638
60	765	627	652
61	792	640	668
62	820	653	678
63	853	668	700
64	885	682	716
65	921	697	728

[a] From ASM *Metals Handbook*, 1939, p. 127.
[b] From Williams, *Hardness and Hardness Measurements*, ASM, 1942, p. 463.

11 HOT HARDNESS

Interest in hot hardness results from the assumption that this property implies hot-wear resistance. Recognizing that even at ordinary temperatures hardness and wear resistance do not necessarily have a direct relationship, at elevated temperatures hardness has even more the status of an unvalidated wear test. Nevertheless, such information is useful as it is the first recourse for qualitatively ranking different alloys in the absence of a hot-wear test with proven validity. This area has been discussed in Reference 14 in considerable detail for hard-facing alloys, and the information here has been condensed from that source.

The Rockwell hardness test, modified to use a diamond indenter set in a superalloy holder and with a time factor included to reflect the influence of creep, was used to survey a variety of hard-facing materials up to 1200°F (649°C). The effect of time becomes increasingly important as temperature rises, and above 1200°F it may dominate the load-carrying and stress-resisting behavior of an alloy. Figure 14.7 shows the negligible effect at 200°F, the quite minor difference at 600°F, the modest but increased sensitivity to time at 800°F, and the

Table 14.7. Condensed Summary of Hard-Facing Alloys Useful for Hot-Wear Applications

A.W.S. Class	Type	W	T	Hot Hardness, °F				Room Temperature		W.S.A.F.
				1000		1200		Hardness		
				0	2	0	2	RC	BHN	
	Air hardening, low-alloy steel	G	X	24	14	−17	−50	50	—	0.70
		G	T	13	−9	−22	−71	35	—	0.80
		G	S	36	26	−5	−44	42	340	0.73
	Air hardening, medium-alloy steel	G	X	36	30	11	−11	60	—	0.68
		G	T	15	0	−20	−61	37	—	0.78
		G	S	40	32	8	−26	54	600	0.68
	Martensitic, medium-alloy iron	G	X	42	33	15	−7	60	—	0.32/0.42
		—	S	46	38	19	−8	63	652	0.37
	Martensitic, high-alloy iron	G	X	46	40	33	14	63	—	0.28 0.38
		—	S	50	44	32	7	62	—	0.33
RFe5-B	High-speed steel	G	X	48	45	31	13	63	—	0.52
		G	T	27	16	5	−22	46	—	0.62
		—	S	45	40	23	−3	52	477	0.58
RFeCr-A1	Austenitic, high-alloy iron	G	X	39	31	23	−9	56	—	0.65/0.85
		G	S	40	36	7	1	57	578	0.67
E/R CoCr-A	Chrome-cobalt-tungsten, No. 6 type	G/A	—	30	29	24	22	43	—	1.10/1.30
		G	S	31	30	24	20	44	387	1.17
E/R CoCr-C	Chrome-cobalt-tungsten, No. 1 type	G/A	—	44	44	39	33	55	—	0.50/0.75
		G	S	45	44	40	34	54	—	0.74

NOTES: W = weld method: G = gas; A = arc. T = treatment before hot testing: X = none; T = tempered 24 hr. at 1200°F; S = stabilized 24 hr. at test temperature. 0 and 2 indicate loading interval in minutes. W.S.A.F. = wet sand abrasion factor vs. 1.00 for S.A.E. 1020 steel. Hardness numbers are Rockwell "C" scale values.

profound time control of apparent hot hardness as loads are sustained at 1000°F. In general the surfacing filler metals are not much affected by temperatures below 800°F. However, at 1000°F most of them are not only softer but the creep effect becomes important. At 1200°F the iron-base alloys lose much of their hardness and cobalt- or nickel-base alloys show marked superiority. For this reason the chromium-cobalt-tungsten grades are popular for hot-wear applications above 1000°F. Table 14.7 provides a convenient summary of the hardness at room temperature, 1000°F (538°C), and 1200°F (649°C) for several important surfacing metals.[14]

A few alloys have the capacity to harden instead of softening in the vicinity of 1100°F. The high-speed steels are of this type, and the secondary hardening they exhibit, superimposed on their primary martensitic hardening, gives them a unique advantage in the temperature range from 800 to 1200°F. They are useful at low red heat for abrasion resistance as well as for cutting tools.

Graphs of hardness versus temperature are shown for a number of the important hard-facing alloys in the section about alloy types (Figs. 14.19–14.21, 14.27–14.29, 14.32).

12   EVALUATION OF ABRASION RESISTANCE

Though many abrasion tests have been described and have been the subjects of experiments, in the fields of interest here there has been no general acceptance of well-standardized procedures. Consequently the data provided herein depend very largely on one laboratory. However, they are based on some forty years of research, hundreds of field tests, and continual study in an attempt to clarify the important relationships between these. It was during this study that the three types of abrasion described briefly in Section 4 were identified as sufficiently different to merit separate classification.

A different laboratory test was developed for each kind of abrasion and then validated so far as possible against field tests. The erosion test uses a stream of sand that is dragged across the metal face by a rubber wheel, thus cushioning the abrasive to some extent and holding the imposed stress to a relatively low level that does not fragment the abrasive grains. The grinding abrasion test uses a copper lap over which the metal specimen is dragged in a pulp or slurry of sand and water. The force imposed is great enough to crush the abrasive and grind it finer as the test proceeds, thus establishing the high-stress conditions. Gouging abrasion is simulated by an Alundum grinding wheel that is prevented from loading during the test by means of air vibrators attached to the metal specimens.

To minimize variability from test to test each apparatus was set up to abrade a standard along with the specimen being evaluated. This did not work out for the rubber-wheel erosion test and the apparatus as finally standardized employed two rubber wheels operating at the same time. It was found that despite close geometrical and procedural similarity the two wheels behaved like separate tests and the standard run on one of them could not be considered as receiving treatment identical with that of the evaluated specimen. The test procedure

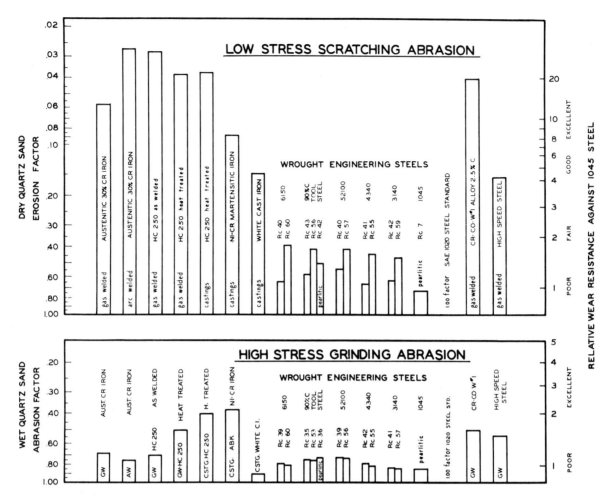

Figure 14.10   Comparative Resistance of Various Hard-facing and Engineering Alloys to Low- and High-stress Abrasion. Note that the merit ranking of certain alloys can be reversed under the two conditions (Ref. 17)

was then modified to run the standard and the specimen being tested in alternation, using the average weight loss of the standard as run before and after the "unknown" for the comparison and calculation of an abrasion factor. These comparisons are based on use of the same rubber wheel in each case. For the other two tests, two specimens (the standard and the "unknown"), are run at the same time under identical conditions and their weight losses can be compared directly with confidence.

For each test the specimens are weighed carefully before and after, weight loss is determined, and the ratio of the two computed. The abrasion factors thus reported are in terms of relative weight loss; a factor of 1.00 indicates weight loss identical with that of annealed 1020 simple carbon steel; a larger factor means more weight loss and thus poorer abrasion resistance than 1020 steel; and the values below 1.00 indicate better abrasion resistance by virtue of less weight loss. If a standard of some metal other than 1020 steel is used its representative weight loss in comparison with 1020 is carefully determined and the results converted to the 1020 basis in reporting on the "unknown" materials being evaluated. The reference material (1020 steel) was selected because of its reproducibility and availability.

To distinguish the several abrasion factors they have been given the following individual descriptions and abbreviations:

Low Stress: Dry Sand Erosion Factor or DSEF
High Stress: Wet Sand Abrasion Factor or WSAF
Gouging: Alundum Wheel Abrasion Factor or AWAF

It should be recognized that each of these tests can be run either wet or dry and that the abrasive can be changed at will. To avoid confusion in the following comparisons only the conditions just described were used and the abrasives were carefully controlled. Graded subangular quartz sand was used for the first two, initially at a grain size of 55 as measured by the AFS method.[16] The Alundum wheel, which is considerably harder than quartz, is of 70 grit aluminum oxide.

The method of reporting whereby merit is shown by low numbers may not be familiar to everyone. For those who think that better performance should be reflected by higher numbers, thus reporting wear resistance rather than wear, the two rating systems are illustrated in Figure 14.10.[17] Note in this figure, which compares the low and high stress abrasion tests on a group of identical materials, that there are some reversals of ranking. This is one of the justifications for different tests.

Note also that hardness seems to influence the behavior rather consistently in the erosion test while it appears less important than carbon content in the high stress test.

Each of these three tests has shown good reproducibility. The erosion test is qualitatively validated against such field service as sand sliding down a chute, plowshares in sand, and parts that dig in fine-grained and loosely consolidated sand or soil. The wet-sand grinding abrasion test has been well validated for ball-mill service provided the impact conditions of the ball mill are not excessive.[13, 18, 19] The Alundum grinding wheel test seems to have some relation to crusher service. The lack of more precise validation for the erosion and Alundum wheel tests is not a deficiency of the laboratory tests but a fault of the field evaluations. A full-scale service installation that will produce statistically acceptable field test data is still lacking for this validation.

To the battery of abrasion tests just described the determination of hardness should be added. It will exhibit partial validation in many cases but it can also be misleading. The engineer should be especially cautious when comparing two

materials of quite different nature. Some examples are instructive. Austenitic manganese steel at about 200 BHN outwears pearlitic white cast iron at 400 BHN or higher in ball-mill service or in the wet-sand abrasion test illustrated by Figure 14.11.[13] This is not because manganese steel work-hardens during the test as is frequently assumed. Its abrasion resistance seems to be inherent, as hardness changes little during the test, and if the steel is prehardened by impact its behavior is little different. Again, if soft 1020 carbon steel at about 110 BHN is compared to an aluminum bronze at 300 BHN, the harder bronze loses weight about four times as rapidly in the wet-sand abrasion test.

However, even though exceptions are possible and also the puzzling occurrence of wear of a hard object from contact with a much softer substance has been occasionally observed,[20] it should not prevent consideration of relative hardness. If the characteristics of the wearing agent are known and it is possible to select a hard-facing alloy of superior hardness, a prediction of good service performance is reasonable. With mineral substances the hardness values on Mohs' scale that are

Table 14.8. Hardness Comparisons[a]

Mohs' No. (1)	Mineral Equivalent (2)	Knoop No. (3)	VHN (4)	BHN (5)	Rockwell (6) B	Rockwell (6) C	Microcharacter (7)	Steel Equivalents
1	Talc	—	—	—	—	—	1– 21	
1.5								
2	Gypsum	32	30	30	20	—	10– 57	
2.5								
3	Calcite	135	135	135	75	—	126– 135	Ferrite (8)
3.5								
4	Fluorite	163	160	160	85	3	138– 145	Spheroidite (9)
4.5								Coarse pearlite (9), austenite (8)
5	Apatite	430	430	410	—	43	870–1740	Fine pearlite (9), coarse bainite (9)
5.5								Fine bainite (9)
6	Orthoclase	560	560	510	—	52	2100–2500	
6.5								Martensite (8)
7	Quartz	790	790	640	—	61	2070–3900	
7.5								Iron carbide (8)
8	Topaz	1250	<1250	—	—	—	2800–4400	Complex alloy carbides (8)
8.5								Complex alloy carbides (8)
9	Corundum	3000	<3000	—	—	—	3900–8300	
9.5								
10	Diamond	8000	<8000	—	—	—	—	

[a] This comparison of the Mohs' mineral scale and Microcharacter scratch scale with the common indentation scales for metals should not be considered a table for conversions, since the individual hardness methods are arbitrary and do not measure the same combination of properties. Conversion from one indentation scale to another is in itself very approximate and generally is not considered reliable except for purposes of making rough estimates of hardness level in terms of the hardness scale to which the conversion is made.

NOTES:

(1) Estimation of Mohs' scale hardness is generally restricted to 0.5 of a scale point.

(2) Minerals used by Mohs to establish the original scale.

(3) Knoop hardness measurements of the Mohs' scale minerals presented in Hoyt's *Metals and Alloys Data Book*, p. 7.

(4) From Knoop according to conversions given in *Trans. ASM* **33** 134 (1944).

(5) From VHN according to conversions, for steel ball, given in Williams' *Hardness and Hardness Measurements*, p. 462.

(6) Same as (5).

(7) Microcharacter hardness measurements of the Mohs' scale minerals given in Williams' *Hardness and Hardness Measurements*, p. 133.

(8) Related to Knoop hardness; unpublished data of the American Brake Shoe Company.

(9) Related to Rockwell C hardness; data presented by M. Gensamer, *Trans. ASM* **30**, 983 (1942).

ABRASION

A surface wear phenomenon caused by the cutting action of abrasives. These are usually non-metallic minerals or hard compounds.

VARIABLES:

1 Compressive stress behind the abrasive — 54 P.S.I.

2 Impact stress associated with abrasive — Negligible

3 Velocity of abrasive — Specimen moves at 119 ft. min.

4 Anchoring of abrasive

5 Loose abrasives

6 Suspension of abrasive — in 30 quarts of water

7 Concentration of abrasive — 40 pounds

8 Abrasive hardness — 7 – Quartz

9 Crushing character of abrasive

10 Sharpness of abrasive

11 Toughness of abrasive

12 Fineness of abrasive

Mohs Scale: 6 5 4 3 2 1

Grain Fineness No. 50 - 55

Figure 14.11   Schematic Diagram Showing the Variables That Are Included in the ABEX Corporation Laboratory Wet-sand Test for Grinding or High-stress Abrasion. This test correlates with ball mill service (Refs. 2, 13, 18, 19, 24)

Martensitic White Cast Iron (Ni–Hard)

Constituents: Very hard carbides (white) and martensite (acicular).

Chemical Analysis:

C%	Ni%	Cr%
3.18	4.58	1.39

| *Hardness* | | Wet-Sand |
Rc	BHN	Abrasion Factor
63	713	0.30

Figure 14.12   Good Structure Wet-sand Abrasion Resistance. Magnification: 1000 ×. (Ref. 2)

Pearlitic Gray Cast Iron

Constituents: Graphite flakes in a pearlite matrix. The pearlite alone (which consists of alternate lamellas of iron carbide and ferrite) has an abrasion factor of about 0.80, but the presence of the graphite increases it to about 1.20.

Chemical Analysis:

C%	Ni%	Cr%
3.02	0.95	1.43

Hardness: BHN	Wet-Sand Abrasion Factor
217	1.20

Figure 14.13   Poor Structure for Wet-sand Abrasion Resistance Magnification: 500× (Ref. 2)

Table 14.9. Classification of Hard-Facing Alloys by Chemical Composition

	Description	C%	Cr%	Other Elements
I	Ferrous Alloys			
IA	Hardenable Alloys			
I-A-1-	Carbon Steels			
a.	Low	0.19 max.		(Refer to AWS Spec. A5.1-48T)
b.	Medium	0.20–0.60		1% maximum
c.	High	0.61 min.		1% maximum
I-A-2-	Low Alloy Grades			
a.	Low Carbon	0.19 max.		(Refer to AWS Spec. A5.5-48T)
b.	Medium Carbon	0.20–0.60		0.4–6.0%
c.	High Carbon	0.61–1.50		0.4–6.0%
d.	Cast Iron Types	1.50 min.		0.4–6.0%
I-A-3-	Medium Alloy Grades			
a.	Medium Carbon	0.21–0.60	4–10	6–12% including chromium
b.	High Carbon	0.61–1.70	4–10	6–12% including chromium
c.	Cast Iron Types	1.70 min.	4–10	6–12% including chromium
I-A-4-	High Alloy Grades			
a.	Low Carbon	0.20 max.	10–14	12–30% including chromium
b.	Medium Carbon	0.21–0.50	10–14	12–30% including chromium
c.	High Carbon	0.51–2.5	10–14	12–30% including chromium
d.	Cast Iron Types	2.50 min.	10–14	12–30% including chromium
I-A-5-	High Speed Steels	0.61 min.	2–6	Tungsten and/or Mo
IB	Austenitic Steels			
I-B-1-	Chromium and Cr-Ni Grades			
a.	Low Carbon			(Refer to AWS Spec. A5.4-62T)
b.	High Carbon, low nickel	0.20–1.70	4.0 min.	4% maximum
c.	High Carbon, high nickel	0.20–1.70	15.0 min.	4% maximum; except for 7.0%-Ni min.
I-B-2-	High Manganese			
a.	Nickel-manganese type	0.70 min.		Minima of 3.5% Ni and 12.0% Mn
b.	Moly-manganese type	0.60 min.		Minima of 1.0% Mo and 11.0% Mn
IC	Austenitic Irons — not usually heat treated			
I-C-1-	High Chromium Iron	2.0 min.	22.0 min.	
I-C-2-	High Alloy Irons			
a.	Medium Carbon	1.7 min.	4–10	6–12% including chromium
b.	High Carbon	2.5 min.	10–14	12–30% including chromium
c.	Very high alloy	2.5 min.		Minimum of 40% iron
IIA	Cobalt Base — medium carbon		24.0 min.	40% Co min. 3.5 W min.
IIB	Cobalt Base — high carbon		24.0 min.	40% Co min. 10.0 W min.
IIIA	Tungsten Carbide Inserts			
IIIB	Composite Tungsten Carbide	1.8 min.		
IIIC	Tungsten Carbide Powder			
IVA	Copper-zinc alloys (brasses)			
IVB	Copper-silicon bronzes			85% Cu min. — 5.0% Si max.
IVC	Copper-aluminum bronzes			80% Cu min. — 8.0% Al min.
VA	Nickel-Copper Alloys			2–30% Ni, 70–98% Cu
VB	Nickel-chromium alloys			
VC	Nickel base, Cr-W-Mo alloys			

included in any standard text on mineralogy[21] are useful. For comparison, a rough correlation with metallurgical hardness scales is included in Table 14.8.

Caution in judging mineral hardness is advised. The important abrasive agent may be concealed from casual observaion. Thus limestone, which is softer than the martensite of hardened steel, may contain siliceous areas or quartz inclusions whose hardness considerably exceeds that of martensite. The harder substances will probably be the determining factor in service behavior.

A similar complication is encountered in abrasion-resistant metals. In heterogeneous structures such as cast irons, the various constituents may differ widely in hardness, ranging from soft graphite, which is deterimental to abrasion resistance, to very hard carbide. Also, the matrix in which the hard particles are imbedded may be a controlling factor. Conventional hardness tests, such as the Brinell or Rockwell, provide an average hardness, masking the effect of both hard and soft constituents (Figs. 14.12 and 14.13).

Carefully determined abrasion factors and hardness values

will be provided under the head of the various types of hard-facing alloys that are described here.

Table 14.1 was prepared for convenience in judging the role of various naturally abrasive and rock-forming minerals.

13 MATERIALS FOR HARD-SURFACING

Almost any metal or alloy that can be melted and cast may be utilized in the form of welding rod. This provides an embarrassing wealth of possibilities. From the most useful of these a number have been generally selected for commercial exploitation as hard-surfacing materials under dozens of trade names. As the concept of surface protection has broadened in recent years, a variety of stainless steels and corrosion-resistant alloys have been added to the list. Hundreds of others cover the multitude of engineering steels and nonferrous alloys that are provided in a form suitable for welding, primarily for joining metals, but that are available for surfacing upon occasion. This makes especially difficult the selection of material in the complex and composition sensitive grades. Manufacturers' recommendations provide a guide, but no guarantee, to optimum structure.

Some of these alloys that have achieved approximate industrial standardization are listed in the American Welding Society Specification A 5.13 (ASTM 399).[11] Its appendix provides a useful description for that group of alloys and is an aid to selection.

A logical classification of the hard-surfacing alloys is desirable. *Chemical composition* is widely used and reflects a fundamental aspect. However, it should be recognized that in many cases composition ranges are so wide that they do not define a fixed set of properties; instead a very great range in behavior is encompassed. The AWS 5.13 specification attempts to avoid this difficulty, and in general those alloys covered are well-defined types.

One chemical basis of classification is shown in Table 14.9, which has served to guide the development of specifications. Another that is much less satisfactory loosely classifies on the basis of total alloy content with the implication that merit increases with the amount of alloy. Since it can be demonstrated that this is a fallacy for many applications,[22] this method should not be used. Among other deficiencies it disregards the potent effect of carbon content. Where chemistry is used for description and as a guide for selection it should be complete, detailed, and narrowly restricted to those applications for which it has been clearly validated.

Metallurgical research has generally established that the constitution and structure of metals are fundamental factors that determine their properties. The sciences of metallography and physical metallurgy have developed around this trend. Composition is important and may be decisive in the production of a desired structure with its attendant properties, but other factors such as thermal history may be more influential. It is also possible to produce a required structure with several compositions or to produce several structures by different heat treatments of the same alloy. For these reasons it is desirable to classify hard-surfacing deposits in terms of structure where possible.

The classification scheme in Table 14.10 combines chemistry and structure. It has proved useful, but as with all convenient and simple generalizations, exceptions can be expected in the applicability sections of the table. It is better to use it as a guide to understanding, supplemented by study in detail of the material of a tentative selection before making a final decision.

The form in which a filler metal is provided may be only incidental. Surfacing alloys are sold in the form of solid wrought wire or cast rods; composite electrodes that have a simple steel-core wire and a coating that contributes essential alloy ingredients; tubes that reverse this scheme by using a mild steel sheath filled with alloys or hard compounds; powders that are applied by spray techniques; fluxes that are fed to the arc zone during welding; or separate ingredients (such as tungsten carbide granules) that the user can apply as he sees fit. The solid wire and the tubular electrodes are adaptable to automatic and semiautomatic as well as ordinary manual arc welding.

Some weldors do not hesitate to remove the coating from an electrode to obtain a bare rod for oxyacetylene deposition. This is an undesirable practice since some of the essential elements of the deposit may be in the coating. Gas welding should be done only with bare rods intended for this purpose.

Each of a number of important types of surfacing alloys are described in the following sections. The coverage is not exhaustive, but most of the important grades are included.

14 TUNGSTEN CARBIDE SURFACING MATERIALS

A number of carbides are outstanding in hardness. Among these are the carbides of tungsten, boron, columbium, tantalum, chromium, and iron. Of these, tungsten carbide is considered the most wear resistant. It has a quality of toughness as well as hardness that makes it superior to the others in abrasion resistance. Chromium and iron carbides, as produced by crystallization in iron-base weld deposits, are softer but have the advantage of being less expensive.

Tungsten Carbide,[23] either in the form of small cast slugs or as sintered shapes made by powder metallurgy, can be readily wetted by many iron and nonferrous alloys. Thus this very hard material can be incorporated into a variety of composites. Cutters for machine tools, such as lathe tool bits, are widely available as steel shanks containing a cutting edge of cemented carbide brazed in place. This is a common example of the carbide-insert method of providing maximum wear resistance at low cost. The principle can be adapted to any surface or edge that does not encounter heavy impact or excessive stress concentration. An important and widespread application is the hard setting of the cutting and wearing parts of oil-well drilling tools. Atomic hydrogen or oxyacetylene welding[8] is usually preferred for this kind of hard setting.

With large inserts the procedure involves first cutting grooves in the steel base metal by means of either an oxyacetylene cutting blowpipe, an arc-air torch, or a forging hammer. The size and spacing of the inserts will vary with the type of tool being protected. Next, the steel in one groove should be melted with an excess-acetylene flame (to avoid oxidation). Then one insert is picked up with the heated end of a high-strength steel welding rod and partially submerged in the molten puddle. Finally the insert is covered over with a deposit from the steel rod. This operation is repeated until

Table 14.10.   Classification of Surfacing Alloys

Classification by Basic Types	Important Features	Successful Applications
Tungsten carbide deposits	Maximum abrasion resistance	Oil well rock drill bits and tool joints
Granules or inserts		A wide range of severely abrasive conditions
Coarse granule tube rods	Worn surfaces become rough	Nonskid horseshoes
Fine granule tube rods	Best performance when gas welded	
High chromium irons	Excellent erosion resistance	
Multiple alloy type	Hot hardness from 800–1200°F with W and Mo	Abrasion by hot coke
Martensitic type	Can be annealed and rehardened	Erosion by (1000°F) catalysts in refineries
Austenitic type	Oxidation resistant	Agricultural equipment in sandy soil
Martensitic alloy irons	Excellent abrasion resistance	General abrasive conditions with light impact
Chromium-tungsten type	High compressive strength	Machine parts subject to repetitive metal-to-metal wear and impact
Chromium-molybdenum type	Good for light impact	A wide variety of abrasive conditions
Nickel-chromium type		
Austenitic alloy irons	More crack-resistant than martensitic irons	General erosion conditions with light impact
Chromium-molybdenum type		
Nickel-chromium types		
Chromium-cobalt-tungsten alloys	Hot strength and creep resistance	
High carbon (2.5%) type	Brittle and abrasion-resistant	Hot wear and abrasion above 1200°F
Medium carbon (1.4%) type		
Low carbon (1.0%) type	Tough and oxidation-resistant	Exhaust valves of gasoline engines. Valve trim of steam turbines
Nickel base alloys	Good hot hardness and erosion resistance	
Nickel-chromium-boron type		Oil well slush pumps
Nickel-chromium-molybdenum-tungsten type	Corrosion resistance but not abrasion resistance	Many corrosive environments
Nickel-chromium-molybdenum type	Resistant to exhaust gas erosion	Exhaust valves of trucks, buses and aircraft
Nickel-chromium type	Oxidation resistant	
Copper base alloys	Anti-seizing; resistant to frictional wear	Bearing surfaces
Martensitic steels		General abrasive conditions with medium impact
High carbon (0.65–1.7%) type	Fair abrasion resistance	
Medium carbon (0.30–0.65%) type	Good resistance to medium impact	Hot working dies
Low carbon (below 0.30%) type	Tough, economical	Tractor rollers
Semi-austenitic steels	Tough, crack resistant	General low-cost hard facing
Pearlitic steels	Crack resistant and low in cost	Base for surfacing or a build-up to restore dimensions
Low alloy steel	Suitable for build-up of worn areas	
Simple carbon steel	A good base for hard facing	
Austenitic steels	Tough; excellent for heavy impact	General metal-to-metal wear under heavy impact
13% manganese — 1% molybdenum type	Fair abrasion and erosion resistance	Railway trackwork
13% manganese — 3% nickel type	Lower yield strength	
13% manganese-nickel-chromium type	High yield strength for austenitic types	Many conditions involving heavy impact, with or without abrasion, and joining austenitic manganese steel parts
High carbon nickel-chromium stainless type	Oxidation and hot wear resistant	Frictional wear at red heat; furnace parts
Low carbon nickel-chromium stainless type	Oxidation and corrosion resistance	Corrosion resistant surfacing of large tanks

the desired number of inserts are set. For additional protection a final surface coat deposited from composite tungsten carbide tube rod may be added.

If a surface is armored with carbide inserts, subsequent wear may cause them to stand in relief. If a smooth surface is required this may become a distinct disadvantage and a homogeneous deposit may therefore be preferable. A very thin deposit with inserts is not usually practicable. Figures 14.14 and 14.15 show the abraded faces of composite materials.

Composite tungsten carbide welding rods are a more convenient arrangement for exploiting the wear resistance of this material. They cost less but the deposits usually have lower

Figure 14.14 Mosaic of Six Cemented Tungsten Carbide Inserts Brazed to a Block of SAE 1020 Steel. The abraded face shown was the result of two standard wet-quartz-sand abrasion tests. The abrasion factors obtained were 0.16 and 0.18 in comparison with the standard of 1.00 for annealed SAE 1020 steel (Ref. 2)

Figure 14.15 Mosaic of Angular Tungsten Carbide Fragments Set in a Matrix of Martensitic Cast Iron (Ni Hard). Note the differential wear. Mechanical loss of the harder fragments may occur when the matrix is deeply worn by loose abrasive. The wet-sand abrasion factor of this face was 0.36, vs. 1.00 for annealed SAE 1020 steel (Ref. 2)

abrasion resistance. Various devices are used to secure the simultaneous deposition of carbide particles and of the matrix alloy, one of the most common being a thin steel tube containing granules of tungsten carbide. The tubes can be coated for electric arc deposition like other rods.

Undissolved particles impart a sluggishness to the molten material that prevents ready flow or placement of the deposit. If the contained carbides are coarse they sink to the bottom; if very fine, they may dissolve in and alloy with the matrix

metal. In the latter case, the properties of tungsten carbide are replaced by those of a modified matrix alloy, and smoother, thinner deposits with somewhat higher impact resistance are obtainable.

These rods are essentially of two types — solid composites and tube composites — both of which are applied in much the same manner. This operation[8] should be performed with the oxyacetylene process, using a flame containing a small excess of acetylene. The application should be made without penetrating

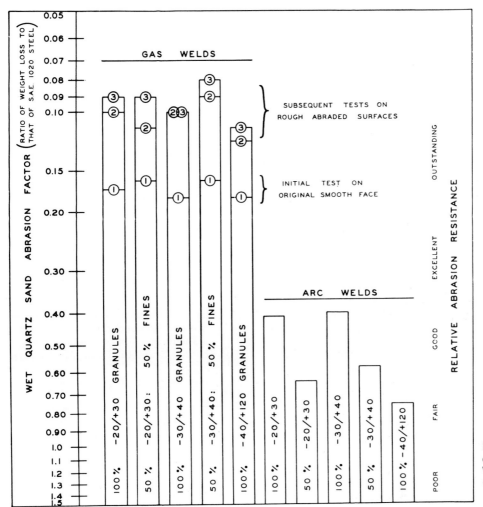

Figure 14.16 Abrasion Resistance of Composite Tungsten Carbide Weld Deposits as Affected by Granule Size and Welding Method (Ref. 24)

GAS WELDS   ARC WELDS

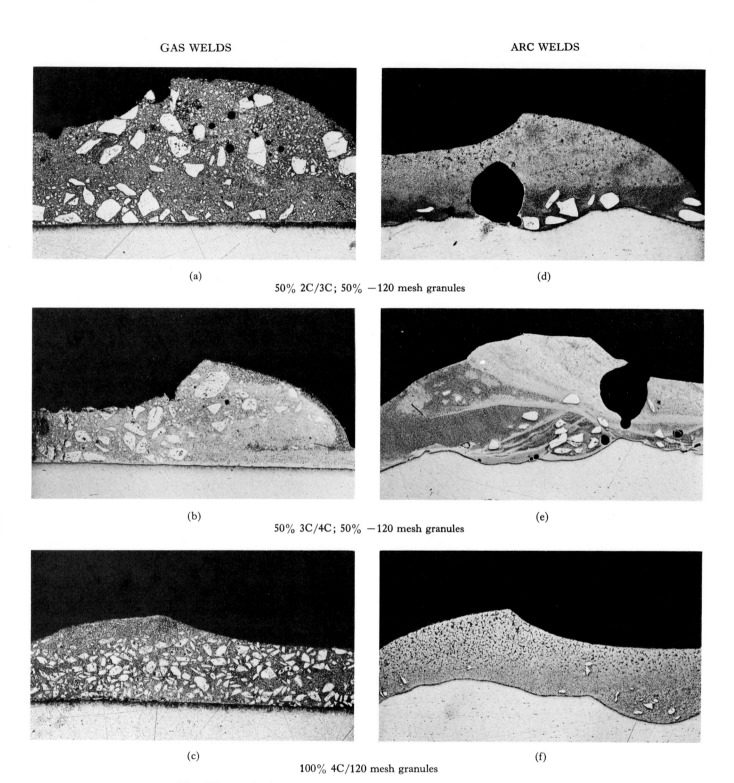

(a)   (d)

50% 2C/3C; 50% −120 mesh granules

(b)   (e)

50% 3C/4C; 50% −120 mesh granules

(c)   (f)

100% 4C/120 mesh granules

The difference in the proportions of undissolved tungsten carbide is significant

Figure 14.17 Composite Tungsten Carbide Weld Deposits — 9× Magnification Cross Sections Through Overlays. Welded Overlays of Tungsten Carbide Composites on a Soft Steel Base, Showing the Effect of Welding Method and Granule Size on the Structure. See also Figure 14.16 (Ref. 24)

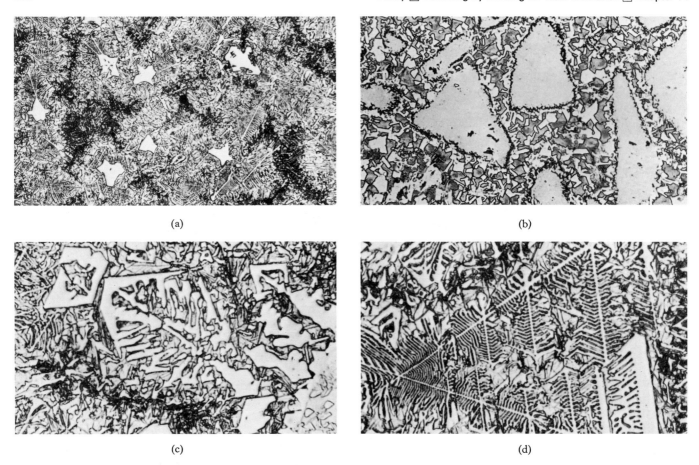

Figure 14.18   Composite Tungsten Carbide Weld Deposits. (a) Arc Weld from Tungsten Carbide Tube Rod Containing 100% 40/120 Mesh Granules, Showing Eta Carbide Crystals and a Eutectiferous Matrix. No primary granules appear because they were dissolved by the molten steel of the tube. Magnification: 250×. (Ref. 24). (b) Oxyacetylene Gas Weld from Tube Rod Containing 50% 30/40 Mesh and 50% Finer than 120 Mesh granules. Large primary tungsten carbide grains appear in a complex matrix of eta carbides and martensite. There is much less solution of the granules during gas welding in contrast to arc welding. Magnification: 250 ×. (c) Complete Matrix Resulting from Reaction of Molten Steel with Tungsten Carbide. Eta carbide (white) is mingled with martensite in a eutectic pattern. Magnification: 1000 ×. (Ref. 24). (d) Eutectic or Lowest-melting-point Composition in the Ternary Iron-Tungsten-Cementite System. The chief constituents here are the eta phase, $(Fe, W)6^c$, and austenite that has partially transformed to martensite. From an arc weld with 100% 40/120 mesh granules. Magnification: 1000×. (Ref. 24)

as deeply into the base metal as in ordinary steel welding. A certain amount of stirring with the rod is necessary to obtain the most even distribution of the deposited metal. These rods do not flow as freely as most ordinary welding rods, because of the presence of the refractory tungsten carbide particles which are not melted during the application. It is best to avoid keeping the deposit molten for too long a period. The welding tip should be large enough to produce a flame that will supply the required amount of heat with the pressure low enough to avoid blowing the molten metal.

The sintered carbide tool bit industry uses the WC type carbide,[23] perhaps combined with titanium or other carbides. In contrast the surfacing material, which is a cast, crushed, and sized product, is a relatively low-melting and intimate mixture of WC and W_2C containing from 2.6 to 4.2 per cent carbon and a minimum of 94 per cent tungsten. It is marketed in bulk but most of it is employed in the form of composite tube rods. Granule sizes are specified by screen mesh sizes; 8–12 mesh, 20–30, 30–40, 40–120, and minus 120 mesh grades are typical of those available. These numbers represent meshes per inch of the classifying screens in the AFS system. The

coarser grades are used for such applications as nonskid horseshoes, while the finer grades are used for producing relatively smooth deposits. The usual tube rod contains 60 per cent by weight of the filler granules.

The granule size of the tube filler not only controls surface roughness (apparent especially after differential wear of the matrix) but also controls matrix character. The matrix is not a mild steel as some think, but is a special product that results from solution of some tungsten carbide in the molten iron. Depending on the temperature, the granule size (which affects the ease of solution in the iron), the method of welding, and the details of welding, the matrix ranges from a medium-carbon tungsten steel to a high-tungsten cast iron. The circumstances have been described in detail.[24] Figure 14.16 from this paper illustrates this graphically and shows the potent effect of granule size and welding method on abrasion resistance. Figure 14.17 shows the marked difference in the amount of undissolved tungsten carbide with the two welding methods, explaining why the abrasion factors in Figure 14.16 are so different. Figure 14.18 illustrates the complexity of the structures that can develop in the matrix. Besides the original carbide there may be a considerable amount of the eta double carbide, $(Fe, W)_6C$, such as occurs in high-speed steel.

Rock drill bits are armored with cast tungsten carbide, applied by several methods. Use of the tube rod is very common, but in some cases the carbide granules are sprinkled on to a surface made sticky with sodium silicate solution and are then fused in place by an atomic hydrogen flame. In use the carbide armored teeth of a rotary cone rock drill are subjected to great pressure; the contact with rock crushes it, and so the drilling is accomplished largely by overstressing and fragmentation. This is believed to be an extreme case of high-stress abrasion. For this reason the materials have been evaluated in the laboratory chiefly by the wet-sand abrasion test as in Figure 14.16.

However, cast tungsten carbide is outstanding in abrasion resistance by all criteria. The low stress test with a rubber wheel hardly affects it. As soon as composite deposits begin to wear into the matrix, the carbide in relief tends to cut the rubber wheel and thus ruin the test. If the gouging abrasion test is attempted with an Alundum grinding wheel here again the wheel suffers! One manufacturer and user of tungsten carbide armored rock bits actually evaluates the carbide quality by the wear that is caused on the Alundum wheel face.

Tungsten carbide is not likely to be selected for properties other than hardness (cast granules average about 2400 VPN with a 25-gram load) and abrasion resistance (wet-sand abrasion factor around 0.10). It is not particularly corrosion resistant, it is vulnerable to oxidation at red heats and above,

and the roughness that results from differential matrix wear tends to make composite deposits unsuitable for friction faces. However, its hot hardness is good up to 1000 or 1200°F (Fig. 14.19),[24] and it thus can serve in some hot-wear applications that involve abrasion. Hot hardness of the composite is determined largely by the amount of tungsten that has been dissolved by the matrix, and because of this the arc-weld deposits seem to retain hardness at 1200°F better than do gas welds.

15   HIGH-CHROMIUM IRONS

These alloys depend for their properties on a combination of 25–30 per cent chromium and several per cent carbon. These combine and crystallize from the molten weld pool (or in castings) chiefly as the chromium carbide, Cr_7C_3. This has a hardness of about 1800 VPN[17] and is thus harder than quartz and the hard silicates that dominate most industrial abrasive situations. However, these carbide crystals are more brittle than tungsten carbide and may perform poorly under high-stress abrasive conditions unless they are very well supported in a quite hard matrix.

There are a number of types of these alloys. Austenitic chromium iron as represented by AWS Specification A 5.13 Grades R FeCr and E FeCr (Table 14.1) has been popular. It has proved to be an excellent material for resisting low-stress

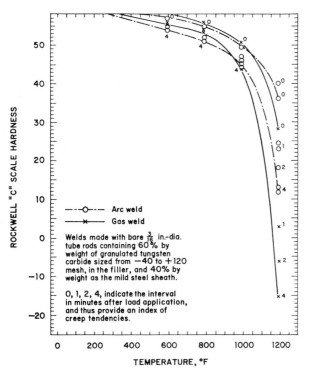

Figure 14.19   The Effect of Temperature on Apparent Hot Hardness of Composite Tungsten Carbide Weld Deposits. Welds made with bare $\frac{3}{16}$-in.- diameter tube rods containing 60 per cent by weight of granulated tungsten carbide sized from −40 to +120 mesh, in the filler, with 40 per cent by weight as the mild steel sheath

Table 14.11   Properties of Austenitic High-Chromium Iron

Type composition	4.0–5.0% C	6% Mn	2% Si	25–30% Cr

	Oxyacetylene Gas Welded	Small Castings[a]	Electric Arc Welded
Resistance to flow under repeated impact	Excellent	Excellent	Good
Tensile toughness	Low	Low	Low
Resistance to cracking under repeated impact	Low	Low	
Oxidation resistance	Excellent	Excellent	Excellent
Maximum temperature considered advisable	2000°F	2000°F	2000°F
Creep resistance	Fair	Fair	Fair
Rockwell hardness at[b]			
70°F	C50–58	C57	C50–C58
600°F	C55	C50	
800°F	C48	C47	
1000°F	C39–C40	C36	C37
1200°F	C23–C25		C6
Compression strength			
0.10% set yield, psi		86–122000	
Ultimate, psi		152–287000	
Elastic deformation, %		0.7–1.1	
Plastic deformation, %		0.4–3.4	
Abrasion resistance	Fair	Fair	Fair
Wet sand abrasion factor	0.67	0.88	0.74
Erosion resistance	Excellent	Excellent	Excellent
Dry sand erosion factor	0.04–0.07		

[a] Properties considered between those of gas and electric weld deposits. Used for some test specimens to eliminate welding variables and defects.

[b] Instantaneous values. See Ref. 14 and Figure 14.20 for effect of creep.

or scratching abrasion. The weld deposit can be placed with precision where it is needed and in use tends to polish to a smooth surface. Thus it is widely used to protect agricultural equipment such as plowshares. Good performance is expected in all low-stress abrasive situations provided impact is absent or is very light.

The austenitic types are not hardenable by heat treatment. The relatively soft matrix does not resist either tensile or compressive stresses well, and probably because it does not support the carbides in the face of high compressive stress the austenitic grades are mediocre against high-stress grinding abrasion. The austenitic matrix is conferred by the several per cent of manganese or nickel that these types carry.

If the alloy content of a high-chromium iron is adjusted to avoid overstabilization of the austenite, upon cooling there may be some transformation to the harder martensite. Such alloys differ in several important respects from the grades listed in Table 14.11. They can be hardened by heat treatment, in some cases they can be softened to permit machining

Table 14.12   Properties of Hardenable High-Chromium Iron

Type composition	2.5% C 1.0% Mn 1.0% Si 28% Cr		
	Oxyacetylene Gas Welds	Small Castings	Electric Arc Welds
Resistance to flow under repeated impact	Excellent		Good
Resistance to cracking	Fair	Fair	Fair
Tensile toughness	Fair	Low	Fair
Compressive properties Yield strength, 0.05% set		145,000 psi[a]	
Ultimate strength		404,000 psi[a]	
Elastic deformation		1.2%[a]	
Plastic deformation		2.5%[a]	
Oxidation resistance	Excellent	Excellent	Excellent
Useful up to		1800–2000°F	
Rockwell hardness at 70°F	C55 as welded	C56 as cast	C48–C57 as welded
70°F annealled at 1400°F	C42		
70°F hardened from 1950°F	C59	C65	C60
800°F hardened from 1950°F		C54	
1000°F hardened from 1950°F		C48	
1200°F hardened from 1950°F		C5	
Resistance to grinding abrasion	Excellent[a]	Excellent[a]	Good[a]
Wet sand abrasion factor	0.36[a]	0.30–0.40[a]	Variable
Resistance to scratching abrasion	Excellent	Excellent	Good
Dry sand erosion factor	0.03–0.15	0.04–0.12	0.03–0.22
Thermal expansion coefficient 70–1500°F		0.0000081 in/in/°F	
1700–2000°F		0.0000143 in/in/°F	

[a] Hardened by air cooling from 2 hours at 2000°F.

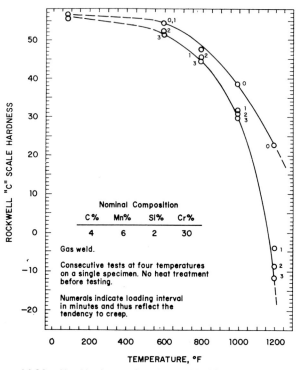

Figure 14.20   Hot Hardness of an Austenitic 28 per cent Chromium Iron (AWS Specification A 5.13-56T. Grade R Fe Cr-A 2). This alloy, which is widely used for surfacing agricultural equipment, has also proved to be good for protecting bar mill guides in a steel mill (Refs. 14, 16)

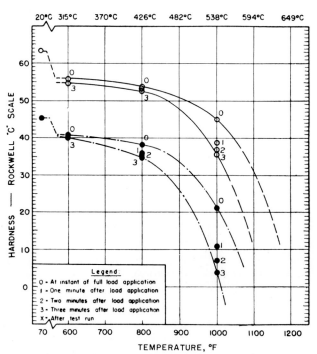

Figure 14.21   Hot Hardness of a Martensitic High-Chromium Iron as Affected by Temperature and Thermal History. The upper curves represent the cast alloy, as-hardened (oil-quenched after 6 hours at 1900°F), while the lower curves reflect tempering (24 hours at 1000°F) of this hardened structure (Ref. 14)

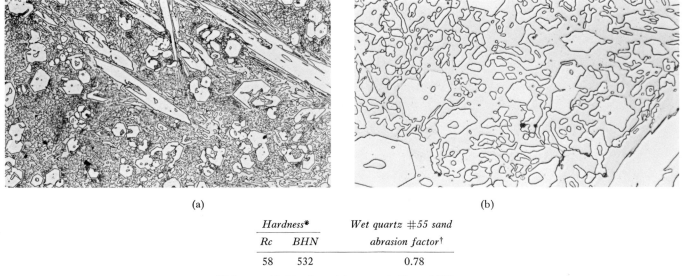

(a)   (b)

Hardness*		Wet quartz #55 sand
Rc	BHN	abrasion factor†
58	532	0.78

*Subsequent hot hardness data appear in Figure 14.20
†Versus 1.00 for SAE 1020 steel

Figure 14.22   Structure of Gas-weld Deposit from 4%C-6%Mn-2%Si-28%Cr Welding Rod. The light trigonal chromium carbides (basically Cr_7C_3) in the darker austenitic matrix have a hardness of about 1750 VPN (25 gm load) and are very resistant to erosion, though they are quite brittle. (a) Magnification: 100×; (b) Magnification: 500× (Ref. 17)

operations, and they are somewhat tougher. The properties of one type are summarized in Table 14.12.[17]

Both grades exhibit the oxidation resistance conferred by 25 per cent or more chromium. They are occasionally used for heat-resisting and hot-wear applications. Their hot hardness (Figures 14.20 and 14.21[14]) is inferior to that of the Cr-Co-W alloys at 1000°F and above, but below 1000°F they may be more economical.

Other alloying elements are sometimes added to the high-chromium iron base. These include molybdenum, tungsten, combinations of these two, and titanium. Little has been published about these modified alloys. The most obvious advantage is an increase in hot hardness because of the secondary hardening conferred by molybdenum and tungsten. One combination features superior abrasion resistance (U.S. Patent 2776208), providing in air-cooled weld deposits a resistance to high-stress abrasion ordinarily achieved only after heat treating the hardenable grade, as in Table 14.12.

Additional details about these alloys are provided in a previous publication[17] and in the appendix to AWS Specification A 5.13.[11]

16   MARTENSITIC IRONS

The martensitic irons consist essentially of a mixture of the microconstituents *carbide, martensite,* and *austenite.* Each of these has distinctive properties and is moreover affected by alloy content. Austenite and its harder transformation products, which are discussed in more detail in Section 18, are also affected by thermal history.

The high-chromium martensitic irons previously mentioned differ from most of these grades in that the carbides tend to crystallize from the melt first, and are thus found in a matrix of the austenite–martensite aggregate. The carbide, Cr_7C_3, is also harder than the more common iron carbide, Fe_3C, and its modifications (Fig. 14.22).

Most of the other martensitic irons are lower in alloy content and exhibit a reversal of the matrix–carbide relation. Ordinarily the austenite forms first from the melt, becomes progressively richer in carbon content as it separates from the fluid portion, and finally the composition reaches the eutectic (or lowest melting) point. The solidification of the alloy occurs rather quickly at a definite temperature after the eutectic composition is reached, and the product is a rather intimate mixture of carbide and austenite that is called ledeburite (Fig. 14.23). After further cooling the austenite transforms, at least partially, to martensite or other hard aggregates. However, if carbon content is high enough, carbides separate initially and the end result may be a mixture of carbide and ledeburite rather than austenite and ledeburite. Most of the martensitic irons, whether as castings or weld deposits, as seen under the microscope tend to exhibit a continuous matrix of

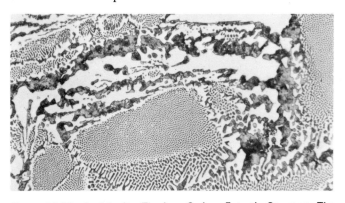

Figure 14.23   Ledeburite, The Iron-Carbon Eutectic Structure. The eutectic is the lowest melting composition of an alloy series. It contains in finely divided form the same constituents that appear elsewhere in coarser form; Cr—Mo—Fe carbides (white) and austenite that has transformed to pearlite (dark grey). Hardness: 600 BHN (3.8% C, 7.6% Cr, 2.0% Mo). The eutectic is formed frequently on the surface by "sweating" with a reducing gas flame preparatory to adding surfacing filler metal. Magnification: 250×

Figure 14.24   Arc-welded Overlay of the Ni-Hard Type, Showing Alloyed Iron Carbide (Cementite), Austenite, and High-carbon Martensite. Hardness: 560 BHN (3.1% C, 3.9% Ni, 1.3% Cr). Magnification: 250×

carbide with large or small islands and masses of partially transformed austenite.

Because of this carbide matrix these materials tend to have quite high compressive strength, good abrasion resistance, and relatively low tensile strength. Their performance is affected considerably by the amount of transformation to martensite that has occurred. When alloy balance and weld cooling rates are optimum these irons can provide excellent and economical protection against abrasion of all kinds.

Probably the best known alloy of this type is the 4.5% Ni — 1.5% Cr grade known as Ni-Hard (Fig. 14.24). It is widely used in the form of castings but has not been very popular for hard-facing. For surfacing, the chromium-molybdenum grade (Fig. 14.25), which has been widely used for some thirty years, seems to be preferred. For hot-wear applications the more highly alloyed chromium-tungsten type may have an advantage (Fig. 14.26). Properties of three grades appear in Table 14.13 and hot hardness is shown in Figures 14.27,

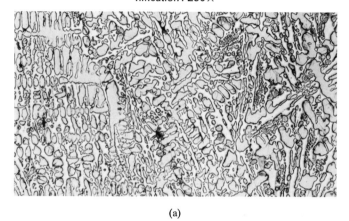

(a)

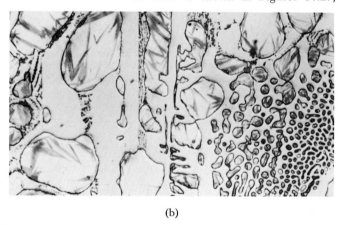

(b)

Figure 14.25   Gas-weld Deposit of a 5% Chromium–4% Molybdenum Martensitic Iron. Hardness: 600 BHN. WSAF: 0.35. This is one of the best of the martensitic irons for general protection against abrasion. The matrix is of alloyed carbides; the islands are austenite partially transformed to high-carbon martensite. (a) Magnification: 100×. (b) Magnification: 500×.

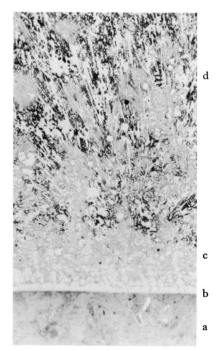

d

c

b

a

Magnification: 100×

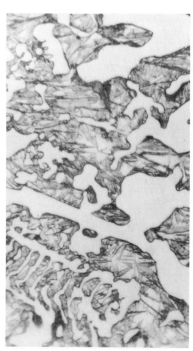

Magnification: 1500×

Figure 14.26   Oxyacetylene Torch Deposit of a Chromium-Tungsten Martensitic Iron on a Soft Steel Base, Showing (a) The Portion of the Base That Has Absorbed Carbon from the Weld Deposit; (b) The Junction Zone of High-carbon Alloy Steel; (c) A Zone of Carbide and Martensite (1500×); and (d) The Fine Mixture of Carbides and Transformation Products in the Bulk of the Overlay. Hardness: 782 BHN. WSAF: 0.36. This alloy has very high compressive strength. It is more expensive than the Cr—Mo iron in Figure 14.25 but has higher hot hardness

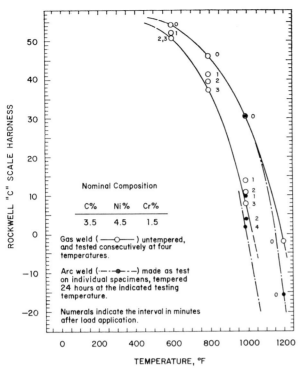

Figure 14.27   Hot Hardness of the Ni-Hard Type Hard-Facing Alloy
(Ref. 14)

14.28, and 14.29. There are other variations available but their characteristics are less well defined.

The martensitic irons also exhibit good metal-to-metal wear resistance, probably because of their high compressive

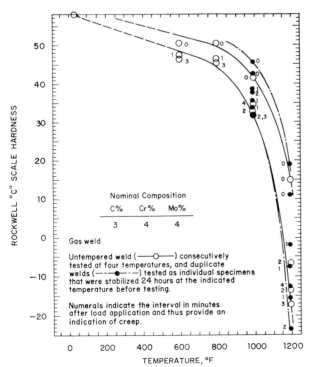

Figure 14.28   Hot Hardness of Gas-weld Deposits from Chromium-Molybdenum Martensitic-iron Hard-facing Alloy; Tested with and without Prior Tempering (Ref. 14)

strength. Thus they qualify as general-purpose hard-facing alloys. Their salient limitation is brittleness; though suitable for light impact they are likely to crack under medium or heavy impact. They also tend to crack from thermal stresses, and where these cannot be avoided, as in some weld deposits, cracking behavior must be considered. This does not mean that cracks are unacceptable; for many applications they are not serious and usually terminate when they reach the tougher

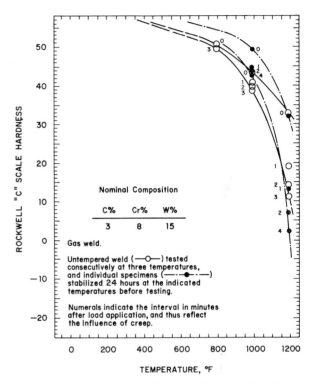

Figure 14.29   Hot Hardness of Gas-weld Deposits of a Chromium-Tungsten Martensitic-iron Hard-facing Alloy; Tested with and without Prior Tempering (Ref. 14)

base metal. Where cracks are not acceptable the alloy choice usually shifts to austenitic grades or to lower-carbon hardenable alloys.

17   AUSTENITIC IRONS

All surfacing by welding is likely to result in residual tensile stresses on the surface as a result of the thermal gradients during welding. These can cause cracking and are very likely to do so with thick deposits, multiple layers, or where large areas are surfaced. Even with single-layer deposits cracks may form in martensitic irons. This is due partly to the brittle carbide matrix of such irons and partly to the volume change as austenite transforms to martensite. The cracks are always disconcerting, sometimes unacceptable, and occasionally detrimental to abrasion resistance.

Cracking can be minimized, if not prevented entirely, by alloying the iron to prevent the martensitic reaction. Such irons have become popular as a result of operator and user appeal; there is a loss of abrasion resistance, but this is not obvious without well-controlled tests. As a result austenitic

Table 14.13　Properties of Some Martensitic Irons

Type Composition	4.5% Ni 1.5% Cr Chrome-Nickel Type			4% Cr 4% Mo Chrome-Molybdenum Type			8% Cr 15% W Chrome-Tungsten Type		
	Gas Welded	Small Castings[a]	Arc Welded	Gas Welded	Small Castings[a]	Arc Welded	Gas Welded	Small Castings[a]	Arc Welded
Resistance to flow under repeated impact		Good			Good			Excellent	
Tensile toughness		Low			Low			Low	
Resistance to cracking under repeated impact		Low			Fair			Fair	
Oxidation resistance		Low			Low			Medium	
Maximum temperature considered advisable		1000°F			1100°F			1200°F	
Creep resistance		Low			Fair			Fair	
Rockwell hardness at[b]									
70°F	C60		C64	C63		C57	C62		C65
600°F	C51	C48		C51–54					
800°F	C51	C42		C52					
1000°F	C27	C17	C42	C46–43		C43	C50		C51
1200°F	C-13		C-8	C19-7		C4	C32		C27
Compression strength									
0.10% set yield, psi		180,000			172–285,000			184–226,000	
Ultimate, psi		260–380,000			310–386,000			292–334,000	
Elastic deformation, %		0.9–1.6			1.2–1.4			0.9–1.0	
Plastic deformation, %		0.6–3.0			0.3–1.4			0.3–0.5	
Abrasion resistance	Good	Good	Fair	Excellent	Excellent	Good	Excellent		Good
Wet sand abrasion factor	0.37/0.57	0.30/0.65	0.60/0.74	0.30/0.46	0.34/0.47	0.38/0.54	0.28/0.50		0.40/0.60
Erosion resistance	Good		Fair	Excellent		Good	Excellent		Good
Tensile strength, psi		22,000–43,000							
Elongation, %		0.001–0.050							
Brinell hardness		500–650							

[a] Properties considered between those of gas and electric weld deposits. Used for some test specimens to eliminate welding variables and defects.
[b] Instantaneous values. See Ref. 14 and Figures 14.27–14.29 for effect of creep.

irons are frequently selected in preference to the more abrasion-resistant martensitic irons. Specifications do not yet cover such grades but one widely used type contains about 15 per cent chromium and 1 per cent molybdenum. Manganese may be around 2 per cent. The high-stress abrasion resistance is likely to be similar to that of martensitic steels and sometimes lower. However, the erosion resistance that depends largely on the volume of hard carbides is expected to be good. The popularity of this hard-facing alloy suggests that it is reasonably satisfactory.

It will be recognized that the material in this chapter, which is intended for engineers, contains much technical detail. While this is important for critical engineering judgments, there are many small operators in the mining, quarry, and construction fields who lack both time and inclination to study this detail. For them the surfacing alloy choice is made easier by the recommendation of just two grades: the 15Cr-1Mo type above and a high-performance austenitic manganese steel that is described later. Both have good operator appeal and the manganese steel is very tough and suitable for heavy-impact service. The iron is used where service involves light impact and usually low-stress abrasion. These three factors usually dominate most situations in earth-handling applications. Of course this pair of alloys does not cope with the need for hot hardness, corrosion resistance, or very high elastic compressive strength, but these properties may not be important in mining and quarrying.

18　MARTENSITIC STEELS

If the carbon content of an alloy is reduced, the volume of hard carbides decreases. At the same time the alloy may be expected to become tougher. In the iron-base alloys carbon can range from around 0.10 per cent up to perhaps 5.0 per cent. This permits a tremendous range in toughness and abrasion resistance. The carbon limit for steels is usually set at about 1.5 or 2.0 per cent carbon, the upper limit of carbon solubility being near 1.7 per cent. Thus even for steels the range in properties is wide. Unfortunately this is frequently neglected or not realized, the tendency being to focus on the nominal alloy content of metals. Recognition of this tendency is especially important when the comparative service performance of wear-resistant materials is being evaluated.

The martensitic steels represent a long step toward toughness from the hard irons. In terms of tonnage they probably dominate the hard-facing field. Also, there is probably more variety in compositions for this group than for any other employed for surfacing. Though Table 14.9 provides some basis for their classification it is difficult to offer simple generalizations, because the same results can be achieved with

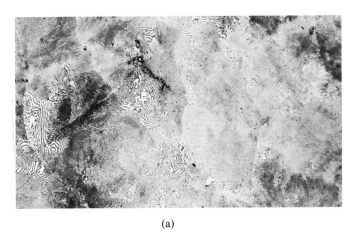 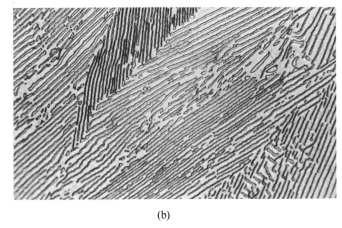

(a)   (b)

Figure 14.30   (a) *Pearlite* as it May Appear in Steels Containing About 0.8% Carbon. The fine structure is not resolved at this magnification, but shows the characteristic pearlite lamellae at higher magnification, as in Figure 14.30 (b). High-carbon unalloyed steels will tend to be pearlitic after air cooling from weld deposition. Magnification: 100×. (b) The lamellar structure of the grey, unresolved areas in Figure 14.30 (a) are shown here. This structure has moderate toughness and abrasion resistance but it is better as a base or a build-up layer than as a surface overlay. Quenching or incorporation of alloying elements is required to produce harder steel microstructures. Magnification: 1500×.

a number of alloy combinations. However, it is possible to offer guidance about carbon content.

Alloys with carbon contents below 0.20 per cent are not much used for hard deposits. In this range most of the structural-steel welding alloys are found, even if they feature rather high strength levels. (Refer to AWS Specifications 5.1[25] and A5.5[26].)

A fully martensitic structure can attain its maximum hardness at around 0.55 to 0.60 per cent carbon. Thus the range from 0.20 to 0.60 per cent carbon is a logical category into which many commercially available steels fall. Figure 14.6 shows this relation between hardness and carbon content.

Attainment of the hard structure known as martensite requires metallurgical control. With simple steels this involves a heat treatment: quenching from above 1400 to 1500°F. However, only in light sections can full hardening be attained thus. Heavy sections or weld deposits that cool with similar

slow rates require additional "hardenability," which is conferred by alloying elements in addition to carbon. Chromium, molybdenum, nickel, tungsten, and vanadium are used for this purpose, with chromium the most common. If the alloying is scant, pearlite may form upon cooling (Fig. 14.30); over-alloying produces too much retained austenite.

There are at present no clear-cut specifications that will ensure optimum alloying for martensitic-steel hard-facing alloys. A number of "self-hardening" or "air-hardening" materials are available, usually in the form of covered electrodes, that exploit alloying for additional hardenability. These can be purchased on the basis of carbon content and deposited hardness. The specific alloy content is less important than the way in which it is balanced for an intended use (Table 14.14).

There are several air-hardening tool steels that produce martensitic weld deposits. Of course they can also be heat-treated like tool steels. If they are used to repair tool-steel

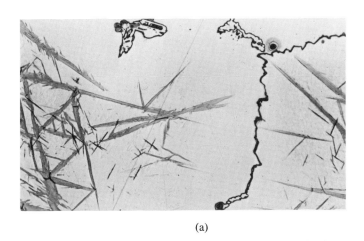 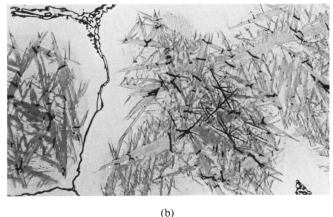

(a)   (b)

Figure 14.31   (a) Martensite That Has Formed in 1%C, 6%Cr, 1%Mo Steel During Air Cooling. The blades are typical of high-carbon martensite. The thick grain boundary and carbide islands also reflect the high carbon content. There is considerable retained austenite in this matrix. Etchant: Hot 4% Picral. Magnification: 250×. (b) The Same Steel As in Figure 14.31 (a) After Air Cooling to Room Temperature and Refrigerating to −20°F. Additional blades of martensite have formed by austenite transformation. Their brittleness is attested by the many micro cracks that have ruptured the blades. Etchant: Hot 4% Picral. Magnification: 250×

Table 14.14　Compositions of Martensitic Steel Surfacing Alloys

AWS-ASTM Class	C%	Mn%	Si%	Cr%	Ni%	Mo%	W%	V%	
E 502	0.10	1.0	0.9	4–6	0.4	0.55	—	—	
E 410	0.12	1.0	0.9	11–13.5	0.6	—	—	—	
E 430	0.10	1.0	0.9	15–18	0.6	—	—	—	
	0.08	0.7	0.8	9	—	1	—	—	
	0.11	2	1.1	0.6	5	4.5	—	—	
	0.12	1.2	0.8	2.1	—	1	—	—	Build-up type
	0.23	0.8	0.5	0.4	1.2	0.6	—	—	Build-up type
	0.23	1.2	0.5	0.4	1.4	0.4	—	—	Build-up type
	0.25	1.8	0.7	2.8	1.3	0.5	—	—	
	0.25	2.0	0.7	4.3	—	0.8	—	—	
	0.30	1.8	0.8	4.0	—	0.5	—	—	
	0.30	0.8	0.8	4.5	—	1.5	1.5	—	
	0.30	1.7	1.2	6.0	—	0.8	—	—	
	0.50	2.0	1.4	9.0	.—	1.8	—	—	
	0.52		0.9	4.8	—	1.4	—	—	
	0.58	1.1	0.6	2.5	—	0.4	—	—	
	0.60	1.2	0.8	5.5	—	0.5	—	—	
	0.60	1.5	0.8	6.0	—	0.5	—	—	
	0.60	1.1	0.6	2.5	—	0.4	—	—	
	0.65	1.0	0.8	5.8	—	0.5	—	—	
	0.75	6.0	1.8	15.5	—	—	—	—	
	0.75	7.0	0.2	16.5	—	—	—	—	
	1.00	1.5	1.3	2.5	—	1.2	—	—	
	1.00			2.0	—	1.0	—	—	
	1.00	1.0	0.8	5.3	—	0.5	—	—	
	1.00			5.0	—	1.5	—	—	
	1.00	0.9	0.3	6.7	—	0.8	—	—	
	1.25	3.0		12.0	—	—	—	—	
	1.50	0.7		5.3	—	1.1	—	0.5	
	2.00	0.3		12.0	—	0.8	—	0.8	
E/R Fe5-A[a]	0.85	0.5 max. ⎫	0.7/0.5	4.0	—	5.0	6.0	1.8	
E/R Fe5-B[a]	0.70	0.5 max. ⎭	max.	4.0	—	8.0	1.8	1.0	
E　Fe5-C[a]	0.40	0.5 max.	0.7 max.	4.0	—	7.0	1.8	1.0	

[a] High-speed steel types from AWS specification A5.13.

NOTES:

(1) Toughness tends to decrease and abrasion resistance to increase as carbon content increases.

(2) There are obviously large differences in the amount of alloying elements and consequently in hardenability of these alloys.

(3) The leaner alloys will require more rapid cooling (perhaps quenching) to develop the martensitic structure.

(4) Some of the more generously-alloyed compositions will retain considerable austenite with rapid cooling.

(5) Inclusion in this table does not provide assurance that weld overlays will be martensitic as-welded. Depending on thermal history some of these can be pearlitic, austenitic, or martensitic.

(6) Reheating below red head (tempering) will soften most of these alloys; this can occur to prior beads of a multiple layer deposit.

(7) The maximum attainable hardness is controlled by carbon content; see text and Figure 14.6.

parts such as hot-working dies, they should be subjected to the same annealing, hardening, and tempering operations as the tool-steel base. Such surfacing operations are highly specialized and should have the benefit of well-informed technicians.

Figure 14.31(a) is a photomicrograph of martensite as it may result from a heat-treating operation on high-carbon steel. Figure 14.31(b) is the structure observed after a deposit from a tool-steel type alloy that has cooled from the heat of welding, with no other heat treatment.

High-speed tool steels are a special type that have earned a respected place in hard-facing. They have ample hardenability to give high hardness levels as welded, good abrasion resistance, superior hot hardness up to about 1100°F (593°C), and clear-cut type compositions. They are among the few steels that have attained industrial standardization sufficient for

inclusion in AWS Specification A 5.13. Table 14.14 shows their compositions and Figure 14.32 provides a typical picture of hot hardness.

The ramifications of austenite transformation are too complex to describe in detail here. However, it should be noted that some steels used for surfacing contain enough alloy to inhibit much of the hardening transformation. Fast cooling (for example, a small deposit that is in effect mass quenched by a heavy base) may retain most of the austenite unchanged, and the deposit will be relatively soft. Slower cooling results in more martensite and higher hardness. An alloy of this type is sometimes termed " semi-austenitic." The austenite has work-hardening qualities but nevertheless is expected to show lower abrasion resistance than hard martensite of the same carbon content. As a compensating advantage it may be tougher.

Worth noting is the fact that even a grade balanced to be martensitic and high in hardness when deposited on a carbon-steel base may be relatively soft and austenitic when welded upon austenitic 13 per cent manganese steel. This effect is due to pickup of manganese from the base and resultant over-alloying of the deposit.

There is need for tough low-carbon martensites that can be deposited crack free in multiple layers. Resurfacing of tractor rollers and idlers is an important example. To fill this need there are available several alloys in the tube-rod form adapted to automatic welding. The arc is usually submerged

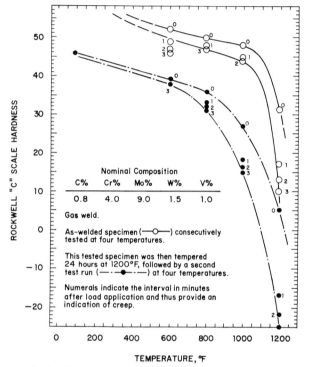

Figure 14.32 Hot Hardness of a High-speed Steel (AWS Specification A 5.13-56T. Grade R Fe 5-B) Gas-weld Deposit Before and After Tempering 24 hours at 1200°F (Ref. 14)

under granulated flux during deposition and high quality deposits can be secured. Such tractor-part reconditioning is a well-established technique and the specialized equipment is commercially available. Adaptation to the surfacing of any cylindrical parts of moderate size should be relatively easy. The carbon level selected is likely to be from 0.20 to 0.35 per cent. This is a case where abrasion resistance has been sacrificed for toughness and freedom from cracks. Higher-carbon steels and the martensitic irons would give higher abrasion resistance, but the thermal stresses from continuously welding on a cylindrical surface would make cracks almost inevitable in high-carbon deposits.

The resistance of several martensitic alloys to high-stress grinding abrasion is shown in Figure 14.33. The base line for this comparison is soft (SAE 1020) mild steel with an abrasion factor of 1.00. The low carbon-martensites used for tractor rollers have factors similar to those of austenitic manganese steel. Figure 14.10 provides additional data on several martensitic steels tempered to different hardness levels. Note that hardness is more important for low-stress abrasion than for high-stress conditions.

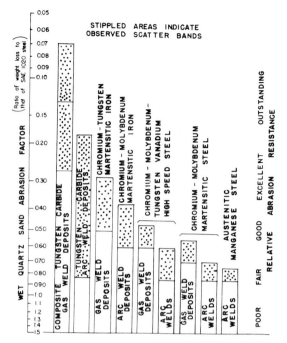

Figure 14.33 Resistance to High-stress Grinding Abrasion of Various Martensitic Alloys, Showing the Contribution of the Volume of Hard Carbides to Wear Resistance (Ref. 12)

Weld deposits of martensitic steels probably represent the best combination of economy in first cost, hardness, strength, abrasion resistance, good impact resistance, and relative toughness. They are the recommended materials for medium-impact conditions. Hardness may range from about 500 to 700 BHN (Rc 48 to 66) if carbon is above 0.60 per cent and retained austenite is not excessive. Of all the hard-facing alloys, their tensile strength is the highest, sometimes exceeding 300,000 psi. Weld deposits usually range from 100,000

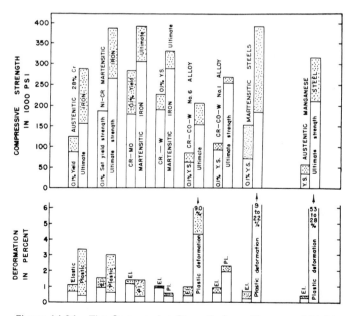

Figure 14.34 The Compressive Strength (as a Measure of Light-impact Resistance) and the Tolerable Deformation (as a Measure of Toughness) of Various Hard-facing Alloys. Stippled areas show observed ranges (Ref. 12)

to 300,000 psi unless brittle constituents, such as free cementite, provide structural discontinuities. Compressive strengths are in the same range. High-stress abrasion resistance exceeds that of pearlitic steels, Cr-Co-W alloys of comparable carbon content and toughness, low-alloy cast irons of comparable hardness, and sometimes that of austenitic manganese steel.

The ductility of high-carbon martensite is low in comparison with that of most engineering steels, but it is usually adequate for hard-overlay requirements where the base is expected to provide the toughness. With good support from a strong base it is flow resistant and thus withstands medium impact well. Even if such welded overlays crack from heavy impact or from cooling stresses it is unusual for the bond between the deposit

and the base to fail, and thus the overlay is not lost even after it cracks. However, a cracked deposit may not develop its maximum abrasion resistance.

Cracking is usually the result of tensile or shear stresses. Thus cracking can be minimized by avoiding these and designing overlays so they are stressed in compression. Comparisons among some of the surfacing alloys is provided by Figure 14.34.[27] Note that the martensitic steels have the highest tolerable plastic flow in compression[27] of the high-yield group. The wide range in strength is attributed to carbon contents.

Martensite may occur with austenite, pearlite, or both in the same deposit. Evaluation of the merits of such mixtures is

Table 14.15   Properties of Austenitic Manganese Steels

Type: ASTM A128-64	Form	Status	Section	Composition C%	Mn%	Si%	Others	Tensile Test Prop. Y. Str. 0.2% set 1000 psi	T. Str. 1000 psi	El %	Red. Area %	Hardness BHN
	Cast	As-C	1 in. φ	1.11	12.7	0.54		52.4	64.8	4	—	—
A	Cast	TQ	1 in. φ	1.15	12.5	0.50		53.0	124	45	—	—
A	Cast ranges	TQ	1 in. φ	1.0–1.4	10–14	0.2–1.0		50–57	100–145	30–65	30–40	185–210
A	Cast	TQ	4 in.	1.11	12.7			52	89.6	25	35	—
A	Cast	TQ	8 in.	1.11	12.7	0.54		47	65.7	18	25	—
	Rolled ranges	TQ	1 in. φ	1.1–1.4	11–14	0.2–0.6		43–67	131–158	40–63	35–50	170–200
C	Cast ranges	TQ	1 in. φ	1.10–1.25	12.5–13.5	0.5	1.8–2.1 % Cr	58–68	96–147	27–59	26–38	205–215
C	Cast	TQ	4 in.	1.17	13.0	0.49	2.0% Cr	53	82	31	29	—
C	Cast	TQ	6 in.	1.17	13.0	0.49	2.0% Cr	56	81	20	19	—
D	Cast ranges	TQ	1 in. φ	0.6–0.9	12.4–14.3	0.48–0.90	3.4–3.6 % Ni	42–49	93–132	40–88	—	174
	Rolled ranges	TQ		0.8–0.9	13.9–15.1	0.9–1.3	2.4–4.0 % Ni	46–56	134–146	74–87	45	179
	Rolled	As-R		0.82	13.9	0.88	2.8% Ni	65.3	154.5	50	36	192
D	Cast ranges	TQ	1 in. φ	1.05–1.17	11.7–15.4	0.52–0.67	3.5–5.0 % Ni	47–53	106–137	35–63	27–43	157–194
E-1	Cast ranges	TQ	1 in. φ	0.75–0.98	12.1–14.1	0.44–0.63	0.9–1.1 % Mo	50–59	106–137	37–67	30–39	179–207
E-1	Cast ranges	TQ	8 in.	0.75–1.0	12.1–14.2	0.44–0.70	0.9–1.3 % Mo	42–55	82–133	27–61	26–60	—
	Rolled	TQ	L / T	0.72	13	—	1.0	54L / 53T	147 / 145	72 / 60	49 / 43	187 / 185
	Rolled	TQ	L / T	0.72	13	—	1.0	53L / 54T	140 / 137	49 / 46	44 / 45	187
E-1	Cast ranges	TQ	1 in. φ	1.15–1.16	12.8–14.3	0.45–0.56	1.0–1.1 % Mo	56.4–73.7	120–144	44.5–53.0	31–37	202–207
E-1	Cast ranges	TQ	8 in.	1.16	13.6–14.3	0.45–0.56	1.0–1.1 % Mo	50–56	76.0–77.2	16.0–33.0	12–92	—

Status Code: As-C = as-cast; As-R = as-rolled; T = transverse section; L = longitudinal section; TQ = toughened by water quenching from above temperature of complete carbide solubility (Acm).

The composition and property ranges are not specifications. They show the actual range of values from the laboratory test group.

difficult; some are quite satisfactory. However, the presence of pearlite usually lowers the abrasion resistance and hardness of a martensitic matrix. Austenite may lower hardness without adversely affecting wear resistance. The tensile properties of mixed structures are usually inferior. The improper application of correctly balanced filler metal compositions, such as the use of arc electrodes for gas welding or vice versa, may produce unintended structures of the mixed type.

19 AUSTENITIC STEELS

Austenitic steels are characteristically high-alloy materials. As a group they are moderately soft, usually very tough, have exceptional work-hardening capacity, and are well adapted to surfacing by welding. The austenitic manganese steels and the Cr-Ni stainless steels are the best-known and generally available groups.

Austenic manganese steel,[28] which is also called Hadfield's manganese steel after its inventor, is an extremely tough non-magnetic alloy in which the usual hardening transformation has been suppressed by a combination of high manganese content and rapid cooling from a high temperature. It is characterized by high strength, high ductility, great work-hardenability, and excellent wear resistance. In the form of castings or rolled shapes it serves many industrial requirements economically and has built an enviable reputation as the outstanding material for resisting severe service that combines abrasion and heavy impact. It usually contains 1.0–1.4 per cent carbon and 10–14 per cent manganese when produced in cast form, and is quench-annealed from 1800°F (982°C) to develop its toughness. It serves as an excellent tough base for hard-facing and it is also widely used in modified form as a surfacing alloy. Many generalizations about the cast alloy are appropriate for weld deposits also. Table 14.15 is a summary of properties.

The salient difference as a welding filler metal is the need for good toughness after air cooling from welding heat. This requires that the carbon be lowered to the range of 0.50 to 0.90 per cent to prevent precipitation of embrittling carbides at grain boundaries. However, lowering carbon reduces strength and toughness unless compensated; welding electrodes therefore contain about 3.5 per cent nickel, 1.0 per cent molybdenum, or other alloy combinations that restore toughness.[28,29] Current specification ranges appear in Table 14.16.[11]

Manganese steel weld deposits usually have an initial hardness below 200 BHN. They can work-harden up to about 500 BHN if subjected to enough deformation. Extreme compression stress or pounding from impact is the usual source of work-hardening, as these factors are able to cause some plastic flow on the surface. Yield strength and ultimate strength both rise as this flow occurs. This behavior is an asset in metal-to-metal wear applications, operating to provide a kind of case-hardening on a very tough base. Work-hardening is also widely credited with making manganese steel abrasion resistant. This cannot be proved experimentally with good abrasion tests, the effect with hard abrasives like quartz being little if any greater than the experimental error. It may well be true with soft abrasives like limestone, where work-hardening changes the steel's hardness from below to above that of the abrasive. However, it should not be assumed that un-work-hardened manganese steel is deficient in abrasion

Table 14.16

(a) Austenitic manganese steel compositions as covered by ASTM A128-64 Specification

ASTM A128-64 Classification	Casting Compositions				
	C%	Mn%	Cr%	Mo%	Ni%
A	1.05–1.35	11.0 min.			
B-1	0.9 –1.05	11.5–14.0			
B-2	1.05–1.2	11.5–14.0			
B-3	1.12–1.28	11.5–14.0			
B-4	1.2 –1.35	11.5–14.0			
C	1.05–1.35	11.5–14.0	1.5–2.5		
D	0.7 –1.3	11.5–14.0			3.0–4.0
E-1	0.7 –1.3	11.5–14.0		0.9–1.2	
E-2	1.05–1.45	11.5–14.0		1.8–2.1	

Silicon: 1.00% max.   Phosphorus: 0.07% max.

(b) Austenitic manganese and chromium-manganese steel weld metal deposits from covered electrodes

AWS A5.13-56T ASTM A399-56T Classification	Weld Metal Deposits from Covered Electrodes					
	C%	Mn%	Cr%	Mo%	Ni%	V%
E FeMn-A	0.5–0.9	11.0–16.0	0.5 max.	—	2.75 min.	
E FeMn-B	0.5–0.9	11.0–16.0	0.5 max.	0.6–1.4	—	

Silicon: 0.3–1.3%   Phosphorus: 0.07% max.

	C%	Mn%	Cr%	Mo%	Ni%	V%	
Proprietary grades	0.75	15.0	4.0	—	3.5		U.S. Pat. 3118760
	0.9	16	16	2.0 max.	2.0 max.		
	0.7	4	19		10		U.S. Pat. 2436867
	0.4	4.1	19.5	1.4	10		
	0.4	15	15		1		
	0.4	14	15	1.7	1	0.60	U.S. Pat. 2711959

resistance. Under gouging abrasion conditions it is better than many harder steels. For grinding abrasion as in a ball mill it is better than medium-carbon steels of similar hardness and pearlitic white cast irons that are harder. It is not quite as good as high-carbon air-hardening steel in such service, and where impact conditions are not too severe it has been largely replaced in some ball mills by 0.50 to 0.90 per cent carbon medium-alloy steels or even by martensitic white cast irons like Ni-Hard. It is seldom appropriate for low-stress abrasion with no impact, not because it does not work-harden but because more abrasion-resistant but brittle alloys usually have enough toughness for the purpose.

Weld surfacing may be required to build up worn areas of manganese steel castings or to provide a wearing surface on some dissimilar base. The latter should be done cautiously, as

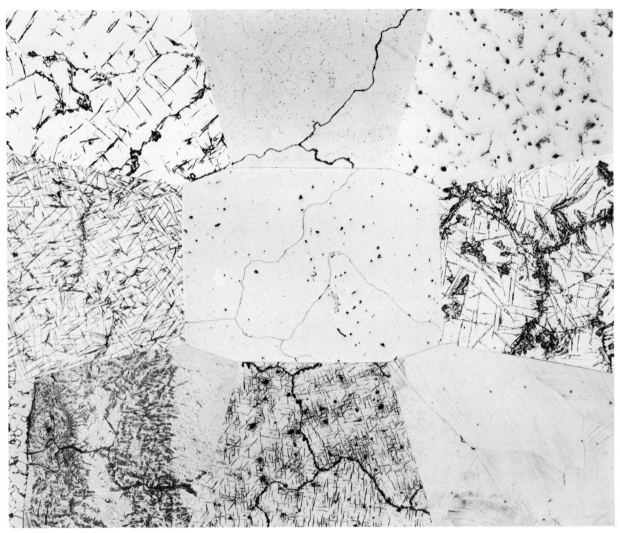

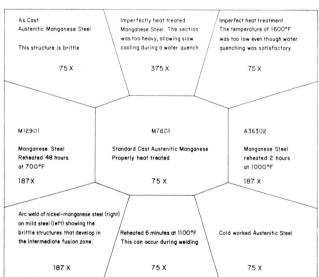

As Cast Austenitic Manganese Steel This structure is brittle 75 X	Imperfectly heat treated Manganese Steel. The section was too heavy, allowing slow cooling during a water quench 375 X	Imperfect heat treatment. The temperature of 1600°F was too low even though water quenching was satisfactory. 75 X
M12901 Manganese Steel Reheated 48 hours at 700°F 187 X	M7801 Standard Cast Austenitic Manganese Properly heat treated 75 X	A36302 Manganese Steel reheated 2 hours at 1000°F 187 X
Arc weld of nickel-manganese steel (right) on mild steel (left) showing the brittle structures that develop in the intermediate fusion zone. 187 X	Reheated 6 minutes at 1100°F This can occur during welding 75 X	Cold worked Austenitic Steel 75 X

Figure 14.35   Photomicrographs of the Structure of Austenitic Manganese Steel

the mixed metal in the fusion zone may be partially martensitic and brittle. Though such surfacing on carbon steels is not generally recommended there are a number of successful applications that are known.[30] If the surfaced parts are subject to compression, the mixed structure zone that is hard but also strong may provide good support. It is excessive tensile stress that should be avoided in the mixed structure area.

Managanese steel electrodes are the accepted filler metals

for joining manganese steel parts. Difficulties that occur are most likely to stem from excessive heating of the base, with consequent embrittlement. Proper welding precautions should be taken,[10],[29] and it is well for the weldor to understand something of the metallurgy involved.

The tough austenite of manganese steel precipitates iron-manganese carbides if reheated in the 700 to 1400°F (371–760°C) range. These form as flat plates or as brittle grain-boundary networks and greatly reduce the ductility and toughness of the steel. The effect depends on time as well as temperature, and is more rapid and severe at high carbon levels. At 800°F (427°C) a day may be required for embrittlement but at 1100°F or 1200°F a few minutes may suffice to reduce the tensile elongation to below 1 per cent. This reheating effect explains why gas welding of manganese steel is unsatisfactory. Once embrittled the toughness can be restored only by reaustenitizing near 1800°F (982°C) or above and water quenching as was originally done in the manufacturing plant. The effect is cumulative; thus, repeated repointing, rebuilding, or hard-facing manganese steel dipper teeth makes them progressively more brittle. This may not preclude good service unless carried too far. Dipper teeth that have served for nine repointing operations have been examined. Considerable embrittlement in the weld-heat-affected zone of the massive tooth body was evident microscopically but the shank that is stressed in tension was still satisfactory. Figure 14.35 is a group of photomicrographs for reference.

Cast manganese steel frogs and crossings in railway track work are widely resurfaced with the electrodes listed in Table 14.16 after the running surface has been battered down by heavy traffic. Crushers in ore-dressing plants are also built

Figure 14.36 Worn Power-shovel Dipper Teeth. The left example retains a sharp cutting edge as a result of hard facing in a careful pattern, even though nearly half its length has worn away. The rounded, inefficient, blunt contour of the right hand example is typical of ordinary unprotected teeth (Ref. 12)

up to replace worn metal. To add metal more economically than is possible with simple weld deposition, special filler bars of manganese steel are sometimes welded in during the surfacing operation.

Austenitic manganese steel is not only a popular wear-resistant alloy in its own right, but it is an excellent base for hard-facing where even greater abrasion resistance is required.

Figure 14.37 The favorable wear pattern in Figure 14.36 can be obtained by using cast manganese steel repointers with the groove design illustrated here. The outer and end grooves are filled with hard-facing alloy as in the right-hand photograph. The crescent backed repointer is then welded to the curved end of the worn tooth, providing an effective replacement that is less expensive than the original, but which may outwear it several times while maintaining an efficient cutting edge (Ref. 12)

A martensitic iron provides an effective overlay for this composite, contributing excellent resistance to low- and high-stress abrasion and to gouging, while the manganese steel base has the toughness to withstand heavy impact.[30] Crusher surfaces and the tips of dipper teeth are examples of successful use of this combination. The best results may be obtained when the martensitic iron is placed in grooves or pockets in the steel, thus largely protecting the brittle high-carbon deposit from the tensile stresses to which it is vulnerable (Fig. 14.36 and 14.37).

If large areas are to be surfaced by automatic or semi-automatic means, the heat input from welding causes high thermal expansion and contraction stresses that may crack brittle deposits. If the base is tough enough, these do no harm in most cases. However, there are some applications where cracks are considered unacceptable. The seating area of a blast furnace bell is one of them. A number of ductile, machinable alloys have been tried for this application but unfortunately there has been a sacrifice of abrasion resistance for ductility to the point that the overlay is little or no better than the 0.30 per cent carbon base upon which it is deposited. Experience has demonstrated that this is a relatively uneconomical procedure. The sliding of iron ore over the surface of a furnace bell is considered erosion or low-stress abrasion; inspection of laboratory results on a number of alloys indicates that a Ni-Cr manganese* overlay (un-work-hardened) has the best erosion resistance of the very tough materials. One trial of this on a bell established that it could be deposited continuously without cracks, machined, and work-hardened by a pressure-roller technique. Service evaluation is incomplete

* U.S. Patent 3118760.

but the bell has operated satisfactorily for over three years at this writing.

This Ni-Cr manganese steel seems to be superior to the nickel-manganese and molybdenum-manganese grades that have been popular for many years. Its deposits are usually crack-free, even under adverse conditions. When welded on a

Table 14.17   Tensile Properties of Nicromang Weld Deposits

Nominal Chemical Composition   0.75% C   15.0% Mn   low% Si
3.5% Ni   4.0% Cr
Structure: Wholly austenitic

Welding Conditions	Yield Strength, psi	Tensile Strength, psi	Elonga-tion % in 2 in.	Red. in Area %	Hard-ness BHN
Manual welds with	74,880	116,500	39.0	33	223
200 amperes DC	74,160	122,000	48.5	35	223
Reverse polarity	77,760	119,700	36.5	30	
Semi-automatic	77,280	121,000	36.5	32	241
welds	80,160	123,500	37.5	31	228
Submerged arc	79,400	120,600	38.0	34	207
welds	74,250[a]	115,200[a]	30.0[a]	26[a]	223[a]

The above represent all-weld-metal specimens.

Composite: SAE 1045 steel joined by Nicromang to hot rolled 13% Mn-1% Mo steel

| Fracture in 1045 steel | 61,000 | 98,000 | 13.5 | 25 | |

Composite: SAE 1045 steel joined by 18 Cr-8 Ni stainless to 13% Mn-1% Mo steel, for comparison

| Fracture in weld metal | 47,500 | 83,500 | 12.0 | 24 | |

[a] Reheated at 1000°F for 48 hours after welding.

Table 14.18   Erosion Resistance of Various Alloys Considered for Blast Furnace Bell Surfacing

Alloy Type		As Weld Deposited	
		Hardness RC	Dry Quartz Sand Erosion Factor[a]
High Chromium Iron	4% C	57	0.08
Cr-V-Mo Iron	2.7% C	49	0.27
Austenitic Iron (15Cr-1Mo)	2.7% C	45	0.47
High Chromium Iron	2% C	45	0.72
Ni-Cr-Austenitic Manganese Steel	0.75% C	15	0.74
Cr-Mo-W Martensitic Steel	0.35% C	53	0.84
Cr-Mo Martensitic Steel	0.25% C	48	0.85
High Alloy Ni-Cr-Mo Martensitic Steel	0.10% C	41	0.87
Semi-austenitic steel 16Cr-6Mn	0.75% C	29	0.89
High-Carbon Stainless 19Cr-9Ni	0.8% C	24	1.01
Nickel-base 15Cr-16Mo-4W	0.12% C	13	1.04

[a] The erosion factor is the *weight loss* of the alloy being evaluated divided by the weight loss of the standard of annealed 1020 steel (at about 107 BHN). The highest factors indicate the poorest abrasion resistance.

carbon-steel base its high alloy contribution may produce a martensitic fusion zone, but this has relatively high strength and is usually quite competent. Tensile tests of cross-weld specimens joining carbon steel to manganese steel characteristically break in the carbon steel, which is the weakest portion of the weldment.

The yield strength of deposits from this grade is higher than that of most cast manganese steels (Table 14.15) as as evidenced by all-weld-metal tensile specimens. Moreover, this alloy is more resistant to the embrittlement that results when standard manganese steel is heated in the 800–1200°F range. Table 14.17 demonstrates its performance. The relative erosion resistance to quartz sand in comparison with a number of other alloys that have been used for blast-furnace-bell surfacing is shown in Table 14.18.

20   STAINLESS STEELS

Chromium-iron and chromium-nickel-iron alloys are widely available as welding electrodes and lend themselves well to application as overlays. The purpose of such surfacing is usually corrosion or heat resistance. Chemical composition is

Table 14.19   Austenitic Stainless Steels Adaptable to Surfacing

AWS-ASTM Class[a]	Nominal Chemical Composition							
	C% max.	Mn% max.	Si% max.	Cr %	Ni %	Mo%	Cb%	Notes
E 308	0.08	2.5	0.90	19	9	—	—	
E 308 L	0.04	2.5	0.90	19	9	—	—	(1)
E 309	0.15	2.5	0.90	25	12	—	—	(3)
E 309 Cb	0.12	2.5	0.90	25	12		0.7–1.0	(2)
E 309 Mo	0.12	2.5	0.90	25	12	2.0–3.0	—	(5)
E 310	0.20	2.5	0.75	25	20	—	—	(3)
E 310 Cb	0.12	2.5	0.75	25	20		0.7–1.0	(2)
E 310 Mo	0.12	2.5	0.75	25	20	2.0–3.0	—	(5)
E 312	0.15	2.5	0.90	29	9			(6)
E 16-8-2	0.10	2.5	0.50	16	8	2		(6)
E 316	0.08	2.5	0.90	18	12	2		(5)
E 316 L	0.04	2.5	0.90	18	12	2		(1) (5)
E 317	0.08	2.5	0.90	19	13	3.0–4.0		(5)
E 318	0.08	2.5	0.90	18	12	2	0.5–1.0	(2)
E 330	0.25	2.5	0.90	15	35	—		
E 347	0.08	2.5	0.90	19	9	—	0.65–1.0	(2)

Phosphorus: 0.03 or 0.04% max.   Sulfur: 0.03% max.

[a] For details refer to AWS A5.4-62T (ASTM A298-62T) specifications, and for Bare Welding Rod or Electrodes to AWS A5.9-62T (ASTM A371-62T). Suffixes to these designations indicate the type of electrode coating: 15 for lime type, adapted to DC Reverse Polarity use; 16 for lime or lime-titania coatings with potassium arc stabilization, adapted for AC welding. Both types are usually "low-hydrogen" coatings.

NOTES:
(1) The low carbon inhibits intergranular corrosion.
(2) Stabilized against intergranular corrosion; higher creep-rupture strength than E 308 L.
(3) Oxidation resistance is superior to that of the 19Cr-9Ni group.
(4) Base metal dilution effects are minimized by the higher alloy reserve.
(5) Enhanced corrosion resistance. E 317 is intended for H_2SO_4 and H_2SO_3 service.
(6) E 312 deposits contain considerable ferrite; E 16-8-2 up to about 5%.

especially important in these cases. This discussion will not attempt to provide comprehensive information about corrosion behavior or heat resistance, but it will provide some guidance.

The Cr-Ni-Fe alloys can be placed in three somewhat similar groups: the wrought stainless steels, the cast stainless steels, and the cast heat-resisting alloys. These are not completely interchangeable, even though the three variants of a type may have the same nominal Cr : Ni ratio. The cast and wrought stainless steels differ chiefly in minor elements (such as silicon) and in metallographic microstructure. Castings are normally more coarse grained than wrought products. Wrought steels usually have marked directional properties as a result of the rolling operation in production. Weld deposits in general are more like castings and the pertinent metallurgy of castings is applicable. The heat-resistant castings of these types are

distinguished chiefly by carbon content, which confers important elevated temperature strength. For corrosion service low carbon is usually an asset; for creep-rupture strength at high temperature it may be a liability.

The well-standardized designations for wrought stainless steels have been adapted to welding filler metals by the addition of prefixes: R for welding rods and E for electrodes. Table 14.19 lists those commonly available. The 18% Cr-8% Ni or 19% Cr-9% Ni grades, as represented by E308, ER308, E308L and ER308L, are the most common and provide effective corrosion resistance under many conditions (Table 14.20). These provide austenitic weld metal and thus have toughness and a considerable capacity for work-hardening. Nonmagnetic behavior, which is sometimes an asset in service, can be achieved. However, in recent years there has been a

Table 14.20   Media at Room Temperature (70°F.) in which 18-8 is Practically Unaffected by Corrosion

Acetic 10%, 33%, 100%	Soda Ash 50% up to 200°F	Potassium Iodide
Acetic Anhydride	Spirits of Nitre	Potassium Nitrate
Benzoic		Potassium Oxalate
Boric	Alum	Potassium Permanganate
Butyric	Aluminum sulphate	Salt and Sea Water
Chloro-sulphonic conc.	Ammonium bromide	Silver Bromide
Citric	Ammonium carbonate	
Hydrocyanic	Ammonium Alum, saturated, slightly acid	Acetone
Lactic	with sulphuric acid, up to 200°F	Benzol
Malic	Ammonium Chloride	Camphor
Nitric, conc. plus 2% HCl	Ammonium Hydroxide	Coffee
Nitric	Ammonium Nitrate	Copal Varnish
Nitrous	Ammonium Sulphate plus 0.5%	Ethyl Alcohol
Oleic	sulphuric acid	Ethyl Ether
Phosphoric conc.	Barium Hydrate	Food Pastes
Phosphoric 10%	Barium Carbonate	Formaldehyde
Picric	Calcium Carbonate	Fruit Juices
Pyrogallic	Calcium Chloride	Gasoline
Stearic	Calcium Hypochlorite made slightly alkaline	Glue
Sulphuric conc.	with NaOH	Inks
Sulphurous	Calcium Hydroxide or Oxide	Lemon Juice
Tannic	Copper Carbonate	Lysol
Tartaric	Copper Nitrate	Methyl Alcohol
	Copper Sulphate plus 2% H2SO4	Milk—fresh or sour
Silver Nitrate	Ferric Nitrate	Mustard
Sodium Bisulphate	Ferrous Sulphate	Naphtha
Sodium Bromide	Glauber's Salt	Oils—mineral or vegetable
Sodium Citrate	Lead Acetate	Paraffin
Sodium Chlorate 10% sol	Lactic Acid Salts	Paregoric Compound
Sodium Chlorate, 25% sol	Magnesium Carbonate	Quinine Sulphate
Sodium Hydroxide	Magnesium Chloride	Soaps
Sodium Hypochlorite, slightly	Magnesium Sulphate	Sodium Salicylate
alkaline with NaOH	Mercurous Nitrate	Soft Soap
Sodium Nitrate	Mercuric Cyanide	Vinegar
Sodium Peroxide at 212°F	Nickel Nitrate	
Sodium Sulphate	Nitrates	Atmosphere of Steam, Carbon Dioxide
Sodium Sulphide	Nitrous Acid Salts	and Air
Sodium Sulphite	Potassium Bromide	Atmosphere of Steam, Sulphur Dioxide,
Sodium Thiosulphate plus 4%	Potassium Chloride	Carbon Dioxide and Air
Potassium Bisulphate	Potassium Cyanide	Sulphur Dioxide
Sodium Thiosulphate 20% plus	Potassium Dichromate	Baking Oven Gases
Acetic Acid 20%	Potassium Ferricyanide	Calcium Chloride
Soda Ash 10% up to 200°F	Same Boiling	Mine Water
		Steam and Air, Refluxed

ᵃ From *The Book of Stainless Steels*, edited by E. E. Thum, American Society for Metals, Metals Park, Ohio, 1935.

tendency to formulate the alloys so a few per cent of ferrite appear in the weld metal and make it slightly magnetic. This is intentional, as it tends to minimize weld cracking and fissuring.[31] Castings may also be balanced to be partially ferritic, enhancing tensile strength and stress-corrosion resistance.[32,33]

Austenitic stainless steels are expected to show their best corrosion resistance when they are free from precipitated carbides.[34] This motivates a heat treatment at a high austenitizing temperature such as 2000°F (1093°C) and water quenching. If this is omitted, or if subsequent processing exposes the metal to the 900 to 1600°F (483–871°C) range there may be carbides precipitated as a film at grain boundaries

as well as within the grains. Selective corrosion attack at such grain boundaries is sometimes called "weld decay" because it may result from welding heat. There may be little or no surface evidence of such intergranular corrosion but in some liquid media it can destroy the integrity of the metal. Types 321, 347, and the very low-carbon grades like 308L have been formulated to minimize intergranular corrosion.

Obviously a welded overlay will involve reheating adjacent weld beads and thus require consideration of intergranular corrosion. Quench annealing, as the carbide solution heat treatment is sometimes called, may be advisable for such weldments if the use of very low carbon or a carbide stabilizing element such as columbium in the weld metal does not give

Table 14.21   Properties of Important Heat Resistant Alloys[a]

Type A.C.I. Designation	28Cr HC	21Cr : 9Ni HF	26Cr : 12Ni HH	26Cr : 20Ni HK	16Cr : 35Ni HT	12Cr : 60Ni HW
C%[b]	0.30	0.25–0.35	0.30–0.40	0.27–0.71	0.35–0.70	0.31–0.77
Ni%	0–3	8–11	10–13	19–21	34–37	59–62
Cr%	27–30	18–23	23–27	23–26	13–17	10–14
Room Temperature Tensile Properties[c] on Small Castings						
Yield strength, psi	—	40–50,000	40–54,000	62–83,000	40–53,000	36–39,000
Ultimate strength, psi	67250	73–95,000	80–95,000	66–100,000	65–71,000	68–78,000
Elongation, %	0.5	27–47	14–40	6–26	5–14	3–13
Red. area, %	0.5	26–40	13–45	7–25	5–18	4–11
Hardness, BHN	238	163–181	160–190	175–202	150–195	160–220
Hot Gas Corrosion as Metal Loss in Inches per Year[d]						
In air at 1600°F	0.019	0.013	0.007		0.0063	0.0055
	0.017	0.001	0.005	0.004	0.004	0.0043
In air at 1800°F	0.048	0.14	0.039	0.025	0.029	0.012
	0.038	0.037	0.030	0.02	0.0097	0.0092
In air at 2000°F	0.063	0.6	0.097	0.07	0.26	0.045
	0.05	0.1	0.071	0.049	0.048	0.037
In oxidizing flue gas						
5 gS per 100 cu. ft. at 1800°F	0.029	0.14	0.036	0.029	0.075	0.038
		0.035	0.029	0.026	0.025	0.024
100 gS per 100 cu. ft. at 1800°F	0.033	0.12	0.033	0.027	0.085	0.05
	0.03	0.033	0.027	0.024	0.024	0.028
5 gS per 100 cu. ft. at 2000°F	0.50	1.0	0.5	0.15	0.30	0.07
	0.15	1.0	0.10	0.05	0.07	0.025
100 gS per 100 cu. ft. at 2000°F	0.17	1.0	0.12	0.07	0.25	0.07
	0.07	0.10	0.07	0.03	0.04	0.02
In reducing flue gas						
5 gS per 100 cu. ft. at 1800°F	0.026	0.34	0.048	0.028	0.056	0.027
	0.025	0.046	0.026	0.022	0.026	0.017
100 gS per 100 cu. ft. at 1800°F	0.034	0.48	0.036	0.025	0.46	1.8
		0.034	0.028	0.022	0.036	0.036
5 gS per 100 cu. ft. at 2000°F	0.25	1.0	0.3	0.10	0.15	0.03
	0.04	0.3	0.035	0.02	0.04	0.02
100 gS per 100 cu. ft. at 2000°F	0.5	1.0	0.5	0.10	1.0	1.0
	0.07	0.4	0.05	0.025	0.10	0.025

[a] In the design of H.R.A. parts, creep and rupture strength are vital properties; in weld overlays they may usually be neglected as the load carrying properties of the base metal overshadow them.

[b] Nominal range to cover tensile properties given here; not part of usual specifications.

[c] Data from research program of The Electro Alloys Div., Elyria, Ohio.

[d] Data from research program sponsored by the Alloy Casting Institute at Battelle Institute; test specimens provided by the Experimental Foundry of the American Brake Shoe Co. The significant elements other than Cr and Ni were: C: 0.31–0.50%; Mn: 0.67–0.90%; Si: 0.76–1.35%; and N: 0.01–0.14%. The ranges given by the two adjacent lines reflect the Cr and Ni extremes.

adequate protection. The 347 grade (columbium stabilized 18-8) may be necessary if service involves heating to near 1200°F (649°C) followed by liquid corrosion, which is a particularly bad combination.

Most stainless overlay applications are served by Type 347, Type 308 (which is the basic 18Cr–8Ni grade); by Types 316 and 317, which contain molybdenum for enhanced resistance to corrosion by sulfuric and sulfurous acids and salts, and to pitting in brine; and by Types 309, (24% Cr-12% Ni) or 310 (25% Cr-20% Ni), which are low-carbon compositions similar to the HH and HK heat-resistant alloys in Table 14.21.

If welding is done by the oxyacetylene process there is little tendency to dilute the deposit with the base metal. However, some pickup of carbon from the gas flame is likely. On the other hand if the corrosion resistance of 18-8 (Table 14.20) is required and is to be obtained by an arc-deposited overlay on ordinary carbon steel, the use of a more highly alloyed filler metal, such as 309 or 310, is advisable to compensate for dilution of the weld by molten base metal. The first weld layer made by arc welding may include perhaps 50 per cent base metal, though this ratio will vary considerably with welding conditions. For large critical applications, such as lining a big tank with a stainless overlay by automatic welding, it is advisable to make preliminary tests, establishing with assurance the maximum dilution that will occur and formulating the filter metal to achieve precisely the end composition selected.

Good corrosion resistance demands unusually high surface quality. Thus the overlay should be carefully inspected for cracks, pits, and slag inclusions. High residual tensile stresses on the surface are likely to follow the welding operation and may even cause delayed cracking. If stress relief[35,36,37] is employed to reduce such stress, it is advisable to avoid the carbide precipitation range mentioned above.

The cast Cr-Ni-Fe heat-resistant alloys are closely allied to the cast and wrought stainless steels. The matching welding electrodes can be employed to produce oxidation and creep-resistant overlays. The important hot-gas corrosion resistant behavior of these grades is detailed in Table 14.21. Because of their higher carbon contents the creep-rupture strengths of the heat-resistant grades may be up to three times those of the corrosion-resistant counterparts. If the distinctive hot strength of higher carbon is not needed it is usually satisfactory to substitute a stainless grade with the same chromium and nickel contents from Table 14.19. However, if carburization resistance is also required the higher silicon of the casting specifications (up to 2.0 or 2.5 per cent) is a distinct asset.

21 NICKEL-BASE ALLOYS

High-nickel surfacing alloys do not fit well into the scheme of Table 14.10. They are shown as midrange but the various types differ so widely in properties that generalizations should be used with caution and each grade studied carefully before it is selected on its merits. Most of the high-nickel alloys were developed for and are exploited for their corrosion resistance. A noteworthy exception is the chromium-nickel-silicon-boron group, which has very good erosion resistance and hot hardness.

Nickel confers both heat and corrosion resistance. However, the alloys depend considerably on chromium for oxidation resistance, which is about three times as effective on a weight percentage basis. The HW alloy in Table 14.21 is a nickel-base type; it has good surface stability in air and oxidizing flue gases up to 2000°F. Like other predominantly nickel alloys it is very susceptible to attack by sulfur and hydrogen sulfide. Thus high-nickel alloys are generally unsuitable for use in reducing atmospheres containing more than about 50 grains of sulfur per cubic foot.

Table 14.22 lists compositions compiled from three AWS specifications for convenience. The three Cr-Ni-Si-B types (E or R NiCr-A, B, and C) are popular for certain kinds of hard-facing because of their fluidity and excellent wetting action. They can be spray coated and then fused in place to provide a thin overlay that can follow irregular contours of the base metal. For the same reason it can be used to make a thin, precise protective coating on a shaft or as the lining of a cylinder. The coating has good hot hardness (Table 14.23) and resists low-stress abrasion well but against high-stress or gouging abrasion poor results may be expected. Where heat and corrosion are not important and the wetting action is not needed the high chromium irons are likely to provide more economical protection against abrasion.

The other high-nickel alloys in Table 14.22 should not be considered for hard facing or abrasion resistance but should be grouped with the stainless steels because they are corrosion resistant and tough. They are more expensive than stainless steels and are appropriately selected when their corrosion resistance in certain media is superior. Hydrochloric acid attacks the steels but the 60% Ni-28% Mo alloy[38] (ER NiMo-4 or E NiMo-1) resists it well and is widely used for the purpose. The 51% Ni-16% Cr-16% Mo-4% W alloy[38] (ER NiMo-5 or E NiMo-2) is said to be the most universally corrosion-resistant alloy available today.

The nickel-copper alloy,[39] which is very widely used in wrought form, is valuable for corrosion resistance in many applications. Sulphuric acid up to 80 per cent, hydrochloric acid up to 5 per cent (if quiet and not heated), cold phosphoric acid, acetic, formic, tartaric, lactic, citric, oxalic, and malic acids, sodium chloride solutions, alkalies, fresh and sea water, food products, and steam are usually resisted satisfactorily. Agitation, aeration, and high temperatures tend to increase the severity of corrosion. Nitric and sulphurous acids, ferric salts, stannous and mercuric salts in acid solution, alkaline hypochlorite solutions, fused sulfur and low-melting metals should not be used with this alloy.

As these metals are relatively expensive, especially with high molybdenum contents, stainless steel should receive first consideration for suitability. For general corrosion resistance, and especially for nitric acid service, the austenitic Cr-Ni-Fe alloys, typified by 18% Cr-8% Ni, are satisfactory and less costly.

22 COBALT-BASE ALLOYS. CHROMIUM-COBALT-TUNGSTEN TYPES

The cobalt-base alloys are an important group, sometimes referred to as nonferrous hard-surfacing alloys. They are usually composed chiefly of cobalt, chromium, and tungsten in

Table 14.22   Nominal Compositions of Nickel-Base Surfacing Alloys

AWS-ASTM Classification	C%	Ni%	Cr%	Cu%	Mo%	W%	Fe%	B%	Si%	Mn%
R NiCr-A[a]	0.40	80	11	—			2.3	2.5	2.3	
R NiCr-B[a]	0.60	76	13	—			4.0	3.0	4.0	
R NiCr-C[a]	0.80	70	15	—			4.5	3.5	4.5	
E NiCr-A[a]	Same as R NiCr-A but as covered electrode deposit									
E NiCr-B[a]	Same as R NiCr-B but as covered electrode deposit									
E NiCr-C[a]	Same as R NiCr-C but as covered electrode deposit									
ER NiCr-3[b]	0.10 max.	67+	20	(Cb 2.5%)			3 max.			3.0
ER NiCrFe-6[b]	0.08 max.	67+	15.5	(Ti 2.5%)			10 max.			2.3
ER NiMo-4[b]	0.08 max.	60	1.0 max.		28		5.0			
ER NiMo-5[b]	0.08 max.	51	15.5		16	4	5.0			
E NiMo-1[c]	0.12 max.	60	1.0 max.		28		5.0			
E NiMo-2[c]	0.12 max.	51	15.5		16	4	5.0			
E NiCr-1[c]	0.15 max.	70+	17.5+	(Cb + Ta: 3%)			4 max.			
E NiCrFe-3[c]	0.10 max.	65	15.0	(Cb 2%)			8			7
E NiCu-1[c]	0.15 max.	65		28	(Cb + Ta: 3.0% max.)		2.5 max.			4 max.
E NiCu-2[c]	0.15 max.	64		28	(Cb + Ta: 2.5% max.)		2.5 max.			6 max.

[a] For details see AWS A5.13-56T (ASTM A399-56T) Specification for Surfacing Welding Rods and Electrodes.
[b] These are the grades from AWS A5.14-64T (ASTM B304-64T) Specification for Nickel and Nickel-Alloy Bare Welding Rods and Electrodes that are most used for surfacing.
[c] These similarly are grades from AWS A5.11-64T (ASTM B295-64T) Specification for Nickel and Nickel-Alloy Covered Welding Electrodes that are used for surfacing.

Table 14.23   Properties of Nickel-Base Alloys

Type Composition	70% Ni 15% Cr 4.5% Si 3.5% B			0.10% C 13% Cr 18% Mo 5% W		
	AWS Class E/R NiCr-C			AWS Class ER NiMo-5/E NiMo-2		
	Gas Welded	Small Castings[a]	Arc Welded	Gas Welded	Small Castings[a]	Arc Welded
Resistance to flow under repeated impact		Fair			Low	
Tensile toughness		Low			High	
Resistance to cracking under repeated impact		Low			High	
Oxidation resistance		Good			Good	
Maximum temperature considered advisable		1500°F			1500°F[c]	
Creep resistance		Good			Good	
Rockwell hardness at[b]						
70°F	C56–62		C42–50	C24–C32	C13–C24	
600°F	C55		C49			
800°F	C52		C46			
1000°F	C48		C39	C16	C1–C8	
1200°F				C7	C0–C3	
Abrasion resistance	Poor		Poor	Poor		
Wet sand abrasion factor	0.90		1.12	0.97		
Erosion resistance	Good			Poor		
Dry sand erosion factor	0.076–0.082			0.73		
Tensile strength, psi					71,500	
Elongation, %					15.0	
Brinell hardness					190	

[a] Properties considered between those of gas and electric weld deposits. Used for some test specimens to eliminate welding variables and defects.
[b] Instantaneous values. See Ref. 14 for effect of creep.
[c] Molybdenum bearing alloys are subject to catastrophic oxidation under some conditions.

various proportions, with carbon as an important minor element. Less frequently other elements such as molybdenum or columbium may replace or accompany the chief components. The Co-Cr-W group has been widely exploited under the trade name "Stellite," which is the forerunner of the variety of currently available materials based on this alloy pattern.

Nominal compositions may range up to 35 per cent chromium and 20 per cent tungsten, with the remainder chiefly cobalt. Largely because of their chromium content they are stainless and nontarnishing. The component metals have high melting points and have conferred a high resistance to heat, which includes two salient factors: oxidation or hot-gas corrosion resistance up to about 2100°F and elevated temperature strength, which is best represented in the hard-facing field by hot hardness.

Current AWS-ASTM specifications recognize three distinct grades, as listed in Table 14.24. There is no sharp line of distinction between these grades and transition compositions are

feasible. However, the standardized grades have the advantage of providing known properties and the weight of considerable industrial experience. The alloys intended for surfacing have carbon in the range of 0.9 to about 3.0 per cent, which confers a considerable spread in mechanical properties and abrasion resistance. Lower-carbon alloys of this type have become prominent as jet engine alloys because of their high creep strength.

Developed by Elwood Haynes, to whom U.S. Patents 1057423 and 1057828 were granted in 1913 and 1915, these materials were among the oldest of the hard-facing alloys. Many of the applications incident to the early experience did not require heat resistance. However, some of these were successful and have persisted. The high-carbon type R CoCr-C in Table 14.24) contains many hard carbides of the Cr_7C_3 type (Fig. 14.38), which usually confer resistance to abrasion. The high-chromium irons also have these chromium carbides and if the distinctive hot hardness of the Cr-Co-W type is not

Table 14.24   Cobalt-Chromium-Tungsten Surfacing Alloys

	AWS-ASTM Classification[a]					
	R CoCr-A E CoCr-A		R CoCr-B E CoCr-B		R CoCr-C E CoCr-C	
Composition						
Carbon, %	0.9/0.7–1.4		1.2/1.0–1.7		2.0/1.75–3.0	
Manganese, %	1.0/2.0 max.		1.0/2.0 max.		1.0/2.0 max.	
Silicon, %	0.4–2.0		0.4–2.0		0.4–2.0	
Cobalt, %	50 min.		46 min.		40 min.	
Chromium, %	26.0/25.0–32.0		26.0/25.0–32.0		26.0/25.0–33.0	
Tungsten, %	3.0–6.0		7.0–9.5		11.0–14.0	
Molybdenum, %	1.0 max.		1.0 max.		1.0 max.	
Iron, %	3.0/5.0 max.		3.0/5.0 max.		3.0/5.0 max.	
Total other elements, %	0.5 max.		0.5 max.		0.5 max.	
	Gas Welded	Arc Welded	Gas Welded	Arc Welded	Gas Welded	Arc Welded
Rockwell hardness at[b]						
70°F	C37–C47	C35–C45			C45–C60	C45–C60
600°F	C37	C37			C50	C53
800°F	C31–C35	C35			C47–C48	C50
1000°F	C27–C31	C28			C44–C46	C35–C47
1200°F	C21–C27	C22			C40–C44	C35–C36
Compressive strength						
0.10% set yield	64–81,000 psi				90–108,000 psi	
Ultimate strength	155–209,000 psi				255–268,000 psi	
Elastic deformation	0.4–1.0%				0.6–0.9%	
Plastic deformation	4.3–9.0%				1.9–2.2%	
Flow resistance under impact	Fair				Good	
Tensile toughness	Moderate				Low	
Oxidation resistance	Excellent		Excellent		Excellent	
Abrasion resistance	Poor	Poor	Poor	Fair	Good	Poor
Wet sand abrasion factor	1.05–1.30	1.15–1.35			0.63–0.85	0.85–1.20
Erosion resistance	Fair				Excellent	
Dry sand erosion factor	0.17				0.05	
Creep resistance	Outstanding		Excellent		Good	

[a] R = Bare rods. E = Electrode deposits. Where compositions differ they are shown thus R/E.
[b] Instantaneous Values. See Ref. 14 and Figures 14.38 and 14.39 for effect of creep. Refer to AWS A5.13-56T or ASTM A399-56T for detailed specifications.

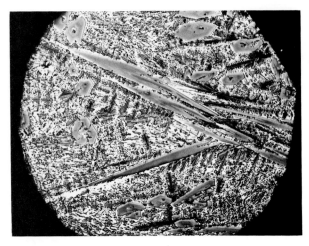

Figure 14.38   Photomicrograph at 250× of an Arc Weld Deposit from a 2.6%C-32%Cr-43%Co-14%W Welding Rod. The structure consists of prominent psuedo-hexagonal spines of the Cr_7C_3 type in a matrix composed largely of an intimate mixture of finer carbides and solid solution. The "fishbone" structure indicates areas of eutectic composition (Ref. 14). (Metal Congress prize winning photograph by R. J. Gray)

required such irons can provide the abrasion resistance more economically.

In the selection of these alloys for surfacing, hardness above 1000°F (538°C) should receive primary consideration. Below this temperature high-speed steel and the tungsten or molybdenum alloyed chromium irons can provide higher hardness; near 1000°F properties may be similar but cost can be unfavorable to the cobalt-base type. At 1200°F and above the Co-Cr-W types seem clearly superior. The choice between the several Co-Cr-W types is based on the level of toughness needed.

For valve facing the 1.0 per cent carbon grade (R CoCr-A) has established a favorable reputation based on years of experience. It is widely used on truck and bus valves, providing a marked increase in useful life. It is also favored for steam valve trim. Chain-saw guides are surfaced with it to protect against frictional wear. These successful examples can serve to suggest other appropriate uses. As conditions become less clearly metal-to-metal wear at red heat and trend toward more abrasion, the tendency is to select the higher carbon grades. The microstructure, for comparison with Figure 14.38 is shown in Figure 14.39.

Scaling and corrosion resistance are secondary properties that may be important but are less often required.[40] Hot-gas attack of the 30 per cent chromium alloys is not usually serious at 1800°F (982°C) and in air they are good up to about 2100°F (1149°C). Corrosion resistance is good in food juices; in nitric, acetic, citric, formic, lactic, sulfuric, sulfurous, and trichloracetic acids; and in many chemicals. Since detailed corrosion data are not given here it must suffice to suggest caution in applying the alloys under severely corrosive conditions. Since minor variables may be decisive a preliminary field test (with an attempt to include all service factors) is advised.

Tarnish resistance up to 500°F (260°C) coupled with a reflectivity of 68 to 83 per cent (versus 91 to 98 per cent for a new silver mirror of highest quality or 68 to 95 per cent for a lacquered silver mirror) has led to the use of Stellite for mirrors in scientific instruments.[41]

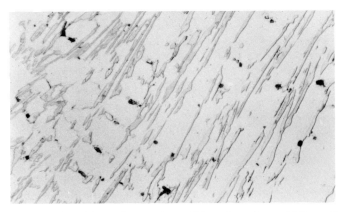

Figure 14.39   Photomicrograph at 500× of an Arc-weld Deposit from a 1%C-28%Cr-60%Co-3.5%W Welding Rod. The structure consists of a moderate volume of carbides in a cobalt rich solid solution (Ref. 14)

Some measure of tensile toughness is occasionally required of these alloys and carbon content is adjusted accordingly. As shown in Table 14.25 cobalt is not the best base for the high-chromium alloys if toughness is intended, but it is better than iron without nickel.[2] However, if corrosion resistance and toughness together are required the austenitic stainless steels should receive first consideration.

Table 14.25   Room Temperature Tensile Properties Chromium-Cobalt Alloys (Ref. 2)

Chemical Analysis				Yield Strength psi	Ult. Tens. Strength psi	Elong. in 2 in. %	Red. Area %	Hardness BHN
C%	Cr%	Co%	Others					
0.50	27.4	—	Fe% = 68.4±		67,250	0.5	nil	238
0.35	26.2	—	Ni% = 70.6	44,000	73,250	15	15	179
0.31	26.6	69.0	—	63,000	82,500	6	7	269
0.50	26.9	69.3	—	69,000	87,500	4	4	297
0.73	26.5	69.3	—	71,000	90,500	2	2	321
0.55	26.3	62.6	Mo% = 6.9	81,000	96,250	4	4	364
0.53	26.3	62.8	W% = 6.9		106,500	1	1	321
0.50	26.4	62.8	Cb% = 7.0	68,000	100,250	nil	nil	364

Table 14.26   Compression Test Properties of Cobalt-Base Alloys

Chemical Composition, %						Yield Strength, psi		Ultimate Strength, psi	Deformation, %			Hardness	
C	Mn	Si	Cr	Co	W	0.05% set	0.10% set		Elastic	Plastic	Total	Rc	BHN
Specimens 0.400 in. diameter × 1.00 in. high machined from small castings													
0.75	0.88	1.92	27.2	58.9	5.1	53,000	64,000	155,500	1.03	7.06	8.09	41	351
0.86	1.03	1.60	25.2	58.8	5.1	67,000	76,000	209,000	0.95	9.76	10.71	41	375
1.07	3.67	1.56	27.9	52.5	5.9	64,000	73,000	227,000	>0.40	7.34+	7.74+	41	364
Specimens 0.400 in. diameter × 1.25 in. high machined from small castings													
0.75	0.88	1.92	27.2	58.9	5.1	70,000	76,000	174,000	0.58	4.35	Buckled	41	351
0.86	1.03	1.60	25.2	58.8	5.1	74,000	81,000	189,200	0.69	6.04	Buckled	41	375
1.07	3.67	1.56	27.9	52.5	5.9	66,000	73,000	185,000	0.51	6.84	7.35	41	364
2.58	3.84	1.62	26.0	39.5	14.3	69,000	90,000	255,000	0.89	1.92	2.81	52	—
2.46	3.67	1.54	30.7	42.4	14.3	88,000	108,000	268,000	0.58	2.15	2.73	49	—

The 1.0 per cent carbon valve facing alloy (R CoCr-A) has a room temperature tensile strength of about 100,000 psi with 1 per cent elongation as determined on small castings. Ductility tends to increase as temperature rises. As carbon is increased ordinary temperature behavior becomes similar to that of cast irons and tensile tests are poorly adapted for comparisons.

Since most service conditions involve compressive loading, compression tests are much more appropriate. Relative elasticity, strength, and plasticity are compared against carbon (and tungsten) by this criterion in Table 14.26. Note that the softer and weaker of the two types deforms more readily in compression, but that once plastic behavior begins, with 1.0% C there is a greater reserve of plasticity before failure occurs.

These data are useful in judging relative performance in applications where experience has not yet clarified the selection problem. The indicated plasticity is valuable for redistribution of concentrated stresses that might otherwise become dangerous, particularly where some impact is involved.

At room temperature the hardness of the 0.5% C alloy is near 300 Brinell. At 1.0% C it approximates 400 BHN and at 2.5% C it is seldom under 500 BHN. Hot hardness follows much the same pattern, with zero time values of Rockwell C27-C31 for 1% C and C35-C47 for 2.5% C at 1000°F; C21-C27 for 1% C and C35-C44 for 2.5% C at 1200°F. Carbon is thus a potent minor component. (Figures 14.40 and 14.41.)

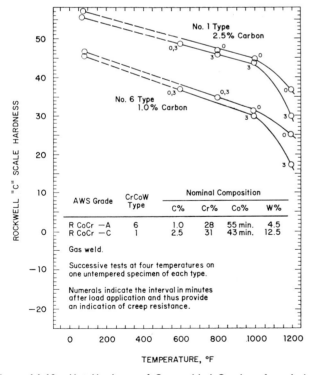

Figure 14.40   Hot Hardness of Gas-welded Overlays from Industrially Standardized Cobalt-base Hard-facing Alloys (AWS Specification A 5.13-56T) Note their superior resistance to creep (Ref. 14)

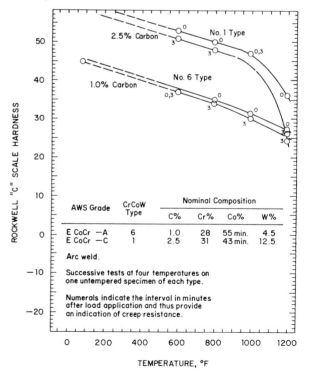

Figure 14.41   Hot Hardness of Arc-welded Overlays from Industrially Standardized Cobalt-base Hard-facing Alloys (AWS Specification A 5.13-56T) (Ref. 14)

Hot hardness alone does not sufficiently differentiate the 1.0 and 2.5% C types, since other qualities, notably toughness, control their selection. The 1% C alloy can be lathe-turned with difficulty and subjected to some other machining operations, but the 2.5% C grade must be shaped by grinding.* It contains very hard carbide spines of the Cr_7C_3 type, which will quickly dull even tungsten-carbide cutting tools. Figures 14.38 and 14.39 illustrate how microstructure changes with carbon content.

23 COPPER-BASE ALLOYS

Various high-copper alloys are used for overlays to cope with frictional wear in bearings, liquid corrosion, and metal-to-metal wear with little or no lubrication. They may be used for bearing inlays. They are not suitable for the abrasive conditions associated with mining, quarrying, and earth-moving operations. The elements alloyed with the copper are usually aluminum, silicon, zinc, tin, or lead. Compositions are outlined in Table 14.27.

Table 14.27   Copper-Base Surfacing Alloys

AWS-ASTM Classification	Phosphor Bronzes		
	R CuSn	R CuSn-A	R/E CuSn-D
Copper, %	94.0 min.	93.5 min.	88.5 min.
Tin, %	4.0–6.0	4.0–6.0	9.0–11.0
Phosphorus, %	0.20–0.30	0.10–0.35	0.10–0.30
Deposit hardness, BHN	70–85	70–85	90–110

AWS-ASTM Classification	Aluminum Bronzes				
	R/E CuAl-A2	R/E CuAl-B	R/E CuAl-C	R/E CuAl-D	R/E CuAl-E
Copper, %	Balance	Balance	Balance	Balance	Balance
Aluminum, %	9–11	11–12	12–13	13–14	14–15
Iron, %	1.5 max.	3.0–4.3	3.0–5.0	3.0–5.0	3.0–5.0
Deposit hardness, BHN	130–150	140–180	180–220	230–270	280–320

AWS-ASTM Classification	Silicon Bronzes	
	R/E CuSi-A	R CuSi-B
Copper, %	94 min.	97 min.
Silicon, %	2.8–4.0	1.0–2.0
Max. Sn, Mn, each, %	1.5	1.5
Zinc, %	1.5 max.	—
Iron, %	0.5	0.5
Deposit hardness, BHN	80–100	

AWS-ASTM Classification	Leaded Bronze R/E CuSn-E	Brass R/E CuZn-E
Copper, %	Balance	56.0 min.
Tin, %	5.0–7.0	2.0–3.0
Lead, %	14.0–18.0	—
Zinc, %	—	Balance
Phosphorus, %	0.30–0.50	—
Deposit hardness, BHN	40–60	130 min.

For additional detail refer to AWS Specifications A5.6-57, A5.7-57, and A5.13-56T.

Abrasion resistance differs considerably. The 2.5% C grade has relatively good resistance to wear from abrasive particles like sand, though a number of other alloys outperform it in such service. In contrast, the 1.0% C material is poor for most abrasive conditions. As a principle, selection of these cobalt-base alloys should be based primarily on their properties above 1000°F.

Table 14.24 permits detailed comparison of the two most popular cobalt-chromium-tungsten alloys.

* While the 2.5% C alloy has been machined at 0.006–0.010 in. feeds with cemented carbide tools, it is unlikely that this would be more economical than grinding.

The phosphor-bronzes are tin bronzes with phosphorus added for deoxidation and hardening. Weld deposits are relatively soft (70-85 BHN for R CuSn and E/R CuSn-A, and 90-110 BHN for R CuSn-D). They are used to prepare bearing surfaces or to provide corrosion resistance. The R CuSn and R CuSn-D grades are best applied with the carbon arc (using covered welding rods), or using bare rods with the gas-tungsten arc or atomic hydrogen welding. The E/R CuSn-A grade is adapted to shielded metal-arc welding or gas-metal arc deposition with bare electrodes. The last method provides better shielding from oxidation and loss of alloying elements and thus may give somewhat harder deposits.

Leaded bronze is a variation of phosphor-bronze that provides droplets of lead dispersed through the deposit. The lead improves machinability but is employed primarily to achieve a self-lubricating quality and some plasticity. This grade is adapted to heavily loaded bearings as in locomotive journals or rolling mills.

The aluminum-bronzes are an important group that permit selection of hardness over the range of 130 to 300 BHN, yield strength from 25,000 to 65,000 psi, and ultimate compressive strength from 120,000 to 171,000 psi. Ductility decreases as hardness increases until at the 300 BHN level it is very low. They are oxidation and corrosion resistant. Seawater attack is resisted well. Corrosion resistance is aided by a film of aluminum oxide that forms on the surface. This film interferes with welding unless it is removed by a powerful flux or is prevented from forming by an inert-gas shield. It is soluble in alkalis and thus alkaline corrosive environments should be avoided. Oxyacetylene welding is not practicable because of the film.

The Al_2O_3 surface film is very effective in preventing "welding wear" phenomena such as seizing and galling. Thus the aluminum bronzes are outstanding for metal-to-metal wear applications where lubrication is scant or lacking. Drawing and forming dies, wearing plates, sheaves, gears, pinions, and cams are parts that have been successfully protected against this kind of wear.[9] However, hard abrasives can cut through the thin film and the metal underneath is not abrasion resistant.

Abrasion and erosion test data appear in Table 14.28. This evidence serves as a warning against using aluminum bronzes in abrasive environments such as those that characterize the earth-moving industries. Practically any steel overlay would be better and less expensive. Note particularly that this relative performance cannot be predicted on the basis of hardness.

Table 14.28   Abrasion and Erosion Resistance of Aluminum-Bronze Weld Deposits

Grade AWS-ASTM	High-Stress Grinding Abrasion			Low-Stress Scratching Erosion	
	Hardness		Abrasion Factor	Hardness Rockwell	Erosion Factor
	Rockwell	BHN			
E CuAl-A	B72	137	3.38	B75	2.18
E CuAl-B	B86	176	3.63	B80	2.20
E CuAl-C	B92	209	4.26	B85	1.83
E CuAl-D	C22	244	4.24	C20	1.85
E CuAl-E	C32	297	4.23		
For comparison SAE 1020 steel		107	1.00		1.00

NOTE: Large factors denote great susceptibility to abrasive wear.

The silicon bronzes in Table 14.27 are not recommended for bearing surfaces; they are used primarily for corrosion resistance.

A convenient reference for detailed information on corrosion is the "Corrosion of Wrought Copper and Copper Alloys" section of the ASM *Metals Handbook*, 8th Edition, Vol. 1, pp. 983–1005, 1961 (or pp. 895–899 of the 1948 Edition).

24   BUILDUP ALLOYS

In the practical application of hard facing against abrasion it is sometimes necessary to build up the dimensions of the previously worn surface before applying the final protective layer. Thick layers of the hard and more abrasion-resistant irons are very likely to crack, and multiple layers are almost certain to do so. To avoid such cracking and yet provide a competent support for the brittle overlay, experience has developed the practice of building up with steels of intermediate hardness. They provide better support than soft mild steel deposits, which also could be used for multilayer buildup, and at the same time they are not so susceptible to cracking as the very abrasion-resistant hard-facing alloys.

Most of the buildup alloys can be typed as pearlitic steels. Simple steels with carbon contents between 0.3 and 0.6 per cent carbon are of this type. Modest amounts of manganese, chromium, or molybdenum may be added to increase deposit hardness and strength but the total alloy content should remain below that which causes martensite to form during cooling from the welding heat. Martensite formation makes the deposit subject to tension cracks on the surface and to underbead cracks that may cause the deposit to spall away from the base. The latter difficulty can be minimized by low-hydrogen electrode coatings as with the E XX15 and E XX16 types in AWS Specifications A 5.1-64T and A 5.5-64T for mild steel and low-alloy steel covered arc-welding electrodes.

For reasons of economy almost all buildup welding is with the electric arc. The usual base is steel. During the surfacing of certain parts such as tractor rollers, it may be expedient to make the buildup and the final layer of the same alloy deposited by submerged arc welding. This is successful if the steel deposited is formulated to give a low-carbon martensite or a very fine pearlite. Such steels and the pearlitic buildup alloys (other than those in the A 5.1 and A 5.5 specifications) are not yet covered by professional society specifications but they may be in the near future.

25   THE SELECTION PROBLEM

An engineer who has read this chapter can probably at this point identify tentatively the important wear factors with which he is concerned, select the most dominant, and by inspection of Table 14.10 decide on one or several prospective alloys. By reviewing again the sections that discuss these, the choice can usually be narrowed to one grade. The element most in doubt may be impact. This problem arises because the severity of impact may vary widely in a single application area, and means for measuring its velocity and force may be lacking. In such cases it is better to err on the side of toughness in alloy selection. An alloy that is too brittle may fail suddenly by fracture, which is more serious than a somewhat shorter life. If the choice has sacrificed abrasion resistance unnecessarily by overestimating the need for toughness to resist impact the result will be shorter life, but under most field conditions this cannot be recognized without carefully controlled tests unless the difference is great.

For abrasive-wear situations where impact is obviously a factor it may be feasible to proceed in steps, starting with austenitic manganese steel as the tough extreme, next trying a high-carbon martensitic steel, then a martensitic or high-chromium iron, and finally a tungsten carbide type. After each step, if no impact failures have resulted, the next step is logical. The step where too many or unacceptable impact fractures result provides the signal to drop back to the preceeding tougher alloy. Table 14.29 is a convenient reference for three of these steps.

Table 14.29   Hard-Facing Alloy Selection for Impact Service (Ref. 13)

Material Suitable for	Martensitic Irons[a] Light Impact	Martensitic Steels [a] Medium Impact	Austenitic Manganese Steel Heavy Impact
Yield strength tensile, psi	60,000–80,000	60,000–80,000	45,000–55,000
Ultimate tensile strength, psi	65,000–90,000	60,000–125,000	100,000–150,000
Tensile elongation, %	0.01–0.04	0.01–0.10	30–70
Brinell hardness	500–750	500–750	160–220
Rockwell hardness	C48-C65	C48-C65	C0-C20
Yield strength compressive 0.10% set	150,000–285,000	76,000–175,000	45,000–55,000
Ultimate compressive strength, psi	265,000–385,000	180,000–385,000	200,000–310,000
Elastic deformation, %	0.8–1.4	0.8–1.5	
Plastic deformation, %	0.2–5.4	1.0–22	30–55
Total deformation, %	1.2–6.5	1.8–23	
Wet sand abrasion factor[b]	0.30–0.50	0.65–0.80	0.75–0.85
Dry sand erosion factor[b]	0.04–0.10	0.15–0.55	0.41–1.00

[a] Untempered, corresponding to the usual status of weld deposits.
[b] Weight loss ratios compared with annealed SAE 1020 steel.

Choices may be less difficult where special properties are needed. An attempt has been made to provide adequate detail within the limitations of space for these cases.

There are many more alloys offered for hard-facing use than are covered here. For many there is limited industrial experience, no standardized specification, and little or no good technical data. Obviously it is difficult to evaluate them properly and their omission here is justified. The coverage has been aimed at those grades covered by specifications that have been studied in some detail by the author or that have been described by good technical publications.

ACKNOWLEDGMENTS

The foundation of this chapter was a book, *Hard Facing by Fusion Welding*, copyright by the American Brake Shoe Company, and Chapter 14 of the 1949 Edition of *Engineering Laminates*. It has been extensively revised using material from American Welding Society publications (*Welding Journal* articles and various volumes of the *Welding Handbook*), the American Society for Metals book *Surface Protection Against Wear and Corrosion*, the *Iron and Steel Engineer* (September 1951) paper on "Hard Facing Alloys for Steel Mill Use," and *The Encyclopedia of Engineering Materials and Processes* (Reinhold Publishing Corporation, New York, 1963). Much of the original data in these papers came from the laboratories of the American Brake Shoe Company (now ABEX Corporation) and its AMSCO Division. The availability of these data and permission to include hitherto unpublished information is acknowledged with appreciation. The author is also grateful to Henry J. Chapin and Chas. E. Ridenour for many years of assistance in evaluating abrasion-resistant and hard-facing alloys.

REFERENCES

1. *Welding Handbook*, Fifth Edition, American Welding Society, New York, 1963, Section 2.
2. Howard S. Avery, "Hard Surfacing by Fusion Welding," American Brake Shoe Company, New York, 1947.
3. Homerberg, "Nitralloy and the Nitriding Process," The Nitralloy Corporation, New York, 1943.
4. *Nitriding Symposium*, Special Edition of the *Trans. Am. Soc. Steel Treating*, **XVI**, No. 5 (October 1929) American Society for Metals, Metals Park, Ohio.
5. Landau, "Wear of Metals," The Nitralloy Corporation, New York, 1945.
6. Robert D. Zimmerman, "Chromium Plating for Wear Resistance," In *The Book of Stainless Steels*, American Society for Metals, Metals Park, Ohio, 1935.
7. F. N. Rhines, "Diffusion Coatings on Metals," Carnegie Institute of Technology Metals Research Laboratory, Contribution 124, Carnegie Institute of Technology, Pittsburgh, Pa., Nov. 1941.
8. "Hard-Facing by the Oxy-acetylene Process," International Acetylene Association, New York, 1936.
9. F. E. Garriott, "Jobs You Can Do With Bronze Electrodes," *Welding Engr.* **36**, 20–24 (July 1951).
10. *Welding Handbook*, Fifth Edition, American Welding Society, New York, 1964, Section 3.
11. American Welding Society Tentative Specification for "Surfacing Welding Rods and Electrodes", AWS Designation A 5.13–56T, A.S.T.M. Designation A 399–56T, 1956, available from American Welding Society, United Engineering Center, New York.
12. Howard S. Avery, "Hard Facing for Impact," *Welding J.* **31**, No. 2, 116–143, (1952).
13. Howard S. Avery, "The Measurement of Wear Resistance," *Wear* **4**, No. 6, 427–449 (Nov.–Dec. 1961).
14. Howard S. Avery, "Hot Hardness of Hard-Facing Alloys," *Welding J.* **29**, No. 7, 552–578 (1950).
15. Littman, "Hardness and Wear Resistance," *Engineering* June 29, 1945, p. 502 (abstracted in *Metal Prog.* January, 1946).
16. *Foundary Sand Handbook*, Sixth Edition, American Foundrymen's Society, Chicago, 1952, pp. 30–41.

17. HOWARD S. AVERY and H. J. CHAPIN, "Hard Facing Alloys of the Chromium Carbide Type," *Welding J.* **31**, No. 10, 917–930 (1952).

18. T. E. NORMAN and C. M. LOEB, "Wear Tests on Grinding Balls," *Trans. Am. Inst. Mining Met. Engrs.* **176**, 490–520 (1948).

19. HOWARD S. AVERY, "Discussion of Wear Tests on Grinding Balls," *Trans. Am. Inst. Mining Met. Engrs.* **176**, 521–523 (1948).

20. H. W. GILLETT, "Considerations Involved in the Wear Testing of Metals," *Symposium on the Wear of Metals* American Society for Testing and Materials, Philadelphia, Pa., 1937.

21. EDWARD S. DANA, *A Text-Book of Mineralogy*, Revised and Enlarged by Wm. E. Ford, John Wiley Sons, Inc., New York, 1922.

22. HOWARD S. AVERY, "Hard Facing Alloys for Steel Mill Use," *Iron and Steel Engr.* **28**, No. 9, 81–106 (1951).

23. SAMUEL L. HOYT, "Hard Metal Carbides and Cemented Tungsten Carbide," *Trans. Am. Inst. Mining Met. Engrs.* (Institute of Metals Division), **89**, 9–58 (1930).

24. HOWARD S. AVERY, "Some Characteristics of Composite Tungsten Carbide Weld Deposits," *Welding J.* **30**, No. 2, 144–160 (Feb. 1951).

25. American Welding Society Tentative Specification for "Mild Steel Covered Arc-Welding Electrodes," AWS Designation A 5.1-64T, A.S.T.M. Designation A 233–64T, 1964.

26. American Welding Society Tentative Specification for "Low-Alloy Steel Covered Arc-Welding Electrodes," AWS Designation A 5.5–64T, A.S.T.M. Designation A 316–64T, 1964.

27. HOWARD S. AVERY, "The Selection of Hard Facing Alloys," *Prod. Eng.* **23**, No. 3, 154–157 (March 1952).

28. HOWARD S. AVERY, "Austenitic Manganese Steel," American Brake Shoe Company, New York, 1949. [A somewhat condensed version appearing in the *American Society for Metals Handbook* (1948), pp. 526–534, includes 33-reference bibliography.]

29. HOWARD S. AVERY and H. J. CHAPIN, "Austenitic Maganese Steel Welding Electrodes," *Welding J.* **33**, No. 5, 459–479 (May 1954).

30. THEODORE GAYNOR, "Discussion by Theodore Gaynor," of "Hard Facing for Impact," (Ref. 12), *Welding J.* **31**, No. 2, 144–145 (Feb. 1952).

31. R. D. THOMAS, JR., "Crack Sensitivity of Chromium-Nickel Stainless Weld Metal," *Metal Prog.* **50**, pp. 474–478 (Sept. 1946).

32. F. H. BECK, E. A. SCHOEFER, J. W. FLOWERS, and M. G. FONTANA, "New Cast High-Strength Alloy Grades by Structure Control," in *Advances in the Technology of Stainless Steels and Related Alloys*, American Society for Testing and Materials Special Technical Publication Number 369, American Society for Testing and Materials, Philadelphia, Pa., 1965.

33. J. W. FLOWERS, F. H. BECK, and M. G. FONTANA, "Corrosion and Age Hardening Studies of Some Cast Stainless Alloys Containing Ferrite," *Corrosion*, **19** No. 5, 186t–198t (May 1963).

34. HELMUT THIELSCH, "Physical Metallurgy of Austenitic Stainless Steels," *Welding J.* (Research Suppl.) **29**, No. 12, 577s–621s (Dec. 1950).

35. GEORGE SANGDAHL, "Stress Relief of Austenitic Stainless Steels," *Metal Prog.* **86**, No. 2, 100–104 (August 1964).

36. EARL R. PARKER, "Stress Relieving of Weldments," *Welding J.* (Research Suppl.) **36**, No. 10, 433s–441s (1957).

37. HOWARD S. AVERY, "Temperatures for Rapid Uniaxial Stress Relief of Heat Resistant Alloys," *Welding J.* (Research Suppl.) **39**, No. 11, 509s–512s (Nov. 1960).

38. B. E. FIELD "Some New Development in Acid-Resistant Alloys," *Trans. Am. Inst. Mining Met. Engrs.* (Institute of Metals Division) **83**, 149–159 (1929).

39. INCO Technical Bulletin T-5, "Engineering Properties of Monel Nickel-Copper Alloys," Huntington Alloys Products Division of the International Nickel Company, New York 1965.

40. H. CORNELIUS, "Scaling Resistance of Hard Facing Alloys with Different Cobalt Contents," *Stahl und Eisen* **64**, 529–532 (1944).

41. W. A. WISSLER, "Haynes Stellite," in *The Book of Stainless Steels*, American Society for Metals, Metals Park, Ohio, 1935.

Chapter 15

Flame-Sprayed Coatings

H. S. Ingham, Jr.

Flame spraying is a process for applying layers of metal or ceramic to a substrate. A "gun" is used in which either a combustion flame or a plasma flame melts and propels the material in finely divided form toward the article to be covered. The sprayed deposit is usually between about 0.002-in. and 0.100-in. thick, but it may be as much as an inch or more. Standard flame-spray materials include numerous steels, aluminum, zinc, copper, hard-surfacing alloys, carbides, oxides such as aluminum oxide and zirconium oxide, and refractory metals including molybdenum and tungsten. Substrates are usually metal but may be ceramic, glass, or carbon, or even wood, plastic, or cloth. Applications for the flame-spray process are as varied as one might guess from the number of different materials that may be used. Occasionally, free shapes are created by flame spraying, but the process is particularly used for the formation of laminates having a substrate and at least one coating material. Very often the laminate is more complex, having several types of flame-sprayed layers, and sometimes paint is also applied.

Thus, the engineer considers flame spraying as a means for designing with several materials, when there is no one material which has all the properties desired.

Flame spraying has grown rapidly from the older but still very important field of metallizing, where the material is in wire form that is atomized and sprayed by the metallizing gun. In the 1930's and 1940's this was used almost entirely as a repair process for replacing metal on worn parts. However, since repaired parts often lasted much longer than new parts, flame spraying in original manufacture became an accepted process. In the 1950's combustion flame-spray guns using powdered materials extended the process to include hard-surfacing alloys, carbides, and oxides. The plasma flame was introduced in the late 1950's. The very high-temperature operation of plasma guns made it possible to spray high-melting-point powders that could not be sprayed by combustion guns and, in addition, produced higher efficiency and better coatings with many materials already sprayable with combustion guns. There are also other kinds of flame-spray guns, but the three basic type mentioned are most widely used.

Flame-sprayed deposits have many properties significantly different from those generally expected of the starting material. This is a particularly important factor to consider in the design of products and structures. Therefore, the intent of this chapter is to acquaint the engineer with the nature of these flame-sprayed layers so that he may be generally familiar with the advantages and scope of a process that is being used for many vastly different applications. After a description of the process, emphasis will be placed on engineering information for those coatings that are particularly useful for design purposes. For detailed information on how to spray or on other uses such as maintenance the reader is referred to available handbooks* and instruction books.

1 THE FLAME-SPRAY PROCESS

The wire metallizing gun (Fig. 15.1), being the oldest type, is the best established and most commonly used equipment for flame spraying production parts. Wire is used in sizes ranging

Figure 15.1 Metallizing Gun Used for Production Spraying. Smaller guns with pistol grip are used for hand operation

from 15 gauge (0.057-in. diameter) or smaller for finer spraying, to $\frac{3}{16}$-in. diameter for high-speed deposition. As illustrated in Figure 15.2, the wire is fed axially through a ring of combustion flame (oxygen, and acetylene, propane, or other fuel gas) which melts the wire tip. An outer sheath of compressed air "atomizes" the molten metal to droplets roughly 10

H. S. INGHAM, JR., is Manager, Research and Development, Metco, Inc., Westbury, N.Y.

* For example, H. S. Ingham and A. P. Shepard *Flame Spray Handbook*, Vol. I: Wire Process; Vol. II: Powder Process; and Vol. III: Plasma Flame Process; Westbury, N. Y., Metco, Inc., 1964 and 1965.

to 100 microns in diameter, and propels these to velocities ranging from 300 to 800 ft/sec. The droplets strike the substrate about 4 to 10 in. from the gun and flatten out while freezing, thus building up a coating. Table 15.1 lists typical

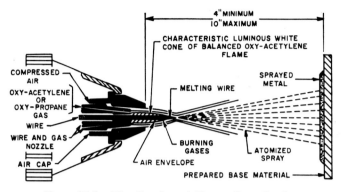

Figure 15.2 Wire Nozzle and Air-cap Cross Section

flame-spray wires. Certain of these spray materials most commonly used for manufacture of laminates will be described in more detail with specific types of applications in later sections.

Powder is sprayed similarly with a combustion flame-spray gun (Fig. 15.3) using, for example, approximately 325 mesh powder. The powder is carried from a hopper in a carrier gas and is fed axially through the ring of flame. No compressed air is required for atomizing, and the flame itself accelerates the molten particles to about 150 ft/sec. In some models air is used for carrying the powder or for providing additional acceleration.

Figure 15.3 Airless, Combustion Flame Spray Gun for Powder

In the plasma flame-spray gun, shown in Figure 15.4, a direct-current arc similar to a welding arc is maintained inside the unit. The arc heats a gas, which may be argon but is usually nitrogen containing some hydrogen. Gas temperatures range above 8000°F where the gas is in a dissociated and partially ionized, or "plasma," state. This plasma issues as a "flame" from the nozzle. Powder is usually fed radially into the flame near this exit point, and the plasma flame accelerates the particles. Velocities between 200 and 600 ft/sec or higher are achieved, depending on choice of gas flow, power, and nozzle.

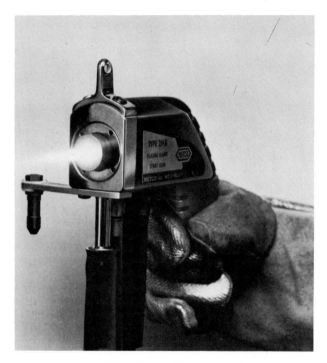

Figure 15.4 Plasma Flame Spray Gun

Powders are listed in Tables 15.2 through 15.5 for both the combustion and plasma guns, with the preferred type of gun indicated for many powders. It may be noted that some powders are suitable for either process. Actually, nearly all of the materials may be sprayed by plasma, even if not specifically indicated in the tables, but it is usually more economical to choose combustion when similar results are obtained with both processes. This economy is changing for some materials, as the potentially high spray speeds and efficiencies of plasma are being realized through continuing developments.

Usually the starting wire or powder is formed simply of the metal, alloy, or compound indicated by the description of the material. However, there is a relatively new class of flame-spray materials which are fabricated as composites to take advantage of a chemical reaction that occurs between the constituents during spraying. "Nickel aluminide" is such a material and is produced in wire and powder form. The wire (labeled 405 in Table 15.1) is an aluminum sheath surrounding a mixture of aluminum and nickel. The powder (labeled 404 in Table 15.2) consists of an aluminum core clad with nickel.

Table 15.1   Flame-Spray Wires

Type Analysis

Alloy	Metco Wire Designation	Carbon	Phosphorus	Sulphur	Iron	Manganese	Nickel	Chromium	Molybdenum	Silicon	Aluminum	Cobalt	Zinc	Tin	Lead	Antimony	Arsenic	Cadmium	Copper
Aluminum	Aluminum	—	—	—	—	—	—	—	—	—	99.0+	—	—	—	—	—	—	—	—
Silicon-aluminum	Aluminum SF	—	—	—	—	—	—	—	—	6	94	—	—	—	—	—	—	—	—
Babbitt	Sprababbitt A	—	—	—	—	—	—	—	—	—	—	—	—	Bal.	0.25	7.5	—	—	3.5
Brass	Sprabrass Y	—	—	—	—	—	—	—	—	—	—	—	34	—	—	—	—	—	66
Aluminum-bronze	Sprabronze AA	—	—	—	1	—	—	—	—	—	9	—	—	—	—	—	—	—	90
Commercial bronze	Sprabronze C	—	—	—	—	—	—	—	—	—	—	—	10	—	—	—	—	—	90
Phosphor-bronze	Sprabronze P	—	—	—	—	—	—	—	—	—	—	—	—	5	—	—	—	—	95
Tobin bronze	Sprabronze TM	—	—	—	0.75	0.25	—	—	—	—	—	—	Bal.	0.8	—	—	—	—	58.2
Cadmium	Cadmium	—	—	—	—	—	—	—	—	—	—	—	—	—	—	—	—	99.9+	—
Copper	Copper	—	—	—	—	—	—	—	—	—	—	—	—	—	—	—	—	—	99.8+
Lead	Lead	—	—	—	—	—	—	—	—	—	—	—	—	—	99.9+	—	—	—	—
Antimony-Lead	Lead 6A	—	—	—	—	—	—	—	—	—	—	—	—	—	94	6	—	—	—
Molybdenum	Sprabond	—	—	—	—	—	—	—	99.9+	—	—	—	—	—	—	—	—	—	—
Monel	Monel	0.15	—	0.01	1.5	1	67	—	—	0.1	—	—	—	—	—	—	—	—	Bal.
Nickel	Nickel	0.25	—	0.04	0.6	—	Bal.	—	—	0.15	0.1	1	—	—	—	—	—	—	0.25
"Nickel aluminide"	405 Wire	—	—	—	Bal.	—	80	—	—	—	20	—	—	—	—	—	—	—	—
Nickel-chromium alloy	Metcoloy 33	—	—	—	Bal.	—	60	16	—	1.5	—	—	—	—	—	—	—	—	—
Carbon steel SAE 1010	Sprasteel 10	0.10	0.04	0.04	Bal.	0.5	—	—	—	—	—	—	—	—	—	—	—	—	—
Carbon steel SAE 1025	Sprasteel 25	0.23	0.04	0.04	Bal.	0.6	—	—	—	—	—	—	—	—	—	—	—	—	—
Carbon steel SAE 1080	Sprasteel 80	0.80	0.04	0.04	Bal.	0.7	—	—	—	—	—	—	—	—	—	—	—	—	—
Steel, low shrink machineable	Sprasteel LS	0.04	0.03	0.03	Bal.	2	4	1.5	1.5	—	—	—	—	—	—	—	—	—	—
Stainless steel Type 202	Metcoloy 5	0.15	0.06	0.03	Bal.	8.5	5	18	—	1	—	—	—	—	—	—	—	—	—
Stainless steel Type 304	Metcoloy 1	0.08	0.03	0.03	Bal.	2	8	18	—	0.75	—	—	—	—	—	—	—	—	—
Stainless steel Type 316	Metcoloy 4	0.08	0.04	0.03	Bal.	2	12	17	2.5	1	—	—	—	—	—	—	—	—	—
Stainless steel Type 420	Metcoloy 2	0.35	0.02	0.02	Bal.	0.35	0.5	13	—	0.5	—	—	—	—	—	—	—	—	—
Tin	Tin	—	—	—	—	—	—	—	—	—	—	—	—	99.8+	—	—	—	—	—
Zinc	Zinc	—	—	—	—	—	—	—	—	—	—	—	99.9+	—	—	—	—	—	—

Table 15.2 Standard Flame-Spray Powders: Metals and Carbides

Alloy	Metco Powder Designation	Recommended Gun(s)[a]	NBS Mesh Size Range	Type Analysis													
				Aluminum	Carbon	Silicon	Nickel	Chromium	Boron	Iron	Cobalt	Copper	Molybdenum	Tantalum	Tungsten	Tungsten Carbide (WC)	Chromium Carbide (Cr_3C_2)
Self-fluxing Ni-Cr-B alloy	12C	Combustion	−120 +325	—	0.15	2.5	Bal.	10	2.5	2.5	—	—	—	—	—	—	—
Self-fluxing Ni-Cr-B alloy	14E	Combustion	−140 +325	—	0.6	3.5	Bal.	14	2.75	4	—	—	—	—	—	—	—
	14F		−270 +15μ	—	—	—	—	—	—	—	—	—	—	—	—	—	—
Self-fluxing Ni-Cr-B alloy (meets AMS 4775)	15E	Combustion	−140 +325	—	1	4	Bal.	17	3.5	4	—	—	—	—	—	—	—
	15F		−230 +15μ	—	—	—	—	—	—	—	—	—	—	—	—	—	—
Self-fluxing Ni-Cr-B alloy	16C	Combustion	−120 +325	—	0.5	4	Bal.	16	4	2.5	—	—	3	—	—	—	—
Self-fluxing Co-Cr-B alloy	18C	Combustion	−120 +270	—	0.2	3.5	Bal.	18	3	2.5	40	3	6	—	—	—	—
Stainless steel type 316	41C	(Both)	−140 +325	—	0.1	1	12	17	—	Bal.	—	—	2.5	—	—	—	—
High chrome stainless steel type 431	42C	(Both)	−140 +325	—	0.2	—	2	16	—	Bal.	—	—	—	—	—	—	—
Nickel-chromium alloy	43C	(Both)	−140 +325	—	—	—	80	20	—	—	—	—	—	—	—	—	—
	43F		−230 +10μ	—	—	—	—	—	—	—	—	—	—	—	—	—	—
Aluminum	54	(Both)	−200 +325	99.0+	—	—	—	—	—	—	—	—	—	—	—	—	—
Copper	55	(Both)	−170 +325	—	—	—	—	—	—	—	—	99.0+	—	—	—	—	—
Nickel	56F	(Both)	−325 +10μ	—	—	—	99.3+	—	—	—	—	—	—	—	—	—	—
Tungsten	61	Plasma	−200 +30μ	—	—	—	—	—	—	—	—	—	—	—	99.5+	—	—
	61F		−325 +15μ	—	—	—	—	—	—	—	—	—	—	—	—	—	—
Tantalum	62	Plasma	−170 +325	—	—	—	—	—	—	—	—	—	—	99.5+	—	—	—
Molybdenum	63	Plasma	−200 +30μ	—	—	—	—	—	—	—	—	—	99.0+	—	—	—	—
Chromium carbide	70C	Plasma	−140 +325	—	—	—	—	—	—	—	—	—	—	—	—	—	99.0+
"Nickel aluminide"	404	(Both)	−170 +270	20	—	—	80	—	—	—	—	—	—	—	—	—	—
Tungsten carbide 12% cobalt aggregate	XP1110	Plasma	−270 +15μ	—	0.1	—	—	—	—	2	12	—	—	—	—	Bal.	—

a "(Both)" indicates either plasma or combustion flame spray gun may be suitable.

Table 15.3   Standard Flame-Spray Powders: Ceramics

Ceramic	Metco Powder Desig-nation	Recom-mended Gun(s)[a]	NBS Mesh Size Range	Type Analysis							
				Aluminum Oxide	Zirconium Oxide	Titanium Oxide	Calcium Oxide	Silicon Dioxide	Iron Oxide	Chromium Oxide	Others
Grey alumina contain-ing titania	101	(Both)	−270 +15μ	94	—	2.5	—	2	1	—	Bal.
White alumina	105	(Both)	−270 +15μ	98.5	—	—	—	1	—	—	Bal.
Lime stabilized zirconia	201	Com-bustion	−270 +10μ	0.5	93	—	5	0.4	—	—	Bal.
	201B	Plasma	−200 +30μ								
Chromium oxide	106	(Both)	−140 +10μ	—	—	—	—	—	—	99.0+	—

[a] "(Both)" indicates either plasma or combustion flame spray gun may be suitable.

Table 15.4   Standard Flame Spray Powders: Carbide Blends

			Type Analysis				
			Carbide			Metal	
Metco Powder Designation	Recommended Gun	NBS Mesh Size Range	Tungsten Carbide 12% Cobalt Aggr. (75F)[a]	Chromium Carbide (70C)[a]	Self-Fluxing Alloy (15E)[a]	Nickel-Chromium Alloy (43C)[a]	"Nickel Aluminide" (404)[a]
31C	Combustion	−120 +325	35	—	65	—	—
32C	Combustion	−120 +325	80	—	20	—	—
34F	Combustion	−230 +15μ	50	—	50	—	—
80	Plasma	−140 +10μ	—	85	—	15	—
81	Plasma	−140 +10μ	—	75	—	25	—
439	Plasma	−170 +15μ	50	—	35	—	15

[a] Metco powder designation is only for purpose of indicating Type analysis, and is not to imply the mesh size is the same.

Table 15.5   Other Flame Spray Powders

Chromium
Cobalt
Nickel (Coarse)
Chromium Carbide (Fine)
60% Chromium Carbide — Cobalt Blend
Rare Earth Oxides
Titanium Oxide
Zirconium Oxide — Hafnium Free
Zirconium Silicate
Magnesium Zirconate
Calcium Zirconate
40% Cobalt — Zirconia Blend
40% Nickel — Alumina Blend
50% Titania — Alumina Blend
75% Tungsten Carbide — Self-fluxing Alloy Blend
Stainless Steel Type 316 (Fine)
Barium Titanate
30% "Nickel Aluminide" — Alumina Blend
70% "Nickel Aluminide" — Alumina Blend

35% "Nickel Aluminide" — Zirconia Blend
65% "Nickel Aluminide" — Zirconia Blend
Columbium
Molybdenum Disilicide
Magnesia Alumina Spinel
Hexaboron Silicide
Chromium Silicide
Tin
Yttrium Zirconate
Silver
Platinum
Iron
Nickel Oxide
Mullite (Aluminum Silicate)
Self-fusing "Nickel Aluminide" Blend
Self-fusing Tungsten Blend
Titanium Hydride
Tantalum Carbide — 20% Zirconium Carbide

When sprayed these composites react exothermically in the flame, forming a nickel aluminide intermetallic compound, and the extra heat from the reaction helps the particles to bond.

In spraying with any of the different guns, bonding con-siderations and surface preparation are necessary preliminary steps. Some flame-spray materials, notably molybdenum and composite nickel aluminide, may be applied directly to a smooth, freshly cleaned surface of many substrate materials including most metals. The majority of other flame-sprayed coatings require more substantial preparation. For example the sub-strate may be roughened by such means as coarse-grit blasting Alternately, a layer of molybdenum or nickel aluminide about 0.002–0.005 in.-thick may be applied first as a bond coat, followed by the desired spray material. Nearly all flame-sprayed coatings adhere very well to such bond-coat layers, and this method of bonding is commonly used. Table 15.6 lists combinations of different bond coats, substrates, and spray materials, and gives typical bond strengths where known. Specific bonding recommendations will be made for those coatings discussed in detail in later sections.

Spraying is done in passes, much like paint spraying, with one or more guns. For example, cylindrical parts are rotated while a gun slowly traverses. Spraying is nearly always done in air, with sufficient booth exhaust facilities to remove dust and smoke. Special precautions are required for spraying toxic materials.

For production purposes the system is often automated using electronic controls for the guns and special work-hand-ling equipment. Extensions are used for spraying inside cylin-ders. Air jets directed at the workpiece are used occasionally for prevention of overheating.

Coatings may be used in the as-sprayed condition or may be finished by grinding or machining.

An important feature of flame spraying is that the workpiece itself is seldom heated above about 600°F and may be kept below 200°F if required. This distinguishes flame spraying from coating processes involving welding, whereby material is essentially both deposited and fused onto the substrate at the same time. Such welding processes are not part of this discussion. Flame spraying is often chosen over other processes in order to avoid tempering, recrystallization, oxidation, and warping problems associated with overheating of substrates.

However, there is one specialized but important branch of flame spraying in which a class of hard-facing nickel and cobalt-base alloys, known as "self-fluxing alloys," is applied with the combustion powder gun. The deposited layers are

subsequently fused by torch or furnace at about 2000°F. These coatings will be discussed in more detail later.

2 GENERAL NATURE OF COATING SYSTEMS

The surface of a sprayed coating typically has a roughness about 200 microinches rms, as measured with a profilometer; such a surface texture is similar to a fine-grade sandpaper. This reflects the fact that discrete particles are applied. When a flame-sprayed deposit is ground or machined, it looks like clean metal with very small, scattered pits visible on the finished surface of most sprayed materials.

The general metallurgical structure of a deposit (Fig. 15.5) has a laminar nature which results from the fact that individual molten particles were sprayed. It can be seen that these

Table 15.6 Bondcoat Materials and Bond Strengths

Bondcoat Material (Metco Designations)	Base Materials[a]	Flame Sprayed Coatings[b]	Typical Tensile Bond Strength (psi)[a]
Molybdenum wire (Sprabond)	Cast iron; steels (except surface-hardened, nitrided, etc.)	Wire sprayed steels	2200
		Wire sprayed aluminum	1850
	Aluminum alloys	Wire sprayed steels	2150
Plasma sprayed molybdenum powder (63)	Steels; other metals	Sprayed steels	3000
Nickel aluminide wire (405 wire)	Cast iron; steels (including surface-hardened, nitrided, etc.)	Wire sprayed steels	3000
		Wire sprayed aluminum	2400
	Aluminum	Wire sprayed steels	2150
	Titanium	Wire sprayed steels	3150
Combustion sprayed nickel aluminide powder (404)	Cast iron; steels (including surface-hardened, nitrided, etc.); titanium, quartz	None; or sprayed steels	2750
		Wire sprayed aluminum	2200
		Combustion sprayed zirconia powder	1100
Plasma sprayed nickel aluminide powder (404)	Cast iron; steels (including surface-hardened, nitrided, etc.); titanium, quartz	None; or sprayed steels	3000
Combustion or plasma sprayed nickel aluminide powder (404)	Aluminum alloys, tantalum alloys, columbium alloys, hard graphites, quartz	Most materials	(Good bond)
	Lightly blasted molybdenum, copper, or copper alloys	Most materials	(Good bond)
Combustion sprayed copper-glass powder (XP1159)	Smooth refractory oxides (e.g. alumina)	None; or copper, aluminum, etc.	(Good bond)
Plasma sprayed tungsten powder (61)	Refractory oxides (e.g. alumina); quartz	None; or copper, aluminum, etc.	(Good bond)
(None)	Rough-grit blasted steels	Wire sprayed steels	2850
		Wire sprayed aluminum	1030
	Rough-grit blasted aluminum	Wire sprayed steels	1150
	Rough-grit blasted molybdenum, copper, or copper alloys	Most materials	(Good bond)

[a] Bond strengths are for smooth substrate surfaces, except where indicated. Somewhat higher values are usually obtained when a bondcoat is applied to a roughened substrate.

[b] Bond strengths have been measured for the specific types of flame sprayed coatings indicated. Good bonding is also obtained for most other sprayed materials applied over indicated bondcoat.

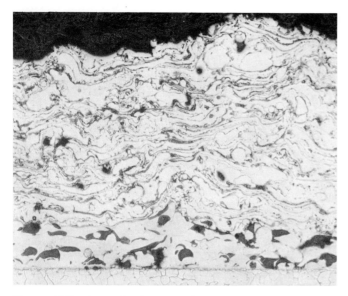

Figure 15.5 Cross Section Showing Flame Sprayed Coating. Sprayed stainless steel (center) is bonded to mild steel base (bottom) with nickel aluminide. Steel etched with Nital. Magnification 133×

particles flattened and individually solidified as they hit, without completely fusing with the rest of the coating. In spite of this, deposited material has excellent coherence as shown by the tensile strengths given in Table 15.7. The cohesion mechanisms are not understood completely, but appear to involve a combination predominantly of van der Waals' and chemical forces as well as mechanical interlocking and possibly some fusion. Oxides play an important role in the coherence of sprayed metals, and the expression " oxide cementation " is sometimes descriptively applied.

For most materials the nature of the bonding is similar to coating cohesion; that is, the forces are believed to be mainly van der Waals' and mechanical interlocking. Those materials that adhere to smooth surfaces and are used for bond-coat

Table 15.7   Tensile Strengths of Deposits (Parallel to Substrate)

Process	Flame Spray Deposit (Metco Designations)	Typical Ultimate Tensile Strength (psi)
Wire	Stainless steel type 420 (Metcoloy 2)	40,000
	Carbon steel SAE 1080 (Sprasteel 80)	27,500
	Molybdenum (Sprabond)	7,500
	Aluminum	19,500
	Zinc	13,000
Combustion powder	Nickel aluminide (404)	5,400
	Fused self-fluxing alloys	70,000
Plasma	Copper (55)	12,500
	Tungsten (61)	5,000
	Fine tungsten (61F)[a]	9,500
	Alumina (101)	5,000
	Nickel aluminide (404)	25,000

[a] Special inert gas shrouding of workpiece used for fine tungsten.

purposes, such as molybdenum, nickel aluminide, and a few others, actually alloy with the substrate in very thin layers. Such a layer is illustrated in Figure 15.6 for nickel aluminide on steel.

The photomicrograph in Figure 15.5 shows the oxide (typically about 5 per cent) that is generally present in the metal layers. Some porosity also exists in most cases, about 5 per cent to 15 per cent depending on material, gun, and spray technique. Advantage is taken of this porosity for oil retention in applications involving wear resistance. If the porosity is a disadvantage where barrier protection against corrosive elements is required, sealing is used, such as vinyl or phenolic paints or wax. Often a coating can be made thick enough to prevent permeation; about 0.030 in. may be sufficient. Sealing is usually combined with anodic coatings such as zinc or aluminum for excellent long-term corrosion protection of steel, and the surface roughness and porosity of these deposits is a big advantage for holding paint.

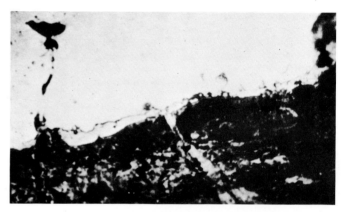

Figure 15.6   Photomicrograph Showing Bonding Alloy Layer. This is about 0.01–0.05 microns thick, between nickel aluminide (light area) and quench hardened, polished SAE 1095 steel base. Section was taken 14½° to interface, to "magnify" thickness of layer. Base etched with Nital. Magnification 1333×

There are several exceptions to the general porosity of flame-sprayed coatings. The fused self-fluxing alloy coatings are nonporous; in fact, their structure is not at all like that in Figure 15.5, but rather is castlike. These alloys contain fluxing elements, boron and silicon, which effectively remove the oxides and trapped gases during fusing. The coatings may contain small slag inclusions, but most of these come to the surface during the fusing operation.

In another exception to the usual porosity, recently developed powder mixtures (for example, 439 in Table 15.4) comprise several constituents that include a material such as a self-fluxing alloy. During spraying, the heat in the other components causes some fusing of the self-fluxing alloy, giving partially " self-fused " coatings with very low porosity.

Other new techniques allow the spraying of finer powders. These yield fine, low-porosity deposits of such materials as carbides and oxides.

Maximum allowable thickness of the coating is usually related to the effects of shrinkage that occurs during the cooling of each particle from its melting point. Stresses develop which are generally greater for the denser and thicker coatings, and which depend on material, spray technique, and other factors. The result could be cracking, lifting, or bending, which thus imposes maximum thickness limitations for some coating-substrate combinations. Normal thickness ranges are included with the coating data given in subsequent sections, but unusual substrates, such as thin sheets, may require thinner layers. On the other hand, careful temperature control of the workpiece sometimes permits coatings thicker than normal to be applied.

Structural strength of a composite should be designed into the substrate, unless requirements are within the capabilities of the deposits. Tensile strengths of some deposits are given in Table 15.7. As mentioned above, shrinkage can leave residual stress in deposited material, and, therefore, a measured tensile strength includes the effect of this stress. The tensile strength in a particular composite depends on factors including coating thickness, part size, shape of the base structure, and spray technique.

The tensile strength figures given in Table 15.7 were obtained parallel to the coating surface. These values are usually higher than for the perpendicular direction because of the laminar nature of sprayed material. The strength perpendicular to the surface is often nearly the same as the bonding

Table 15.8 Wear Resistant Materials (Non-fused)

Flame Spray Material		Flame Spray Gun	Thickness Range (Inches)	Bond Method (see notes)	Coating Hardness Rockwell	Particle Micro-hardness (Knoop 50 gm, and Rockwell Conversion)	Silica Slurry Wear Resistance (mil/hr loss)	Remarks	Typical Applications
Metco Designation	Material								
72F	Tungsten carbide-12% cobalt aggregate	Plasma	0.003 to greater than 0.030	a	Rc45–50	2500 (Rc75)	0.5	Best wear resistance in this list	Turbine vane separators, slurry pump parts, well drilling guides
439 Tungsten carbide-self-fusing blend		Plasma	{0.0015–0.005 {0.002–0.015	a} b}	Rc65 (conversion from VHN 300)	Carbide: 2615 (Rc75) Matrix: 960	3.3	Good finish, excellent for sliding parts	Seal areas in engines and fuel pumps
70C, 80 or 81	Chromium carbide or Chromium carbide blends	Plasma	0.003 to greater than 0.030	b				For wear resistance at elevated temperature (e.g. 1500°F)	Seal areas in higher temperature portions of turbine engines
101 or 105 (sealed)	Aluminum oxide	Plasma or combustion	0.003–0.030	c	Rc55	(Knoop 200 gm 1636)	17	For corrosive environments	Pump shafts, seal rings, pump impellers
Sprabond	Molybdenum	Wire	0.003–0.035	a	Rc40	1162–1800 (Rc65–75)		Excellent for sliding parts	Piston rings, lathe ways, extrusion dies, synchronization rings
Metcoloy 2	Stainless steel Type 420	Wire	0.005–0.160	d	Rc33	604–868 (Rc55–59)	18	Economic and moderately corrosion resistant	Clutch plates, paper and cloth rolls, transmission parts
Sprabronze AA	Aluminum-bronze	Wire	0.005–0.240	d	Rb82	356–449 (Rc31–36)		Marine environment, or combine wear with electrical	Ship propellor shafts, motor and generator armatures

Bond notes:
[a] Smooth clean surface is usually satisfactory for bonding. Roughen surface if desired to increase bond strength.
[b] For bond, use plasma flame sprayed nickel aluminide (404) or molybdenum (63) 0.005 in. thick on smooth clean surface. In addition use roughened surface if desired to increase bond strength.
[c] For bond, use plasma or combustion flame sprayed nickel aluminide

(404) 0.005 in. thick, on clean surface. Alternately, for high temperature (above about 1000°F), use nickel-chromium alloy (43C powder) on roughened surface.
[d] For bond, use wire sprayed nickel aluminide (405) or molybdenum (Sprabond) 0.005 in. thick on smooth clean surface. Use roughened surface if desired to increase bond strength.

strength, and, in fact, some of the "bond strengths" in Table 15.6 actually represent perpendicular tensile strengths of the bond-coat material.

3 MATERIALS FOR WEAR AND ABRASION RESISTANCE

One of the most common areas of usefulness for flame-sprayed deposits is in providing resistance to wear, abrasion, or erosion. It is advantageous to devide coatings for this into two groups, those used in their as-sprayed condition, and those that are fused after spraying. Table 15.8 lists typical as-sprayed coatings that are excellent for wear, abrasion, or erosion resistance. These are applied to base structures that may be of mild- or high-strength steel, aluminum, or almost any metal or alloy. Ordinarily, deposits are finished by grinding, but for some applications finishing is not necessary.

Table 15.9 is a similar list of self-fluxing alloys and carbide mixtures thereof. These are fused by torch or furnace at about 2000°F in air after spraying and are nonporous. Fusing does not permit the use of certain materials for substrates, such as low melting-point materials like aluminum, steels that might lose temper, or parts that might warp. No bond-coat materials are used; base preparation is by grit-blasting or rough-cutting, and the fusing results in a metallurgical bond. A combustion gun for powder is used for economic reasons, although plasma-flame equipment produces about the same coating results. Most fused coatings are finished by grinding, but the softer ones may be machined.

The listings in Tables 15.8 and 15.9 are approximately in decreasing order of wear resistance. However, such ordering should be cautiously interpreted, as the actual degree of wear can vary highly between different jobs. Also, there are other considerations that are just as important, such as compatability with the material of a mating part. This includes galling tendencies, if any, and relative wear resistance of the mating part. Whether or not there is lubrication is of obvious importance. Cost is another major consideration; the more wear-resistant coatings often tend to be higher in initial cost.

For the "slurry wear resistance" results in Table 15.8, dimensional loss was determined on samples rubbed in a silica slurry for a definite time period in a special machine. It should be emphasized that, as for any accelerated wear test, the results can only be used as a very rough guide. For any new application it is advantageous to run several coatings in wear tests that more closely simulate the actual application.

The sprayed particles in many deposits are very hard. This is to be expected for carbides and ceramics, but the hardness of metal coatings is sometimes surprising. Rapid quench rates and the oxides both contribute to high microhardness. It should be noted that there is a very significant difference between the microhardness of the particles that make up the coating, and the Rockwell hardness of the coating itself. The coating hardness is influenced not only by particle microhardness but also by the effects of porosity, cohesion, and the form of the oxides. Both types of hardness must be taken into account when selecting a coating for a particular job. For example, if the application is for sliding frictional wear resistance, the chief requirement is for particle microhardness. On the other hand, if the application involves erosion or heavy loads that might tear out particles, a coating Rockwell hardness that reflects the best interparticle bonding becomes important.

Most fused self-fluxing alloys have fair-to-good resistance to attack by various acids, salt water, and similar corrosive media. They are also oxidation resistant in air at elevated temperatures and are useful up to about 1500°F. Chemically these alloys behave like 80 : 20 nickel-chromium alloy (which is better known and may be used for guidance). This chemical

Table 15.9   Fused Self-fluxing Alloys. (Applied by combustion powder gun, then fused at 2000°F.)

Metco Powder Designations	Thickness (inches)	Hardness (Rockwell C Scale)	Silica Slurry Wear Resistance (0.001 in./hr. loss)	Remarks	Typical Applications
32C System[a]	0.010–0.025	Matrix 62 Carbide 75	0.4	Highest carbide content of fused coatings	Grit-blast plates and blades, abrasive wheel molds and plungers, extrusion mandrels
31C	0.005–0.025	Matrix 62 Carbide 75		Carbide, simpler system than 32C	Oil well joints, wire drawing capstans, mixing blades
34F	0.003–0.020	Matrix 62 Carbide 75	1.1	Carbide, fine texture, excellent finish	Buffing fixtures, plug gauges, punch guides
15E	0.006–0.070	62	3.2	Alloy conforms to AMS 4775	Textile spindles, engine valves and stems, oil well seal areas
16C	0.010–0.125	60		Similar to 15E but easier to fuse	Heat treating rolls, pump plungers, extrusion rams
12C	0.010–0.050	30		Machineable	Glass molds, valve plugs

[a] For 32C, use undercoat of 15E (0.002–0.015 in.) apply 32C, then overcoat with 15E (0.005–0.010 in.) before fusing.

behavior and the lack of porosity make the fused coatings especially useful where protection against wear and corrosion is required.

Sealed coatings of inert aluminum oxide or other ceramics are used for more severe chemical environments such as in pumps.

In milder conditions, the wire-sprayed Type 420 stainless steel applied thick enough, or sealed, is quite suitable and may be more economical. The wire-sprayed coatings listed in Table 15.8 are especially useful for lubricated service. Here the coating porosity with its oil retention is an important advantage. Of these coatings, molybdenum has been particularly successful and is rapidly growing in use, as, for example, in replacing chrome plate on piston compression rings in engines. An added advantage of molybdenum is its bonding ability, which simplifies production.

In some applications advantage is taken of the surface roughness of as-sprayed material. For example, the Type 420 stainless steel is being used to provide a high-friction surface on rolls used for handling cloth.

Tables 15.1 through 15.5 include many materials for wear resistance that are not listed in Table 15.8. These are considered for unusual cases that cannot be covered in detail here. They include other types of carbides and carbide mixtures, finer grades of powders or smaller wires for denser coatings, other self-fluxing alloys, and quite a selection of oxide ceramics. Stainless steel powders as well as wire can be sprayed by combustion or plasma guns. Softer steels and other alloys are usually used for their machinability and are particularly important in maintenance for replacing worn metal.

4 CORROSION PROTECTION

4.1 Anodic coatings

The protection of iron or steel with anodic (sacrificial) coatings of zinc or aluminum is well known, and flame spraying is a good method of applying these coatings. Generally the wire process is used. Aluminum powder can produce denser coatings with less oxide than wire, but such attributes are not ordinarily needed for corrosion protection, and the wire coatings are substantially lower in cost. Best results are obtained when a sealer or paint system is applied to the sprayed coatings. Excellent protection for 30 years or much longer often can be expected from such sealed coatings. No repainting is needed for the lifetime of the structure unless appearance must be maintained, and then it can be done as infrequently as at 5- to 10-year intervals.

In the past zinc was much more commonly used than aluminum, both with flame spraying and with other coating methods such as galvanizing. However, experience and test results now indicate that aluminum protects longer in the more severe environments including industrial, very humid, or salt atmospheres. This is attributed to the higher "activity" of zinc, which results in more rapid corrosion of the zinc. In the less severe environments the strong anodic effect of zinc is advantageous because it protects sizable areas of bare iron or steel which may result from skips during spraying or from subsequent damage. Zinc vapors during spraying are sufficient to give some coverage by condensation on areas usually shadowed from the direct spray, such as in crevices and behind projec-

tions. These considerations, and the fact that the surface preparation for zinc is less critical than for aluminum, must be compared with the lower material cost for the same thickness of aluminum and with the value of the longer life of aluminum coatings.

As mentioned, a paint system is generally recommended for sealing the sprayed metal, except that thin unsealed coatings of zinc are sometimes used where lowest cost protection is required. Unsealed aluminum sometimes shows flaking or discoloration. The main purpose of the sealer is to close the pores of the coating to reduce the surface area of sprayed material exposed to the environment. Surface coverage of paint is not important except for appearance. The paint can be scraped or worn off without significantly shortening the life of the structure.

Many sealers have been tested with flame-sprayed zinc and aluminum. Certain types have proved much more satisfactory than others. Important factors include chemical compatibility. penetration into the coating, ease of application, and viscosity. Very volatile or very thick paints are undesirable, as penetration is poor. An inhibitive primer such as a "wash-type" primer of polyvinyl-butyral resin containing zinc chromate inhibitor is usually applied first. For overcoating, a good vinyl varnish or pigmented vinyl paint that is easy to apply and penetrates well should be used. Phenolic epoxy is good and may be used without a primer. Such premium sealers as these have proved successful, but there are many other paints, including cheaper ones, that should do well.

Paints adhere very well to sprayed zinc and aluminum, which is not usually true of metal coatings applied by plating or dipping. Consideration must also be given to the fact that often most of the cost of a complete coating system is in the preparation of the base and application of the paint. Many objects are being finished today by sandblasting and then painting. At a relatively low additional cost, a thin coating of zinc can be sprayed on the blasted surface before painting, thus increasing the life of the product many times.

Table 15.10 suggests various coating systems for the protection of iron and steel in different environments. The thicknesses of flame-sprayed layers applied for service may be different from those indicated, where desired.

Surface preparation is by sandblasting (for large areas, especially outdoors) or grit blasting. Bonding requirements for corrosion protection are not stringent so no bond coat is necessary.

4.2 Impermeable coatings

Sprayed coatings that are cathodic, or inert, with respect to the base must be impermeable if corrosion is involved and thus must be sufficiently thick, or be sealed, or be fused self-fluxing alloys. Layers produced from certain "self-fusing" powder mixtures need not be as thick as other as-sprayed coatings.

Useful sealers for the unfused coatings include phenolic varnishes for most organic or aqueous environments except alkalies, or a good crystalline wax for acidic or basic solutions. Again, sealer penetration is important.

Sprayed cathodic or inert materials are not usually used to provide corrosion protection only, since anodic coatings are better. However, such protection is often needed in addition to other attributes such as wear resistance. Therefore, reference should be made to materials covered in other sections, particularly Section 3.

5 HIGH-TEMPERATURE OXIDATION PROTECTION

Table 15.11 lists systems that are applied specifically for protection against oxidation at high temperatures. Not included

Fused self-fluxing alloys are oxidation resistant, but are usually considered where there is also wear, or at least some erosion from hot flames and combustion products. There are other oxidation-resistant materials, some of which are listed in Table 15.5, that have been considered for protecting at very high temperatures, but coating porosity is a problem. A few (such as those containing silicon) have shown promise because of the formation of self-protective oxides that seal the pores.

Table 15.10 Systems for Corrosion: Protection of Steel

| General Environment | Types of Applications | System (on Iron or Steel) | | | Minimum Expected Life (Years) | | | |
		Sprayed Metal	Thickness (inches)	Sealers[a]	10	15	20	30
Salt water immersion or heavy spray, including tide level	Ship hulls, above or below water line; bilges, bridges (low parts); pilings	Aluminum Aluminum	0.003 0.004–0.006	primer + 2 coats vinyl[b] primer + 2 coats vinyl[b]		×	×	
Salt atmosphere (no direct spray), or very humid	Mild conditions Extreme conditions, e.g. decks	Aluminum Aluminum	0.003 0.006	primer + 2 coats vinyl primer + 2 coats vinyl		× ×		
Industrial atmosphere	Mild conditions Extreme conditions	Aluminum Aluminum	0.003 0.006	primer + 2 coats vinyl primer + 2 coats vinyl			× ×	
Rural atmosphere	Poles, fixtures, outdoor furniture, nuts and bolts	Aluminum Zinc Zinc Zinc	0.003 0.003 0.010 0.003	primer + 2 coats vinyl (none) (none) primer + 1 coat vinyl	×			× × ×
Cold fresh water	pH less than 6.5 pH 5–10 pH over 6.5	Aluminum Aluminum Zinc	0.006 0.006 0.010	primer + 2 coats vinyl 2 coats phenolic-epoxy (none)	× × ×			
Hot fresh water or steam (400°F) or food processing		Aluminum	0.006	2 coats phenolic epoxy	c			

[a] If desired to maintain appearance, re-apply top coat approximately every 5 to 10 years, preparing surface by cleaning only.
[b] If antifouling is desired, add coating of "organo-tin" vinyl type of antifouling paint.
[c] Limited information indicates at least 5 years life expected.

here are thermal-insulation coatings and coatings to resist wear at the high temperatures. These are covered in other sections. The systems in the table were established to protect iron-base parts, but they may be considered or adapted for other oxidizable metals. Sealers are used because of the coating porosity. Wire is usually used because it is cheaper than powder. Coarse-grit blasting is used to prepare the base for all materials in the table except nickel aluminide, which requires only light preparation.

The short-term oxidation protection by nickel aluminide at higher temperatures (up to about 2400°F) is useful for easily oxidized materials (including titanium) during processing such as extruding. In this case the coating is removed and the final products are not laminates, but the engineer has used the flame-spray process for intermediate stages.

The problems of protection at high temperature not only involve choice of oxidation-resistant materials and minimizing coating porosity, but also require consideration of diffusion between the coating and substrate. Thus aluminum protects steel by forming iron aluminide layers which help protect the steel by creating complex oxide layers. For such reasons, it becomes increasingly important to design coating systems specifically for a particular substrate in a particular environment.

6 ELECTRICAL APPLICATIONS
6.1 Conductors

Copper sprayed by any of the three processes has a resistivity of about 5 ohm-cm, which is about three times that of wrought copper. It is sprayed up to $\frac{1}{8}$-in. thick on steel, for example, to

Table 15.11   Systems for Protection against High-Temperature Oxidation

| Max. Temp. (°F) | Flame Spray Material | | Thickness (inches) | Sealer (see notes) | Suggested Applications |
	Type[a] Metco Designation	Material			
900	Wire or Powder (54)	Aluminum	0.006	b	Smokestacks, exhaust stacks, hot oil refinery lines
1600	Wire or Powder (54)	Aluminum	0.008	c, d	Damper plates, annealing covers, exhaust manifolds
1800	Wire (33) or Powder (43C)	Nickel-chromium alloy	0.010	c	Pyrometer tubes and supports, wire annealing retorts, carburizing boxes
2100	Wire (33) or Powder (43C)	Nickel-chromium alloy	0.015		Furnace structure parts, furnace conveyers
	overcoated with: Wire or Powder (54)	Aluminum	0.004	c	
1200 (or 2200[e])	Wire (405) or Powder (404)	Nickel Aluminide	0.010	(none)	Protecting billets during processing, protecting molybdenum stirring rods for glass during warmup

[a] Powder coatings may be sprayed by either combustion or plasma guns; except for 404, denser coatings are obtained with plasma.

[b] For example, a silicone-alkyd resin combination containing aluminum flake, with 6–8 per cent silicone content.

[c] For example, a bituminous coal tar containing aluminum flake. This burns out at high temperature, but protects long enough for protective diffusion layers of alloy to develop.

[d] Heat treat at 1450°F for 5 minutes.

[e] Short term protection above 1200°F.

form laminates with strength and good electrical conductivity for such applications as spark electrodes and locomotive rail contacts. Thin layers (about 0.005 in.) of sprayed copper are used to provide points for soldering. It is applied to ceramic or glass to provide conducting paths or heating elements; however, flame spraying has been less practical for the very small printed circuits.

Other metals used in electrical applications include aluminum, zinc, babbit, and tin. Most of these have low melting points and good ductility. These are useful properties for spraying onto cloth or paper as, for example, in the production of capacitors.

Surface preparation of metals may be by grit blasting (for thinner coatings) or a bond-coat material may be used. Bonding on glass or enamel is obtained by preheating the base to about 1000°F, maintaining temperature during spraying, and then cooling slowly to prevent cracking. If a conducting layer is required on a refractory ceramic substrate and the surface is rough, a good bond can be obtained directly with most sprayed metals. However, if the surface is smooth, combustion- or plasma-sprayed copper-glass mix adheres well and may be used to provide a bond coat. The resultant coating is essentially metallic for making solder connections and similar applications.

6.2   Insulators

Flame-sprayed aluminum oxide (101 or 105 in Table 15.3) is commonly used for providing electrical insulation. A rule of thumb is to use 50 volts per mil in determining suitability and required thickness, but a slightly lower value should be used for thicker coatings since the relationship is not quite linear.

Slightly higher break-down strength can be obtained by sealing with, for example, a phenolic varnish.

Barium titanate is another ceramic used for electrical insulation because of its dielectric properties, which are approximately the same as for aluminum oxide in flame-sprayed coatings. Other ceramics for special problems may be chosen from Tables 15.3 and 15.5. Special techniques for applying denser ceramic coatings with higher dielectric strength are under development.

Ceramic coatings are usually applied between about 0.003- and 0.050-in. thick. It is best to use a bond-coat metal, although bonding of the thinner coatings may be satisfactory on a rough-grit-blasted surface.

7   REFRACTORY MATERIALS

This section discusses refractory materials for use in very high temperatures. Such materials are usually sprayed with plasma, although the combustion processes will handle most that have melting points up to and including that of zirconium oxide (4650°F). Table 15.12 gives data for some of the refractory metal and ceramic deposits.

Tungsten, molybdenum, tantalum, and columbium are the more common refractory metals and are easily plasma flame-sprayed. In one application tungsten about 0.1-in. thick is bonded to a graphite substrate with a thin coating of sprayed tantalum, and the laminate is used in a rocket exhaust system. The tungsten may be sprayed with normal procedures. Finer powder may be sprayed with special techniques of shrouding and cooling the workpiece to yield coatings with higher tensile strength. Sometimes bonding of refractory metals requires

Table 15.12   Refractory Coatings[a]

Metco Powder Designation	Material	Melting Point (°F)	Coating Density (gm/cc)	Porosity	Coating Hardness (Rockwell)	Particle Microhardness (Knoop 50 gm)
61, or 61F[b]	Tungsten	6170	16.9	12.5%	Ra 50	500
63	Molybdenum	4760	9.06	11%	Rc 25–34	786
62	Tantalum	5430	14.15	15%	Ra 65	1585
XP 1132	Columbium	4532			Rc 61	1912
201B	Stabilized Zirconium Oxide	4650	5.0	11%	Rb 96	
201	Stabilized Zirconium Oxide (fine)	4650	5.3	5%	Rc 47	
101	Grey Aluminum Oxide	3700	3.3		Rc 56	1636 (Knoop 200gm)
105	White Aluminum Oxide	3700	3.3		Rc 60	

[a] See Tables 15.7 and 15.8 for other data.
[b] Spray 61F at close distance with heavy shrouding of workpiece with inert gas, or with cooling with CO_2 jets.

unusual considerations. For example, the tantalum bond coat mentioned above is used to minimize the formation of brittle tungsten carbides.

Of the oxides used in refractory applications, zirconium oxide has one of the lowest thermal conductivities and has been used as a heat barrier in applications similar to that described above for tungsten. Usually the part is cooled in use, perhaps by circulating the liquid fuel of the rocket engine.

Figure 15.7   Cross Section of "Gradated" Coating of Zirconia and Nickel Aluminide. Upper part is all zirconia, lower part of deposit is all metal. Steel substrate is at bottom. Unetched. Magnification 166×

Aluminum oxide, plasma- or combustion-sprayed, is a good heat barrier and also affords good erosion resistance. An example tested is a piston dome in a combustion engine. It is resistant to wetting by molten metals and is used for protecting such items as brazing fixtures and pouring troughs. Since the coatings exhibit some porosity, the base material must be protected against high-temperature oxidation by such techniques as spraying an intermediate layer of a nickel-chromium alloy. If there are fluxes involved, the substrate must also be protected against these, but some fluxes (such as for zinc) are extremely severe and are a serious problem in this type of application.

Plasma-flame spraying is suitable for producing "gradated" coatings of metal and ceramic by applying mixtures of powders in different proportions in intermediate layers (Fig. 15.7). Such a laminated system is useful for extreme thermal shock or temperature-gradient conditions, where a simple coating of ceramic on metal might crack. No one system can be universally recommended, as each set of conditions is different and requires its own laminate design. A necessary design consideration is that although the coating system is basically for heat barrier some heat flow may be needed to prevent melting and failure of the ceramic on the high-temperature side.

Carbides, without metal binder, may be plasma-sprayed, but most of them decarburize and oxidize too much during spraying. An exception is chromium carbide. Recently, excellent deposits have been obtained with "solution" carbides of tantalum carbide containing about 20 per cent zirconium carbide or hafnium carbide. These have extremely high melting points and appear to be more stable than simple carbides. They are considered for conditions of heavy erosion in neutral or reducing atmosphere, as in some solid-fuel-rocket exhausts.

Borides and silicides such as those listed in Table 15.5 are sprayable and may be used in special refractory applications. Porosity has limited their use in protecting substrates against oxidation. Materials high in boron are of interest for design of nuclear installations.

8   OTHER APPLICATIONS

There are many individual uses that cannot be categorized. Rather than present these here, it is hoped that the reader has gained sufficient basic understanding of flame-sprayed laminates to apply his own ingenuity in solving new design problems.

Chapter 16

Glassed Steel

William B. Crandall

1 INTRODUCTION

Glassed steel is an engineering laminate. Actually, it is a family of laminates, each member of which consists of a glass structure (which may vary in formulation and number of layers) applied and then fused by high-temperature firing to a base metal, usually mild steel. Unlike most laminates, glassed steel is formed into its desired size and shape before the laminate, in this case glass, is applied to the base material. Glass thickness is built up by repeating the cycle of application and fusion.

Because of the special conditions required to produce a bond between the glass and base metal to which it is applied, there are some distinct differences in composition and character between the formulations used for glassed steel and the more commonly known glasses. However, the properties obtained with glassed-steel compositions are comparable in most respects, and in some cases are a decided improvement to high-grade laboratory glasses.

One of the most striking aspects of glassed-steel composites is their variety. As there is no one type of glass, there is no one type of glassed steel. This is an extremely important point to remember for this type of laminate can be tailored to a variety of uses with the judicious selection of formulation and base metal to which it is applied.

Coatings of amorphous glass have been available for many years. They have been noteworthy mainly because of their use in making corrosion-resistant process equipment. They have taken the form of storage and reaction vessels, valves, piping, pumps, heat exchangers, columns and other types of process equipment. In addition to their corrosion resistance, they have been used because the glass surface is inert and smooth. Sizes vary from small pump parts to vessels having over 40,000 gallons capacity.

Crystallized-glass coatings are also used as an engineering laminate that falls into this category of composites. Coatings of this type retain the corrosion resistance, inertness, and smoothness of amorphous glass coatings, but are superior in most other engineering properties. They provide superior abrasion resistance, impact resistance, thermal shock, and can be used at much higher temperatures. The electrical insulation properties of this composite have also found practical use. The process for manufacturing the crystallized glass-metal composites is closely related to that for amorphous glass composites, but the former is more critical in terms of high-temperature firing. The crystallized glasses can be applied to a variety of base metal. Since a number of crystallized-glass formulations are available as well, the composites can be produced from the different combinations of base metal and coating from a family of materials each with different properties. Both of these composites are competitive in cost with specialty alloys.

2 SUBSTRATE

A number of metals and alloys may be used as the base material for glassed-steel composites (see flow diagram Fig. 16.1). The standard is low-carbon steel to ASTM A-285 Grade B with special surface conditions. Few base metals can be arbitrarily dismissed as potential substrates for at least some type of glass coating. There is great latitude in the selection of glass formulations, and manufacturing methods are sophisticated enough to permit matching of the base-metal characteristics with those of some of the glass formulations available.

In practice, almost any metal or alloy can be used as a substrate for the application of some type of glass composition. Practically, however, there are a number of metals which are not suitable because either they melt below firing temperature, are thermally or chemically incompatible, or a combination of

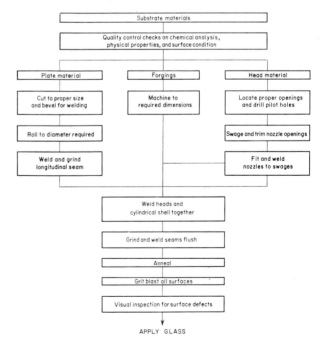

Figure 16.1 Flow Diagram 1. Manufacture of Glassed-steel: Substrate Preparation

WILLIAM B. CRANDALL is Vice President, Research, The Pfaudler Company, a division of Sybron Corporation, Rochester, N.Y.

these reasons. A few common materials that fall into this category are brass, Monel, nickel, and ductile iron.

When glassed steel takes the form of a vessel, the base metal ordinarily is approved under the ASME Boiler and Pressure Vessel Code and is of flange or firebox quality.

The best steelmaking practices must be used during smelting, pouring and mill fabrication to produce plates and other shapes having the desired metallurgical characteristics for application of the glass. Minor surface irregularities, too slight to warrant rejection for ordinary applications, can result in excessive rework and/or rejection during the coating operation.

2.1 Fabrication

For the most part fabrication of the base metal is aimed at producing vessels, columns, and other cylindrical, tanklike objects. Fabrication is carried out with rectangular steel plates ¼- to 2-in. thick, which are sized and edge-beveled by oxy-acetylene torch cutting. These plates are then rolled into cylindrical shells and tack welded. Top and bottom heads for vessels are formed from circular plates by hot spinning or hydraulic pressing. All openings in the heads and shells — for example, agitator, baffle, thermowell and sight glass ports as well as flanged pipe connections — are completed prior to assembly. Such openings may be either radial or vertical and require a gradual transition between the appendage and the vessel proper. Such a configuration is obtained by swaging or "flueing."

Agitators, baffles, thermowells, and similar items that require a coating on the outer surface are usually constructed from pipe, tubing, bar, and other shapes more suitable to the end product than plate.

2.2 Welding

Circumferential seams (joining shells to each other axially or heads to shells) and longitudinal welds (completing fabrication of the shells) are done by submerged arc automatic welding. Where this process is not practical (as in nozzles or other flanged openings which require joining a flanged piece to the outside of a swaged opening) a combination of semiautomatic gas metal-arc (short circuiting transfer) and the manual shielded metal-arc welding techniques is used.

Welds are an integral and very important part of the pressure vessel. All welding is performed by operators qualified under the ASME code. In addition to the code-required X-ray inspection of welds, other nondestructive tests, such as visual, magnetic particle, and dye penetrant methods, are made. Any defects revealed are cut out and these areas rewelded and rechecked.

2.3 Surface preparation

After fabrication and initial surface inspection are completed, the entire unit is usually given a normalizing heat treatment. Grain refinement and burn-off of surface impurities result. After cooling, the surface to be coated is cleaned by grit blasting. In this operation, minute steel defects which have been shown by experience to cause difficulty during coating are often revealed. All defects are repaired by grinding and, if

necessary, rewelding. The unit is then given a final grit blast prior to application of the coating.

3 APPLICATION OF GLASS

3.1 Frit manufacture

The first step is to mix the raw materials in the proper proportions for the formulation selected (see flow diagram Fig. 16.2). Ingredients used are of reasonably high purity and quality, checked by periodic chemical analysis. Accurate weighing of raw components followed by thorough mixing is necessary to ensure intimate contact of aggregates during melting.

Raw materials are converted to a molten glass in a smelter. This operation usually requires 1–3 hours at a temperature of 2200–2600°F. The values are determined by the composition of the glass. Variations in time and/or temperature can radically change the operating properties of the glass. There are essentially two types of smelters used, one in which the glass is agitated during smelting by flowing, rocking, or other means. The operation may be continuous or batch. The second type is a batch smelter, and the glass is not agitated. Gas, oil, or electricity is used as fuel.

After the glass has been smelted, it is poured into water from the smelter. The sudden cooling results in fracture of the molten stream into small particles called frit. In some cases the frit is produced by pouring the molten stream onto cooled metal rolls; this is called roll quenching.

The frit is ball-milled with water and mill additions to form a slurry called slip. The mill additions are essentially electrolytes and suspending agents to keep the glass particles in suspension. Normal particle-size range of glass after milling is 10–100 microns.

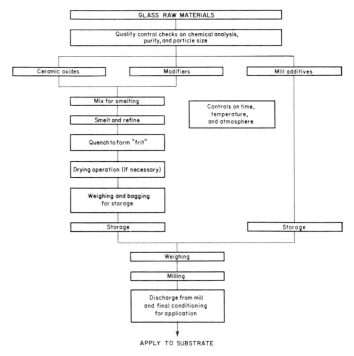

Figure 16.2   Flow Diagram 2. Manufacture of Glassed-steel: Glass Preparation

3.2 Application of glass to substrate

The application of the glass to the substrate is a multistep process involving both a ground (primer) coat and a cover (finish) coat (see flow diagram, Fig. 16.3). The ground coat is specially formulated to possess the necessary properties of adherence, oxide solubility, and "fit" (thermal expansion) for best mating with the base metal. Since there are numerous base metals used, each requires a ground coat with specific properties. The cover coat chosen depends not only on the base metal but the requirements of the service. For example, a cover coat for best chemical corrosion resistance would not necessarily be the best for abrasion resistance.

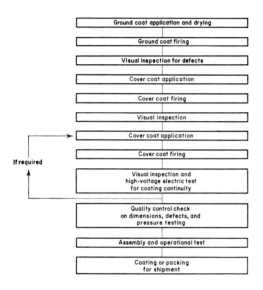

Figure 16.3   Flow Diagram 3. Manufacture of Glassed-steel: Application of Glass to Substrate

Another consideration in the selection of a coating is whether to use a conventional glass or one of a family of crystallized glasses. The conventional glasses are amorphous and remain so throughout the complete processing cycle. The crystallized glasses are initially amorphous but are formulated to develop controlled crystal growth *in situ* during a special heat-treating of the composite.

Two methods are commonly used to apply a coating to a base metal:

1. *Spray application.* The coating in the form of slip (slurry) is sprayed onto the item to be coated. Spraying equipment is similar to that used for paint application.

2. *Dipping or Slushing.* The item to be coated is submerged in the slip, removed, and allowed to drain. A thin coating of the slip remains on the metal.

For both methods the ground coat is dried and the item is fired in a furnace, which may vary in size from small laboratory types to giants capable of handling 50,000 gallon vessels. After the ground coat is fused to the base metal, the vessel is taken out of the furnace and allowed to cool at room temperature. The first cover coat is then applied, dried and fired. These steps are repeated until the desired thickness is obtained. For conventional glass the coating process is now complete. For crystallized glass, the equipment is returned to the furnace and

held at a temperature below the cover coat firing temperature. This allows crystallization to occur.

3.3 Types of coating

Because of the vast diversity of glassed coatings available, classification is somewhat complex. Suitable types for specific applications can best be determined by referring conditions to the manufacturer for recommendations. However, there are certain broad categories into which glass-lined coatings fall.

The coatings are classified according to composition. They are also classified in terms of the integrity of the coating.

The composition of the coating relates most directly to the properties it will ultimately exhibit. The degree of corrosion resistance, abrasion resistance, impact strength, as well as a combination of other properties are determined by the composition selected. As with most materials there is no single formulation that provides optimum performance in all respects. Certain types may provide the best acid-alkali resistance, while still others are best for thermal shock or some other characteristic.

The integrity of the coating refers to the quality of the glass coverage. Certain services are mild and it is not necessary for all metal areas to be completely covered by glass of a certain thickness. Pinholes may exist without detriment to the service or longevity of the equipment. Other services demand complete glass coverage. The integrity of the coating is specified by the inspection method employed at the factory. There are three designations of this type:

1. *Visual inspection* ensures negligible metal exposure. Recommended field test: visual.

2. *Low voltage* ensures complete glass coverage. Recommended field test: visual.

3. *High voltage* ensures 100 per cent glass coverage and a minimum of thickness to resist severe corrosives. Recommended field test: Pfaudlertron(R) Glass Tester.

Crystallized glasses create another classification condition. In these types of coatings, the degree, type, and dispersion of crystals grown within the glassy matrix are indicated in classifications. Since physical and chemical properties vary considerably, depending on the degree, type and dispersion of crystals, these composites are classified accordingly.

4   PROPERTIES
4.1   A summary

The properties of glass-to-metal composites illustrate their versatility and serve as a guide to their use.

Corrosion resistance

Glassed steel offers excellent resistance to all acids (except HF) at all concentrations to the boiling point, in many cases to 350°F, and much higher under certain conditions. Under alkaline conditions, the normal recommendation is a maximum operating limit of *p*H 12 at 212°F. This limitation on *p*H concentration varies with temperature and frequency of exposure.

Crystallized-glass composites are capable of withstanding corrosives at higher operating temperatures. They have been used for handling metal chlorides, molten salts, and molten metals such as selenium, gallium, and zinc. Depending on the base metal used and the environment, operating temperatures up to 1450°F have been achieved.

Abrasion resistance

Glassed steel, including the crystallized glass composites, has a hardness of 6 to 7 on the Mohs scale. While both are in the same hardness range, the crystallized glasses offer superior abrasion resistance.

Heat transfer

While ceramic materials are generally thought of as insulators, glassed steel is comparable to other materials having higher heat-conductivity factors. The resistance to product film build-up or surface fouling of glassed steel affects the over-all co-efficient of heat transfer, making the over-all conductivity improvement possible.

Product adherence

Product adherence or buildup is minimized, resulting in im-proved heat transfer and minimum down-time for cleaning.

Purity protection

The inert characteristics of these materials result in freedom from objectionable effects on the purity, flavor or taste, and color of product being processed in glassed-steel equipment.

Electrical properties

Glassed steel has demonstrated excellent electrical resistance even at elevated temperatures. At room temperature, glassed steel has a dielectric strength of approximately 700 volts per mil of glass thickness.

Other properties

In addition, these materials exhibit unusual tolerance for impact and thermal shock. They may also be ground or honed to close dimensional tolerances.

4.2 Resistance to chemical corrosion

Because of the importance of glassed steel's corrosion resis-tance to its use in industry, this aspect of the material is to be emphasized. Glassed steel is particularly resistant to corrosion by aqueous acid solutions at elevated temperatures. It exhibits this characteristic because the glasses used are composed en-tirely of oxides. The major constituent of glass, silica (SiO_2),

Table 16.1.   Corrosion Resistance for #3300 Glass

Acid	Conc. % wt.	Temp. °F	Corrosion Rate	
			mdd	mpy
HCl	5	325	26.0	14.6
HCl	20	325	47.5	26.7
HCl	30	325	31.3	17.6
HNO_3	5	300	8.9	5.0
HNO_3	10	300	14.2	8.0
HNO_3	30	300	22.2	12.5
HNO_3	50	300	15.6	8.8
HNO_3	60	300	8.4	4.7
HNO_3	70	300	3.6	2.0
H_2SO_4	5	300	10.1	5.7
H_2SO_4	20	325	28.8	16.2
H_2SO_4	50	375	6.9	3.9
H_2SO_4	50	425	35.2	19.8
H_2SO_4	90	450	10.7	6.0
H_3PO_4	5	300	7.1	4.0
H_3PO_4	10	300	9.1	5.1
H_3PO_4	40	300	25.1	14.1
H_3PO_4	50	300	40.2	22.6
H_3PO_4	60	275	21.8	12.3
H_3PO_4	80	266	29.9	16.8
H_3PO_4	86	250	31.4	17.7
H_3PO_4	100	176	7.6	4.3
H_3PO_4	115	176	3.4	1.9
CH_3COOH	5	350	21.5	12.1
CH_3COOH	10	350	19.0	10.7
CH_3COOH	20	350	17.6	9.9
CH_3COOH	40	350	11.2	6.5
CH_3COOH	50	375	29.9	16.8
CH_3COOH	70	400	15.5	8.7
CH_3COOH	70	450	23.3	13.1
CH_3COOH	90	425	10.8	6.1
CH_3COOH	90	450	10.7	6.0

Table 16.2.   Collective Properties

	Crystallized Glass	Glassed Steel
Maximum operating temperature	1450°F	450°F
Corrosion rate		
20% HCl, boiling, vapor phase	4–5 mpy	4–5 mpy
Dist. H_2O, boiling, vapor phase	1 mpy	7–8 mpy
NaOH, pH12, 180°F, liquid phase	7–8 mpy	8–10 mpy
Thermal shock		
Max. recommended ΔT at 250°F	260°F	260°F
Abrasion rate		
P.E.I. test, mg. loss/in.^{2}/15 min. cycle	1.6	7.6
Impact resistance		
In.-lb. to cause chipping	130	55
Hardness		
Mohs' scale	6–7	6–7
Rockwell, approximate	50	N.A.
Density		
Sp. gr., ceramic component	2.6	2.6
Thermal conductivity		
Ceramic component only, BTU/hr/sq. ft/°F/in.	8	6
Composite, 0.05 in. ceramic plus 11/16 in. steel, k/l	123	98
Typical chemical service, (liquid product, steam jacketed vessel), overall "U"	87	74
Dielectric strength		
Volts per mil, room temp., 60 cycle	735	700
Dielectric constant		
$f = 10^6$ cps	8.24	8.02

forms an ionic network throughout which the various oxides of sodium, calcium, aluminum, and cobalt are dispersed.

There are three basic types of corrosion mechanisms affecting glassed steel. The rate of attack by all mechanisms is highly dependent on both concentration and temperature.

A standard test procedure is used to obtain basic corrosion data on all attack mechanisms for both amorphous and crystallized glasses subjected to a selected number of very destructive and commonly encountered corrosives. This is a basic procedure in corrosion evaluation, and is backed by corrosion testing related to actual service conditions when required for evaluating performance of glassed steel for a specific application. Values for No. 3300 glass are shown in Table 16.1.

Stainless steel test disks, glassed on one side and weighed before and after exposure, are used. The heat source is a hot plate. A unit of the same dimensions made from stainless steel, but without the condensing arm, is used for alkali testing. The heat source is a stagnant air oven. Rate of attack in mils (inches/1000) penetration for year (mpy) is calculated from the weight lost.

Three test solutions are used:

1. 20% HCl at 230°F (condensing vapors);
2. 5% NaOH at 150°F (liquid phase);
3. Distilled water at 212°F (condensing vapors).

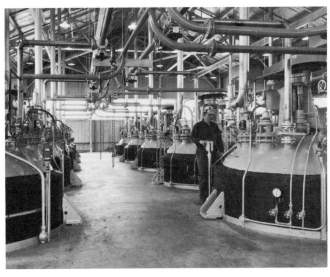

Figure 16.5   The Nonadherent Surface of Glassed-steel Accounts for Its Wide Use in Plastics Manufacturing. Shown is a typical polymer plant in which 3700 gallon glassed-steel polymerizers are used

Figure 16.4   A 150-gallon Ceramic-Metal Evaporator Pan Used in the making of Ceric Ammonium Nitrate. Operator is using a metal spade to scrape the material off the abrasion resistant ceramic metal surface

Figure 16.6 Largest Bottom-entering Furnace in the USA for Firing Glassed-steel Reactors. Shown is a 22,000 gallon reactor about to be fired.

An HCl solution (near the azeotropic concentration) is used because it is commonly encountered and is very destructive to most materials of construction, especially at this concentration and temperature; NaOH (at pH 14) is the most common and most corrosive of the alkalis; water of high purity (distilled) at elevated temperatures and small quantities such as condensing steam, can cause quite rapid attack on some glasses. Tests are usually run in duplicate for periods of 48 hours to 15 days.

A comparison of amorphous glassed steel and crystallized glassed steel points out the improvement in a number of properties that can be obtained with the crystallized glass. The figures shown in Table 16.2 represent the best attainable performance from among all available formulations in each property covered. While a specific composite may display dominance in one or more of the properties, it will not be outstanding in all areas. The data for crystallized glasses are not based solely on performance of composites having a steel base. The crystallized glasses can be applied to a variety of metals and alloys. Indeed, application at temperatures (over 900°F) usually requires the use of metals such as Inconel or some other high-temperature substrate.

4.3 Applications

Glassed steel is used for processing and storing materials in the chemical, petrochemical, pharmaceutical, plastic, brewery, biochemical, and other process industries. It is offered in the form of reactors, heat exchangers, dryers, evaporators (Figure 16.4), storage tanks, as well as various types of pumps, valves, and pipe lengths.

Versatility of glassed steel is demonstrated by the variety of standard and custom-built vessel designs available (Figure 16.5). To illustrate, storage vessels can be furnished in sizes up to 20,000-gallon capacity for corrosive storage, or to 50,000-gallon capacity for mild service where noncontamination and easy cleaning are the principal requirements. The design pressure and size of a glassed-steel reactor are limited only by the handling capabilities of furnace-charging cranes and by furnace dimensions (Figure 16.6).

Crystallized glass composites are filling traditional end uses of glassed steel and have branched into a number of new types of equipment applications. In process vessels, for example, crystallized glasses can withstand severe corrosion and abrasion at much higher temperatures than glassed steel. Slurries of aqua regia and sandlike abrasion at temperatures over 200°F have been handled.

Items such as pump parts and textile machinery parts which are subject to severe abrasive conditions receive protection from crystallized glasses which can be applied to them. Instrument parts such as thermocouple tubes subjected to molten metals have shown improvement when protected by crystallized-glass compositions. The electrical properties of various types of glassed steel (amorphous and crystalline) are also being put to use in several applications. Rolls made from crystallized glass-metal composites have been used as supports for plastic films being subjected to high-voltage, high-frequency corona discharge. The process is designed to make the plastic films suitable for printing.

Index